S0-ALJ-753

FACTORING FORMS

1. Common Monomial Factor $AB + AC = A(B + C)$
2. Difference of Squares $A^2 - B^2 = (A + B)(A - B)$
3. Difference of Cubes $A^3 - B^3 = (A - B)(A^2 + AB + B^2)$
4. Sum of Cubes $A^3 + B^3 = (A + B)(A^2 - AB + B^2)$
5. Perfect Square Trinomial $A^2 + 2AB + B^2 = (A + B)^2$
    $A^2 - 2AB + B^2 = (A - B)^2$

THEOREMS ON EXPONENTS

For x and y real numbers, m and n rational numbers,

1. Product of Two Powers $x^n x^m = x^{n+m}$
2. Power of a Power $(x^n)^m = x^{nm}$
3. Quotient of Two Powers $\dfrac{x^n}{x^m} = x^{n-m}$ if $x \neq 0$
4. Power of a Product $(xy)^n = x^n y^n$
5. Power of a Quotient $\left(\dfrac{x}{y}\right)^n = \dfrac{x^n}{y^n}$ if $y \neq 0$

THE QUADRATIC FORMULA

If $ax^2 + bx + c = 0$ and $a \neq 0$, then $x = \dfrac{-b \pm \sqrt{b^2 - 4ac}}{2a}$

EXPONENTIAL FUNCTION — DEFINITION

$y = b^x$, and x is any real number, $b > 0$, $b \neq 1$

LOGARITHMIC FUNCTION — DEFINITION

$y = \log_b x$ if and only if $x = b^y$, and $b > 0$, $b \neq 1$

LOGARITHMIC OPERATIONS THEOREMS

For x, y, and b any positive numbers, $b \neq 1$, and k any positive integer:

1. $\log_b xy = \log_b x + \log_b y$
2. $\log_b \dfrac{x}{y} = \log_b x - \log_h y$
3. $\log_b x^y = y \log_b x$
4. $\log_b \dfrac{k}{x} = \dfrac{1}{k} \log_b x$

MODERN
INTERMEDIATE
ALGEBRA
FOR COLLEGE
STUDENTS

THIRD EDITION

MODERN
INTERMEDIATE
ALGEBRA
FOR COLLEGE
STUDENTS

Susanne M. Shelley

Sacramento City College

SAUNDERS COLLEGE

Philadelphia

PREFACE

Copyright © 1980 Saunders College/Holt, Rinehart and Winston
Copyright © 1974, 1969 by Rinehart Press
All rights reserved

Library of Congress Cataloging in Publication Data

Shelley, Susanne M.
 Modern intermediate algebra for college students

 Second ed. by V. S. Groza and S. M. Shelley
published in 1974.
 Includes index.
 1. Algebra. I. Groza, Vivian Shaw. Modern
intermediate algebra for college students. II. Title.
QA154.2.G76   1980   512.9′042   79-23598
ISBN 0-03-042581-6

Printed in the United States of America
0 1 2 3   032   9 8 7 6 5 4 3 2

As a result of teaching from the Second Edition of *Modern Intermediate Algebra for College Students* for four years, and with the many helpful suggestions from numerous users of the text, this Third Edition has been prepared. While maintaining the successful features of the previous edition, some major changes were also made.

The sequence of topics has been changed in order to introduce topics from elementary algebra as they are needed for better comprehension. For many students who take intermediate algebra, there has been a time lag since learning elementary algebra, and for this reason some review at each stage, where necessary, should be helpful. If students in a particular class have just completed elementary algebra, much of the material in Chapter 1 can be covered very rapidly. However, for the student who needs to have these concepts available for review, the chapter should be extremely useful.

Chapter 2, on polynomials and factoring, builds on the basic concepts introduced in elementary algebra. Chapter 3, which deals with rational expressions, reviews the fundamental operations of fractions, and extends the basic concepts to include synthetic division, the remainder and factor theorems, complex fractions, and ratio, proportion, and variation.

Exponents and radicals, which were the topic of Chapter 1 in the Second Edition, are now the topic of Chapter 4. Students seemed to respond somewhat negatively to the early introduction of exponents and radicals, so their placement at their present point in the text should ease this difficulty. Included in Chapter 4 is an increased emphasis on rational exponents in decimal form for computer and calculator use.

Chapter 5—quadratic equations—and Chapter 6—inequalities and absolute values in one variable—are essentially unchanged except for an increase in illustrative examples, and a deletion of some theorems and proofs for which students at this level do not seem to be ready. Chapter 7—relations and functions —has been streamlined in a similar manner.

Chapter 8—exponential and logarithmic functions—has undergone some major adjustments. The section on inverse relations and functions has been expanded, simplified, and many examples and suitable exercises have been added. There are more exercises on solutions of logarithmic and exponential equations, and there is less emphasis on the use of logarithms for computation, owing to the availability of electronic calculators.

Chapter 9—systems of equations—deals with the solutions of linear systems and quadratic systems, and presents determinants. Matrices have been eliminated from this edition.

Chapter 10 consists solely of the binomial theorem and its applications, and may be omitted without loss of continuity.

The A and B exercise sets of the previous editions have been incorporated into one exercise set, with the even numbered exercises paralleling the odd numbered exercises. All exercise sets are graded and frequent references are made to illustrative examples to help students solve the problems. The answers to all odd numbered exercises are in the back of the book, while a separate manual contains the answers to all even numbered exercises. Each chapter has a set of review exercises with all answers in the back of the book. Each chapter also has a diagnostic test with complete solutions.

The Instructors Manual, in addition to the answers to the even numbered exercises, also contains a set of tests; there are two forms of one-hour tests for each chapter, as well as answers for the tests.

Tables of powers and roots, and logarithms, as well as frequently used formulas, are inside the covers of the book for convenient reference.

I would like to express my appreciation to the following persons who reviewed the Second Edition and provided many excellent suggestions for improving this Third Edition. Richard DeTar, Sacramento City College, California; Arthur Dull, Los Medanos College, California; Ronald D. Faulstich, Moraine Community College, Illinois; B. Wells Gorman, Southwestern College, California; Richard E. Linder, Southwestern College, California; Louis S. Perone, State University of New York at Farmingdale, New York; Alan Leigh Sawyer, Orange Coast College, California; Richard Ringeisen, Indiana University–Purdue University at Fort Wayne, Indiana; Mary Carter Smith, Laney College, California; John Spellman, Pan American University, Texas.

Sacramento, California
September, 1979   S. M. S.

CONTENTS

CONTENTS

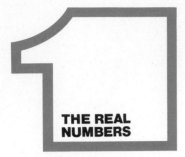

CONTENTS

RELATIONS AND FUNCTIONS

EXPONENTIAL AND LOGARITHMIC FUNCTIONS

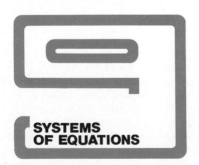

SYSTEMS OF EQUATIONS

CONTENTS

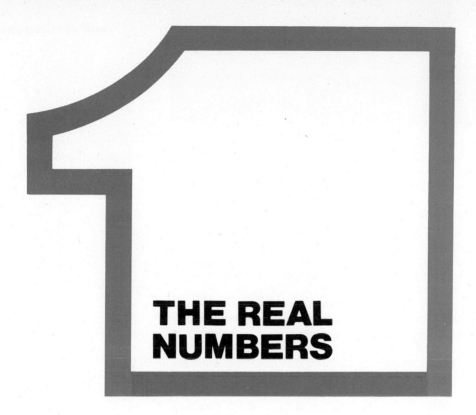

THE REAL NUMBERS

Algebra is a study of numbers and their properties. Most of the discussion in this course will deal with the real numbers, so we start with a description of the real numbers, some basic concepts, the symbolism to be used, and the relationship between the abstract character of numbers and their visual geometric representation.

1.1 BASIC CONCEPTS: SETS, SYMBOLISM, AND THE NUMBER LINE

A **set** is a collection of objects or things. The objects in the set may be any well-defined collection, such as the set of students in Professor Smith's 9 o'clock algebra class or the items in John's backpack. In algebra the objects in a set are usually numbers.

Symbolically, a set is described by three general methods: the **listing method,** the **description method,** and the **set-builder method.**

In the listing method the set is defined by listing or stating the names of the objects it contains (called **elements** or **members**) enclosed by braces and separated by commas. A capital letter is often designated as the name of the set.

For example, the set of the first five counting numbers may be called S and *listed* as follows:

$S = \{1, 2, 3, 4, 5\}$

By the description method S is the set of the first five counting numbers. This *describes* the set or states a property possessed by each member of the set.

The set-builder method is very similar to the description method and is symbolized as follows:

S is the set of all numbers x such that $x = 1, 2, 3, 4,$ or 5

$S = \{ \quad x \quad | \quad x = 1, 2, 3, 4, $ or $5\}$

Written compactly:

$S = \{x | x$ is one of the first five counting numbers$\}$

The letter to the left of the vertical bar is a "dummy" or symbol used to describe a typical object in the set. The vertical bar, read "such that," is followed by a description of the objects in the set.

A set that contains no elements is called the **empty set** and is symbolized as $\varnothing$. For example, the set of women who were presidents of the United States prior to 1980 is the empty set, and the set of all triangles with four sides is also the empty set.

In the example just given, S (the set of the first five counting numbers), 1, 2, 3, 4, and 5 are elements of S; 6 is *not* an element of the set, nor is 7, 8, or 9.

The most fundamental set underlying the set of real

numbers is the set of **counting numbers.** These are also called the **natural numbers** and are called *N* in this course.

$$N = \{1, 2, 3, 4, 5, \ldots\}$$

The three dots indicate that the set continues with the pattern that is established by the listed numbers, and since there is no last element listed, there are infinitely many numbers in this set. The set is said to be **infinite.** Set *S*, as discussed above, for example, is **finite** because we know exactly how many elements it contains.

If we add the number zero to the set of natural numbers, we obtain the **whole numbers,** which we call *W*.

$$W = \{0, 1, 2, 3, 4, \ldots\}$$

We could also write the set of whole numbers in set-builder notation as

$$W = \{x \,|\, x \text{ is zero or } x \text{ is a natural number}\}$$

Notice that *W* is also an infinite set and that it contains *all* the elements of *N*; that is, every natural number is also a whole number. A set that is completely contained in another set is called a **subset.** The set of natural numbers *N* is a subset of the whole numbers *W*.

The negatives of the natural numbers, combined with the whole numbers, form the set of **integers,** *I*.

$$I = \{\ldots, -5, -4, -3, -2, -1, 0, 1, 2, 3, 4, \ldots\}$$

This set is infinite in both directions; we have no "first" number and no "last" number in this set.

The symbols used to represent numbers are **numerals,** such as 3, −2, or 0, and letters of the alphabet. These letters are called **variables** if they stand for an unspecified number and **constants** if they stand for a specified number. For example, 3, −17, and 59 are numerals; *x*, *y*, and *p* are variables; and π is a constant.

The set-builder notation is used to define the set of **rational numbers,** *Q*, which consists of fractions whose numerator and denominator are integers, but the denominator is never zero. Symbolically, the set of rational numbers is

$$Q = \left\{\frac{a}{b} \,\middle|\, a \text{ and } b \text{ are integers, and } b \neq 0\right\}$$

The symbol $\neq$ means "is not equal to."

Examples of rational numbers, or elements of *Q*, are $\frac{3}{5}$, $\frac{-2}{7}$, and $\frac{4}{9}$. Note also that all natural numbers can be written in this form, $5 = \frac{5}{1}$ or $\frac{10}{2}$, and so on, so that *N* is a subset of *Q*; that is, the set of rational numbers contains all the natural numbers. By the same reasoning, the set of integers *I* is a subset of the set of rationals *Q*.

The saying that one picture is worth a thousand words applies equally well to mathematics, and a number line is used to that end.

A line is an infinite set of points, so any point may be selected as the starting point. This point is called the **origin** and is given the name 0 (zero).

A unit of measurement and a direction, called the positive direction, are selected. Using this unit of length, points are marked off in succession in the positive direction, which is to the right in Figure 1.1.

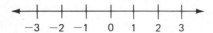

FIG. 1.1

The arrow at the right indicates that the line continues indefinitely in this direction.

The negative integers $\{-1, -2, -3, \ldots\}$ are assigned to the points in the opposite direction in a similar manner. This direction is to the left in Figure 1.1.

The arrow at the left indicates that the line continues indefinitely in this direction.

A line whose points are named by using numbers is called a **number line.**

The number that names a point is called the **coordinate** of the point.

The point that is given a number name is called the **graph** of this number.

All the rational numbers can be identified as coordinates on a number line by considering subdivisions of the basic unit and by using the concept of opposites (Figure 1.2).

FIG. 1.2

We can see that each rational number can be made to correspond to exactly one point on a number line.

One might be tempted to believe that the rational numbers account for all the points on the number line. However, this is not the case. There are many numbers that are not rational. For example, $\sqrt{2}$ is *not* a rational number.

The real number $\sqrt{2}$ is called an **irrational number.** Some other examples of irrational numbers are $\sqrt{3}$, $\sqrt{5}$, $\sqrt[3]{2}$, $-\sqrt{2}$, $-\sqrt{3}$, and π.

All the numbers that can be associated with points on a number line form a set called the set of **real numbers.** Thus a number line is called a **real number line.**

It follows that the set of irrational numbers, Ir, consists of all the real numbers that are *not* rational.

$$Ir = \{x \mid x \text{ is a real number but } x \text{ is not rational}\}$$

Each point on a real number line can be made to correspond to exactly one real number, and each real number can be made to correspond to exactly one point on a real number line. Thus there is a *one-to-one correspondence between the set of real numbers and the set of points on a real number line.* This important property of the set of real numbers is called the **axiom of completeness.** The set of real numbers is "complete" in the sense that all the real numbers are "used up" in naming the points on a number line, and every point has exactly one real number name.

The numbers associated with points to the right of 0 are called the **positive real numbers,** and those to the left of 0 are called the **negative real numbers** (Figure 1.3).

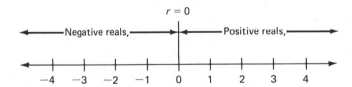

FIG. 1.3 Real Number Line

A positive number and its corresponding negative are called **additive inverses.** For example, 3 and -3 are additive inverses; so are $-\dfrac{1}{2}$ and $\dfrac{1}{2}$, and $\sqrt{5}$ and $-\sqrt{5}$.

Any point, r, on the number line, and its additive inverse, r, are the same *distance* from the origin, but the two points are located on opposite sides of the origin. The distance, without regard to the location of the point to the right or left of the origin on a horizontal number line, is called the **absolute value** of the number r and is symbolized $|r|$. Any nonzero real number without a sign is assumed to be positive, so it follows that the absolute value of a number is positive unless the number is zero. The absolute value of zero is zero. (A formal definition of absolute value will be given in a later chapter.)

EXAMPLES

$$|5| = 5, \; |-3| = 3, \; |\sqrt{2}| = \sqrt{2}, \; |0| = 0, \; \left|-\frac{1}{5}\right| = \frac{1}{5}.$$

But

$$-|3| = -3 \quad \text{and} \quad -|-4| = -4.$$

In summary:

The **natural numbers** $N = \{1, 2, 3, 4, \ldots\}$
The **whole numbers** $W = \{0, 1, 2, 3, 4, \ldots\}$
The **integers** $I = \{\ldots, -2, -1, 0, 1, 2, 3, \ldots\}$

The **rational numbers** $Q = \left\{\dfrac{a}{b} \mid a \text{ and } b \text{ are integers} \right.$

$$\left. \text{and } b \neq 0 \right\}$$

The **irrational numbers** $Ir = \{x \mid x$ is a real number but x is not rational$\}$

All the above sets put together form the set of **real numbers.**

EXAMPLE 1   Use the listing method to describe each of the following sets.
(a) The natural numbers *less than* 5
(b) The integers between -3 and 2
(c) The whole numbers less than 0

Solution
(a) $\{1, 2, 3, 4\}$
(b) $\{-2, -1, 0, 1\}$
(c) $\varnothing$. This is the empty set; there are no negative whole numbers.

EXAMPLE 2   Use the set-builder method to describe each of the following sets.
(a) $\{2, 4, 6\}$
(b) $\{0, 1, 2, 3, \ldots\}$
(c) The set of fractions that have 1 as the numerator and any natural number as the denominator.

Solution
(a) $\{x \mid x$ is one of the first three even natural numbers$\}$
(b) $\{w \mid w$ is a whole number$\}$
(c) $\left\{\dfrac{1}{n} \mid n \text{ is a natural number}\right\}$

EXAMPLE 3   Given: $S = \left\{-3, -\dfrac{7}{3}, -\sqrt{2}, 0, \dfrac{3}{4}, \dfrac{\sqrt{2}}{3}, 5, 8\right\}$

List each of the following subsets of S.
(a) the natural numbers
(b) the whole numbers
(c) the integers
(d) the rational numbers
(e) the irrational numbers
(f) the real numbers

Solution   (a) $\{5, 8\}$

(b) $\{0, 5, 8\}$

(c) $\{-3, 0, 5, 8\}$

(d) $\left\{-3, -\dfrac{7}{3}, 0, \dfrac{3}{4}, 5, 8\right\}$

(Remember that all whole numbers are also rational.)

(e) $\left\{-\sqrt{2}, \dfrac{\sqrt{2}}{3}\right\}$

Even though $\dfrac{\sqrt{2}}{3}$ is a fraction, it is *not* the ratio of two integers and is therefore irrational.

(f) S; every number in set S is a real number.

EXAMPLE 4 Find the value of each of the following. (This is also called "simplifying.")

(a) $|-3|, \left|-\dfrac{1}{2}\right|, -|3|, -\left|-\dfrac{2}{5}\right|$

(b) $|2| + |-2|, |4 - 4|, |-5| + |-2|, |-5| - |-2|$

Solution   (a) $|-3| = 3, \left|-\dfrac{1}{2}\right| = \dfrac{1}{2}, -|3| = -3,$

$-\left|-\dfrac{2}{5}\right| = -\dfrac{2}{5}$

(b) $|2| + |-2| = 4$

$|4 - 4| = 0$

$|-5| + |-2| = 7$

$|-5| - |-2| = 3$

EXAMPLE 5 State the additive inverse of each of the following.
(a) -2
(b) 3
(c) $\dfrac{1}{5}$
(d) 0

Solution   (a) 2

(b) -3

(c) $-\dfrac{1}{5}$

(d) 0 (zero is its own additive inverse)

EXERCISES 1.1

In Exercises 1–20 use the listing method to describe each set. (Example 1)

1. The natural numbers between 8 and 10.
2. The natural numbers between −1 and 4.
3. The integers between 0 and 1.
4. The integers between $\frac{1}{2}$ and $\frac{5}{2}$.
5. The integers greater than 4.
6. The integers less than −2.
7. The whole numbers between −2 and 2.
8. The whole numbers less than 5.
9. An **even number** is a natural number that is exactly divisible by 2. List the even numbers between 9 and 15.
10. The even numbers greater than 13.
11. The **odd numbers** are the natural numbers that are not even. List the odd numbers between 2 and 8.
12. The odd numbers greater than 100.
13. The odd numbers that are also even.
14. The even positive integers.
15. A **prime number** is a natural number other than 1 that can be divided exactly *only* by itself and 1. Examples of prime numbers are 2, 3, 5, 7, 11. Note that 9 is not prime because it can also be divided exactly by 3. List the prime numbers between 20 and 30.
16. The prime numbers between 40 and 60.
17. All the even prime numbers.
18. The negative integers.
19. The natural numbers between −1 and 1.
20. The integers between −1 and 1.

In Exercises 21–30 use the set-builder method to describe each set. (Example 2)

21. The prime numbers.
22. The even natural numbers.
23. The fractions that have −1 as a numerator and a natural number as a denominator.
24. The fractions that have any integer as the numerator and 3 as the denominator.
25. Let $S = \{2, 6, 19, 28\}$. Describe this set using the set-builder notation.

In Exercises 26–30 use the set-builder notation to describe the set.

26. $P = \{-3, -\sqrt{6}, 13, \sqrt{53}, 205\}$.
27. $T = \left\{\frac{1}{2}, \frac{1}{4}, \frac{1}{6}, \frac{1}{8}, \cdots\right\}$

28. $A = \left\{ \dfrac{-5}{1}, \dfrac{-5}{2}, \dfrac{-5}{3}, \dfrac{-5}{4}, \ldots \right\}$

29. $S = \{a, b, c, d, e, f, \ldots, z\}$

30. $V = \{a, e, i, o, u\}$

In Exercises 31–36 let $S = \left\{ -7, -\sqrt{11}, -\dfrac{3}{5}, 0, \dfrac{2}{3}, \sqrt[3]{5}, 3, \dfrac{\sqrt{10}}{4} \right\}$.

List each of the following subsets of S. (Example 3)

31. The natural numbers.

32. The whole numbers.

33. The integers.

34. The rational numbers.

35. The irrational numbers.

36. The real numbers.

Simplify Exercises 37–46. (Example 4)

37. $|4|$

38. $|-4|$

39. $\left| -\dfrac{2}{3} \right|$

40. $|-4| + |-2|$

41. $|-4| - |-2|$

42. $|12| - |-5|$

43. $|12| + |-5|$

44. $|8 - 8|$

45. $-|8 - 8|$

46. $|-|-3||$

In Exercises 47–56 state each additive inverse. (Example 5)

47. 4

48. -4

49. $-\dfrac{3}{5}$

50. $\dfrac{2}{3}$

51. $\sqrt{3}$

52. $\sqrt[3]{5}$

53. 0

54. $|-5|$

55. $-|-5|$

56. $-(-5)$

Determine whether each statement in Exercises 57–62 is true or false.

57. Every natural number is a rational number.

58. Every whole number is a natural number.

59. All prime numbers are natural numbers.

60. The set of irrational numbers is a subset of the set of real numbers.

61. The set of rational numbers is a subset of the set of irrational numbers.

62. The totality of the set of even numbers and the set of odd numbers is the set of natural numbers.

1.2 SOME AXIOMS: EQUALITY AND INEQUALITY

Now that we have defined the set of real numbers, we may begin to discuss the relationship between any two numbers.[1] Two numbers are either equal to each other or they are not equal. Considering the possibility that the numbers are equal, we call this the **equal relation** and state certain properties of this relation.

AXIOMS OF EQUALITY

For all real numbers r, s, and t:

1. $r = r$ Every real number is equal to itself. This is the **reflexive property.**
2. If $r = s$, then $s = r$.   This is the **symmetric property.**
3. If $r = s$ and $s = t$, then $r = t$.   This is the **transitive property.**

The terms "axiom" and "property" are used interchangeably. An **axiom** is an *assumed* property, contrasted with a **theorem,** which must be *proved.*

EXAMPLES By the reflexive axiom, $3 = 3$, $-2 = -2$, and $x = x$.
By the symmetric axiom, if $2 = y$, then $y = 2$; and if $3 = 4 - 1$, then $4 - 1 = 3$.
By the transitive axiom, if $x = y$ and $y = 3$, then $x = 3$.

Another axiom that is closely related to the transitive axiom is the substitution axiom.

SUBSTITUTION AXIOM

If r and s are real numbers, and if $r = s$, then r may replace s or s may replace r in an algebraic expression without changing the value of the expression, or in an algebraic statement without changing the truth or falsity of the statement.

EXAMPLE 1 If $y = x + 3$ and $x = 7$, then by the substitution axiom we can replace x by 7 and obtain $y = 7 + 3$ or $y = 10$.

[1] Unless otherwise indicated, "number" in this book will mean "real number."

Now consider the second possible relation between two numbers—that they are *not* equal. If two numbers, r and s, are not equal, then r is either greater than s or r is less than s. The number line provides a means to visualize this **inequality relation:** if r is *less* than s, then on a horizontal number line the point corresponding to r is to the *left* of the point corresponding to s; and if r is *greater* than s, then the point corresponding to r is to the *right* of s.

The symbols $<$ and $>$ are used to mean "less than" and "greater than," respectively. The number relations, their symbolic statement, their algebraic meaning, and their geometric meaning are shown in Table 1.1.

TABLE 1.1   NUMBER RELATIONS

SYMBOLIC STATEMENT	ALGEBRAIC MEANING	GEOMETRIC MEANING (for horizontal number line)
1. $r < s$	r is less than s	r is to the left of s
2. $r = s$	r is equal to s	r is the same point as s
3. $r > s$	r is greater than s	r is to the right of s

EXAMPLE 2   In Figure 1.4 the corresponding locations for the numbers -5, -1, $-\dfrac{1}{2}$, 0, 3, and 6 are shown on a number line. Insert the correct symbol, $=$, $<$, or $>$, between each of the following pairs of numbers by observing the relative positions of their graphs on the number line.

(a) 0, 3
(b) 6, 0
(c) -5, 0
(d) $-\dfrac{1}{2}$, -5
(e) $-\dfrac{1}{2}$, 3

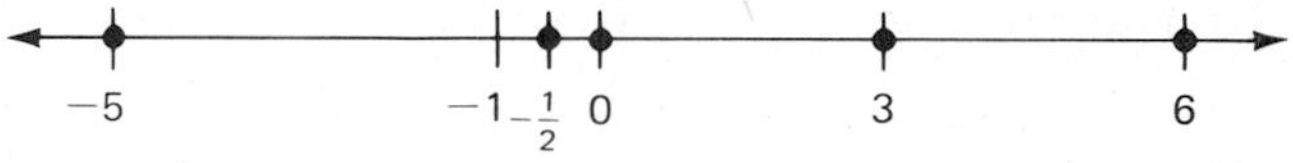

FIG. 1.4

Solution
(a) $0 < 3$
(b) $6 > 0$
(c) $-5 < 0$
(d) $-\dfrac{1}{2} > -5$
(e) $-\dfrac{1}{2} < 3$

The three possible relations between numbers as stated in Table 1.1 are summarized in the trichotomy axiom.

TRICHOTOMY AXIOM

For any two real numbers r and s, one and only one of the following statements is true.

1. $r < s$
2. $r = s$
3. $r > s$

A combination of the symbols $<$ and $=$ may be used as a shorthand notation to say that r and s have the relationship that either r is less than s *or* r is equal to s.

We write $r \leq s$ to mean $r < s$ *or* $r = s$ and say "r is less than or equal to s."

We write $r \geq s$ to mean $r > s$ *or* $r = s$ and say "r is greater than or equal to s."

A slash drawn through any symbol has the effect of the word "not." Table 1.2 gives a summary of all the relation symbols.

TABLE 1.2 SUMMARY OF RELATION SYMBOLS

SYMBOL	VERBAL TRANSLATION	EXAMPLES
$=$	Equals, is equal to	$3 + 4 = 7$
$\neq$	Does not equal	$3 + 4 \neq 5$
$<$	Is less than	$3 < 7$
$>$	Is greater than	$7 > 3$
$\nless$	Is not less than	$7 \nless 3$
$\ngtr$	Is not greater than	$3 \ngtr 7$
$\leq$	Is less than *or* is equal to	$x \leq 7$
$\geq$	Is greater than *or* is equal to	$x \geq 3$

The following statements are self-evident from the number line.

If x is a positive number, then $x > 0$.

If x is a negative number, then $x < 0$.

The transitive axiom applies to the relations "less than" and "greater than."

TRANSITIVE AXIOM—Inequalities

For all real numbers r, s, and t:
If $r < s$ and $s < t$, then $r < t$; and if $r > s$ and $s > t$, then $r > t$.

1.2 SOME AXIOMS: EQUALITY AND INEQUALITY

The reflexive and symmetric axioms do not apply to inequalities. For example, no number is less than itself, and if 5 is less than 7, then 7 is *not* less than 5, but is greater than 5.

EXAMPLE 3 By the transitive axiom, if $-2 < 3$ and $3 < 5$, then $-2 < 5$. Also, if $x < y$ and $y < 3$, then $x < 3$.

We can also use the number line to illustrate the set of all real numbers less than a given number, less than or equal to a given number, and so forth, in the following manner.

EXAMPLE 4 Use a graph (number line) to illustrate $\{x \mid x > 4\}$.

Solution Figure 1.5 is this graph. Since it is difficult to differentiate between the original number line and the graph, the graph is drawn above the number line for clarity. The circle above the numeral 4 indicates that 4 is *excluded* from the solution set; the solution set is indicated by the half-line starting at 4 (but not including 4) and all values greater than 4 as shown by the direction of the line.

FIG. 1.5

EXAMPLE 5 Use a graph to illustrate $\{x \mid x \leq 3\}$.

Solution The graph is shown in Figure 1.6. The filled-in circle above the numeral 3 indicates that this number is *included,* and the direction of the half-line indicates that all the numbers less than 3 are also included in the graph. The graph represents all numbers that are less than or equal to 3.

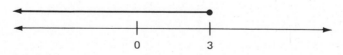

FIG. 1.6

The statement $a < x < b$ means that x is a number *between* a and b. Again, on the graph an open circle above a or b means that number is to be *excluded* from the permissible values for x, and a filled-in circle above the a or b means that number is to be *included*.

EXAMPLE 6   Draw the graph of $\{x \mid -1 \le x < 2\}$.

Solution   Figure 1.7 is the graph.

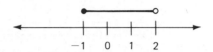

FIG. 1.7

EXERCISES 1.2

In Exercises 1–16 insert the symbol $<$, $=$, or $>$ between each pair of numbers so that the resulting statement is true. (Example 2)

1. 23, 18
2. 15, 22
3. $-3, 2$
4. $-2, 3$
5. $-4, -6$
6. $-3, -4$
7. 4, 0
8. 0, -2

9. $|3|, |-4|$
10. $|-3|, |4|$
11. $|-6|, |6|$
12. $6, |-6|$
13. $0, |-3|$
14. $0, -|3|$
15. $-|-3|, 0$
16. $-|4 - 4|, 0$

Graph each of the sets in Exercises 17–30 on a number line and describe the set in words. (Examples 4–6)

17. $\{x \mid x < 3\}$
18. $\{x \mid x > 2\}$
19. $\{x \mid x > -1\}$
20. $\{x \mid x < -2\}$
21. $\{x \mid x \le 5\}$
22. $\{x \mid x \ge 1\}$
23. $\{x \mid x \ge 0\}$

24. $\{x \mid x \le -4\}$
25. $\{x \mid x > -2\}$
26. $\{x \mid x < 1\}$
27. $\{x \mid -1 < x < 2\}$
28. $\{x \mid 2 < x < 5\}$
29. $\{x \mid 0 \le x < 3\}$
30. $\{x \mid -3 < x \le -1\}$

Each statement in Exercises 31–40 illustrates one of the axioms discussed in this section. Match each exercise with one of these axioms, as listed below.

(a) *Reflexive axiom* $r = r$

(b) *Symmetric axiom*   If $r = s$, then $s = r$.

(c) *Transitive axiom*   If $r = s$ and $s = t$, then $r = t$.

(d) *Substitution axiom*   If $r = s$, then r may replace s or s may replace r in any expression without changing its value.

(e) *Trichotomy axiom*   Either $r < s$ or $r = s$ or $r > s$.

31. If x is less than 3 and 3 is less than y, then x is less than y.

32. If $z = 5$, then $5 = z$.

33. Every number is equal to itself.

34. Every real number is either positive, negative, or equal to zero.

35. If $x = 2$ and $x + 3 = y$, then $2 + 3 = y$.

36. If $x = y$ and $y = 2$, then $x = 2$.

37. If $x + 2 = y$, then $y = x + 2$.

38. If $x + 2 = y + 2$ and $y = 3$, then $x + 2 = 5$.

39. If a is less than 4 and 4 is less than b, then a is less than b.

40. If the absolute value of -6 is 6, then 6 is equal to the absolute value of -6.

1.3 PROPERTIES OF THE REAL NUMBERS

In Section 1.2 we discussed how numbers are *related.* Any two numbers are either equal or unequal, and the axioms of equality and inequality formalize these relationships. Our next concern is what we can do with numbers. From arithmetic we recall that we can add numbers, subtract them, multiply them, or divide them. These are the four basic **operations** on numbers. Later we will extend these to six operations by including raising to a power and root extraction.

   Since we later define subtraction in terms of addition, and division in terms of multiplication, we first look at the properties of the real numbers for addition and multiplication.

   It is first necessary to note that these are **binary** operations; that is, you can add only two numbers at a time and you can multiply only two numbers at a time. In order to add three or more numbers, you must always add two numbers first, and then add the third number to that sum. The same is true for multiplication.

The following properties are all axioms. They are assumed to be true for all real numbers.

PROPERTIES OF THE REAL NUMBERS

Let r, s, and t be any real numbers.

Axioms

1. Closure

$r + s$ is a real number — The sum of any two real numbers is always a real number.

rs is a real number — The product of any two real numbers is always a real number.

2. Commutative Axioms

$r + s = s + r$ — In adding two numbers, the order of addition does not matter.

$rs = sr$ — In multiplying two numbers, the order of multiplication does not matter.

3. Associative Axioms

$(r + s) + t = r + (s + t)$ — When adding three numbers it does not matter which two are added first.

$(rs)t = r(st)$ — When multiplying three numbers it does not matter which two are multiplied first.

4. Distributive Axiom

$r(s + t) = rs + rt$ — This axiom links the operations of addition and multiplication.

5. Identity Axioms

There exists a unique real number 0 such that $r + 0 = 0 + r = r$ — 0 is the additive identity. Adding 0 to a number does not change its value.

1.3 PROPERTIES OF THE REAL NUMBERS

	There exists a unique real number 1 such that $r \cdot 1 = 1 \cdot r = r$	1 is the multiplicative identity. Multiplying a number by 1 does not change its value.
6. Inverse Axioms	For each real number r there exists a unique real number $-r$ called the additive inverse of r, such that $r + (-r) = 0$	Additive inverse. The sum of a number and its additive inverse is 0.
	For each *nonzero* real number r there exists a unique real number $\frac{1}{r}$ called the **reciprical** of r such that $r \cdot \frac{1}{r} = 1$	Multiplicative inverse (reciprocal). The product of any real number other than zero and its reciprocal is 1. Zero has no reciprocal.

By using combinations of the axioms of equality and the properties of the real numbers just given, we obtain useful variations of the axioms. For example, the distributive axiom

$r(s + t) = rs + rt$

can also be written

$rs + rt = r(s + t)$

by the symmetric axiom and

$r(s + t) = (s + t)r$

by the commutative axiom for multiplication.

EXAMPLE 1 Use the distributive axiom to simplify
$5x + 7x$

Solution
$$5x + 7x = (5 + 7)x$$
$$= 12x$$

Other useful properties of the real numbers may be obtained by using a combination of the axioms, called **theorems.** They are presented without proof.

MORE PROPERTIES OF THE REAL NUMBERS

Let r be any real number.

Theorems

1. Multiplication by zero	$r \cdot 0 = 0 \cdot r = 0$	Zero times any number is always zero.
2. Negative of a negative	$-(-r) = r$	The negative of a negative number is positive.
3. Reciprocal of a reciprocal	If $r \neq 0$, then $\dfrac{1}{\frac{1}{r}} = r$	The reciprocal of the reciprocal of a nonzero number is the number itself.

EXAMPLE 2

(a) $5 \cdot 0 = 0;\ 0 \cdot 0 = 0;\ -\dfrac{1}{2} \cdot 0 = 0$

Theorem 1, zero times any number is always zero

(b) $-(-3) = 3;\ -(-7) = 7$

Theorem 2, negative of a negative

(c) $\dfrac{1}{\frac{1}{3}} = 3;\ \dfrac{1}{-\left(\frac{1}{3}\right)} = -3$

Theorem 3, reciprocal of a reciprocal

(d) $3 + (-27)$ is a real number

The closure axiom for addition

(e) $3(-27)$ is a real number

The closure axiom for multiplication

(f) $x + 3 = 3 + x$

The commutative axiom for addition

(g) $(3)(-27) = (-27)(3)$

The commutative axiom for multiplication

(h) $(16 + 40) + 10 = 16 + (40 + 10)$

The associative axiom for addition

(i) $(5 + 7) + (-7) = 5 + (7 + [-7])$

The associative axiom for addition

$\qquad = 5 + 0$

The additive inverse axiom

$\qquad = 5$

The additive identity axiom

$(5 + 7) + (-7) = 5$

The transitive axiom of equality

The distributive axiom has many useful applications, so it is convenient to recognize the following immediate consequences of this axiom.

1. $a(b - c) = ab - ac$
2. $ab + ac = a(b + c)$
3. $ac + bc = (a + b)c$
4. $ac - bc = (a - b)c$
5. $ad + bd + cd = (a + b + c)d$
6. $-(b + c) = -b - c$
7. $-(b - c) = -b + c$

1.3 PROPERTIES OF THE REAL NUMBERS

One of the uses of the distributive axioms is to "collect **like terms.**"

DEFINITIONS

Like terms are terms with identical literal factors. A **literal factor** is a factor denoted by a letter.

A factor is one of the numbers that is multiplied to form a product. For example, since 3 times 4 equals 12, 3 and 4 are factors of 12. Similarly, in the expression $3x$, 3 and x are multiplied, so that 3 is a numerical factor and x is a literal factor.

EXAMPLE 3 Use the distributive axiom to collect like terms in each of the following:
(a) $3x + 5x$
(b) $7x^2 - 4x^2$
(c) $xy + 7xy$
(d) $2z + 3 + 4z$
(e) $xy + 3x - 2y$
(f) $x^2 + 5x + 6x + 4$

Solution
(a) $3x + 5x = (3 + 5)x = 8x$
(b) $7x^2 - 4x^2 = (7 - 4)x^2 = 3x^2$
(c) $xy + 7xy = 1xy + 7xy = (1 + 7)xy = 8xy$ From the fact that 1 is the multiplicative identity, $xy = 1xy$
(d) $2z + 3 + 4z = 2z + 4z + 3$ Using the commutative axiom
$2z + 4z + 3 = (2 + 4)z + 3$ for addition
$= 6z + 3$
Note that $6z$ and 3 are *unlike* terms.
(e) $xy + 3x - 2y$ cannot be simplified because there are *no like terms.*
(f) $x^2 + 5x + 6x + 4 = x^2 + (5 + 6)x + 4$
$= x^2 + 11x + 4$

In order to make the comparison of the student's answer with those listed in the back of the book more consistent, the following convention for writing answers will be used whenever appropriate.

CONVENTIONS

The terms of a sum are written so that:

1. A numerical term is to the right of all literal terms.

ILLUSTRATIONS

1. $x + 2$; $y - 3$

2. The letters involved are written in alphabetical order whenever possible.

2. $3x + 2y + 5$;
$4a - 2b + c$

The factors of a product are written so that:

1. A numerical factor is to the left of all literal factors.

1. $3xyz$

2. The letters involved are written in alphabetical order whenever possible.

2. $5x^2yz^3$

In the expression x^3, x is the **base** of the term and 3 is an **exponent.** We say that "x is raised to the third power" and recognize that $x^3 = x \cdot x \cdot x$. Similarly, y^2 is "y raised to the second power," or $y \cdot y$, and in general z^n is a number z (the base) raised to the nth power (n is the exponent and n is a natural number). Again, $z^n = z \cdot z \cdot z \cdot \ldots \cdot z$ for n factors of z. More discussion and applications of exponents will be found in this text as we go along. At this time we need to recognize these terms in the context of like or unlike terms.

An expression such as $3x^4 + 2x^3 + x^2 + 5x$ is said to be arranged in **descending powers of x** because the exponents on the variable x (the base) are arranged from the largest to the smallest. This arrangement is often used to make the reading of an answer less confusing.

EXERCISES 1.3

Each of the statements in Exercises 1–30 may be justified by one or more of the following axioms or theorems. Match each exercise with the correct axioms or theorems. Assume all variables stand for non-zero real numbers and refer to Example 2.

(a) Closure: *The sum of any two real numbers is always a real number.*

(b) Commutative: *In adding two numbers, the order of addition does not matter. In multiplying two numbers, the order of multiplication does not matter.*

(c) Associative: *When adding three numbers, it does not matter which two are added first. When multiplying three numbers, it does not matter which two are multiplied first.*

(d) Distributive: *$r(s + t) = rs + rt$. Multiplication "distributes" over addition.*

(e) Identity: *Adding 0 to a number does not change its value. Multiplying a number by 1 does not change its value.*

(f) *Inverse:*   *The sum of a number and its negative is 0. The product of a number and its reciprocal is 1.*

(g) *Zero times any number is always zero.*

(h) *The negative of a negative number is positive.*

(i) *The reciprocal of the reciprocal of a nonzero number is the number itself.*

1. $-2 + 3 = 3 + (-2)$

2. $2(x + 5) = 2x + 10$

3. $-3 + 3 = 0$

4. $5\left(\dfrac{1}{5}\right) = 1$

5. $4x + 7x = 11x$

6. $4 + 0 = 4$

7. $0 + 2 = 2 + 0$

8. $0 \cdot 2 = 2 \cdot 0$

9. $-5 + [-(-5)] = 0$

10. $2(-3 + 4)$ is a real number

11. $4(-2 + 2) = 0$

12. $4y + 12y = 16y$

13. $\dfrac{1}{x} \cdot x = 1$

14. $-2 + (3 + 4) = (-2 + 3) + 4$

15. $(x + 3) + y = y + (x + 3)$

16. $2(x + 4) = (x + 4)2$

17. $(x + 3) + y = (3 + x) + y$

18. $2(x + 4) = 2(4 + x)$

19. $\left(-\dfrac{1}{2}\right)(-2)$ is a real number

20. $-(-3) - 3 = 0$

21. $[-(-3) - 3] \cdot 15 = 0$

22. $14(2x) = 28x$

23. $6(5y) = 30y$

24. $5(2x + 3) = 10x + 15$

25. $4ax + 3x = (4a + 3)x$

26. $\dfrac{1}{\frac{1}{7}} = 7$

27. $3(8 + x) = (x + 8) \cdot 3$

28. $3(8 + x) = 3x + 24$

29. $(x + 3) + y = x + y + 3$

30. $(x + 3) + (-3) = x$

Simplify Exercises 31–40 by using one or more of the properties of the real numbers. Name the properties you use. (Assume all variables stand for nonzero real numbers.)(Example 2)

31. $(x + 3) + (-3)$

32. $5z + 7z$

33. $(-7)\left(-\dfrac{1}{7}\right)$

34. $\dfrac{1}{\frac{1}{9}}$

35. $(57)(3) + (57)(2)$

36. $(25)(9)(4)$

37. $(64 + 75) + (25 + 36)$

38. $(-8 + 5) + 3$

39. $6n + 3n + 5n$

40. $4x + (3 + 5x)$

In Exercises 41–50 simplify each expression by collecting like terms and arranging the answer in alphabetical order, or in descending powers of the variable whenever possible. (Example 3)

41. $4y + 5y + 2y$

42. $8z^2 + 12z^2$

43. $3x + 5 + 2x + x^2$

44. $2xy + x + y + xy$

45. $x^2y + 2x^2y + xy^2$

46. $2x^2 + 3x - x + 4$

47. $8xz - z$

48. $xy + 2x + 5x$

49. $12x^2y - 5x^2y + 2$

50. $12x^2y + 5xy^2 + 2xy^2$

1.4 SUMS, DIFFERENCES, PRODUCTS, AND QUOTIENTS OF THE REAL NUMBERS

As stated in the previous section, the four basic operations of arithmetic are addition, subtraction, multiplication, and division. These four operations are extended to the real numbers in the following manner.

SUM OF TWO REAL NUMBERS—Addition

The sum of two *positive* real numbers is *positive*.
If $r > 0$ and $s > 0$, then $r + s > 0$.
The sum of two *negative* real numbers is *negative*.
If $r < 0$ and $s < 0$, then $r + s < 0$.

In general the sum of two real numbers having the *same* sign is obtained by adding the absolute values of the numbers and prefixing their common sign.

In general the sum of two real numbers having *opposite* signs and *unequal* absolute values is obtained by finding the difference of their absolute values (the larger minus the smaller) and prefixing the sign of the number that has the larger absolute value to the answer.

For example, $15 + (-2) = 13$ because $|15| - |-2| = 15 - 2 = 13$, and the answer is positive because $|15| > |-2|$. Similarly, $-15 + 2 = -13$ because $|-15| - |2| = 15 - 2 = 13$, but the answer is negative because 15 is negative, and $|-15| > |2|$.

EXAMPLE 1 Perform the indicated additions:
(a) $(+18) + (+2)$
(b) $(-18) + (-2)$
(c) $(+18) + (-2)$
(d) $(-18) + (+2)$

Solution
(a) $(+18) + (+2) = 18 + 2 = 20$
(b) $(-18) + (-2) = -(18 + 2) = -20$
(c) $(+18) + (-2) = +(18 - 2) = +16 = 16$
(d) $(-18) + (+2) = -(18 - 2) = -16$

EXAMPLE 2 Perform the indicated additions:
(a) $8 + 7$
(b) $-8 + 7$

(c) $8 + (-7)$
(d) $(-8) + (-7)$

Solution
(a) $8 + 7 = 15$
(b) $-8 + 7 = -(8 - 7) = -1$
(c) $8 + (-7) = +(8 - 7) = 1$
(d) $(-8) + (-7) = -(8 + 7) = -15$

EXAMPLE 3 Find the indicated sum: $18 + [(-33) + 15]$

Solution
$$18 + [(-33) + 15] = 18 + (-18) = 0$$

EXAMPLE 4 Find the indicated sum: $[4 + (-4)] + (-3)$

Solution
$$[4 + (-4)] + (-3) = 0 + (-3) = -3$$

EXAMPLE 5 Find the indicated sum: $10 + (-5) + (-4)$

Solution
$$10 + (-5) + (-4) = [10 + (-5)] + (-4)$$
$$= 5 + (-4)$$
$$= 1$$

Since the associative axiom for addition of real numbers states that the way in which the numbers are grouped does not affect the sum, we may check the answer by regrouping the original numbers.

Check
$$10 + [(-5) + (-4)] = 10 + (-9) = 1$$

The difference of two real numbers is defined in terms of addition by the following rule.

DIFFERENCE OF TWO REAL NUMBERS—Subtraction

The subtraction of two real numbers is performed by changing the sign of the number to be subtracted and then adding according to the rules of addition of real numbers. In symbols, if r and s are real numbers, then

$$r - s = r + (-s) \quad \text{and} \quad r - (-s) = r + s$$

EXAMPLE 6 Perform the indicated subtractions:
(a) $8 - 5$
(b) $-8 - 5$
(c) $-8 - (-5)$
(d) $8 - (-5)$

Solution
(a) $8 - 5 = 8 + (-5) = 3$
(b) $-8 - 5 = -8 + (-5) = -13$
(c) $-8 - (-5) = -8 + 5 = -3$
(d) $8 - (-5) = 8 + 5 = 13$

EXAMPLE 7   Subtract and check:
(a) $15 - 8$
(b) $15 - (-8)$
(c) $-15 - 8$
(d) $-15 - (-8)$

Solution
(a) $15 - 8 = 7$ (The answer is the same as it is in arithmetic.)
(b) $15 - (-8) = 15 + 8 = 23$
(c) $-15 - 8 = -15 + (-8) = -23$
(d) $-15 - (-8) = -15 + 8 = -7$
Check
$$23 + (-8) = 15$$
$$-23 + 8 = -15$$
$$-7 + (-8) = -15$$

EXAMPLE 8   Subtract:
(a) $0 - 10$
(b) $0 - (-10)$
(c) $-10 - 10$
(d) $-10 - (-10)$

Solution
(a) $0 - 10 = 0 + (-10) = -10$
(b) $0 - (-10) = 0 + 10 = 10$
(c) $-10 - 10 = -10 + (-10) = -20$
(d) $-10 - (-10) = -10 + 10 = 0$

EXAMPLE 9   Write as a single integer: $-(5 - 12)$
Solution
$$-(5 - 12) = -[5 + (-12)] = -(-7) = 7$$
(Note the use of the negative of a negative property.)

PRODUCT OF TWO REAL NUMBERS—Multiplication

The product of two nonzero real numbers having the *same* sign is *positive.*
The product of two nonzero real numbers having *different* signs is *negative.*
In symbols, if r and s are any nonzero real numbers,

1. $(r)(s) = rs$ 3. $(-r)(s) = -rs$
2. $(-r)(-s) = rs$ 4. $(r)(-s) = -rs$

The numbers 1 and 0 are such special numbers that their properties need to be restated.

PRODUCTS INVOLVING 1 AND 0

Identity Property of Multiplication

The product of any real number and 1 is that number. 1 is the multiplicative identity, and multiplying a real number by 1 does not change the value of that number. In symbols, if r is any real number, then

$$1 \cdot r = r \cdot 1 = r$$

Zero Property of Multiplication

The product of any real number and 0 is 0. In symbols, if r is any real number, then

$$0 \cdot r = r \cdot 0 = 0$$

EXAMPLE 10 Find the indicated products:

(a) 5(8)
(b) 5(−8)
(c) (−5)(8)
(d) (−5)(−8)

Solution

(a) $5(8) = 40$
(b) $5(-8) = -40$
(c) $(-5)(8) = -40$
(d) $(-5)(-8) = 40$

EXAMPLE 11 Multiply:

(a) 4(−1)
(b) −4(1)
(c) 4(0)
(d) (0)(−4)

Solution

(a) $4(-1) = -4$
(b) $-4(1) = -4$
(c) $4(0) = 0$
(d) $(0)(-4) = 0$

EXAMPLE 12 Find the product of (25)(−3)(4) in two different ways.

Solution

$$(25)(-3)(4) = (-75)(4) = -300$$
$$(25)(-3)(4) = (-3)(25)(4) = (-3)(100)$$
$$= -300$$

We defined subtraction of real numbers in terms of addition. We will now define division in terms of multiplication.

QUOTIENT OF TWO REAL NUMBERS—Division

If r and s are any real numbers, and $s \neq 0$, then $\dfrac{r}{s} = t$ if and only if there is exactly one real number t so that $r = st$.

The expression "if and only if" means that the definition works two ways:

1. If $\dfrac{r}{s} = t$, then $r = st$; and

2. If $r = st$, then $\dfrac{r}{s} = t$

EXAMPLE 13 Find the indicated quotients:

(a) $\dfrac{20}{4}$

(b) $\dfrac{-20}{4}$

(c) $\dfrac{20}{-4}$

(d) $\dfrac{-20}{-4}$

Solution

(a) $\dfrac{20}{4} = 5$ because $20 = (4)(5)$

(b) $\dfrac{-20}{4} = -5$ because $-20 = (4)(-5)$

(c) $\dfrac{20}{-4} = -5$ because $20 = (-4)(-5)$

(d) $\dfrac{-20}{-4} = 5$ because $-20 = (-4)(5)$

The quotient of two nonzero real numbers having the *same* sign is *positive*.

The quotient of two nonzero real numbers having *different* signs is *negative*.

The quotient $\dfrac{0}{s} = 0$ for any real number s different from zero. *The quotients $\dfrac{s}{0}$ and $\dfrac{0}{0}$ are undefined.*

EXAMPLE 14 Find the indicated quotients, if they exist:

(a) $\dfrac{0}{52}$

(b) $\dfrac{0}{-52}$

(c) $\dfrac{52}{0}$

(d) $\dfrac{0}{0}$

Solution

(a) $\dfrac{0}{52} = 0$ because $0 = (52)(0)$

(b) $\dfrac{0}{-52} = 0$ because $0 = (-52)(0)$

(c) $\dfrac{52}{0}$ is undefined. There is no number t such that $52 = 0 \cdot t$ since $0 \cdot t = 0$ for all real numbers t.

(d) $\dfrac{0}{0}$ is undefined. By definition, if $\dfrac{0}{0} = t$, then there must be exactly one real number t such that $0 = 0 \cdot t$. In this case, there are many possibilities since $0 = 0 \cdot t$ for all real numbers t.

EXAMPLE 15 Perform the indicated operations, if possible:

(a) $\dfrac{7 - 22}{-5}$

(b) $\dfrac{6 - 6}{3}$

(c) $\dfrac{4 - (-4)}{4 - 4}$

(d) $\dfrac{8 - 2}{2 - 8}$

Solution

(a) $\dfrac{7 - 22}{-5} = \dfrac{-15}{-5} = 3$

(b) $\dfrac{6 - 6}{3} = \dfrac{0}{3} = 0$

(c) $\dfrac{4 - (-4)}{4 - 4} = \dfrac{4 + 4}{4 + (-4)} = \dfrac{8}{0}$, undefined

(d) $\dfrac{8 - 2}{2 - 8} = \dfrac{6}{-6} = -1$

EXERCISES 1.4

Perform the indicated additions in Exercises 1–20. (Examples 1–5)

1. $-12 + (-18)$

2. $-12 + 18$

3. $14 + (-9)$

4. $15 + (-25)$

5. $-5 + 5$

6. $-5 + (-5)$

7. $-7 + 0$

8. $0 + (-7)$

9. $(-6) + (-9) + (-5)$
10. $-12 + 4 + (-2)$
11. $-9 + (-6 + 6)$
12. $[7 + (-7)] + 8$
13. $[20 + (-15)] + (-10)$
14. $20 + [(-15) + (-10)]$

15. $-3 + [(-5) + (-2)]$
16. $[(-4) + (-5)] + (-6)$
17. $[10 + (-20)] + (-30)$
18. $10 + [(-20) + (-30)]$
19. $(-3 + 3) + (-2)$
20. $-3 + [3 + (-2)]$

Perform the indicated subtractions in Exercises 21–40. (Examples 6–9)

21. $7 - 3$
22. $3 - 7$
23. $7 - (-3)$
24. $-3 - 7$
25. $-6 - 4$
26. $-6 - (-4)$
27. $6 - 6$
28. $-4 - (-4)$
29. $-7 - 7$
30. $8 - (-8)$

31. $-20 - 8$
32. $-15 - (-10)$
33. $22 - 12$
34. $17 - 19$
35. $0 - 2$
36. $0 - (-5)$
37. $-8 - (-15)$
38. $8 - 15$
39. $-14 - 9$
40. $-12 - 7$

Perform the indicated operations in Exercises 41–60.

41. $(-5 + 5) - 9$
42. $(-1) - [(-2) - (-3)]$
43. $2 + (8 - 20)$
44. $0 - (6 - 12)$
45. $27 + [59 + (-27)]$
46. $(-38) + (38 - 26)$
47. $(-65) + [46 - (-65)]$
48. $(-6) + (7 - 10)$
49. $50 - (20 - 20)$
50. $[-(-8) - 2] - 3$

51. $20 - (10 - 5)$
52. $(20 - 10) - 5$
53. $-12 - (14 - 17)$
54. $-8 - (-4 + 4)$
55. $35 - [(-20) - (-20)]$
56. $[-65 - (-65)] - 40$
57. $-3 - [4 + (-2)]$
58. $-3 - [4 - (-2)]$
59. $(-3 - 4) + (-2)$
60. $(-3 - 4) - (-2)$

Find the indicated products in Exercises 61–80. (Examples 10–12)

61. $(5)(-4)$
62. $(-5)(4)$
63. $(-3)(-7)$
64. $(5)(-20)$
65. $(-3)(-2)(4)$
66. $(-3)(-2)(-4)$
67. $(3)(-2)(4)$
68. $(-5)(2)(-3)$
69. $-(-3)(5)$
70. $(0)(-3)$

71. $(10 - 6)(6 - 10)$
72. $(3 - 8)(6 - 6)$
73. $(-5 + 4)(-7 + 7)$
74. $-7(8 - 5)$
75. $-7(5 - 8)$
76. $[(-3)(5)] - [4(-2)]$
77. $(-3)(-3)(-3)$
78. $(-2)(-2)(-2)(-2)$
79. $-(7 - 5)(7 - 4)(7 - 3)$
80. $(8 - 1)(8 - 2)(8 - 3)(8 - 4)$

Find the indicated quotients in Exercises 81–110. (Examples 13–15)

81. $\dfrac{-56}{7}$

82. $\dfrac{56}{-4}$

83. $\dfrac{21}{-7}$

84. $\dfrac{-21}{7}$

85. $\dfrac{-21}{-7}$

86. $\dfrac{21}{7}$

87. $\dfrac{0}{-6}$

88. $\dfrac{-6}{0}$

89. $\dfrac{-56}{8}$

90. $\dfrac{63}{-7}$

91. $\dfrac{-55}{-11}$

92. $\dfrac{42}{-7}$

93. $\dfrac{7-2}{2-7}$

94. $\dfrac{4-12}{-4}$

95. $\dfrac{-120}{(-4)+6}$

96. $\dfrac{8-8}{8-(-8)}$

97. $\dfrac{120}{5-17}$

98. $\dfrac{10-30}{40-45}$

99. $\dfrac{25-15}{25-35}$

100. $\dfrac{0-9}{0+(-9)}$

101. $\dfrac{-132}{-22}$

102. $\dfrac{8-8}{8}$

103. $\dfrac{8-(-8)}{-8}$

104. $\dfrac{-3}{-3-(-3)}$

105. $\dfrac{18-25}{25-18}$

106. $\dfrac{96-39}{39-96}$

107. $\dfrac{(-5)+(-5)}{(-5)-(-5)}$

108. $\dfrac{(-5)+(-5)}{(-5)+(-5)}$

109. $\dfrac{8-(-8)}{-8}$

110. $\dfrac{-4}{4-(-4)}$

1.5 COMBINED OPERATIONS; EVALUATION

In all the preceding exercises and examples, whenever the possibility of an ambiguous answer arose, grouping symbols such as parentheses or brackets were used to indicate how the problem was to be worked. For example, if the problem $8 - 9 + 7$ were written as $8 - (9 + 7)$, the answer is $8 - 16 = -8$. However, if the grouping were $(8 - 9) + 7$, the answer is $-1 + 7 = 6$. In order to avoid such ambiguities and at the same time to keep grouping symbols to a minimum, the following convention is adopted.

CONVENTION

Unless the grouping symbols indicate otherwise, the operations are to be performed in the following order:

1. Take any roots as read from left to right.
2. Raise to indicated powers as read from left to right.
3. Multiply and/or divide as read from left to right.
4. Add and/or subtract as read from left to right.

Taking roots and raising to powers higher than second or third will be discussed in subsequent chapters, but it is a good idea to establish the convention for the order of operations in its entire form.

Recall that raising to the second power is the same as squaring a number; that is, multiplying the number by itself. Thus $3^2 = 3 \cdot 3 = 9$, $(-2)^2 = (-2)(-2) = 4$, and in general, $x^2 = x \cdot x$. Similarly, raising a number to the third power is the same as cubing the number. Thus $3^3 = 3 \cdot 3 \cdot 3 = 27$ and $(-2)^3 = (-2)(-2)(-2) = -8$, and in general $x^3 = x \cdot x \cdot x$. Notice that when a number is squared or cubed, if a negative sign is to be included it must be in the parentheses. For instance, $(-3)^2 = (-3)(-3) = 9$, whereas $-3^2 = -(3)(3) = -9$.

TABLE 1.3   SUMMARY OF THE OPERATION SYMBOLS

| | | NUMBER SYMBOLS | |
OPERATION	TWO NUMERALS	NUMERAL AND LETTER	TWO LETTERS
Addition (sum)	$3 + 4$	$x + 4$	$x + y$
Subtraction (difference)	$7 - 3$	$x - 4$	$x - y$
Multiplication (product)	$3 \cdot 4$ or $3(4)$	$3x$	xy
Division (quotient)	$\dfrac{12}{3}$	$\dfrac{x}{3}$	$\dfrac{x}{y}$
Squaring	3^2	x^2	
Cubing	5^3	x^3	Not applicable
Square root	$\sqrt{9}$	$\sqrt{x}$	
Cube root	$\sqrt[3]{27}$	$\sqrt[3]{x}$	

A **square root** of a number is a number whose square is the given number.

A **cube root** of a number is a number whose cube is the given number.

Thus a square root of 9 is 3 because 9 is the square of 3. A square root of 9 is also -3 because $(-3)(-3) = 9$. Similarly, a cube root of 27 is 3 because $(3)(3)(3) = 27$, and a cube root of -8 is -2 because $(-2)(-2)(-2) = -8$.

Table 1.3, a summary of the operation symbols, is provided for quick reference.

EXAMPLE 1 Perform the indicated operation: $(8 - 5)^2 - 13$

Solution

1. Simplify the expression inside parentheses: $8 - 5 = 3$. Thus

$$3^2 - 13$$

2. Raise to power: $3^2 = 9$. Thus

$$9 - 13$$

3. Perform subtraction: $9 - 13 = -4$. Therefore

$$(8 - 5)^2 - 13$$
$$3^2 - 13$$
$$9 - 13 = -4$$

EXAMPLE 2 Simplify $3 + 2(3)^2 - (5 - 2)$.

Solution

$$3 + 2(3)^2 - (5 - 2)$$
$$= 3 + 2(3)^2 - 3 \qquad \text{(Simplifying parentheses)}$$
$$= 3 + 2 \cdot 9 - 3 \qquad \text{(Raise to power)}$$
$$= 3 + 18 - 3 \qquad \text{(Multiply)}$$
$$= 18 \qquad \text{(Add and subtract from left to right)}$$

EXAMPLE 3 Simplify $\dfrac{8^2 + 8 \cdot 6}{2 \cdot 8 + 2 \cdot 6}$.

Solution The bar used in division is a grouping symbol, so the operations in the numerator and in the denominator must be done before the division:

$$\frac{8^2 + 8 \cdot 6}{2 \cdot 8 + 2 \cdot 6} = \frac{64 + 8 \cdot 6}{2 \cdot 8 + 2 \cdot 6} \qquad \text{(Squaring first)}$$
$$= \frac{64 + 48}{16 + 12} \qquad \text{(Multiply)}$$
$$= \frac{112}{28} \qquad \text{(Add)}$$
$$= 4 \qquad \text{(Divide)}$$

With the aid of the properties of the real numbers, especially the substitution axiom, the operations defined on the real numbers, and the convention for the order of operations, we are now ready to evaluate algebraic expressions and formulas.

EXAMPLE 4   Evaluate $3(x + 4) - \dfrac{x}{2}$ for $x = 6$.

Solution   In the expression just given re-place x with 6 wherever x occurs (justified by the substitution axiom). It is a good idea to use parentheses when making this substitution, especially when the replacement is a negative number.

$$3(x + 4) - \frac{x}{2} = 3[(6) + 4] - \frac{(6)}{2}$$
$$= 3(10) - 3$$
$$= 30 - 3$$
$$= 27$$

EXAMPLE 5   Evaluate $2x^2 - 3xy - y^2$ for $x = 3$ and $y = -4$.

Solution

$$2x^2 - 3xy - y^2 = 2(3)^2 - 3(3)(-4) - (-4)^2$$
$$= 2(9) - 3(-12) - (16)$$
$$= 18 + 36 - 16$$
$$= 38$$

EXAMPLE 6   The area of a circle is given by the formula $A = \pi r^2$, where A is the area and r is the length of the radius of the circle. π is a constant, which may be approximated by 3.14. Find the area of the circle whose radius is 4 centimeters (cm).

Solution

$$A = \pi r^2; \quad A = 3.14(4)^2$$
$$A = 3.14(16)$$
$$A = 50.24 \text{ square}$$
$$\text{centimeters (cm}^2)$$

Note that this is an approximation be-cause we used the approximate value 3.14 for π.

EXAMPLE 7   A formula for converting degrees Fahrenheit to de-grees Celsius is

$$C = \frac{5(F - 32)}{9}$$

where F stands for the temperature in degrees Fahrenheit and C in degrees Celsius.
Find C for $F = -4$ degrees.

Solution

$$C = \frac{5(F - 32)}{9}$$
$$C = \frac{5[(-4) - 32]}{9} = \frac{5(-36)}{9}$$
$$= -20 \text{ degrees}$$

EXERCISES 1.5

Evaluate in Exercises 1–50:

1. $x^2 + 10x + 35$ for $x = 3$

2. $2x^3 + 4x - 48$ for $x = 4$

3. $2x^3 - 4x^2 + 8x - 10$ for $x = 5$

4. $x^3 + x^2 - x$ for $x = -2$

5. $(x + 2)(x - 3)$ for $x = 3$

6. $(x + 4)(8 - x)$ for $x = 2$

7. $2(x - 3)^2$ for $x = 4$

8. $2(3 - x)^2$ for $x = 4$

9. $x(x + 1)(x - 2)$ for $x = 6$

10. $x^2(4 - x)^2$ for $x = 2$

11. $\dfrac{y(y - 1)}{2}$ for $y = 10$

12. $\dfrac{a(a + 1)}{3}$ for $a = 3$

13. $(n + 1)(n - 1)$ for $n = -3$

14. $(y + 3)(y - 3)$ for $y = -7$

15. $x^2 + 2xy + y^2$ for $x = 5$, $y = 3$

16. $x^2 + 3xy + 5y^2$ for $x = 5$, $y = 2$

17. $x^2 + y^2$ for $x = 5$, $y = 3$

18. $x^3 + y^3$ for $x = 2$, $y = 3$

19. $c^2 - a^2$ for $c = 3$, $a = -2$

20. $\sqrt{x^2 - y^2}$ for $x = 25$, $y = 24$

21. $\dfrac{a^3 - b^3}{a - b}$ for $a = 5$, $b = 2$

22. $\dfrac{a^2 - b^2}{a + b}$ for $a = 3$, $b = 2$

23. $\dfrac{3x}{5(x - 7)}$ for $x = 10$

24. $ab(a + b^2)$ for $a = 3$, $b = 5$

25. $a^3 - \dfrac{a}{2}$ for $a = 6$

26. $\dfrac{6x}{2x + 4}$ for $x = 4$

27. $x + y + z$ for $x = 7$, $y = -5$, $z = -9$

28. $x - y - z$ for $x = 12$, $y = 3$, $z = -8$

29. $x - y + z$ for $x = -6$, $y = -7$, $z = 8$

30. $x + y - z$ for $x = -15$, $y = 19$, $z = 21$

31. $a^2 + b^2 - c^2$ for $a = 2$, $b = -2$, $c = -1$

32. $a^2 - ab + b^2$ for $a = 7$, $b = -6$

33. $\dfrac{rs}{r - s - 1}$ for $r = 10$, $s = 7$

34. $\dfrac{r^2 + s^2 + 1}{r - s + 1}$ for $r = 4$, $s = -2$

35. $\dfrac{x + y + z}{3}$ for $x = 20$, $y = -17$, $z = -15$

36. $2x + 2y - 1$ for $x = 5$, $y = -4$

37. $\dfrac{c}{a} + \dfrac{d}{b} + \dfrac{a}{b}$ for $a = -6$, $b = -3$, $c = 18$, $d = -15$

38. $c^2 - a^2 - b^2$ for $c = 11$, $a = 2$, $b = 6$

39. $\dfrac{ac + bd}{a + b}$ for $a = 8$, $c = 3$, $b = 10$, $d = -6$

40. $a^2 - ab + b^2$ for $a = -4$, $b = -5$

41. $\dfrac{p + q - w}{2}$ for $p = 20$, $q = 0$, $w = -25$

42. $\dfrac{xyz}{x - y - z}$ for $x = 5$, $y = 6$, $z = -2$

43. $xy + xz + yz$ for $x = 5$, $y = -4$, $z = -2$

44. $a^2 + b^2 + c^2$ for $a = 2$, $b = -3$, $c = -4$

45. $(n + 1)^2 - (n + 1)(n - 1) + (n - 1)^2$ for $n = -5$

46. $b^2 - 4ac$ for $a = -3$, $b = -5$, $c = -4$

47. (Temperature scales: Fahrenheit to Celsius)

$$C = \frac{5(F - 32)}{9}$$

Find C for $F = 23$ degrees.

48. (Temperature scales: Celsius to Fahrenheit)

$$F = \frac{9C}{5} + 32$$

Find F for $C = -15$ degrees.

49. (Center of mass)

$$x = \frac{MD + md}{M + m}$$

Find x for $M = 150$ g, $D = 40$ cm, $m = 50$ g, $d = -20$ cm.

50. Find x in Exercise 49 for $M = 25$ lb, $D = -6$ ft, $m = 15$ lb, $d = -2$ ft.

In Exercises 51–52 (mirrors),

$$f = \frac{pq}{p + q}$$

where

f = local length of mirror (positive for concave mirrors and negative for convex mirrors)

p = object distance from mirror (positive when object is in front of the mirror)

q = image distance from mirror (positive when image is in front of the mirror and negative when image is behind the mirror)

51. Find f for $p = 45$ cm, $q = -180$ cm.

52. Find f for $p = 6$ ft, $q = -3$ ft.

53. (Thermodynamics)

$E = Q - W$

where

E = change in internal energy

Q = change in heat (positive when added to the system and negative when removed)

W = work done (positive when done by the system and negative when done on the system)

Find E for each of the following cases:

(a) $Q = 0$, $W = -6$ ft-lb

(b) $Q = 0$, $W = 50$ joules

(c) $Q = 400$ joules, $W = -175$ joules

(d) $Q = -20$ Btu, $W = 0$

(e) $Q = 810$ Btu, $W = 470$ Btu

54. (Corrective lenses) $P = \dfrac{100(p + q)}{pq}$

where

P = power in diopters (positive for farsighted persons and negative for nearsighted)

p = object distance from lens (positive) in centimeters

q = image distance from lens (positive when on the opposite side of the lens as the object and negative when on the same side) in centimeters

Find P for each of the following cases:

(a) $p = 20, q = -50$ (b) $p = 300, q = -75$

55. (Flow of liquids) $P = \dfrac{d}{2}(V^2 - v^2)$

where

P = change in pressure (positive for increase and negative for decrease)

V = original speed of the liquid

v = new speed of the liquid

d = density of the liquid

Find P for each of the following cases:

(a) $d = 62.5, V = 3$ ft/sec, $v = 6$ ft/sec

(b) $d = 50, V = 4$ ft/sec, $v = 2$ ft/sec

56. (Acoustics) $F = \dfrac{fV}{V + v}$

where

f = observed frequency (pitch) of a moving source of sound waves

V = speed of sound in air

v = speed of the moving source (positive if moving away from the observer and negative if moving toward the observer)

Find F for $f = 460$ cycles/sec, $V = 1100$ ft/sec, $v = -88$ ft/sec.

In Exercises 57–58 (chemistry: capillary tubes),

$$h = \frac{10SC}{49dr}$$

where

h = *height in millimeters that the liquid rises* (+) *or is depressed* (−)

S = *surface tension in dynes per centimeter*

C = *constant depending on the liquid and the material of the tube*

d = *density of the liquid in grams per cubic centimeter*

r = *radius of the tube in millimeters*

57. Find h for $S = 490, C = -0.68, d = 13.6, r = 2$ and state whether the liquid rises or is depressed (mercury in a glass tube).

58. Find h for $S = 72.8, C = 1, d = 1, r = 4$ (water in a glass tube).

1.6 SOLUTION OF FIRST-DEGREE EQUATIONS IN ONE VARIABLE

One of the main purposes of the study of algebra is to learn the techniques for solving problems—problems arising from the natural sciences, the physical sciences, business, economics, and in recent times, the behavioral sciences. Equations and inequalities are the major tools used to solve problems from these areas. In this section we present the simplest type of equation: the **first-degree equation in one variable.** Techniques for the solution of other types of equations and inequalities, as well as some applications, are presented in later chapters.

An **equation** is a mathematical sentence stating that two expressions are equal. For example, $5 = 3 + 2$, $3x + 2 = 7$, and $C = 2\pi r$ are all equations.

An **open equation** is an equation containing one or more variables. An open equation is not classified as true or false, but becomes true or false when each variable is replaced by a numerical value.

The equation $2 + 4 = 6$ is called a **true equation** since it does not contain any variables and since it is known that the sum of 2 and 4 is 6.

The equation $7 - 3 = 8$ is called a **false equation** since it contains no variables and since it is known that the difference between 7 and 3 is 4 and not 8.

The open equation $x + 5 = 5 + x$ becomes true for all replacements of the variable. Such an equation is also called an **identity.**

The open equation $x + 5 = x + 1$ becomes false for all replacements of the variable. Equations such as these are sometimes called **contradictions.**

The open equation $x + 5 = 9$ is neither true nor false, but becomes true or false when the variable is replaced by a numerical value.

For example, if $x = 2$, then $x + 5 = 9$ becomes $2 + 5 = 9$, a false equation.

If $x = 4$, then $x + 5 = 9$ becomes $4 + 5 = 9$, a true equation.

An open equation that is sometimes true and sometimes false is also called a **conditional equation** because the equation becomes true on the condition that the variable is replaced by a certain numerical value. The conditional equation $x + 5 = 9$ is true on the condition that x is replaced by 4.

The number 4 is called a **solution** or **root** of the equation $x + 5 = 9$ because $x + 5 = 9$ becomes true when x is replaced by 4. The number 4 is also said to **satisfy** the equation $x + 5 = 9$.

An important concern of algebra is the process of finding all the solutions of an open equation. This process is called **solving the equation.**

DEFINITIONS

A **solution** or **root** of an open equation in one variable is a number that makes the equation true when the variable is replaced by this number.

The **solution set** of an open equation is the set of all solutions of the equation.

To **solve an equation** means to find the solution set of the equation.

Equivalent equations are equations that have the same solution set.

For example, all the following equations are equivalent because each has {5} for its solution set—that is, 5 is the only solution of each equation:

$$x + 3 = 8$$
$$6(x - 4) = 6$$
$$\frac{x}{7} = \frac{5}{7}$$
$$x = 5$$

To solve an equation in one variable, say x, we replace the given equation by successive equivalent equations until one side of the equation is reduced to x. For example, it will be shown that the equation $3x + 2 = 14$ can be solved by using the equivalent equations

$$3x + 2 = 14$$
$$3x = 12$$
$$x = 4$$

In the last equation the left side has been reduced or simplified to x, and we now say that the equation is solved. The following theorems are used to solve equations.

THE EQUIVALENCE THEOREMS

1. Addition Theorem

If the same number is added to each side of an equation, the resulting sums are again equal, and the two equations are equivalent.

2. **Subtraction Theorem**

If the same number is subtracted from each side of an equation, the resulting differences are equal, and the two equations are equivalent.

3. **Multiplication Theorem**

If each side of an equation is multiplied by the same nonzero real number, the resulting products are equal, and the two equations are equivalent:

4. **Division Theorem**

If each side of an equation is divided by the same nonzero real number, the resulting quotients are equal, and the two equations are equivalent.

The following examples illustrate the use of the equivalence theorems.

EXAMPLE 1   Solve $x - 6 = 4$.

Solution

$x - 6 = 4$ if and only if

$x - 6 + 6 = 4 + 6$ (Addition theorem—add 6 to each side)

$x + 0 = 10$

$x = 10$

Thus the solution is $x = 10$ and the solution set is {10}.

Check   For $x = 10$, $x - 6$ becomes $10 - 6 = 4$, and the statement is true.

EXAMPLE 2   Solve $x + 5 = 8$.

Solution

$x + 5 = 8$ if and only if

$x + 5 - 5 = 8 - 5$ (Subtraction theorem— subtract 5 from each side)

$x + 0 = 3$

$x = 3$

Thus $x = 3$ is the solution and {3} is the solution set.

Check   For $x = 3$, $x + 5 = 3 + 5 = 8$, a true statement.

EXAMPLE 3   Solve $\dfrac{x}{5} = 7$.

Solution $\dfrac{x}{5} = 7$ if and only if

$$5\left(\frac{x}{5}\right) = 5(7) \qquad \text{(Multiplication theorem—}$$
$$\qquad\qquad\qquad \text{multiply each side by 5)}$$
$$x = 35$$

Thus $x = 35$ is the solution and $\{35\}$ is the solution set.

Check   For $x = 35$, $\dfrac{x}{5} = \dfrac{35}{5} = 7$, a true statement.

EXAMPLE 4   Solve $8x = 24$.

Solution

$8x = 24$ if and only if

$$\frac{8x}{8} = \frac{24}{8} \qquad \text{(Division theorem—}$$
$$\qquad\qquad\quad \text{divide each side by 8)}$$
$$x = 3$$

The solution is $x = 3$ and $\{3\}$ is the solution set.

Check   For $x = 3$, $8x = 8(3) = 24$.

Equations such as $2x - 9 = 5$ and $x + 7 = 20$ are called first-degree equations in one variable.

DEFINITION

A **first-degree equation in one variable,** x, is an equation that can be expressed in the form $ax + b = 0$, where $a \neq 0$.

By using the equivalence theorems, any first-degree equation can be solved by transforming the given equation into a simpler equation.

EXAMPLE 5   Solve $2x - 6 = 8$.

Solution

$$2x - 6 = 8$$
$$2x - 6 + 6 = 8 + 6 \qquad \text{(Addition theorem—add 6 to both sides)}$$
$$2x = 14$$
$$\frac{2x}{2} = \frac{14}{2} \qquad \text{(Division theorem—divide each side by 2)}$$
$$x = 7$$

$x = 7$ is the solution and $\{7\}$ is the solution set.

Check   For $x = 7$, $2x - 6 = 8$ becomes $2(7) - 6 = 8$
$$14 - 6 = 8$$
$$8 = 8, \text{ true}$$

Since $2x - 6 = 8$ is true for $x = 7$, 7 is the solution of $2x - 6 = 8$.

EXAMPLE 6 Solve $3 - y = 6$.

Solution

$$3 - y = 6$$
$$-3 + 3 - y = -3 + 6 \qquad \text{(Add } -3 \text{ to each side)}$$
$$-y = 3$$
$$(-1)(-y) = (-1)(3) \qquad \text{(Multiply each side by } -1)$$
$$y = -3$$

The solution set is $\{-3\}$

Check If $y = -3$, then $3 - y = 3 - (-3) = 3 + 3 = 6$. Thus $3 - y = 6$ is true for $y = -3$.

Note in Example 6 that both sides were multiplied by -1 so that the final equivalent equation would have the form $y = a$ and *not* $-y = a$.

EXAMPLE 7 Solve $4(k + 3) = 10$.

Solution

$$4(k + 3) = 10$$
$$4k + 12 = 10 \qquad \text{(Distributive axiom)}$$
$$4k + 12 - 12 = 10 - 12 \qquad \text{(Subtraction theorem—subtract}$$
$$\qquad\qquad\qquad\qquad\qquad \text{12 from each side)}$$
$$4k = -2$$
$$\frac{4k}{4} = \frac{-2}{4} \qquad \text{(Division theorem—divide each}$$
$$\qquad\qquad\qquad \text{side by 4)}$$
$$k = -\frac{1}{2}$$

The solution is $-\dfrac{1}{2}$ and $\left\{-\dfrac{1}{2}\right\}$ is the solution set.

Check

$$4(k + 3) = 10$$
$$4\left(-\frac{1}{2} + 3\right) = 10$$
$$4\left(\frac{5}{2}\right) = 10$$
$$10 = 10, \text{ true}$$

EXAMPLE 8 Solve $2 = 2 - 5p$.

Solution

$$2 = 2 - 5p$$
$$2 + 5p = 2 - 5p + 5p \qquad \text{(Addition theorem—add } 5p \text{ to}$$
$$\qquad\qquad\qquad\qquad\qquad \text{each side)}$$
$$2 + 5p = 2$$
$$5p = 0 \qquad \text{(Subtraction theorem—subtract 2}$$
$$\qquad\qquad\qquad \text{from each side)}$$
$$\frac{5p}{5} = \frac{0}{5} \qquad \text{(Division theorem—divide each}$$
$$\qquad\qquad\qquad \text{side by 5)}$$
$$p = 0$$

The solution is 0. The solution set is $\{0\}$. Note that this is *not* the empty set.

Check
$$2 = 2 - 5p$$
$$2 = 2 - 5(0)$$
$$2 = 2 - 0$$
$$2 = 2, \text{ true}$$

EXAMPLE 9 Solve $x + 2 = x + 5$.

Solution

$$x + 2 = x + 5$$
$$x + 2 - 2 = x + 5 - 2$$
$$x = x + 3$$
$$x - x = x - x + 3$$
$$0 = 3$$

But 0 can never equal 3. Therefore there is no value of x which makes the equation true, and the solution set is the empty set, $\varnothing$.

EXAMPLE 10 Solve $5(x - 4) = 5x - 20$.

Solution

$$5(x - 4) = 5x - 20$$
$$5x - 20 = 5x - 20$$

Since the left side and the right side of the equation are identical, the statement is true regardless of the value of x; in other words, all real numbers satisfy the equation and the solution set is R, the set of real numbers.

EXAMPLE 11 Solve $3x + 25 + 4x = 1 + x$.

Solution

It is usually desirable to simplify each side as much as possible before attempting the solution.

The left side of the equation is $3x + 25 + 4x$.

Combine like terms to obtain $7x + 25$. Thus

$$3x + 25 + 4x = 1 + x$$

$7x + 25 = 1 + x$	(Combine like terms of left side)
$7x + 25 - 25 = 1 + x - 25$	(Subtract 25 from each side)
$7x = -24 + x$	(Simplify right side)
$7x - x = -24 + x - x$	(Subtract x from each side)
$6x = -24$	
$\dfrac{6x}{6} = \dfrac{-24}{6}$	(Divide each side by 6)
$x = -4$	

Check

$$3x + 25 + 4x = 1 + x$$
$$3(-4) + 25 + 4(-4) = 1 + (-4)$$
$$-12 + 25 - 16 = 1 - 4$$
$$-28 + 25 = -3$$
$$-3 = -3, \text{ true}$$

The solution set is $\{-4\}$.

EXAMPLE 12 Solve $y + 7 - 4(y - 8) = 2(y + 2)$.

Solution

$y + 7 - 4(y - 8) = 2(y + 2)$	
$y + 7 - 4y + 32 = 2y + 4$	(Distributive axiom to remove parentheses)
$-3y + 39 = 2y + 4$	(Combine like terms on left side)
$-3y + 39 - 39 = 2y + 4 - 39$	(Subtract 39 from each side)
$-3y = 2y - 35$	(Simplify)
$-3y - 2y = 2y - 2y - 35$	(Subtract $2y$ from each side)
$-5y = -35$	(Simplify)
$\dfrac{-5y}{-5} = \dfrac{-35}{-5}$	(Divide each side by -5)
$y = 7$	

Thus $y = 7$ and $\{7\}$ is the solution set.
The check is left for you to do.

EXAMPLE 13 Solve $\dfrac{2t - 4}{4} + 5 = t - (3 + t)$.

Solution

$\dfrac{2t - 4}{4} + 5 = t - (3 + t)$	
$\dfrac{2t - 4}{4} + 5 = t - 3 - t$	(Remove parentheses on right side)
$\dfrac{2t - 4}{4} + 5 = -3$	(Simplify right side)
$\dfrac{2t - 4}{4} + 5 - 5 = -3 - 5$	(Subtract 5 from each side)
$\dfrac{2t - 4}{4} = -8$	(Simplify)
$4\left(\dfrac{2t - 4}{4}\right) = 4(-8)$	(Multiply each side by 4)
$2t - 4 = -32$	(Simplify)
$2t - 4 + 4 = -32 + 4$	(Add 4 to each side)
$2t = -28$	(Simplify)

$$\frac{2t}{2} = \frac{-28}{2} \qquad \text{(Divide each side by 2)}$$

$$t = -14$$

The solution set is $\{-14\}$. Verify this solution.

All the examples just given are worked in great detail showing every step to illustrate the justification for arriving at the solution. Many of these steps can be done mentally with practice.

All the equations in this section are **first-degree equations in one variable.**

The equivalence theorems for the solution of equations apply for all equations, even if more than one unknown is involved. This is particularly important when a formula is not in the form desired or convenient for the solution of a problem. Since a formula is an equation, it can be treated as such and changed to a desired equivalent equation by the same methods as shown in the preceding examples.

An equation involving more than one letter, some of which are constant, is called a **literal equation.**

EXAMPLE 14 Solve the formula for the circumference of a circle, $C = 2\pi r$ for r.

Solution

$$C = 2\pi r$$
$$2\pi r = C$$
$$\frac{2\pi r}{2\pi} = \frac{C}{2\pi} \qquad \text{(Divide each side by } 2\pi)$$
$$r = \frac{C}{2\pi}$$

Thus the radius of any circle is equal to the circumference of the circle divided by 2π.

EXAMPLE 15 $C = \frac{5}{9}(F - 32)$ is the formula for converting Fahrenheit temperature to degrees Celsius. Solve the formula for F—that is, rewrite it so that it can be readily used to convert Celsius temperature to Fahrenheit.

Solution

$$C = \frac{5}{9}(F - 32)$$

$$9C = 9\left(\frac{5}{9}\right)(F - 32) \qquad \text{(Multiply each side by 9)}$$

$$9C = 5(F - 32) \qquad \text{(Simplify)}$$
$$9C = 5F - 160 \qquad \text{(Distributive axiom)}$$
$$9C + 160 = 5F - 160 + 160 \qquad \text{(Add 160 to each side)}$$
$$5F = 9C + 160 \qquad \text{(Symmetric axiom)}$$

$$F = \frac{9C + 160}{5} \qquad \text{(Divide each side by 5)}$$

$$F = \frac{9}{5}C + 32 \qquad \text{(Further simplification of right side)}$$

Thus $F = \frac{9}{5}C + 32$.

EXAMPLE 16 If $mx - y + b = 0$, solve for each of the following:
(a) x
(b) y
(c) m
(d) b

Solution   In each instance, consider all letters constant except the letter for which you are solving.

(a) Solve for x:
$$mx - y + b = 0$$
$$mx - y + y + b = 0 + y \qquad \text{(Add } y \text{ to each side)}$$
$$mx + b - b = y - b \qquad \text{(Subtract } b \text{ from each side)}$$
$$mx = y - b \qquad \text{(Simplify)}$$
$$x = \frac{y - b}{m} \qquad \begin{array}{l}\text{(Divide each side by } m \\ [m \neq 0])\end{array}$$

(b) Solve for y:
$$mx - y + b = 0$$
$$mx - y + y + b = 0 + y \qquad \text{(Add } y \text{ to each side)}$$
$$mx + b = y \qquad \text{(Simplify)}$$
$$y = mx + b \qquad \text{(Symmetric axiom)}$$

(c) Solve for m:
$$mx - y + b = 0$$
$$mx - y + b - b = 0 - b \qquad \text{(Subtract } b \text{ from each side)}$$
$$mx - y + y = -b + y \qquad \text{(Add } y \text{ to each side)}$$
$$mx = y - b \qquad \text{(Commutative axiom, addition)}$$
$$m = \frac{y - b}{x} \qquad \text{(Divide each side by } x[x \neq 0])$$

(d) Solve for b:
$$mx - y + b = 0$$
$$mx + b = y \qquad \text{(Add } y \text{ to each side)}$$
$$b = y - mx \qquad \text{(Subtract } mx \text{ from each side)}$$

EXAMPLE 17   Solve $2x + y = 10$ for y.
Solution
$$2x + y = 10$$
$$-2x + 2x + y = 10 - 2x$$
$$y = 10 - 2x$$
or
$$y = -2x + 10$$

EXAMPLE 18   Solve $4x - y = 12$ for y.

Solution

$$4x - y = 12$$
$$-4x + 4x - y = -4x + 12$$
$$-y = -4x + 12$$
$$(-1)(-y) = (-1)(-4x + 12)$$
$$y = 4x - 12$$

EXERCISES 1.6

Solve and check the equations in Exercises 1–60.

1. $x + 5 = 7$

2. $x + 5 = -7$

3. $3x + 2 = 11$

4. $5 - 2y = 15$

5. $3x + 2 = x + 4$

6. $2x + 3 = x - 5$

7. $3 + 4x - 2 = 4 + 3x$

8. $x + 4 = 2x - 3$

9. $y + 1 - y$

10. $y + 1 = y - 1$

11. $5t + 7t - 2 = 5 + 7 - 2t$

12. $4 - (3x + 8) = 17$

13. $2(x - 5) = 2x - 10$

14. $2(x + 5) = 10$

15. $4 = 2z + 3$

16. $3 - x = x - 3$

17. $7x - 5(x - 2) = 20$

18. $9 - 6(2 - x) = 7x$

19. $y - 2 = 2 - y$

20. $x + 2 = x + 3$

21. $4 - 3(t + 2) = 7$

22. $3(s + 2) = 3s + 6$

23. $0 = 2(x + 5)$

24. $2(x - 3) - 3(x + 1) = 0$

25. $2(x - 3) + 3(x + 2) = x + 8$

26. $5(z - 1) - (1 - 4z) = 30$

27. $2x = 2(x - 2) + 4$

28. $2p - (6 - p) = 4p - 13$

29. $x + 2(3x - 5) - 4 = 0$

30. $8x + (5 - x) + 30 = 0$

31. $2(m + 3) = 3(2 - m)$

32. $3(x - 2) - (x + 2) = 4$

33. $5(3 - y) = 121 - 6(25 - y)$

34. $5t - (7t - 4) - 2 = 5 - (3t + 2)$

35. $x + (x + 1) + (x + 2) = 33$

36. $3r - 6 = 3(r - 2)$

37. $5(4p - 3) + 3(2p + 2) = 7p + 29$

38. $5(12 - x) = 7 - 2(6 - 4x)$

39. $30 - 3(2x - 3) = 5(3 - x)$

40. $12 = 8z - 3(7 - z)$

41. $4(2y - 5) - 8(y + 5) = 7$

42. $7 - 3x - (5 + 2x) = 12$

43. $p - 7 = 7 - p$

44. $p - 7 = p + 7$

45. $x - [1 - (x - 1)] = 3x$

46. $x - 2[x - 2(x - 2)] = 4$

47. $4t + 2(t - 6) = 0$

48. $4 = 2x + 10 + 4x$

49. $4x + (3 - 2x) = (4x + 3) - 2x$

50. $6(3y + 4) - 3(10y + 16) = 8(2y - 17)$

51. $6(y - 3) - 5(2y - 6) = 0$

52. $-2(t + 1) = 4(1 - 2t)$

53. $x + 4 = \dfrac{4(x - 1)}{4}$

54. $\dfrac{2 - 3z}{2} = 3 - (z + 2)$

55. $x + 2 = \dfrac{3x - 2}{5}$

56. $\dfrac{8y + 7}{5} = 2y - 3$

57. $\dfrac{4(y - 2)}{5} = y - 3$

58. $8 - \dfrac{7t + 3}{4} - 2(3 - t) = 0$

59. $\dfrac{2 - 5x}{4} = 1 - (x + 1)$

60. $\dfrac{5x + 2}{3} = 3x - 2$

Solve the equations in Exercises 61–80 for the indicated variables. (Examples 14–18)

61. $V = \dfrac{1}{3}\,bh$ (Volume of a cone); b

62. $A = a(a + 2s)$ (Surface area of a square pyramid); s

63. $P = I^2 r$ (Electric power); r

64. $F = \dfrac{9}{5}\,C + 32$ (Temperature conversion); C

65. $A = \dfrac{1}{2}\,h(a + b)$ (Area of a trapezoid); h

66. $A + B + C = 180$ (Sum of angles of a triangle); B

67. $P = a + b + c$ (Perimeter of a triangle); b

68. $D = A(n - 1)$ (Physics: law of small prisms); n

69. $D = A(n - 1)$ (Physics: law of small prisms); A

70. $Q = \dfrac{100M}{C}$ (Psychology); C

71. $y - 5x = 10$; y

72. $4x + y = 12$; y

73. $x + 4y = 12$; x

74. $x - 6y = 18$; x

75. $x + y + 1 = 0$; x

76. $x - 2y + 6 = 0$; x

77. $3x + 2y = 6$; y

78. $2x - 5y = 10$; x

79. $ax + by + c = 0$; y

80. $ax + by + c = 0$; x

DIAGNOSTIC TEST

In Problems 1–20 determine if each statement is true or false. If it is false, correct the statement.

1. Zero is a natural number.
2. The set of integers contains all the natural numbers, their additive inverses, and zero.
3. 15 is a prime number.
4. $|-3| = |3|$
5. $|-3| > 0$
6. $-3 - (-3) = 0$

7. $\dfrac{3 - (-3)}{3 + (-3)} = 0$

8. $\dfrac{3 + (-3)}{3 - (-3)} = 0$

9. $(-5)^2 = 25$

10. $-5^2 = -25$

11. $-(5^2) = -25$

12. If $x = 2$, then $3x^2 - 2x + 4 = 12$.

13. If $x = -2$, then $3x^2 - 2x + 4 = 20$.

14. If $x = 3$, then $x(x - 4)^2 = 9$.

15. If $x = 2$ and $y = -3$, then $x^2 + 2xy + y^2 = 1$.

16. If $a = 2$, $b = -1$, and $c = 3$, then $3a^2 + 2ab - bc^2 = 23$.

17. If $F = \dfrac{9}{5} C + 32$ and $C = 25$, then $F = 77$.

18. If $F = \dfrac{9}{5} C + 32$ and $C = -5$, then $F = 23$.

19. If $C = \dfrac{5}{9} (F - 32)$ and $F = 41$, then $C = 5$.

20. If $a = 5$ and $b = 9$, then $\dfrac{a - b}{b - a} = 1$.

Solve and check the equations in Problems 21–24.

21. $3x + 8 = 20$

22. $3(y + 2) = 4 + 2y$

23. $z - 9 = 9 - z$

24. $7 - \dfrac{x + 4}{3} = 2$

Solve the equations in Problems 25–27 for the indicated variable.

25. $x = 2y + 3$; y

26. $V = \dfrac{1}{3} bh$; h

27. $A = P + Prt$; r

Graph the sets in Problems 28–30 on a number line.

28. $\{x \,|\, x > 5\}$

29. $\{x \,|\, x \le 2\}$

30. $\{x \,|\, -2 < x \le 4\}$

REVIEW EXERCISES

For Exercises 1–25 refer to Section 1.1.
In Exercises 1–10 use the listing method to describe each set.

1. The natural numbers between 12 and 18.

2. The integers between $-\dfrac{1}{2}$ and $\dfrac{3}{2}$.

3. The even numbers between 11 and 15.
4. The first six prime numbers.
5. The even prime numbers between 1 and 50.
6. The positive integers.
7. The whole numbers between -5 and 5.
8. The first five odd prime numbers.
9. The negative whole numbers.
10. The prime numbers between 20 and 30.

In Exercises 11–16 let $S = \left\{-2, -\sqrt{2}, -\frac{1}{2}, 0, \frac{3}{5}, \sqrt{8}, 4, \frac{12}{\sqrt{2}}, 20\right\}$.

List each of the following subsets of S.

11. The whole numbers.
12. The natural numbers.
13. The integers.
14. The irrational numbers.
15. The rational numbers.
16. The real numbers.

Simplify Exercises 17–21.

17. $|-3|$
18. $-|2|$
19. $|13| + |-4|$

20. $-|3 - 3|$
21. $|-|-4||$

In Exercises 22–25 state each additive inverse.

22. 6
23. $-\dfrac{1}{2}$

24. $-(-2)$
25. $\sqrt{5}$

For Exercises 26–37 refer to Section 1.2.
In Exercises 26–30 insert the symbol $<$, $=$, *or* $>$ *between each pair of numbers so that the resulting statement is true.*

26. $0, -\dfrac{3}{5}$
27. $|-2|, |2|$
28. $-3, -|3|$

29. $-3, |-3|$
30. $7, -1$

Graph each of the sets in Exercises 31–37 on a number line and describe the set in words.

31. $\{x \mid x > -1\}$
32. $\{x \mid x < 3\}$
33. $\{x \mid x \le 4\}$
34. $\{x \mid x \ge 0\}$

35. $\{x \mid -2 < x < 3\}$
36. $\{x \mid 2 \le x < 4\}$
37. $\{x \mid -4 < x \le 0\}$

REVIEW EXERCISES

For Exercises 38–48 select the theorem or axiom that justifies each exercise from the list below.

(a) Distributive axiom

(b) Commutative axiom

(c) Associative axiom

(d) The sum of a number and its additive inverse is 0

(e) The product of a number and its reciprocal is 1.

(f) Reciprocal of a reciprocal

(g) Negative of a negative

(h) Closure; the sum of two real numbers is always a real number; the product of two real numbers is always a real number.

Assume all variables stand for nonzero real numbers.

For Exercises 38–48 refer to Section 1.3.

38. $3(x + 4) = 3x + 12$

39. $x \cdot \dfrac{1}{x} = 1$

40. $3(5x) = 15x$

41. $2(x - 5) = (x - 5)2$

42. xy is a real number

43. $\dfrac{1}{\dfrac{1}{x}} = x$

44. $(a + b) + c = a + (b + c)$

45. $x(2y) = 2xy$

46. $x(3 + y) = 3x + xy$

47. $(x + 2) + (-2) = x$

48. $-2 + [-(-2)] = 0$

For Exercises 49–81 refer to Sections 1.4 and 1.5.
Perform the indicated operations in Exercises 49–66.

49. $3[-2 - (-5)]$

50. $-2(3 - 10)$

51. $-3(4) - 5$

52. $-3(4 - 5)$

53. $4(3) + 3^2 - 2(6)$

54. $\dfrac{2 - 6}{6 - 2}$

55. $\dfrac{(-2) - (-2)}{(-2) + (-2)}$

56. $\dfrac{2 - (-2)}{2 + (-2)}$

57. $-3(2)(-1) + (-2)(3)$

58. $(-3)^2 + (-3)^3$

59. $(4 - 7)(7 - 4)$

60. $[4 - (-4)](8)$

61. $2(6 + 3)^2$

62. $2(6) + 3^2$

63. $-(-3)^2$

64. $-(3^2)$

65. $\dfrac{3^2 + 3(4)}{2(3) + 2(4)}$

66. $8 - (3 - 7)^2$

Evaluate in Exercises 67–77.

67. $(x + 2)(x - 3)$ for (a) $x = 5$, (b) $x = 0$, (c) $x = -5$

68. $\dfrac{4x}{6 - x}$ for (a) $x = 8$, (b) $x = 0$, (c) $x = -2$

69. $\dfrac{x^2 - 16}{x - 4}$ for (a) $x = 2$, (b) $x = 0$, (c) $x = -5$

70. $3x - (7 - x)$ for (a) $x = 4$, (b) $x = 0$, (c) $x = -2$

71. $x - [x - (x - 2)]$ for (a) $x = 7$, (b) $x = 0$, (c) $x = -1$

72. $10 - 2[x - (5 - 3x)]$ for (a) $x = 1$, (b) $x = 3$, (c) $x = -2$

73. $\dfrac{(x - 1)(x - 2)(x - 3)}{(-1)(-2)(-3)}$ for (a) $x = 9$, (b) $x = 1$, (c) $x = -3$

74. $\dfrac{x^2 + 3xy - 10y^2}{x - 2y}$ for (a) $x = 3$, $y = 2$

(b) $x = -4$, $y = -5$

75. $x - y + z$ for (a) $x = 7$, $y = 9$, $z = 11$

(b) $x = 5$, $y = -2$, $z = -6$

76. $x - y - z$ for (a) $x = 25$, $y = 15$, $z = 35$

(b) $x = -5$, $y = -10$, $z = 15$

77. xyz for (a) $x = 7$, $y = -4$, $z = -25$

(b) $x = -4$, $y = -3$, $z = -2$

78. (Slope of a line)

$m = \dfrac{Y - y}{X - x}$

Find m for $Y = 7$, $y = 10$, $X = -1$, $x = -7$.

79. (Flow of liquids)

$P = p + \dfrac{d}{2}(v^2 - V^2)$

Find P for $p = 50$, $d = 1$, $v = 20$, $V = 40$.

80. (Medication for child)

$C = \dfrac{WA}{150}$

Find C for $W = 60$ lb, $A = 200$ mg.

81. (Radio and television)

$r = \dfrac{R(E - G)}{E}$

Find r for $R = 2$ megohms, $E = -12$ volts, $G = 9$ volts.

For Exercises 82–120 refer to Section 1.6.
Solve and check the equations in Exercises 82–100.

82. $6x - 4 = 26$

83. $2 - 3x = 20$

84. $6y + 9 - 5y = 3$

85. $2(y + 3) + 5 = 3y - 4$

86. $6 - \dfrac{x + 2}{3} = 4$

87. $2x + 6 = 0$

88. $4z - (z - 5) = 4 - (z + 7)$

89. $3(p + 2) = 3p + 6$

90. $x - 2 = 2 - x$

91. $12 - 6(3 - x) = 7x$

92. $\dfrac{x - 7}{3} = 0$

93. $5 - \dfrac{x + 3}{2} = 6$

94. $x + 4 = x - 4$

95. $2(x - 3) - 3(x + 1) = 0$

96. $\dfrac{m + 7}{2} = 5$

97. $\dfrac{5x - (x - 2)}{3} = 6$

98. $2(x - 3) + 3(x + 2) - x = 8$

99. $2(x + 3) = 3(x + 2)$

100. $x - 3[x - 3(x - 3)] = 1$

In Exercises 101–120 solve for the indicated variable.

101. $3x - y = 5$; y

102. $3x - y = 5$; x

103. $2x + 5 + y = 0$; y

104. $2x + 5 + y = 0$; x

105. $5x - 2y = 10$; x

106. $5x - 2y = 10$; y

107. $y = 3x + 2$; x

108. $x = 3y + 2$; y

109. $ax + by = c$; y

110. $ax + by = c$; x

111. $C = \pi d$; d

112. $A = P + Prt$; t

113. $P = x + y + z$; y

114. $S = 2a + 2b$; b

115. $R = \dfrac{g}{T - t}$; t

116. $M = n(n + 2d)$; d

117. $B = 15 - \dfrac{A}{2}$; A

118. $W - 5H + 190 = 0$; H

119. $A + 2B - 3C = 20$; B

120. $A + 2B - 3C = 20$; C

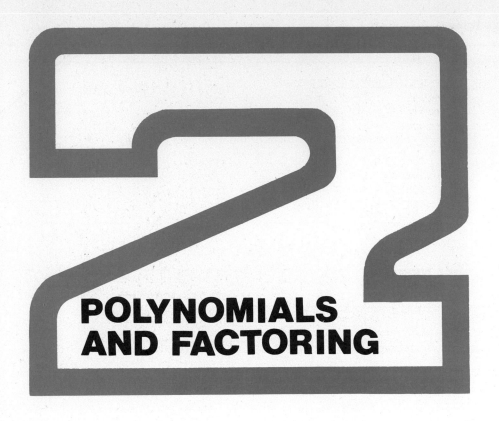

POLYNOMIALS AND FACTORING

I n Chapter 1 we discussed the basic operations of addition, subtraction, multiplication, division, and raising to a power. Some definitions involving these operations are needed before we can deal with the important class of algebraic expressions known as polynomials.

2.1 ALGEBRAIC EXPRESSIONS

DEFINITION

An **algebraic expression** is a combination of numerals, constants, and variables, and their indicated sums, differences, products, quotients, powers, and roots.

Examples of algebraic expressions are $2x + 3$, xy^2, $\dfrac{y - 1}{z}$, and $x^2 - 5xy + 2$.

An important algebraic expression that will lead to the definition of a polynomial is the monomial.

DEFINITION

A **monomial** is:

1. A constant;
2. A term of the form cx^n, where c is a constant, x is a variable, and n is a natural number; or
3. A product of terms as described in (2).

Each of the following is an example of a monomial:

$$3, \; x^2, \; 5x^3, \; \sqrt{2}xy, \; x^2y^9, \; \frac{1}{2}\,xyz^5$$

$\sqrt{x}$ is *not* a monomial; neither is $\dfrac{1}{x}, \dfrac{5}{x^2}$, or any other term that does not fit the description of being a constant or of the form cx^n, where c is constant and n is a natural number.

A monomial that has only one variable is called a **monomial in one variable.** For example, $3x$, $15x^3$, and $-2y^2$ are each a monomial in one variable. The monomial $4x^2y^3$ is not a monomial in one variable but a **monomial in two variables.**

A definition is needed here to explain what is meant by x^n, where n is a natural number. In Chapter 1 the operation "raising to a power" was briefly discussed. For example,

$$x^2 = x \cdot x$$
$$x^3 = x \cdot x \cdot x$$

DEFINITION OF x^n

If x is a real number and n is a natural number, then

$$n \text{ factors}$$

$$x^1 = x \quad \text{and} \quad x^n = \overbrace{x \cdot x \cdot x \cdot \ldots \cdot x} \quad \text{if} \quad n > 1$$

EXAMPLE 1

$$x^2 \cdot x^3 = (x \cdot x) \cdot (x \cdot x \cdot x)$$
$$= x \cdot x \cdot x \cdot x \cdot x$$
$$= x^5$$

EXAMPLE 2

$$8x^2 \cdot 2x^3 = (8 \cdot x \cdot x) \cdot (2 \cdot x \cdot x \cdot x)$$
$$= (8 \cdot 2) \cdot (x \cdot x \cdot x \cdot x \cdot x) \qquad \text{(Associative and commutative}$$
$$= 16x^5 \qquad\qquad\qquad\qquad\qquad \text{axioms, multiplication)}$$

The preceding examples lead to the following theorem.

THE FIRST THEOREM OF EXPONENTS

The product of two powers having the same base is a power whose base is this common base and whose exponent is the sum of the exponents. In symbols,

$$x^m x^n = x^{m+n}$$

EXAMPLE 3

$$x^2 \cdot x^3 = x^{2+3} = x^5$$

EXAMPLE 4

$$8x^2 \cdot 2x^3 = (8 \cdot 2) \cdot (x^2 \cdot x^3) = 16(x^{2+3}) = 16x^5$$

EXAMPLE 5

$$5^2 \cdot 5^3 = 5^{2+3} = 5^5 = 3125$$

EXAMPLE 6

$$2^3 \cdot 3^2 \cdot 2^4 = 2^{3+4} \cdot 3^2 = 2^7 \cdot 3^2$$
$$= 128 \cdot 9$$
$$= 1152$$

It should be emphasized that the variable need not be the letter x but can be any letter that designates a real number.

EXAMPLE 7

$$a^3 \cdot a^7 = a^{3+7} = a^{10}$$

EXAMPLE 8

$$3y^2 \cdot 4y^3 = 3 \cdot 4 \cdot y^2 \cdot y^3 = 12y^{2+3} = 12y^5$$

Now consider raising a power to a power. For example, $(x^3)^2$.

By the definition of x^n,
$$(x^3)^2 = x^3 \cdot x^3$$
and by the first theorem of exponents,
$$x^3 \cdot x^3 = x^{3+3} = x^6$$
Therefore
$$(x^3)^2 = x^6$$
Also,
$$(x^2)^4 = x^2 \cdot x^2 \cdot x^2 \cdot x^2 = x^{2+2+2+2} = x^{4\cdot2} = x^8$$

The problem of raising a power to a power can be generalized by the following theorem of exponents:

THE SECOND THEOREM OF EXPONENTS

The power of a power is a power having the same base and an exponent that is the product of the exponents. In symbols,
$$(x^m)^n = x^{mn}$$

EXAMPLE 9
$$(3^2)^3 = 3^{2\cdot3} = 3^6 = 729$$

EXAMPLE 10
$$(x^3)^4 = x^{3\cdot4} = x^{12}$$

Note: Because multiplication is commutative, $x^{mn} = x^{nm}$; therefore $(x^m)^n = (x^n)^m$.

EXAMPLE 11

$(y^5)^m = y^{5m}$; m is a natural number.

Be careful when applying the first theorem of exponents. For example, $5x^2 \cdot 7y^3 = 35x^2y^3$. The theorem does not apply to the variables x and y since they are different. However, the following example illustrates that the theorem *may* be used for monomials in more than one variable when it applies.

EXAMPLE 12
$$(5x^2yz^3)(3xy^4) = 15xx^2yy^4z^3$$
$$= 15x^3y^5z^3$$

The third theorem applies to the simplification of the quotient of two powers that have the same base.

THE THIRD THEOREM OF EXPONENTS

The quotient of two powers having the same base is:

1. A power having the same base and an exponent equal to the difference (larger minus smaller) of the exponents, if the exponent of the numerator is larger than the exponent of the denominator; or

2. A reciprocal of a power having the same base and an exponent equal to the difference (larger minus smaller) of the exponents, if the exponent of the numerator is smaller than the exponent of the denominator; or

3. The number 1 if the exponents are equal.

In symbols,

1. $\dfrac{x^m}{x^n} = x^{m-n}$ if $m > n$ and $x \neq 0$

2. $\dfrac{x^m}{x^n} = \dfrac{1}{x^{n-m}}$ if $n > m$ and $x \neq 0$

3. $\dfrac{x^m}{x^n} = \dfrac{x^n}{x^n} = 1$ if $m = n$ and $x \neq 0$

EXAMPLE 13 $\dfrac{x^5}{x^2} = x^{5-2} = x^3;$ note that $5 > 2$. To verify, we can show that

$$\frac{x^5}{x^2} = \frac{x^3 \cdot x^2}{x^2} = x^3$$

EXAMPLE 14

$$\frac{2^7}{2^3} = 2^{7-3} = 2^4 = 16$$

EXAMPLE 15

$$\frac{x^2}{x^5} = \frac{1}{x^{5-2}} = \frac{1}{x^3};$$ note that $2 < 5$.

EXAMPLE 16

$$\frac{2^3}{2^7} = \frac{1}{2^{7-3}} = \frac{1}{2^4} = \frac{1}{16}$$

EXAMPLE 17

$$\frac{x^5}{x^5} = 1$$

The following two theorems are very closely related: Theorem 4 deals with the power of a product and Theorem 5 with the power of a quotient.

THE FOURTH THEOREM OF EXPONENTS

A power of a product is a product of two powers whose bases are factors of the product and each of whose exponents is the same as the exponent on the product. In symbols,

$$(xy)^n = x^n y^n$$

EXAMPLE 18 $(xy)^3 = x^3 y^3$; we can verify this result by showing
$$(xy)^3 = (xy)(xy)(xy) = (xxx)(yyy) = x^3 y^3$$

EXAMPLE 19 $(2z)^5 = 2^5 z^5 = 32 z^5$

EXAMPLE 20 $(-3x^2)^3 = (-3)^3(x^2)^3$ By Theorem 4
$$= -27x^6 \qquad \text{By Theorem 2}$$

Remember that by the symmetric axiom of equality,
     if $(xy)^n = x^n y^n$, then $x^n y^n = (xy)^n$
The following example uses this property.

EXAMPLE 21 Simplify $2^6 5^6$.

Solution   By Theorem 4,
$$2^6 5^6 = (2 \cdot 5)^6 = 10^6 = 1{,}000{,}000$$

THE FIFTH THEOREM OF EXPONENTS

A power of a quotient is a quotient of two powers whose bases are the numerator and denominator of the quotient and each of whose exponents is the same as the exponent on the quotient. In symbols,

If $y \neq 0$, then $\left(\dfrac{x}{y}\right)^n = \dfrac{x^n}{y^n}$ and $\dfrac{x^n}{y^n} = \left(\dfrac{x}{y}\right)^n$

EXAMPLE 22
$$\left(\frac{x}{y}\right)^4 = \frac{x^4}{y^4}$$

EXAMPLE 23
$$\left(\frac{-2}{x}\right)^6 = \frac{(-2)^6}{x^6} = \frac{64}{x^6}$$

EXAMPLE 24
$$\frac{(98)^5}{(49)^5} = \left(\frac{98}{49}\right)^5 = 2^5 = 32$$

Following is a summary of the five theorems of exponents. x and y stand for any real numbers except when they occur in the denominator of a fraction, in which case they cannot equal zero. n and m are natural numbers.

THE EXPONENT THEOREMS

Theorem 1 $x^n x^m = x^{n+m}$ Product of two powers

Theorem 2 $(x^n)^m = x^{nm}$ Power of a power

Theorem 3 $\dfrac{x^n}{x^m} = x^{n-m}$ if $n > m$ and $x \neq 0$

                $\dfrac{x^n}{x^m} = \dfrac{1}{x^{m-n}}$ if $m > n$ and $x \neq 0$ Quotient of two powers

                $\dfrac{x^n}{x^m} = 1$ if $m = n$ and $x \neq 0$

Theorem 4 $(xy)^n = x^n y^n$ Power of a product

Theorem 5 $\left(\dfrac{x}{y}\right)^n = \dfrac{x^n}{y^n}$ if $y \neq 0$ Power of a quotient

EXERCISES 2.1

Simplify Exercises 1–108 by using one or more of the exponent theorems. (Assume all exponents are natural numbers and all variables are nonzero real numbers.)

1. $x^5 x^3$

2. $y^4 y^2$

3. $3y(y^3)$

4. $7x(2x^2)$

5. $(2x^5)(3x^7)$

6. $(3x^3)(2x^2)(-4x)$

7. $\dfrac{x^7}{x^3}$

8. $\dfrac{c^3}{c^6}$

9. $(t^4)^3$

10. $\dfrac{x^2}{x^5}$

11. $(-5t)^4$

12. $a^5 a^7$

13. $\left(\dfrac{-n}{2}\right)^5$

14. $5^6 2^6$

15. $\left(\dfrac{-3}{y}\right)^4$

16. $\dfrac{x^4}{x^8}$

17. $2^4 5^4$

18. $-5(2^3)^2$

19. $(3ab)(-2a)(-4b)$

20. $(rst)(rs)(st)$

21. $(-z)(-z^2)(-z^3)$

22. $(xy^2)(3xy)$

23. $(5xz)(3x^2y)$

24. $(abc)(a^2b^2c^2)$

25. $(x^2)^4$

26. $(x^4)^2$

27. $(x^3)^2$

28. $(x^2)^3$

29. $5(z^2)^3$

30. $x(x^2)^3$

31. $(y^3)^4$

32. $(x^3)^2 \cdot x^2$

33. $xy(y^2)^5$

34. $(-x)^2$

35. $(-x^2)^3$

36. $(2xy)(xz^2)(-yz)$

37. $ab(b^4)^5$

38. $(-y^2)^2$

39. $-(y^2)^2$

40. $2(-3x)(xy)(y^2)^4$

41. $2^3 \cdot 2^2$

42. $3^3 \cdot 3^2$

43. $(2^2)^3$

44. $(3^3)^2$

45. $(3^2)^3$

46. $(-2^2)^3$

47. $[(-2)^2]^3$

48. $(-2)^3(-2)^2$

49. $4^2 \cdot 4$

50. $4^2 \cdot 4^3$

51. $3 \cdot 3^2$

52. $5^2 \cdot 5$

53. $4^2 \cdot 2^3$

54. $9 \cdot 3^3$

55. $\dfrac{(100)^6}{(50)^6}$

56. $-7(x^2)^5$

57. $\dfrac{(24)^7}{(48)^7}$

58. $(-a)^4(-a)^6$

59. $(-5x^2)^3$

60. $(-2y^3)^4$

61. $-2(y^4)^3$

62. $-(5x^3)^2$

63. $(a^2b^3)(ab^5)$

64. $(ab^2)^3(a^2b)^4$

65. $\dfrac{6y^6}{2y^2}$

66. $\dfrac{3^8x^7}{3^5x^{10}}$

67. $\dfrac{12y^{12}}{4y^4}$

68. $\dfrac{a^3b^5}{a^4b^2}$

69. $\left(\dfrac{a^3b^5}{ab^3}\right)^4$

70. $\dfrac{(8^2)^3}{(4^3)^2}$

71. $(-x^2)(-x)^2$

72. $(-x^3)(-x)^3$

73. $\dfrac{(10^4)^3}{(10^2)^5}$

74. $\dfrac{(10a^2)^3}{(10a^3)^2}$

75. $(5xy^2)^3$

76. $(-3x^2y^3z^4)^4$

77. $-3(a^2b^3c^4)^2$

78. $(x^4y^2)(x^3y^4)$

79. $\left(\dfrac{12a^2bc}{10ab^3}\right)^2$

80. $(c^3d^5)(cd^3)$

81. $\dfrac{5x^5}{2y^4} \cdot \dfrac{3y^6}{10x^3}$

82. $(-2a^2b^5)^5$

83. $\dfrac{(-y^4)^4}{-(y^4)^4}$

84. $\dfrac{-(x^3)^2}{(-x^3)^2}$

85. $\dfrac{10^6}{10^2 \cdot 10^8}$

86. $\dfrac{10^5 \cdot 10^7}{10^4}$

87. $\dfrac{(10^5)^2}{10^2 \cdot 10^5}$

88. $\dfrac{(10^6)^3}{10^3 \cdot 10^6}$

89. $\dfrac{(2^3 \cdot 3^4)^3}{(2^2 \cdot 3^2)^5}$

90. $\dfrac{2^6 \cdot 5^4}{2^3 \cdot 5^6}$

91. $(2^2)^3(5^2)^2$

92. $\dfrac{(2^4)^3(5^6)^2}{(10^3)^2}$

93. $\dfrac{9^6 \cdot 4^6}{6^6}$

94. $\dfrac{3^5 \cdot 3^{12}}{3^7 \cdot 3^{10}}$

95. $x \cdot x^n$

96. $\dfrac{x}{x^n},\ n > 1$

97. $\dfrac{3^{2n}}{3^n}$

98. $\dfrac{x^n}{x},\ n > 1$

99. $(y^n)^4$

100. $\left(\dfrac{x^3}{y^2}\right)^n$

101. $(x^2y)^n$

102. $\left(\dfrac{x^2}{5^2}\right)^n$

103. $\dfrac{x^n x}{x^{n+1}}$

104. $\dfrac{(x^n)^2}{(x^2)^{n+1}}$

105. $x(x^2)^n$

106. $\dfrac{(x^{n+1})^2}{x^2 x^n}$

107. $x^{n+1}x^n$

108. $\left(\dfrac{x^n}{y^n}\right)^n$

2.2 POLYNOMIALS: SUMS AND DIFFERENCES

In the preceding section a monomial was defined as an algebraic expression of the form cx^n where c is a constant, x is a variable, and n is a natural number; or a product of terms so described. A constant is a special form of monomial. We are now ready to define a polynomial as the algebraic sum of monomials.

DEFINITION

A **polynomial** is a monomial or an algebraic sum of monomials.

EXAMPLE 1 Each of the following is an example of a polynomial:
(a) $x^2 + 3x + 2$ is a polynomial in one variable.
(b) $15y^9 - y$ is a polynomial in one variable.
(c) $x^2 - 3xy + 2y^2$ is a polynomial in two variables.
(d) $9x^5 - 3x^4 + 2x^2 + 5x - 1$ is a polynomial in one variable.
(e) $x^2 + y^2 + z^2 - 4xyz$ is a polynomial in three variables.

Some common types of polynomials have special names. A **monomial** is a polynomial of one term, a **binomial** is a polynomial of two terms, and a **trinomial** is a polynomial of three terms.

Each of the following is an example of a binomial:
$x + 1, 3y^2 + y, -5x^2 - 2, a + b, x^2 - y^2$
Each of the following is an example of a trinomial:
$x^2 + 2x + 1, 5y^3 - 3y + 2, a + b + c, x^2 - 6xy + 9y^2$
Any indicated factor of a monomial is called a **coefficient** of the product of the other factors.

For example, in $5xy$, 5 is the coefficient of xy, $5x$ is the coefficient of y, and x is the coefficient of $5y$.

A **numerical coefficient** of a monomial is its constant factor.

For example, 2 is the numerical coefficient of $2x^3$, and -5 is the numerical coefficient of $-5xy$.

The **degree of a monomial** is the number of times a variable occurs as a factor.

For example, the degree of $7x^2$ is 2, the degree of $-4y^3$ is 3, the degree of $6x^2y^3$ is 5, and the degree of $3xy$ is 2.

The **degree of a polynomial** is the degree of the term of highest degree.

For example, the degree of $4x^3 - 5x^2 - 6x + 7$ is 3. The following table shows the general pattern of first-, second-, and third-degree polynomials in one variable and in two variables.

Assume a, b, c, d, e, and f are constants, and x and y any real numbers.

$ax + b$	$a \neq 0$	First-degree polynomial in one variable
$ax + by + c$	a and b *not* *both* zero	First-degree polynomial in two variables
$ax^2 + bx + c$	$a \neq 0$	Second-degree polynomial in one variable
$ax^2 + bxy + cx^2 + dx + ey + f$	a, b, and c not all zero	Second-degree polynomial in two variables
$ax^3 + bx^2 + cx + d$	$a \neq 0$	Third-degree polynomial in one variable

A polynomial is said to be **arranged in descending powers of a variable** when the term of greatest degree in this variable is written first, at the left, the term of next greatest degree is written second, and so on, with the term of least degree written last, at the right.

If the preceding order is reversed, the polynomial is said to be **arranged in ascending powers of a variable.**

For example, $x^3 + 5x^2 - 7x + 4$ is arranged in descending powers of x, and $7 + 3y - y^2$ is arranged in ascending powers of y.

Similar terms or **like terms** are terms whose literal factors are the same.

For example, $3x$ and $5x$ are like terms, $7x^2$ and $-4x^2$ are like terms, and $\frac{1}{2}xy$ and $-6xy$ are like terms. On the other hand, $3x$ and $3x^2$ are not like terms. Also, $5x$ and $5y$ are not like terms.

Polynomials are added or subtracted by using the commutative and associative axioms to rearrange the terms and by using the distributive axiom to combine similar terms.

EXAMPLE 2 Add $3x^2 + 4x - 7$ to $4x^2 + x + 2$.

Solution

$$(3x^2 + 4x - 7) + (4x^2 + x + 2)$$
$$= (3x^2 + 4x^2) + (4x + x) + (-7 + 2)$$
$$= (3 + 4)x^2 + (4 + 1)x + (-7 + 2)$$
$$= 7x^2 + 5x - 5$$

Most of these steps can be performed mentally, and as a rule, the final result can be written immediately after looking at the problem.

EXAMPLE 3 Simplify by performing the indicated operations:
$(x^2 + 2xy + 3y) + (3x^2 - 5xy + 4y)$

Solution

$$(x^2 + 2xy + 3y) + (3x^2 - 5xy + 4y)$$
$$= 4x^2 - 3xy + 7y$$

To subtract one polynomial from another, the subtraction problem must be changed to an addition problem by using the definition of subtraction: $r - s = r + (-s)$.

It is important to remember that the sign of each term of the subtrahend must be changed, as shown in the following example.

EXAMPLE 4 Subtract $4x^2 + xy - 6y^2$ from $2x^2 + 3xy - 9y^2$.

Solution

$$(2x^2 + 3xy - 9y^2) - (4x^2 + xy - 6y^2)$$ (Recalling that "subtract a from b" is translated as "$b - a$")

$$= (2x^2 + 3xy - 9y^2) + (-4x^2 - xy + 6y^2)$$ (Using the definition of subtraction)

$$= (2x^2 - 4x^2) + (3xy - xy) + (-9y^2 + 6y^2)$$ (Using the associative and commutative axioms several times)

$$= (2 - 4)x^2 + (3 - 1)xy + (-9 + 6)y^2$$ (Using the distributive axiom)

$$= -2x^2 + 2xy - 3y^2$$

Again, it is usually possible to write the final result immediately after inspection of the problem. Thus, by performing much of the work mentally, the solution would be written as illustrated in Example 5.

EXAMPLE 5 Subtract $5x - 2y + 4$ from $-3x + 2y + 4$.

Solution

$$(-3x + 2y + 4) - (5x - 2y + 4)$$
$$= (-3x + 2y + 4) + (-5x + 2y - 4)$$
$$= -8x + 4y$$

It is quite simple to check the addition or subtraction of polynomials by assigning a value to each variable.

EXAMPLE 6 Check the result of Example 5 by letting $x = 3$ and $y = 4$.

Solution   If $x = 3$ and $y = 4$,

$$5x - 2y + 4 = 5(3) - 2(4) + 4$$
$$= 15 - 8 + 4 = 11$$
$$-3x + 2y + 4 = -3(3) + 2(4) + 4$$
$$= -9 + 8 + 4 = 3$$

Now

$$(-3x + 2y + 4) - (5x - 2y + 4)$$
$$= 3 - 11 = -8$$

and

$$-8x + 4y = -8(3) + 4(4)$$
$$= -24 + 16 = -8$$

Therefore for $x = 3$ and $y = 4$,

$$(-3x + 2y + 4) - (5x - 2y + 4)$$
$$= -8x + 4y$$

From the check in Example 6 it is reasonable to conclude that $-8x + 4y$ is the correct answer for Example 5. However, there is a danger involved. Special values of the variables may produce a "check" even though the answer is wrong. This can usually be prevented by avoiding the use of the special numbers 0, 1, and -1 as replacement values.

Since the value of a polynomial changes when the variable is replaced by different numbers, the notation $P(x)$ (read: P of x, or the polynomial in x) is used to indicate the desired replacement for x and the value of the polynomial.

EXAMPLE 7 If $P(x) = x^3 + 2x^2 - 3x + 1$, find $P(0)$, $P(2)$, and $P(-1)$.

Solution $P(0)$ indicates that x is replaced by 0 in the given polynomial.

$$P(0) = (0)^3 + 2(0)^2 - 3(0) + 1 = 1$$

Similar replacements are made for $P(2)$ and $P(-1)$.

$$P(2) = (2)^3 + 2(2)^2 - 3(2) + 1$$
$$= 8 + 8 - 6 + 1 = 11$$
$$P(-1) = (-1)^3 + 2(-1)^2 - 3(-1) + 1$$
$$= -1 + 2 + 3 + 1 = 5$$

Thus $P(0) = 1$, $P(2) = 11$, and $P(-1) = 5$.

All letters used are purely arbitrary. $Q(x)$, $R(x)$ may also designate polynomials in x, $P(y)$ might represent a polynomial in one variable, y. This notation may also be used for polynomials in more than one variable; for instance, $P(x, y)$ may stand for a polynomial in two variables, x and y.

EXERCISES 2.2

Simplify the expressions in Exercises 1–20 and arrange the answer in descending powers of the variable whenever appropriate. (Examples 2–5)

1. $(7x^2 - 5x + 2) + (x^2 + x - 6)$
2. $(3a + 5b - 7) + (2a - 5b + 7)$
3. $(-2x - 3y - 4) + (2x - 3y - 4)$
4. $3(2a - 5) + 4(7 - 3a)$
5. $(5y^2 - 6y + 9) + (5y - 8 - 4y^2)$
6. $(2a - b - c) - (a + b + c)$
7. $(-r - s + t) - (r + s - t)$
8. $(8 - 2x + y) - (9 - 2x + y)$
9. $(1 + x - x^2) - (2 - x^2 + x^4)$
10. $3(2a - b) - 4(b - 2a)$

11. $(2x - 3y + 4z) + (5x + 7y - 9z)$
12. $(x^2 - xy + y^2) + (x^2 + xy - y^2)$
13. $7(x - 2) - 3(2 - x)$
14. $-5(2y + 1) + 6(2y - 3)$
15. $(a^2 - a^2b^2 + 2b^2) - (-a^2 + a^2b^2 - 2b^2)$
16. $(3t^2 + 4t - 7) + (-5t^2 + t - 1)$
17. $(x + y - 1) - (x - y + 1)$
18. $(3x - 1) - (1 - 3x)$
19. $(x^2 - x + 1) - (x^2 - x + 1)$
20. $-5(t + 5) + 2(5t + 1)$

In Exercises 21–40 perform the operations as directed. (Example 6)

21. Add $3x^2 + 5 - x$, $x - 8$, $x^2 + 4$. Check, using $x = 6$.
22. Subtract $2a^2 - 5b^2 - 4$ from $a^2 - b^2 + 1$. Check, using $a = 3$, $b = 2$.
23. From $-t^2 - 3t - 5$ subtract $-t^2 - 5t + 4$. Check, using $t = 5$.
24. Find the sum of $c^3 - 3c^2 - c - 7$ and $4c - 3c^2 - 6c^3 + 6$. Check, using $c = 1$.
25. Find the sum of $x^3 - 3x^2 + x - 5$ and $x^2 - 3x + 4$. Check, using $x = 2$.
26. From the sum of $3x^2 + 5x + 9$ and $x^2 - 7x + 14$ subtract $2x^2 - 3x + 5$. Check, using $x = 2$.
27. Subtract $3a^2 + 5ab - 6b^2$ from the sum of $2a^2 - 3ab + 4b^2$ and $5ab - 6b^2$. Check, using $a = 2$ and $b = 3$.
28. Add $4 - 3x^2$, $2 + 3x + 8x^2$, $3 - 2x^2 - 5x$. Check, using $x = 3$.
29. From $4a + 2b - 4c$ subtract $3a - 3b - 3c$. Check, using $a = 2$, $b = 3$, $c = 5$.
30. Subtract $2y^3 - 5y^2 - 3y + 9$ from $-4y^3 - 3y^2 - 4$. Check, using $y = 4$.
31. Subtract $2x^2 - 6$ from the sum of $-3x^2 + 5x$ and $4 - 7x$. Check, using $x = 1$.

32. Find the number that must be added to $2x^2 - 3xy - y^2$ to obtain $y^2 + xy + x^2$. Check, using $x = 5$, $y = 1$.

33. Subtract $x^3 - 17x + 1$ from the sum of $x^4 + 3x^3 - x^2 + 2$ and $5x^4 - x^3 + 4x - 7$. Check, using $x = 1$.

34. From the sum of $a^2 + 3ab + 4b^2$ and $2a^2 - 4ab + 5b^2$ subtract $3a^2 - 7ab + 8b^2$.

35. Subtract the sum of $x^4 + 3x^2 + 2$ and $2x^4 - 5x^3 + x^2 - 1$ from $6x^3 - 7x^2 + x$.

36. From $2x^3 - 4x^2y + 8xy^2 + 11y^3$ subtract the sum of $x^3 + y^3$ and $3x^2y - 5xy^2 - 7y^3$.

37. Find the sum of $a^3 + b^3 - a^2b + ab^2$, $a^2b + b^3 - 3a^3$, and $b^3 - 5ab^2 + a^2b$.

38. From $x^3 - x^2 - 3x + 2$ subtract the sum of $x^4 + 5x - 1$ and $x^3 + 3x^2 - 6x + 2$.

39. Subtract the sum of $x^3 + 2x^2y - xy^2 + y^3$ and $5x^2y + 6xy^2 - 2y^3$ from $4x^3 - 3x^2y + 3xy^2 + 4$.

40. Find the sum of $a^3 - 3a^2b + 5ab^2 - b^3$, $2a^3 + 5a^2b - 2ab^2$, and $6a^2b + 15ab^2 - 9b^3$.

In Exercises 41–50 find the value of each polynomial for $P(0)$, $P(-1)$, and $P(2)$. (Example 7)

41. $P(x) = 3x^2 + 5x - 4$

42. $P(x) = 2x^2 - 5x + 4$

43. $P(x) = 4x^2 - x + 7$

44. $P(x) = x^2 + 6x + 9$

45. $P(x) = x^3 - 3x^2 + x - 2$

46. $P(x) = 2x^3 + 4x^2 + 1$

47. $P(x) = 5x^3 - 4x^2 + 2x$

48. $P(x) = x^3 - 2x^2 + x$

49. $P(x) = x^4 + 3x^2 - 2$

50. $P(x) = x^5 - x^4 + x^3 - x^2 + x$

2.3 POLYNOMIALS: PRODUCTS

A polynomial has been defined as the algebraic sum of monomials, so the simplest product of two polynomials is the product of two monomials. These products were demonstrated in Section 2.1 by using the theorems of exponents.

EXAMPLE 1 Find the indicated product of monomials.

$(3x^2y)(2xy^2)$

Solution $(3x^2y)(2xy^2) = (3)(2)(x^2)(x)(y)(y^2) = 6x^3y^3$ by application of the commutative and associative principles for multiplication, and the product of powers theorem, exponent theorem 1.

In order to find the products of polynomials of more than one term, the distributive axiom is used. Following is a restatement of the axiom.

THE DISTRIBUTIVE AXIOM

If r, s, and t are real numbers, then
$$r(s + t) = rs + rt$$

This axiom can be applied for multiplying a monomial by a binomial, and it can be extended to multiplying a binomial by a binomial, a monomial by a trinomial, and so on.

When parentheses are removed in the multiplication of a polynomial by a monomial, it is important to remember that each term of the polynomial must be multiplied by the monomial.

EXAMPLE 2 Multiply $2x^3(3x^2 + 4x + 1)$.

Solution
$$2x^3(3x^2 + 4x + 1)$$
$$= 2x^3 \cdot 3x^2 + 2x^3 \cdot 4x + 2x^3 \cdot 1$$
$$= 6x^5 + 8x^4 + 2x^3$$

The product of a polynomial multiplied by a polynomial is obtained by applying the distributive axiom several times.

EXAMPLE 3 Multiply $(2x + 5)(3x + 2)$.

Solution

$(2x + 5)(3x + 2)$

$= (2x + 5)(3x) + (2x + 5)(2)$ (Distributive axiom)

$= 3x(2x + 5) + 2(2x + 5)$ (Commutative axiom)

$= 6x^2 + 15x + 4x + 10$ (Distributive axiom)

$= 6x^2 + 19x + 10$

EXAMPLE 4 Write $(x^2 + 2x)(3x^2 + x + 1)$ as a single polynomial (multiply).

Solution

$(x^2 + 2x)(3x^2 + x + 1)$

$= (x^2 + 2x)(3x^2) + (x^2 + 2x)(x) + (x^2 + 2x)(1)$

$= 3x^4 + 6x^3 + x^3 + 2x^2 + x^2 + 2x$

$= 3x^4 + 7x^3 + 3x^2 + 2x$

Note that the answer is stated in descending powers of the variable—that is, the variable is arranged so that the first term has the largest exponent, the second term the next largest exponent, and so on, with the constant term last. This rearrangement can be done because of the commutative and associative properties of addition of real numbers, and it makes the answer easier to check.

Sometimes it is convenient to arrange the polynomials to be multiplied in vertical order rather than horizontal order, as shown in the following example.

EXAMPLE 5 Multiply $3x^2 + x + 1$ by $x^2 + 2x$.

Solution

$$
\begin{array}{l}
3x^2 + x + 1 \\
\underline{x^2 + 2x} \\
3x^4 + x^3 + x^2 \qquad \text{(Multiply top row by } x^2) \\
 \underline{6x^3 + 2x^2 + 2x} \quad \text{(Multiply top row by } 2x) \\
3x^4 + 7x^3 + 3x^2 + 2x \qquad \text{(Combine like terms)}
\end{array}
$$

Note that the result of Example 5 checks with the result of Example 4.

Example 3 illustrates the product of two binomials. Since this is a problem that is met very frequently, it is worth looking at more closely. By omitting the second and third steps of the solution, we note that

$$(2x + 5)(3x + 2) = 6x^2 + 15x + 5x + 10$$

or
$$(2x)(3x) + (5)(3x) + (2x)(2) + (5)(2)$$

In other words, the first term is the product of the first terms of each binomial, the second term is the product of the "inner" terms, the third term is the product of the "outer" terms, and the last term is the product of the last terms of the binomials.

$$
\overset{\displaystyle 6x^2}{(2x + 5)\underset{+10}{(3x + 2)}} \quad \text{and} \quad \overset{\displaystyle 15x}{(2x + \underset{4x}{5)(3}x + 2)}
$$

The like terms are then combined:

$$6x^2 + 15x + 4x + 10 = 6x^2 + 19x + 10$$

EXAMPLE 6 Use the method just discussed to find the product $(3x - 2)(x + 4)$.

Solution

The product of the first terms is $(3x)(x) = 3x^2$.
The product of the "inner" terms is $(-2)(x) = -2x$.
The product of the "outer" terms is $(3x)(4) = 12x$.
The product of the two last terms is $(-2)(4) = -8$.
Combining like terms,

$$(3x - 2)(x + 4) = 3x^2 + 10x - 8$$

Certain products of binomials occur so frequently as to deserve special attention. Following is a list of these products.

SPECIAL PRODUCTS

1. The Square of a Binomial: $(A + B)^2 = A^2 + 2AB + B^2$
$(A - B)^2 = A^2 - 2AB + B^2$

2. The Cube of a Binomial: $(A + B)^3 = A^3 + 3A^2B + 3AB^2 + B^3$
$(A - B)^3 = A^3 - 3A^2B + 3AB^2 - B^3$

3. The Difference of Squares: $(A - B)(A + B) = A^2 - B^2$

These special products can be verified by multiplication.
$$(A + B)^2 = (A + B)(A + B) = A^2 + AB + AB + B^2$$
$$= A^2 + 2AB + B^2$$
$$(A - B)^2 = (A - B)(A - B) = A^2 - AB - AB + B^2$$
$$= A^2 - 2AB + B^2$$
$$(A + B)^3 = (A + B)(A + B)(A + B)$$
$$= (A + B)^2(A + B)$$
$$= (A^2 + 2AB + B^2)(A + B)$$

Using the vertical method of multiplication,

$A^2 + 2AB + B^2$
$\underline{A\ +\ B}$
$A^3 + 2A^2B +\ \ AB^2$
$\underline{\ \ \ \ \ \ \ A^2B + 2AB^2 + B^3}$
$A^3 + 3A^2B + 3AB^2 + B^3$

Therefore
$$(A + B)^3 = A^3 + 3A^2B + 3AB^2 + B^3$$

$$(A - B)^3 = (A - B)(A - B)(A - B)$$
$$= (A - B)^2(A - B)$$
$$= (A^2 - 2AB + B^2)(A - B)$$

Again, using the vertical method of multiplication,

$A^2 - 2AB + B^2$
$\underline{A\ -\ B}$
$A^3 - 2A^2B +\ \ AB^2$
$\underline{\ \ \ \ -\ \ A^2B + 2AB^2 - B^3}$
$A^3 - 3A^2B + 3AB^2 - B^3$

Therefore
$$(A - B)^3 = A^3 - 3A^2B + 3AB^2 - B^3$$

$$(A - B)(A + B) = A^2 + AB - AB - B^2 = A^2 - B^2$$

When a problem matches one of the "special product" forms, it can be worked very quickly by substitution, as shown in the examples below.

EXAMPLE 7 Find the product $(x + 5)^2$.

Solution This fits the form of a square of a binomial where $A = x$ and $B = 5$. Thus
$$(A + B)^2 = A^2 + 2AB + B^2$$
$$(x + 5)^2 = x^2 + 2(x)(5) + 5^2$$
$$= x^2 + 10x + 25$$

EXAMPLE 8 Expand $(3x - 1)^2$.

Solution To "expand" means to multiply. Consider

$$(A - B)^2 = A^2 - 2AB + B^2$$

The expanded result is a trinomial. The first term, A^2, is the square of the first term of the binomial $A - B$. The middle term of the trinomial, $2AB$, is equal to twice the product of the first and second terms of the binomial $A - B$. The third term, B^2, is the square of the last term of the binomial $A - B$. Thus

$$(3x - 1)^2 = (3x)^2 - 2(3x)(1) + (1)^2$$
$$= 9x^2 - 6x + 1$$

This result can be checked by multiplying:

$$(3x - 1)(3x - 1) = 3x(3x - 1) - 1(3x - 1)$$
$$= 9x^2 - 3x - 3x + 1$$
$$= 9x^2 - 6x + 1$$

EXAMPLE 9 Expand $(y - 2)^3$.

Solution Using the form of a cube of a binomial where $A = y$ and $B = 2$.

$$(A - B)^3 = A^3 - 3A^2B + 3AB^2 - B^3$$
$$(y \quad 2)^3 = y^3 - 3y^2(2) + 3y(2)^2 - (2)^3$$
$$= y^3 - 6y^2 + 12y - 8$$

EXAMPLE 10 Expand $(2b + 1)^3$.

Solution This time the form of a cube of a binomial is used with $A = 2b$ and $B = 1$:

$$(2b + 1)^3 = (2b)^3 + 3(2b)^2(1) + 3(2b)(1)^2 + (1)^3$$
$$= 8b^3 + 12b^2 + 6b + 1$$

EXAMPLE 11 Expand $(x - 3)(x + 3)$.

Solution The binomials to be multiplied are identical except for the middle sign. The form to recognize is the difference of squares form, $A^2 - B^2$, with $A = x$, $B = 3$. Thus

$$(A - B)(A + B) = A^2 - B^2$$
$$(x - 3)(x + 3) \quad = x^2 - (3)^2$$
$$= x^2 - 9$$

EXAMPLE 12 Expand $(2y - 5)(2y + 5)$.

Solution Again the difference of squares form may be recognized, with $A = 2y$ and $B = 5$. Thus

$$(2y - 5)(2y + 5) = (2y)^2 - (5)^2$$
$$= 4y^2 - 25$$

By learning these forms and applying them, special products can be obtained more rapidly than by the longer multiplication method used previously.

A resulting product can be quickly checked by assigning a value to each variable.

EXAMPLE 13 Check that $(2x + 5)^2 = 4x^2 + 20x + 25$ by letting $x = 3$.

Solution For $x = 3$,
$$(2x + 5)^2 = (2[3] + 5)^2 = (6 + 5)^2$$
$$= (11)^2 = 121$$

Check
$$4x^2 + 20x + 25 = 4(3)^2 + 20(3) + 25$$
$$= 36 + 60 + 25 = 121$$

EXERCISES 2.3

In Exercises 1–20 multiply and arrange the answer in descending powers of the variable where appropriate. (Examples 1–5)

1. $x(x^2 + 3)$
2. $3x(4x^2 + 2x - 1)$
3. $y^2(2y^2 - 3y + 1)$
4. $-x^3(x^2 - 3x + 2)$
5. $x^2y(3x^3 - 2x^2y + xy^2 - y^3)$
6. $5x^2(x^3 - 3xy + y^2 + 1)$
7. $(x + 3)(x - 5)$
8. $(x - 3)(x + 5)$
9. $(2x + 1)(3x - 2)$
10. $(2x - 1)(3x + 2)$
11. $(x + 2)(x^2 + 3x - 4)$
12. $(x - 3)(x^2 + 2x - 3)$
13. $x(x + 6)(2x - 3)$
14. $y(2y + 1)(3y + 2)$
15. $(x + 6y)(x^2 - 6xy + 36y^2)$
16. $(2x - y)(3x^2 + 2xy - y^2)$
17. $(t - 5)(2t - 9)(2t + 9)$
18. $(2x^2 - x + 3)(3x^2 + 1)$
19. $(x^2 + 5x + 6)(x^2 - 3x + 1)$
20. $(x^2 + xy - y^2)(x^2 - xy + y^2)$

In Exercises 21–40 use the short-cut method as shown in Example 6 to multiply the indicated binomials.

21. $(x + 2)(x - 5)$
22. $(x - 2)(x + 5)$
23. $(x + 3)(x + 4)$
24. $(x - 3)(x - 4)$
25. $(x + 3)(x - 4)$
26. $(x - 3)(x + 4)$
27. $(2x + 1)(3x - 2)$
28. $(2x - 1)(3x + 2)$
29. $(2x + 1)(3x + 2)$
30. $(2x - 1)(3x - 2)$
31. $(3x - 5)(3x + 5)$
32. $(5x - 3)(5x + 3)$
33. $(x + 3)(x + 3)$
34. $(x - 2)(x - 2)$
35. $(2x - 3)(3x - 2)$
36. $(2x + 3)(3x + 2)$
37. $(x^2 + 3)(x^2 - 3)$
38. $(4x^2 + 1)(4x^2 - 1)$
39. $(x - 5)(2x - 3)$
40. $(x + 5)(2x - 3)$

In Exercises 41–70 use the formulas for special products and Examples 7–12. Check by the method shown in Example 13.

41. $(4x + 5)^2$
42. $(4x - 5)^2$

43. $(3 - x)^2$

44. $(5 - 2x)^2$

45. $(2 + 7x)^2$

46. $(8 - b)^2$

47. $(4 + 3a)^2$

48. $(3 - 2y)^2$

49. $(x + y)^2$

50. $(x - y)^2$

51. $(2x + 5y)^2$

52. $(5x + 2y)^2$

53. $(2x - 5y)^2$

54. $(3a - 2b)^2$

55. $(x + 3)^3$

56. $(y + 8)^3$

57. $(x - 3)^3$

58. $(y - 8)^3$

59. $(2a + 5)^3$

60. $(2a - 5)^3$

61. $(9 + 2y)^3$

62. $(1 - 3b)^3$

63. $(y + 1)(y - 1)$

64. $(2z + 5)(2z - 5)$

65. $(3x + 2)(3x - 2)$

66. $(x + 2y)(x - 2y)$

67. $(4a + 5b)(4a - 5b)$

68. $(x^2 + 1)(x^2 - 1)$

69. $(2x^2 + 3)(2x^2 - 3)$

70. $(3a^2 + 2)(3a^2 - 2)$

In Exercises 71–80 use any of the methods shown in this section to expand and simplify.

71. $x(5x - 3)(5x + 3)$

72. $2x(3x + 2)(3x - 2)$

73. $3x(4x + 1)^2$

74. $x^2(3x - 2)^2$

75. $(x - 3)(2x + 5)^2$

76. $[(x + 1) + y][(x + 1) - y]$

77. $[x + (y + 2)][x - (y + 2)]$

78. $(x^n + 1)(x^n - 1)$

79. $(x^n + 2)^2$

80. $(x^n + 1)(x^{2n} - x^n - 2)$

2.4 FACTORING I: GREATEST COMMON FACTORS

Since the symmetric property of equality justifies exchanging the sides of an equation, an alternate form of the distributive axiom is

$$AB + AC = A(B + C)$$

The product $A(B + C)$ is the factored form of $AB + AC$. **To factor an expression means to write it as a product of factors.**

EXAMPLE 1 Factor $4x + 4y$.

Solution

Using the form $AB + AC = A(B + C)$ with $A = 4$, $B = x$, and $C = y$, then

$$4x + 4y = 4(x + y)$$

EXAMPLE 2 Factor $3x^2 + 5xy$.

Solution   Note that each term has x as a factor.

$$3x^2 + 5xy = x(3x) + x(5y) = x(2x + 5y)$$

It is customary to write the common monomial factor, such as x in Example 2, at the *left* in the factored form.

It is desirable to express a polynomial as a product of the *greatest* common monomial factor and another polynomial. Thus although

$$6x^4 + 12x^2 = 2x(3x^3 + 6x)$$

$2x$ is not the greatest common factor. The greatest common factor is $6x^2$, so the desired factored form is

$$6x^4 + 12x^2 = 6x^2(x^2 + 2)$$

Factoring polynomials can be thought of as the inverse or opposite process of multiplying polynomials. For example, the instructions "simplify," "multiply," "express as a single polynomial," or "expand"

$$x(2x + 1)(3x - 1)$$

mean the following:

First multiply $(2x + 1)(3x - 1)$, and then multiply the product by x:

$$x[(2x + 1)(3x - 1)] = x[6x^2 + x - 1]$$
$$= 6x^3 + x^2 - x$$

To factor the polynomial $6x^3 + x^2 - x$, reverse the steps, observing that x is the greatest common factor:

$$6x^3 + x^2 - x = x(6x^2 + x - 1)$$

It is evident from the problem that $6x^2 + x - 1$ is the product of two binomials, $2x + 1$ and $3x - 1$, but it is rather difficult to tell by just looking at it. This problem will be explored in greater detail in the next sections. It is presented at this time to develop an awareness of the inverse relationship involved in multiplying and factoring polynomials.

The following example illustrates the procedure for finding the greatest common monomial factor of a polynomial.

EXAMPLE 3 Factor $abx^2 - ab^2x^2 - a^2bx^2$.

Solution   Writing each term of the polynomial in factored form, it reads

$$abxx - abbxx - aabxx$$

Now it is easy to identify the greatest common factor:

$$abx^2 - ab^2x^2 - a^2bx^2 = abxx - abbxx - aabxx$$
$$= abxx(1 - b - a)$$
$$= abx^2(1 - b - a)$$

Note the necessity for writing 1 as the first term of the second factor. If this term had been omitted, the product of these two expressions would not have yielded the original polynomial.

EXAMPLE 4 Express $6ax^2 + 4a^2x - 8ax$ as a product of polynomials.

Solution Again, it is helpful to express each term in factored form:
$$2 \cdot 3ax^2 + 2^2a^2x - 2^3ax$$
The common factors are 2, a, and x; thus the greatest common factor is $2ax$, and
$$6ax^2 + 4a^2x - 8ax = 2ax(3x + 2a - 4)$$

When the leftmost term of a polynomial written in descending powers of a variable begins with a minus sign, it is customary to include -1 as one of the factors of the greatest common monomial factor. The next example demonstrates this convention.

EXAMPLE 5 Find the greatest common monomial factor of $-x^3 - x^2 + x$.

Solution
$$\begin{aligned}
-x^3 - x^2 + x &= (-1)x^3 + (-1)x^2 + (-1)(-x) \\
&= (-x)x^2 + (-x)x + (-x)(-1) \\
&= (-x)(x^2 + x - 1) \\
&= -x(x^2 + x - 1)
\end{aligned}$$

A factoring problem can be checked by multiplication.

EXAMPLE 6 Check that $-x^3 - x^2 + x = -x(x^2 + x - 1)$ by multiplication.

Solution
$$\begin{aligned}
-x(x^2 + x - 1) &= (-x)(x^2) + (-x)(x) + (-x)(-1) \\
&= -x^3 - x^2 + x
\end{aligned}$$

A factoring problem can also be checked by assigning a value to each variable and showing that the value of the expanded form is equal to the value of the factored form.

EXAMPLE 7 Factor $8ax^3 - 6ax^2 + 12ax$ and check by letting $a = 5$ and $x - 2$.

Solution
$$\begin{aligned}
8ax^3 - 6ax^2 + 12ax &= 2ax(4x^2) - 2ax(3x) + 2ax(6) \\
&= 2ax(4x^2 - 3x + 6)
\end{aligned}$$
Check For $a = 5$ and $x = 2$,
$$\begin{aligned}
8ax^3 - 6ax^2 + 12ax &= 8(5)(2^3) - 6(5)(2^2) + 12(5)(2) \\
&= 40(8) - 30(4) + 12(10) \\
&= 320 - 120 + 120 = 320 \\
2ax(4x^2 - 3x + 6) &= 2(5)(2)(4 \cdot 2^2 - 3 \cdot 2 + 6) \\
&= 2(10)(16 - 6 + 6) \\
&= 20(16) = 320
\end{aligned}$$

Because it is so easy to make a mistake in algebraic work, good algebraic techniques must involve some type of checking.

Each of the examples just given involves a common **monomial** factor. Often the common factor is a **binomial,** and by making a simple substitution, the problem can be reduced to the form for factoring a common monomial.

EXAMPLE 8   Factor $x(x - 3) - 5(x - 3)$.

Solution   The binomial $x - 3$ is a common factor in this expression. Let $N = x - 3$ and make the following substitution

$$x(x - 3) - 5(x - 3) = xN - 5N$$

Use the distributive axiom to write

$$xN - 5N = N(x - 5)$$

Now substitute for N to obtain

$$N(x - 5) = (x - 3)(x - 5)$$

EXAMPLE 9   Factor $p^2(p + 1) + (p + 1)$.

Solution   Let $N = p + 1$; then $p^2(p + 1) + (p + 1) = p^2N + N$.
Factor: $p^2N + N = N(p^2 + 1)$
Substitute: $N(p^2 + 1) = (p + 1)(p^2 + 1)$
Thus $p^2(p + 1) + (p + 1) = (p + 1)(p^2 + 1)$

From the first theorem of exponents, recall that for appropriate values for x and n (x any real number, n a natural number), $x \cdot x^n = x^{n+1}$, and $x^n \cdot x^n = x^{n+n} = x^{2n}$. These results as well as further application of this theorem, enable us to factor expressions such as $x^n + x^{n+2}$.

EXAMPLE 10   Find the largest common monomial factor of $x^n + x^{n+2}$ and write in factored form.

Solution

$$x^{n+2} = x^n \cdot x^2 \qquad \text{From the first exponent theorem and the symmetric property of equality}$$

$$x^n + x^{n+2} = x^n + x^n \cdot x^2$$
$$= x^n(1 + x^2)$$

EXERCISES 2.4

In Exercises 1–40 factor by finding the largest common monomial factor. (Examples 1–5)

1. $4x + 12$

2. $5x - 20$

3. $6x - 3$

4. $14x + 7$

5. $3x^2 + 6x + 9$

6. $5x^2 + 15xy + 5y^2$

7. $8ab - 12a + 4$

8. $6x + 3y - 3$

9. $4y^3 - 100y^2$

10. $4x^2 + 2x$

11. $ay - a$

12. $by + by^2$

13. $x^4 - x^3$

14. $x^3 - x^4$

15. $x^4 - x^5$

16. $x^5 - x^4$

17. $24x^3 - 30x^2 + x^2y$

18. $a^2n^2 - 3an^2 + 3a^2n$

19. $c^2n^2 + cn^3$

20. $x^2y^2 - 3xy^2$

21. $ax + ay - az$

22. $x^3 - x^2 + x$

23. $a^2x - ax^2$

24. $ax^3 + a^2x^2 - a^3x$

25. $24x^2 + 12x + 6$

26. $ay^2 + aby + ab$

27. $c^4 - c^3 + c^2 - 2c$

28. $x^2 - x + xy$

29. $-25x^3 - 15x$

30. $a^3b^3 + a^2b^4 - a^2b^3$

31. $-xy^3 - xy^2 - xy$

32. $-xyz - xy - yz$

33. $r^2s - rs^2 - 4rs$

34. $6ax^2y - 12axy^2 + 6axy$

35. $18ab + 27a^2b^2 - 63$

36. $-ab - ac$

37. $-6x^2 + 2x^3 - x^4$

38. $-56mn + 72m^2n^2$

39. $3p^2q + 9p^2q^2 - 12p^2q^3$

40. $15x^2 - 10x^3 - 15x^4$

In Exercises 41–55 factor by finding the largest common binomial as shown in Examples 8 and 9.

41. $3y(x - 3) + x(x - 3)$

42. $x(x - 1) + 2(x - 1)$

43. $x^2(y + 2) - (y + 2)$

44. $(y + 2) + x^2(y + 2)$

45. $b^2(b + c) - t(b + c)$

46. $x(x - 3) - y(x - 3)$

47. $a(x + y) + 3(x + y)$

48. $x^2(x - 4) - 5(x - 4)$

49. $x^2(x + 5) - 2y^2(x + 5)$

50. $y(y^2 - 3) + (y^2 - 3)$

51. $5a^2(t^2 + 4) - (t^2 + 4)$

52. $y^2(y + 1) + 4(y + 1)$

53. $t(t^3 + 1) - (t^3 + 1)$

54. $u^2(u - v) + (u - v)$

55. $4(a + b)x^2 + 2(a + b)x + (a + b)$

In Exercises 56–60 factor by finding the largest common monomial. Assume n is a positive integer. (Example 10)

56. $x^n + x^{n+1}$

57. $x^{2n} - x^n$

58. $x^{3n} + x^n$

59. $y^{4n} - y^nz$

60. $x^{n+2} + x^{n+1} + x^n$

2.5 FACTORING II: TRINOMIALS

In the preceding section the relationship between factoring and multiplication was discussed. Let us look again at the product of two binomials, $(x + 2)(x + 3)$. In applying the "short-cut" method for expanding this product,

$$(x + 2)(x + 3) = x^2 + 2x + 3x + 6 = x^2 + 5x + 6$$

The expanded form of the product of the two binomials is a trinomial. Thus $x^2 + 5x + 6$ is expressed in **factored** form as $(x + 2)(x + 3)$. The first term of the trinomial is the product of the first terms of the two binomials, and the third or last term of the trinomial is the product of the second terms of the binomials.

$$\overset{x^2}{\overbrace{(x + 2)(x}} + 3)$$
$$\underbrace{\qquad}_{+6}$$

The middle term of the trinomial, $5x$, is the algebraic sum of the outer product and inner product of the binomials:

Outer product
$$\overbrace{(x + 2)(x} + 3)}$$
$$\underbrace{\quad}$$
Inner product

Outer product $= 3x$; inner product $= 2x$; algebraic sum of outer and inner products $= 3x + 2x = 5x$. Thus
$$x^2 + 5x + 6 = x^2 + (2 + 3)x + (2 \cdot 3) = (x + 2)(x + 3)$$

In general, the product of two binomals of the form $X + A$ and $X + B$ is
$$(X + A)(X + B) = X^2 + (A + B)X + A \cdot B$$

EXAMPLE 1   Express $x^2 + 3x + 2$ in factored form.

Solution   If $x^2 + 3x + 2$ is factorable over the integers—that is, if it can be factored into binomials whose coefficients are integers—then the product of the first terms of the binomials must be x^2:
$$(x \qquad)(x \qquad)$$
and the factors will be in the form
$$(x + A)(x + B)$$
where $A + B = 3$ and $A \cdot B = 2$. Since the only factors of 2 are 2 and 1, and -2 and -1, and the sum $A + B$ is a positive number, $A = 2$ and $B = 1$. (The same result would have been obtained for $A = 1$ and $B = 2$, since addition and multiplication are each commutative.)
Thus
$$x^2 + 3x + 2 = (x + 2)(x + 1)$$
Also
$$x^2 + 3x + 2 = (x + 1)(x + 2)$$

The answer to Example 1 can be checked by multiplication or by substitution of a numerical value for x.

EXAMPLE 2   Factor $x^2 + 3x - 4$ and check by (a) multiplication and (b) substituting 5 for x.

Solution   Again, the factors will be in the form $(X + A)(X + B)$ with $A + B = 3$ and $A \cdot B = -4$. The factors of -4 are $(-1)(4)$, $1(-4)$, and $2(-2)$. If $A = -1$ and $B = 4$, then $A + B = -1 + 4 = 3$. Thus

$$x^2 + 3x - 4 = (x - 1)(x + 4)$$

Check

(a) Multiplying,

$$\begin{aligned}(x - 1)(x + 4) &= x(x + 4) + (-1)(x + 4) \\ &= x^2 + 4x - x - 4 \\ &= x^2 + 3x - 4\end{aligned}$$

(b) For $x = 5$,

$$\begin{aligned}x^2 + 3x - 4 &= 5^2 + 3(5) - 4 \\ &= 25 + 15 - 4 \\ &= 36\end{aligned}$$

$$\begin{aligned}(x - 1)(x + 4) &= (5 - 1)(5 + 4) \\ &= (4)(9) \\ &= 36\end{aligned}$$

If all the numerical coefficients and constants of a factored expression are integers, the polynomial is said to be **factored over the integers.** All the examples worked thus far have been factored over the integers.

EXAMPLE 3   Factor $y^2 + 4y - 12$ over the integers, if possible, and check.

Solution   The integral factors of -12 are $(-1)(12)$, $(1)(-12)$, $(-2)(6)$, $(2)(-6)$, $(-3)(4)$, $(3)(-4)$. The only pair of factors whose sum is 4 is 6 and -2. Therefore

$$y^2 + 4y - 12 = (y + 6)(y - 2)$$

Since multiplication is commutative, these factors can also be written $(y - 2)(y + 6)$.

Check

$$\begin{aligned}(y + 6)(y - 2) &= y^2 + 6y - 2y - 12 \\ &= y^2 + 4y - 12\end{aligned}$$

If we let $y = 2$ and make a numerical check,

$$y + 6 = 2 + 6 = 8; \qquad y - 2 = 2 - 2 = 0$$
$$(y + 6)(y - 2) = (8)(0) = 0$$
$$\begin{aligned}y^2 + 4y - 12 &= 2^2 + 4(2) - 12 \\ &= 4 + 8 - 12 = 0\end{aligned}$$

EXAMPLE 4   Factor $x^2 + 3xy - 10y^2$ over the integers and check.

Solution

1. Comparing the forms:

$$X^2 + (A + B)X + AB = (X + A)(X + B)$$
$$x^2 + 3xy - 10y^2 = (x \; ? \; y)(x \; ? \; y)$$

It is seen that A and B must have the forms py and qy where $p + q = 3$ and $pq = -10$.

2. The pairs of factors of -10 are $(-1)(10)$, $(10)(-1)$, $(-2)(5)$, and $(2)(-5)$. Since $(-2) + (5) = 3$, select $p = -2$ and $q = 5$.

3. Then $x^2 + 3xy - 10y^2 = (x - 2y)(x + 5y)$. Also, by the commutative axiom for multiplication,
$$x^2 + 3xy - 10y^2 = (x + 5y)(x - 2y)$$

Check   Let $x = 3$ and $y = 2$. Then
$$\begin{aligned}
x^2 + 3xy - 10y^2 &= 3^2 + 3(3)(2) - 10(2^2) \\
&= 9 + 18 - 40 \\
&= 27 - 40 = -13 \\
(x - 2y)(x + 5y) &= (3 - 2 \cdot 2)(3 + 5 \cdot 2) \\
&= (3 - 4)(3 + 10) \\
&= (-1)(13) = -13
\end{aligned}$$

EXAMPLE 5   Factor $x^2 + x + 1$ over the integers, if possible.

Solution   Comparing the forms:
$$X^2 + (A + B)X + AB = (X + A)(X + B)$$
it is seen that $A + B = 1$ and $AB = 1$. The only pairs of factors of 1 that are integers are $(1)(1)$ and $(-1)(-1)$; in other words, $A = 1$ and $B = 1$ or $A = -1$ and $B = -1$. But
$$A + B = 1 + 1 = 2$$
or
$$A + B = -1 + (-1) = -2$$
and according to the problem, $A + B = 1$. Therefore $x^2 + x + 1$ cannot be factored over the integers.

A polynomial that cannot be factored over the integers is said to be **prime** over the integers.

In all the trinomials in Examples 1 through 5, the coefficient of the x^2 term is 1. What happens if this coefficient is not 1? Let us begin by examining the product of two binomials.
$$\begin{aligned}
(2x + 1)(5x + 3) &= 2x(5x) + 2x(3) + 1(5x) + 1(3) \\
&= 10x^2 + (6 + 5)x + 3 \\
&= 10x^2 + 11x + 3
\end{aligned}$$
To factor $10x^2 + 11x + 3$, it is necessary to retrace these steps. This principle is illustrated in the following example.

EXAMPLE 6   Factor $10x^2 + 11x + 3$.

Solution   If this trinomial has two binomial factors, the product of their first terms must be $10x^2$. Because we prefer that the leading terms be positive, the possibilities are $2x$ and $5x$ or $10x$ and x.

$$(2x \quad)(5x \quad)$$

or

$$(10x \quad)(x \quad)$$

Since the product of the second terms must be 3, this produces the following possibilities:

$$(2x + 1)(5x + 3) \qquad (10x + 1)(x + 3)$$
$$(2x + 3)(5x + 1) \qquad (10x + 3)(x + 1)$$

Each of these yields the correct first and third terms of the given trinomial, but only one of them produces the correct middle term.

$$(2x + 1)(5x + 3)$$

$5x$	(Inner product)

$6x$	(Outer product)

$11x$	(Sum of inner and outer products)

Therefore $10x^2 + 11x + 3 = (2x + 1)(5x + 3)$.

EXAMPLE 7 Factor $8y^2 - 10y - 7$ over the integers, if possible, and check the result.

Solution The factors of 8 are $(1)(8)$ and $(2)(4)$. Therefore $8y^2 - 10y - 7$ has the form $(y + b)(8y + d)$ or $(2y + b)(4y + d)$. Since $bd = -7$, the possibilities for bd are $(-1)(7)$, $(7)(-1)$, $(1)(-7)$, and $(-7)(1)$.

The possibilities to be tried are:

$$(y - 1)(8y + 7) \qquad (2y - 1)(4y + 7)$$
$$(y + 7)(8y - 1) \qquad (2y + 7)(4y - 1)$$
$$(y + 1)(8y - 7) \qquad (2y + 1)(4y - 7)$$
$$(y - 7)(8y + 1) \qquad (2y - 7)(4y + 1)$$

After several trials,

$$(2y + 1)(4y - 7)$$

$4y$	(Inner product)

$-14y$	(Outer product)

$-10y$	(Sum of inner and outer products)

Thus

$$8y^2 - 10y - 7 = (2y + 1)(4y - 7)$$

Check   Let $y = 5$.

$$8y^2 - 10y - 7 = 8(5^2) - 10(5) - 7$$
$$= 2 \cdot 4(25) - 50 - 7$$
$$= 200 - 57$$
$$= 143$$
$$(2y + 1)(4y - 7) = (2 \cdot 5 + 1)(4 \cdot 5 - 7)$$
$$= (11)(13)$$
$$= 143$$

EXAMPLE 8   Factor $5x^2 + 6x - 1$ over the integers, if possible.

Solution   If the trinomial can be factored, the possibilities are

$$(x + 1)(5x - 1) \quad \text{and} \quad (x - 1)(5x + 1)$$

Trying these,

$$(x + 1)(5x - 1) = x(5x - 1) + (1)(5x - 1)$$
$$= 5x^2 - x + 5x - 1$$
$$= 5x^2 + 4x - 1$$
$$(x - 1)(5x + 1) = x(5x + 1) + (-1)(5x + 1)$$
$$= 5x^2 + x - 5x - 1$$
$$= 5x^2 - 4x - 1$$

Since none of these possibilities works and since they are the only possibilities, $5x^2 + 6x - 1$ *cannot* be factored over the integers; it is prime.

It should be pointed out that factoring often involves much trial and error, and practice will reduce the "error" part considerably.

When attempting to factor a trinomial or any other polynomial, always look for any common monomial factors first.

EXAMPLE 9   Factor $4x^3 + 10x^2 - 6x$ over the integers.

Solution   The given trinomial has a common monomial factor, $2x$.

$$4x^3 + 10x^2 - 6x = 2x(2x^2 + 5x - 3)$$

After factoring the common monomial term, we factor the reduced trinomial in the usual manner.

$$4x^3 + 10x^2 - 6x = 2x(2x - 1)(x + 3)$$

EXERCISES 2.5

In Exercises 1–80 factor over the integers, if possible. Always look for common monomial factors first. Check the answers.

1. $x^2 - 6x + 8$ **2.** $x^2 - 2x - 8$

3. $x^2 - x - 6$

4. $x^2 - 5x + 6$

5. $y^2 + 9y + 8$

6. $x^2 - 8x + 12$

7. $z^2 + 3z + 4$

8. $a^2 + 8a - 20$

9. $x^2 - 9x + 20$

10. $x^2 - 10x + 21$

11. $r^2 + 4r - 32$

12. $p^2 - 6p - 40$

13. $x^2 + 5x + 6$

14. $p^2 - 5p + 6$

15. $y^2 - 7y + 12$

16. $x^2 - 2x - 3$

17. $x^2 - 2x - 4$

18. $y^2 - y - 2$

19. $x^2 + 2xy + y^2$

20. $x^2 - 2xy - 15y^2$

21. $3x^2 - 9x + 6$

22. $35b^2 - 30b - 5b^3$

23. $10x^2 - 20x - 350$

24. $4x^2 + 32x + 60$

25. $150x^4 - 1050x^3 + 1500x^2$

26. $5x^2 + 30x - 35$

27. $4a^3y + 4a^2y - 48ay$

28. $100x^3 - 200x^2 - 800x$

29. $20y^2 + 140y - 120$

30. $6y^4 - 24y^3 - 72y^2$

31. $8a^2 - 50a + 33$

32. $8a^2 + 38a - 33$

33. $4x^2 + 12x + 9$

34. $9b^2 - 30b + 25$

35. $8p^2 + 2p - 21$

36. $8p^2 + 22p - 21$

37. $2x^2 - 3x + 1$

38. $4x^2 - 20x + 25$

39. $12x^2 - 23x + 5$

40. $12x^2 - 17x - 5$

41. $12x^2 - 16x + 5$

42. $3a^2 - 4a - 7$

43. $3a^2 + 4a - 7$

44. $3a^2 + 21a + 7$

45. $3a^2 - 20a - 7$

46. $3a^2 + 20a - 7$

47. $5x^2 + 5x - 2$

48. $15x^2 - 7x - 2$

49. $9y^2 + 24y + 16$

50. $25x^2 - 30x + 9$

51. $10x^2 + 33x - 7$

52. $10x^2 - 9x - 7$

53. $3x + 2x^2 + 1$

54. $-4a + 1 + 4a^2$

55. $15 + 16x + 4x^2$

56. $8 - 21a - 9a^2$

57. $5x + 4x^2 + 1$

58. $-3y - 1 + 4y^2$

59. $6 - 11x + 4x^2$

60. $-5 - 11x + 16x^2$

61. $6a^2 + 8a + 2$

62. $12y^2 - 12y + 3$

63. $16y^2 - 2y - 5$

64. $16y^2 - 16y - 5$

65. $9ax^2 - 21ax - 8a$

66. $4a^2x + ax - 5x$

67. $3a^2 - 7ab + 2b^2$

68. $8x^3 - 20x^2 - 12x$

69. $2m^2 + 5mn - 3n^2$

70. $4p^2 + 5pq + q^2$

71. $15x^2 - 2xy - y^2$

72. $6x^2 + 17xy - 14y^2$

73. $6a^2x + 12ax - 21x$

74. $4x^2y + 2xy - 6y$

75. $3x^5 + 20x^4 - 7x^3$

76. $6x^4 - 10x^3 - 4x^2$

77. $2x^4y^2 + 6x^3y^2 - 20x^2y^2$

78. $x^2 + 4x^3 + 3x^4$

79. $30x^2 - 78x - 36$

80. $2xy + 2x^2y - 60x^3y$

81. For what integers k is $x^2 + kx + 4$ factorable over the integers?

82. For what integers k is $x^2 + kx - 12$ factorable over the integers?

83. For what integers p is $2x^2 + px + 34$ factorable over the integers?

84. For what integers p is $5x^2 + px - 9$ factorable over the integers?

2.6 FACTORING III: SPECIAL POLYNOMIALS

Some polynomials have factors that are readily recognizable. You should familiarize yourself with these.

The polynomial $x^2 + 2ax + a^2$ is called a **perfect square trinomial** because it is the result of the square of a binomial. Note that

$$(x + a)^2 = (x + a)(x + a) = (x + a)x + (x + a)a$$
$$= x^2 + ax + ax + a^2 = x^2 + 2ax + a^2$$

When a perfect square trinomial is written in descending powers of the variable x, the first term and the last term are positive and perfect squares, such as x^2, 25, 16, a^2, and so on. **The middle term is equal to twice the product of the first and second terms of the binomial factor, because the outer product and the inner product are the same.**

PERFECT SQUARE TRINOMIAL FORM

$$x^2 + 2ax + a^2 = (x + a)^2$$

EXAMPLE 1   Find the term which when added to $x^2 + 8x$ will make it a perfect square trinomial.

Solution   The factors of this perfect square trinomial must be of the form $(x + a)(x + a)$ and their product is $x^2 + 2ax + a^2$. The middle term of this perfect square trinomial is $2ax$, and the middle term of the trinomial we wish to form is $8x$; therefore let $2a = 8$ and $a = 4$. Since $a = 4$, $a^2 = 16$. The missing term is a^2, so we must add 16 to $x^2 + 8x$ to make it a perfect square trinomial:

$$x^2 + 8x + 16 = (x + 4)^2$$

Note: You should recognize that the solution to Example 1 could have been found very simply and mechanically by taking half of the coefficient of x, $\frac{1}{2}(8) = 4$, and squaring this number, $4^2 = 16$. This is known as *completing the square.*

EXAMPLE 2   Find the term which when added to $x^2 + 40x$ will make it a perfect square trinomial and thus complete the square.

Solution   The middle term of the desired trinomial is 40x. The coefficient of x is 40; $\frac{1}{2}(40) = 20; 20^2 = 400$. Therefore the missing term is 400 and
$$x^2 + 40x + 400 = (x + 20)^2$$

EXAMPLE 3   Express $x^2 + 10x + 25$ as the square of a binomial.

Solution   Using the form
$$x^2 + 2ax + a^2 = (x + a)^2$$
$$x^2 + 10x + 25 = x^2 + 2(5x) + 5^2$$
$$= (x + 5)^2$$

EXAMPLE 4   Express $4y^2 - 12y + 9$ as the square of a binomial.

Solution

$$
\begin{aligned}
x^2 + 2ax \quad\quad + a^2 \quad &= (x + a)^2\\
4y^2 - 12y + 9 = (2y)^2 + 2(-6y) \quad + 3^2\\
= (2y)^2 + 2(-3)(2y) + (-3)^2 &= (2y + [-3])^2\\
&= (2y - 3)^2
\end{aligned}
$$

EXAMPLE 5   Is $x^2 + 14x - 49$ a perfect square?

Solution   No, because the third term is *not* positive.

EXAMPLE 6   Is $y^2 + 3y + 9$ a perfect square?

Solution   The only possibility is $(y + 3)^2$. But
$$(y + 3)^2 = y^2 + 2(3)y + 9 = y^2 + 6y + 9$$
Thus $y^2 + 3y + 9$ is not a perfect square.

EXAMPLE 7   Is $x^2 + 16 - 8x$ a perfect square?

Solution   First rearranging the polynomial in descending powers of x,
$$x^2 - 8x + 16 = x^2 - 8x + 4^2 = (x - 4)^2$$
The answer is yes.

EXAMPLE 8   Fill in the missing term so that $x^2 + ($ $) + 9y^2$ is a perfect square trinomial.

Solution   Using the form
$$x^2 + 2ax + a^2 = (x + a)^2$$
$$x^2 + (\quad) + 9y^2 = x^2 + (\quad) + (3y)^2$$
$$= (x + 3y)^2$$
$a = 3y$, $2a = 6y$, and $2ax = 6xy$
The missing term is $6xy$ and
$$x^2 + 6xy + 9y^2 = (x + 3y)^2$$

The product of the two binomials $x + a$ and $x - a$ is not a trinomial, but another binomial.
$$(x + a)(x - a) = (x + a)x - (x + a)a$$
$$= x^2 + ax - ax - a^2$$
$$= x^2 - a^2$$

Notice that the outer and inner products are alike except for sign, which makes them additive inverses whose sum is always zero. Since the resulting binomial is the difference of the squares of a and x, this special product is called the **difference of squares.**

DIFFERENCE OF SQUARES FORM

$$A^2 - B^2 = (A + B)(A - B)$$

EXAMPLE 9 Factor $x^2 - 36$.

Solution Since x^2 and 36 are both perfect squares and the binomial is of the form $A^2 - B^2$, it factors as $(A + B)(A - B)$, where $A^2 = x^2$ and $B^2 = 36 = 6^2$.
$$A^2 - B^2 = (A + B)(A - B)$$
$$x^2 - 36 = x^2 - 6^2 = (x + 6)(x - 6)$$

EXAMPLE 10 Write $25x^2 - 4$ in factored form.

Solution $25x^2$ is the square of $5x$ since $(5x)(5x) = 25x^2$. 4 is the square of 2. Therefore
$$25x^2 - 4 = (5x)^2 - (2)^2$$
$$= (5x + 2)(5x - 2)$$

EXAMPLE 11 Factor $2x^2 - 32$.

Solution The two terms have a common factor, 2, so extract this common factor first:
$$2x^2 - 32 = 2(x^2 - 16)$$
$$= 2(x + 4)(x - 4)$$

EXAMPLE 12 Factor $4x^2 - 25y^2$.

Solution $4x^2$ is the square of $2x$. $25y^2$ is the square of $5y$. Therefore
$$4x^2 - 25y^2 = (2x)^2 - (5y)^2$$
$$= (2x + 5y)(2x - 5y)$$

EXAMPLE 13 Factor $x^2 + 1$, if possible.

Solution x^2 is the square of x. 1 is the square of 1. However, these squares are *not* separated by a minus sign, and they are therefore *not the difference* of two squares but their sum. Note that *none* of the following possibilities work:
$$(x + 1)(x + 1) = x^2 + 2x + 1$$
$$(x - 1)(x - 1) = x^2 - 2x + 1$$
$$(x + 1)(x - 1) = x^2 - 1$$
Therefore $x^2 + 1$ is prime and cannot be factored.

2.6 FACTORING III: SPECIAL POLYNOMIALS

In general, **a binomial of the form $x^2 + a^2$ cannot be factored over the integers.**

Two other factorable binomials are useful to know.

DIFFERENCE OF CUBES FORM

$$A^3 - B^3 = (A - B)(A^2 + AB + B^2)$$

SUM OF CUBES FORM

$$A^3 + B^3 = (A + B)(A^2 - AB + B^2)$$

These two factorable binomials can be verified by direct multiplication.

EXAMPLE 14 Factor $x^3 - 8$.

Solution $x^3 - 8$ fits the difference of cubes form, with $A^3 = x^3$ and $B^3 = 8 = 2^3$.
$$A^3 - B^3 = (A - B)(A^2 + AB + B^2)$$
$$x^3 - 8 = x^3 - 2^3 = (x - 2)(x^2 + 2x + 4)$$

EXAMPLE 15 Factor $27x^3 + 125y^3$.

Solution   Since $27x^3 = (3x)^3$ and $125y^3 = (5y)^3$, and
$$A^3 + B^3 = (A + B)(A^2 - AB + B^2)$$
$$27x^3 + 125y^3 = (3x)^3 + (5y)^3$$
$$= (3x + 5y)([3x]^2 - [3x][5y] + [5y]^2)$$
$$= (3x + 5y)(9x^2 - 15xy + 25y^2)$$

EXAMPLE 16 Check by multiplication that
$$x^3 - 8 = (x - 2)(x^2 + 2x + 4)$$
Solution

$$
\begin{array}{r}
x^2 + 2x + 4 \\
x - 2 \\
\hline
x^3 + 2x^2 + 4x \\
-2x^2 - 4x - 8 \\
\hline
x^3 \qquad\qquad - 8
\end{array}
$$

EXAMPLE 17 Check by multiplication that
$$27x^3 + 125y^3 = (3x + 5y)(9x^2 - 15xy + 25y^2)$$
Solution

$$
\begin{array}{r}
9x^2 - 15xy + 25y^2 \\
3x + 5y \\
\hline
27x^3 - 45x^2y + 75xy^2 \\
+ 45x^2y - 75xy^2 + 125y^3 \\
\hline
27x^3 \qquad\qquad\qquad\qquad + 125y^3
\end{array}
$$

Listed for quick reference are the special polynomials and their factored forms.

PERFECT SQUARE TRINOMIAL

$$A^2 + 2AB + B^2 = (A + B)^2$$
$$A^2 - 2AB + B^2 = (A - B)^2$$

DIFFERENCE OF SQUARES

$$A^2 - B^2 = (A + B)(A - B)$$

DIFFERENCE OF CUBES

$$A^3 - B^3 = (A - B)(A^2 + AB + B^2)$$

SUM OF CUBES

$$A^3 + B^3 = (A + B)(A^2 - AB + B^2)$$

EXERCISES 2.6

In Exercises 1–10 fill in the missing terms so that the result is a perfect square trinomial. (Examples 1, 2, and 8)

1. $x^2 + 6x + ($ $)$ **6.** $p^2 + 12p + ($ $)$

2. $x^2 - 20x + ($ $)$ **7.** $x^2 + ($ $) + 9$

3. $x^2 - 14x + ($ $)$ **8.** $t^2 - ($ $) + 36$

4. $x^2 + 36x + ($ $)$ **9.** $x^2 - 12xy + ($ $)$

5. $a^2 - 10a + ($ $)$ **10.** $a^2 - 16ab + ($ $)$

In Exercises 11–62 factor the polynomials over the integers, if possible. Remove all common monomials first and check each problem. Exercises 11–20 refer to Examples 9–13.

11. $x^2 - 81$ **16.** $y^2 - x^2$

12. $9x^2 - 1$ **17.** $16 - 9a^2$

13. $4x^2 - 49$ **18.** $9y^2 - 100z^2$

14. $x^2 + 9$ **19.** $x^2 - 2$

15. $36x^2 - 25y^2$ **20.** $x^2 + 1$

Exercises 21–30 refer to Examples 14–17.

21. $x^3 + 1$ **26.** $8y^3 - 1$

22. $x^3 - 1$ **27.** $p^3 + 27$

23. $x^3 - a^3$ **28.** $y^5 - 8y^2$

24. $x^3 + a^3$ **29.** $x^4 + 64x$

25. $8y^3 + 1$ **30.** $64x^3 + 125y^6$

For Exercises 31–62 use all your factoring knowledge.

31. $x^2 + 8x + 16$

32. $y^2 + 2y + 1$

33. $a^2 + 2ab - b^2$

34. $9y^2 - 18y + 36$

35. $x^2 + 49y^2 - 14xy$

36. $x^2 + 36 - 12x$

37. $n^2 + 4ny + 16y^2$

38. $x^2 + 4$

39. $u^2 + 81 + 9u$

40. $a^2 - b^2$

41. $x^2 + 4x + 4$

42. $x^2 - 2x + 1$

43. $x^2 + 18x + 81$

44. $x^2 - 10x + 25$

45. $y^2 + 16y + 64$

46. $9a^2 + 6a + 1$

47. $16p^2 - 8p + 1$

48. $36 - 12x + x^2$

49. $64y^2 - 48y + 9$

50. $x^2 + 14x + 49$

51. $64x^3 - 125y^6$

52. $4a^3 - 121a$

53. $64x^3 - 9x$

54. $y^2 - 144$

55. $y^2 + 144$

56. $16y^2 - 4y^4$

57. $8y^4 - 16y^2$

58. $a^3 - 16a$

59. $x^4 + 216x$

60. $x^4 - 343x$

61. $2y^4z - 54yz^4$

62. $3ax^4 + 81a^4x$

2.7 FACTORING IV: SUBSTITUTION AND GROUPING

Many factorization problems that at first glance seem difficult or unfactorable may be simplified by making an appropriate substitution.

EXAMPLE 1 Factor $3(x + 2)^2 - (x + 2) - 2$.

Solution If the substitution $y = x + 2$ is made, the expression can be written
$$3y^2 - y - 2$$
which factors as
$$(3y + 2)(y - 1)$$
Replacing y by $x + 2$ yields
$$[3(x + 2) + 2][(x + 2) - 1]$$
which simplifies as
$$(3x + 8)(x + 1)$$
Therefore
$$3(x + 2)^2 - (x + 2) - 2 = (3x + 8)(x + 1)$$

EXAMPLE 2 Factor $4(a + n)^2 - x^2$.

Solution Recognizing the difference of squares form,
$$A^2 - B^2 = (A + B)(A - B)$$
and substituting $Y = a + n$,
$$4(a + n)^2 - x^2 = 4Y^2 - x^2$$
$$= (2Y + x)(2Y - x)$$
$$= (2[a + n] + x)(2[a + n] - x)$$
$$= (2a + 2n + x)(2a + 2n - x)$$

Up to this point we have not factored a polynomial with more than 3 terms. When attempting to factor a polynomial with 4 or more terms, always consider *grouping* and maybe rearranging the terms to see if, with the aid of a substitution, the polynomial may be factored.

EXAMPLE 3 Factor $x^3 + 5x^2 - 4x - 20$.

Solution Since a basic factoring form cannot be recognized readily, the terms are grouped to try to find a form that can be recognized. Since no two terms are perfect squares, the terms are grouped in pairs:

$$x^3 + 5x^2 - 4x - 20 = (x^3 + 5x^2) - (4x + 20) \quad \text{(Grouping in pairs)}$$
$$= x^2(x + 5) - 4(x + 5) \quad \text{(Factoring each group)}$$
$$= x^2(N) - 4(N) \quad \text{(Letting } N = x + 5\text{)}$$
$$= (x^2 - 4)(N) \quad \text{(Using the distributive axiom)}$$
$$= (x^2 - 4)(x + 5) \quad \text{(Replacing } N \text{ by } x + 5\text{)}$$
$$= (x - 2)(x + 2)(x + 5) \quad \text{(Difference of squares)}$$

EXAMPLE 4 Factor $x^2 - y^2 + 4y - 4$ over the integers, if possible.

Solution The polynomial has 4 terms, so a grouping will be considered. After mentally checking the possibilities for a grouping of two and two terms, we recognize that the last three terms are preceded by a minus sign and, if grouped, read $y^2 - 4y + 4$, a perfect square trinomial! This grouping will lead to a difference of squares form, and we can now factor the polynomial.

$$x^2 - y^2 + 4y - 4 = x^2 - (y^2 - 4y + 4)$$
$$= x^2 - (y - 2)^2$$
$$= x^2 - N^2 \quad \text{(Letting } N = y - 2\text{)}$$

$$= (x + N)(x - N) \qquad \text{(Difference of squares)}$$

$$= (x + [y - 2])(x - [y - 2]) \qquad \text{(Replacing } N \text{ by } y - 2)$$

$$= (x + y - 2)(x - y + 2)$$

It can be seen that factoring may involve several steps before a polynomial is completely factored. Unless otherwise indicated, the instructions "factor completely" mean "factor completely over the integers." To summarize the factorization procedure, the following example is given, followed by a "checklist" for assuring that the process has been followed.

EXAMPLE 5   Factor completely $3ax^4 - 3ay^4$.

Solution   First remove the greatest common monomial factor by the distributive axiom:
$$3ax^4 - 3ay^4 = 3a(x^4 - y^4)$$
Now observe that $x^4 - y^4$ is a difference of squares,
$$x^4 - y^4 = (x^2 + y^2)(x^2 - y^2)$$
By examining the factors obtained, we notice that $x^2 - y^2$ also represents a difference of squares,
$$= x^2 - y^2 = (x + y)(x - y)$$
Now the factorization is complete because we cannot factor any further over the integers. Writing these steps sequentially,
$$3ax^4 - 3ay^4 = 3a(x^4 - y^4)$$
$$= 3a(x^2 + y^2)(x^2 - y^2)$$
$$= 3a(x^2 + y^2)(x + y)(x - y)$$
Since the factors of a product can be written in any order, it follows that
$$3ax^4 - 3ay^4 = 3a(x + y)(x - y)(x^2 + y^2)$$

PROCEDURE FOR COMPLETE FACTORIZATION

1. Remove the greatest common monomial factor.
2. Factor any binomial, if present, using one of the following forms.
3. Factor any trinomial, if present, using one of the following forms.
4. Factor by grouping, if necessary.

FACTORING FORMS

1. Common Monomial Factor
$$AB + AC = A(B + C)$$
$$AB + AC + AD = A(B + C + D)$$

2. Difference of
Squares $\qquad$ $A^2 - B^2 = (A - B)(A + B)$
3. Difference of Cubes $\quad$ $A^3 - B^3 = (A - B)(A^2 + AB + B^2)$
4. Sum of Cubes $\qquad$ $A^3 + B^3 = (A + B)(A^2 - AB + B^2)$
5. Perfect Square $\qquad$ $A^2 + 2AB + B^2 = (A + B)^2$
Trinomial $\qquad$ $A^2 - 2AB + B^2 = (A - B)^2$
6. Simple Trinomial $\qquad$ $X^2 + (a + b)X + ab = (X + a)(X + b)$
7. General Trinomial $\qquad$ $aX^2 + bX + c = (rX + s)(kX + n)$
$\qquad\qquad$ where $rk = a$, $sn = c$, and $rn + sk = b$

EXERCISES 2.7

Factor completely over the integers in Exercises 1–54.

1. $x(x + 4) + 5(x + 4)$

2. $y(y - 2) - (y - 2)$

3. $(x + 7)^2 - 6(x + 7) + 5$

4. $3(y - 4)^2 + 14(y - 4) - 5$

5. $(t + 3)^2 - 3(t + 3) - 4$

6. $(x^2 + 6)^2 - y^2$

7. $x^4 - (y - 9)^2$

8. $(x - 4)^2 - (y + 6)^2$

9. $x^2(x + 2) + 3(x + 2)$

10. $4x^2(x - 8) - (x - 8)$

11. $(r - 6)^2 + 3(r - 6) - 18$

12. $5(y + 8)^2 - 3(y + 8) - 2$

13. $(s - 10)^2 - 20(s - 10) + 100$

14. $(s + 3t)^2 - (s - t)^2$

15. $(y - 4)y^2 - (y - 4)$

16. $(x^2 - 4)^2 - 4x^2$

17. $3x^3 - 2x^2 - 12x + 8$

18. $x - x^2 - x^3 + x^4$

19. $x^3 + 2x^2 - 25x - 50$

20. $r^2 - 6r + 9 - s^2$

21. $x^4 - x^3 + 27x - 27$

22. $x^2 + 12x + 36 - 4y^2$

23. $x^2 - y^2 - 16y - 64$

24. $x^4 - x^2 + 18x - 81$

25. $y^4 + 4y^2 + 4 - 25x^2$

26. $(x + 1)^3 - (x - 1)^3$

27. $x^4 - 15x^2 - 250$

28. $(t - 1)^4 - 35(t - 1)^2 - 36$

29. $c^2 + 6c + 9 - 16x^2$

30. $3x^3 + 2x^2 - 3xy^2 - 2y^2$

31. $4a^3 - 12a^2 - a + 3$

32. $(x + 2)^2 - (x - 7)^2$

33. $a^3(x - 3)^2 + b^3(x - 3)^2$

34. $(x^2 - 1)^3 + 1$

35. $-(x + 2)^2 + 14(x + 2) - 49$

36. $2p^3 - 3p^2 + 2p - 3$

37. $x^8 - y^8$

38. $t^6 - 64$

39. $p^3(r - s)^3 + p^3$

40. $(2x + 1)^2 - 3(2x + 1) - 10$

41. $x(x + 3)(4x + 8) - 5(x + 3)$

42. $r^2 + 4rs + 4s^2 - r - 2s$

43. $x^6 - 1$

44. $1 - x^6$

45. $1 - x^2 - 2xy - y^2$

46. $9 - 4a^2 - 4a - 1$

47. $-x^2 - y^2 + 2xy + z^2$

48. $(x^2 - 1)^2 + 3(x^2 - 1) + 2$

49. $(a^2 + 2a - 3)^2 - 4$

50. $2x^4 - x^3 - 4x + 2$

51. $x^2 - 9y^2 + 5x + 15y$

52. $a^4 - 9(2a - 3)^2$

53. $2x^2 + y^2 + 3xy + 3x + 3y$

54. $25x^2 - x^2y^2 + 10xy - a^2 - 2axy + y^2$

2.8   FACTORING   V: APPLICATIONS

If the trinomial $ax^2 + bx + c$ can be factored over the set of integers, then there is an easy way to solve the equation $ax^2 + bx + c = 0$. The method depends on the following theorem.

THE ZERO-PRODUCT THEOREM

A product of two real numbers is zero if and only if one or both of the two numbers is zero.

In symbols, $rs = 0$ if and only if $r = 0$ or $s = 0$.

EXAMPLE 1   Given $8x = 0$. Then by the zero-product theorem $8 = 0$ or $x - 0$. Since $8 \neq 0$, x must $= 0$.

EXAMPLE 2   Solve the equation $(x - 5)(x + 7) = 0$.

Solution

$$(x - 5)(x + 7) = 0$$

By the zero-product thereom,

$$x - 5 = 0 \quad \text{or} \quad x + 7 = 0$$
$$x = 5 \quad \text{or} \quad x = -7$$

The solution set is $\{5, -7\}$. Check *both* answers.

Check

For $x = 5$, $(5 - 5)(5 + 7) = (0)(12) = 0$.

For $x = -7$, $(-7 - 5)(-7 + 7) = (-12)(0) = 0$.

The polynomial $ax^2 + bx + c$ is a second-degree or **quadratic** polynomial if $a \neq 0$, and an equation of the form $ax^2 + bx + c = 0$ is called a **quadratic equation in one variable.** The solution of a quadratic equation by the methods of this section depends on (1) the possibility of factoring the trinomial over the integers; (2) the application of the zero-product theorem; and (3) the solution of the resulting first-degree equations. Solutions of quadratic equations that are not readily factorable over the integers will be treated in a later chapter.

EXAMPLE 3   Solve the equation $x^2 + 15x + 54 = 0$.

Solution   In factored form,

$$x^2 + 15x + 54 = (x + 6)(x + 9)$$

and the given equation is equivalent to

$$(x + 6)(x + 9) = 0$$

By the zero-product theorem,
$$x + 6 = 0 \quad \text{or} \quad x + 9 = 0$$
$$x = -6 \quad \text{or} \quad x = -9$$

Check For $x = -6$,
$$x^2 + 15x + 54 = 0$$
$$(-6)^2 + (15)(-6) + 54 = 0$$
$$36 - 90 + 54 = 0$$
$$0 = 0$$

For $x = -9$,
$$x^2 + 15x + 54 = 0$$
$$(-9)^2 + (15)(-9) + 54 = 0$$
$$81 - 135 + 54 = 0$$
$$0 = 0$$

Therefore the solution set is $\{-6, -9\}$.

EXAMPLE 4 Solve for x: $x^2 + 8x = 0$.

Solution The quadratic polynomial $x^2 + 8x$ has a common factor, x. An equivalent equation is
$$x(x + 8) = 0$$
By the zero-product theorem:
$$x = 0 \quad \text{or} \quad x + 8 = 0$$
$$x = 0 \quad \text{or} \quad x = -8$$
and the solution set is $\{0, -8\}$.

EXAMPLE 5 Solve for x: $2x^2 + 4x = x^2 + 5$.

Solution Before attempting a solution, express the given equation as an equivalent equation whose right side is zero:
$$2x^2 + 4x = x^2 + 5$$
$$x^2 + 4x - 5 = 0$$
Now factor:
$$(x + 5)(x - 1) = 0$$
By the zero-product theorem:
$$x + 5 = 0 \quad \text{or} \quad x - 1 = 0$$
$$x = -5 \quad \text{or} \quad x = 1$$

Check For $x = -5$,
$$2x^2 + 4x = x^2 + 5$$
$$(2)(-5)^2 + (4)(-5) = (-5)^2 + 5$$
$$50 + (-20) = 25 + 5$$
$$30 = 30$$

For $x = 1$,
$$2x^2 + 4x = x^2 + 5$$
$$(2)(1)^2 + (4)(1) = (1)^2 + 5$$
$$2 + 4 = 1 + 5$$
$$6 = 6$$

Since both answers check, the solution set is $\{-5, 1\}$.

2.8 FACTORING V: APPLICATIONS

Polynomial equations of degree greater than 2, which are readily factorable over the integers, may also be solved by applying the zero-product theorem.

EXAMPLE 6 Solve the equation $x^3 - x = 0$.

Solution $x^3 - x$ factors as $x(x^2 - 1) = x(x + 1)(x - 1)$. Therefore

$$x^3 - x = 0$$
$$x(x + 1)(x - 1) = 0$$
$$x = 0 \quad \text{or} \quad x + 1 = 0 \quad \text{or} \quad x - 1 = 0 \qquad \text{By the zero-}$$
$$\text{product theorem}$$

$$x = 0, \quad x = -1, \quad \text{or} \quad x = 1$$

The solution set is $\{0, -1, 1\}$.

A word of caution is necessary for the solution of equations of the type $x^2 - a^2 = 0$. These equations are often stated in the form $x^2 = a^2$. Before applying the zero-product theorem, the polynomial must be written on one side of the equation, with the other side of the equation equal to 0. **Do not divide each side of the equation by a variable.** There is always the possibility that the variable may be equal to zero, and division by zero is not permitted since the operation is undefined. Thus division by a variable restricts the variable to a nonzero value.

EXAMPLE 7 Solve $x^2 = 25$.

Solution

$$x^2 = 25$$
$$x^2 - 25 = 0$$
$$(x + 5)(x - 5) = 0$$
$$x + 5 = 0 \quad \text{or} \quad x - 5 = 0$$
$$x = -5 \quad \text{or} \qquad x = 5$$

and the solution set is $\{-5, 5\}$.

There are many real-life applications of problems involving the solution of quadratic equations, and these applications will be discussed in greater detail in Section 5.9. Some simple problems, however, can be done now, as shown in the following example.

EXAMPLE 8 The sum of the square of a number and twice the number is 15. Find the number.

Solution Let x = the number. Then x^2 = the square of the number and $2x$ = twice the number.

$$x^2 + 2x = 15$$

This is a quadratic equation because it involves x^2, and x is the variable for which we are solving.

$$x^2 + 2x = 15$$
$$x^2 + 2x - 15 = 0$$
$$(x + 5)(x - 3) = 0$$

Applying the zero-product theorem,

$$x + 5 = 0 \quad \text{or} \quad x - 3 = 0$$
$$x = -5 \quad \text{or} \quad x = 3$$

The problem asked us to find "the number," but since we were not restricted to a positive number, both answers are correct and satisfy the stated conditions.

If the number is -5, then $(-5)^2 + 2(-5) = 25 - 10 = 15$.

If the number is 3, then $(3)^2 + 2(3) = 9 + 6 = 15$.

Therefore the number is either -5 or 3.

EXERCISES 2.8

In Exercises 1–50 solve each equation by factoring, if possible. Check all answers.

1. $(x + 2)(x - 3) = 0$
2. $(x - 1)(x + 4) = 0$
3. $(2y - 3)(y + 2) = 0$
4. $(3a + 1)(5a - 2) = 0$
5. $(x + 5)(x + 2) = 0$
6. $(t - 3)(t - 9) = 0$
7. $(4s - 1)(2s + 3) = 0$
8. $(3x - 4)(6x - 1) = 0$
9. $2y^2 + 7y + 6 = 0$
10. $x^2 + 6x = 0$
11. $x^2 - 3x + 2 = 0$
12. $2p^2 - 3p = 0$
13. $z^2 - 5z = 0$
14. $2n^2 - 32 = 0$
15. $x(x + 2) = 0$
16. $x^2 - 8x + 12 = 0$
17. $x^2 + 6x = 27$
18. $8k - k^2 = 0$
19. $24x + 99 = 3x^2$
20. $15z + 3z^2 = z - 8$
21. $a^2 + 4a + 12 = 8a + a^2$
22. $x^2 - 2x = 0$
23. $x^2 - 2x = 3$
24. $(y + 2)^2 = 3y^2 - 2y + 4$
25. $(x + 1)(x - 2) = 4$
26. $(x - 1)^2 = x^2 - 9$
27. $y^2 + 7y = -10$
28. $x^2 = 6 + x$
29. $a^2 - 12a - 30 = 15$
30. $x^2 + 8x - 20 = 4 - 2x$
31. $(m + 3)(m + 4) = 0$
32. $(m + 3)(m + 4) = 6$
33. $(x + 3)(x - 3) = 8x$
34. $(x + 2)(x - 1) = x^2 + 1$
35. $x(x + 2)(x - 3) = 0$
36. $x(x - 1)(x + 4) = 0$
37. $x^2 = 9$
38. $15 - 2x = x^2$
39. $u^2 = 100$
40. $u^2 = 64$
41. $x^2 = 16$
42. $x^2 = 25$
43. $y^2 = 36$
44. $y^2 = 49$
45. $(x - 5)^2 = (x + 5)^2$
46. $(x + 1)(x^2 - 2x) = (x^2 + x)(x - 2)$
47. $(y + 1)^3 - y^3 = 1$
48. $(y + 1)(y - 1) = 2y^2 - 5$
49. $x^3 = 8 + x^3 - 2x^2$
50. $(x + 3)(x + 4)(x + 5)(x + 6) = 0$

51. The square of a number is 91 more than 6 times the number. Find the number.
52. The sum of a number and its reciprocal is 2. Find the number.
53. The product of two consecutive odd integers is 255. Find the integers.
54. The product of two consecutive even integers is 528. Find the integers.
55. If 18 is subtracted from 10 times a number, the result is one-half the square of the number. Find the number.
56. $A = \pi R^2 - \pi r^2$ is the formula for the area between two concentric circles.
    (a) Express the right side of the formula in factored form.
    (b) Using the factored form, calculate the area between two circles, where $R = 43$ and $r = 41$. Use $\pi = \dfrac{22}{7}$.

DIAGNOSTIC TEST

For Problems 1–15 select the best answer. If the answer you select is "none of these," write the correct answer.

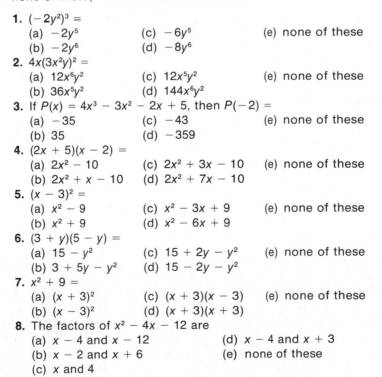

1. $(-2y^2)^3 =$
   (a) $-2y^5$ (c) $-6y^5$ (e) none of these
   (b) $-2y^6$ (d) $-8y^6$
2. $4x(3x^2y)^2 =$
   (a) $12x^6y^2$ (c) $12x^5y^2$ (e) none of these
   (b) $36x^5y^2$ (d) $144x^6y^2$
3. If $P(x) = 4x^3 - 3x^2 - 2x + 5$, then $P(-2) =$
   (a) -35 (c) -43 (e) none of these
   (b) 35 (d) -359
4. $(2x + 5)(x - 2) =$
   (a) $2x^2 - 10$ (c) $2x^2 + 3x - 10$ (e) none of these
   (b) $2x^2 + x - 10$ (d) $2x^2 + 7x - 10$
5. $(x - 3)^2 =$
   (a) $x^2 - 9$ (c) $x^2 - 3x + 9$ (e) none of these
   (b) $x^2 + 9$ (d) $x^2 - 6x + 9$
6. $(3 + y)(5 - y) =$
   (a) $15 - y^2$ (c) $15 + 2y - y^2$ (e) none of these
   (b) $3 + 5y - y^2$ (d) $15 - 2y - y^2$
7. $x^2 + 9 =$
   (a) $(x + 3)^2$ (c) $(x + 3)(x - 3)$ (e) none of these
   (b) $(x - 3)^2$ (d) $(x + 3)(x + 3)$
8. The factors of $x^2 - 4x - 12$ are
   (a) $x - 4$ and $x - 12$ (d) $x - 4$ and $x + 3$
   (b) $x - 2$ and $x + 6$ (e) none of these
   (c) x and 4

9. If $x - 2$ is one of the factors of a binomial of the form $A^3 - B^3$, then the other factor is
   (a) $x^2 - 2x + 4$
   (b) $x^2 + 2x + 4$
   (c) $x^2 - 4x + 4$
   (d) $x^2 + 4x + 4$
   (e) none of these

10. Referring to Problem 9, the original binomial is
    (a) $x^3 - 2$
    (b) $x^3 - 4$
    (c) $x^3 - 8$
    (d) $x^3 - 16$
    (e) none of these

11. Which of the following statements is true?
    (a) $x^2x^3 = x^6$
    (b) $(-y)^2 = -y^2$
    (c) $(-x)^3 = -x^3$
    (d) $x^2 + y^2 = (x + y)^2$
    (e) none of these

12. If $(x + 3)(x - 1) = 0$, then $x =$
    (a) 3 or -1
    (b) 3 or 1
    (c) -3 or 1
    (d) -3 or -1
    (e) none of these

13. If $2x^2 + x - 3 = 0$, then $x =$
    (a) 0 or $\dfrac{3}{2}$
    (b) 1 or $\dfrac{3}{2}$
    (c) -1 or $\dfrac{3}{2}$
    (d) 1 or $-\dfrac{3}{2}$
    (e) none of these

14. If $(x - 4)(x + 3) = -6$, then
    (a) $x - 4 = -6$ or $x + 3 = -6$
    (b) $x - 4 = 3$ or $x + 3 = -2$
    (c) $x - 4 = -1$ or $x + 3 = 6$
    (d) $x - 4 = -3$ or $x + 3 = 2$
    (e) none of these

15. If the product of two consecutive odd integers is 195, then an equation to solve this problem is
    (a) $x(x + 1) = 195$
    (b) $x(x + 2) = 195$
    (c) $(x + 1)(x + 2) = 195$
    (d) $x + (x + 2) = 195$
    (e) none of these

REVIEW EXERCISES

Simplify Exercises 1–20 by using one or more of the exponent theorems. (Assume all exponents are natural numbers and all variables are nonzero real numbers.) Refer to Section 2.1.

1. $x^5 \cdot x^3$
2. $3y(2x^2)^3$
3. $(a^2b^3)(a^3b^2)^3$
4. $3^3 \cdot 3^2$
5. $-(x^2)(-x)^2$
6. $\dfrac{2^5x^7}{2^3x^8}$

REVIEW EXERCISES

7. $(3x^2y)^3$

8. $\left(\dfrac{2x^3}{y^2}\right)^4$

9. $\dfrac{5^3 \cdot 5^5}{5^7}$

10. $\dfrac{(14)^5}{(42)^5}$

11. $\dfrac{9^2 \cdot 4^2}{6^2}$

12. $\left(\dfrac{6x^2}{3x}\right)^5$

13. $3(-2x)(xy^2)^3$

14. $(-3^2)^3$

15. $(-3^3)^2$

16. $-(3^3)^2$

17. $[(-3)^2]^3$

18. $x^{n+1} \cdot x^{n-1}$

19. $\dfrac{x}{x^{2n}}$

20. $\left(\dfrac{x^2y}{x^n}\right)^n$

Simplify the expressions in Exercises 21–25, and arrange the answer in descending powers of the variable whenever appropriate. Refer to Section 2.2.

21. $(4x^2 - 3x + 12) + (x^3 - 3x^2 + 5)$
22. $(3 - x - x^2) + (4x - 2x^2 + 3)$
23. $(a^2b + 3ab - b^2) - (2a^2b - 4ab + ab^2 - 4b^2)$
24. $4(xy + x^2y - 2) - 2(3xy + xy - 2)$
25. Subtract $5x^3 - 3x^2 + x - 2$ from the sum of $x^3 - 7x + 1$ and $4x^2 + 5x - 3$. Check, using $x = 2$.

Exercises 26 and 27 refer to Section 2.2.

26. If $P(x) = 2x^3 - 3x^2 + 4x - 1$, find $P(-2)$, $P(1)$, and $P(2)$.
27. If $R(x) = x^4 - 2x^2 + 7x - 6$, find $R(-1)$, $R(0)$, and $R(3)$.

Expand the indicated products in Exercises 28–60. Refer to Section 2.3.

28. $x(x^2 + 3x + 2)$
29. $x(x^3 + 2x - 1)$
30. $2x(3x^2 - 4x - 1)$
31. $4x(ax^2 - bx + c)$
32. $(x - 2)(x - 3)$
33. $(y + 1)(y - 4)$
34. $(2a + 3)(3a - 1)$
35. $(a + 5)^2$
36. $(5x + 1)(5x - 1)$
37. $(2y - 3)^2$
38. $x(x + 2)(2x - 7)$
39. $x^3(2x^2 + x - 2)$
40. $-x^4(3x^3 + 2x - 4)$
41. $(2x + y)(3x - 16y)$
42. $x(x + 4)(2x - 5)$

43. $2x(3x + 2)^2$
44. $(x + 2)(x + 3)^2$
45. $(x^2 + 3x + 2)(x - 4)$
46. $(2x^2 - x + 3)(3x + 1)$
47. $(x^2 + 5x + 6)(x^2 - 3x + 1)$
48. $(2x + 4)^2$
49. $(3x - 2)^2$
50. $(a + 2)^3$
51. $(a - 2)^3$
52. $(3y + 1)^3$
53. $(2 + 5x)^3$
54. $(2x - 5)(2x + 5)$
55. $(3x + 2)(3x - 2)$
56. $3x(2x + 5)^2$
57. $(a^n + 1)(a^n - 1)$

58. $(x^n - 3)^2$

59. $(b^n + 1)(b^{2n} + b^2 - 1)$

60. $(x^n + 1)^3$

In Exercises 61–100 factor completely over the integers if possible. Refer to Sections 2.4–2.7.

61. $3x^2 - 6x + 24xy$
62. $t^2 + 11t + 24$
63. $x^2 + 10x + 25$
64. $49y^4 - 1$
65. $x^3 - 1$
66. $a^3 + 1$
67. $-3x^3 + 27x$
68. $6a^2 - 7a - 3$
69. $36x + 3x^3 - 3x^5$
70. $100p^2 - 25q^2$
71. $64y^3 - 36y$
72. $x^6 - y^6$
73. $12x^2y - 22xy - 20y$
74. $8a^2 + 5ab - 3b^2$
75. $2y^5 - 162y$
76. $15ax^6 + 42ax^5 - 9ax^4$
77. $25r^2 + 10r + 1$
78. $36x^2 - 12x + 1$
79. $4p^2 - 20pq + 25q^2$
80. $x(y + 2) + 3(y + 2)$

81. $5(x - 3) + p(x - 3)$
82. $3x - x^2(x + 4)$
83. $y^2z(a - 1) + yz$
84. $x(a + 2)^2 + y(a + 2)$
85. $3(x + y) - 2(x + y)^2$
86. $x^2 + 60x + 900$
87. $x^3y^2 + y^2$
88. $7(x + a)^2 - 11y(x + a) - 6y^2$
89. $r^2 - 25s^2 + r - 5s$
90. $ax + cx + a^2b + abc$
91. $x^3 - 27$
92. $x^4 - x^2 + 1 - x$
93. $x^3 - 2x^2 - 5x + 6$
94. $(b - 1)^3 + 8$
95. $x^3 + 125 + x^4 + 125x$
96. $(m^2 + 2m + 1)^2 - 16$
97. $5x^{2n} - 5$
98. $x^{3n} - x^n$
99. $y^{n+2} + 2y^{n+1} - 15y^n$
100. $x^{6n+1} - x$

Solve the equations in Exercises 101–114 by factoring, if possible. Refer to Section 2.8.

101. $2x^2 - 3x - 14 = 0$
102. $x^2 + 7x = 0$
103. $x^2 + 7x = 60$
104. $45y - 15y^2 = 0$
105. $t(t - 11) = -18$
106. $12z^2 + 24z + 12 = 0$
107. $2x - (x + 2) = 3x$

108. $x^2 + 6 = 7x$
109. $x^2 = 3x$
110. $x^2 = 64$
111. $(x - 3)(x + 2)(x - 1) = 0$
112. $x^2 = 9(x - 1)^2$
113. $11x + 26 = 2 - x^2$
114. $x^2 + 36 = 12x$

115. The product of two consecutive even integers is 18 more than 15 times the larger number. Find the two integers.

116. Five times the square of a number added to 10 times the number equals 315. Find the number.

117. $b^2 = c^2 - a^2$ is a formula for finding side b of a right triangle when the hypotenuse c and side a are given.
   (a) Rewrite the formula, expressing the right side in factored form.
   (b) Using the result of (a), calculate b for
       1. $c = 85$ and $a = 84$
       2. $c = 37$ and $a = 35$

118. $A = 2\pi rh + 2\pi r^2$ is a formula for the total area A of a right circular cylinder having radius r and height h.
    (a) Rewrite the formula, expressing the right side in factored form.
    (b) Using the result of (a), calculate A for $r = 15$ and $h = 20$. Use $\pi = \dfrac{22}{7}$.

119. $D = ckwL^2 - 4c^2kwL + 4c^3kw$ is a formula that gives the maximum deflection for a certain beam of length L.
    (a) Rewrite the formula, expressing the right side in factored form.
    (b) Using the result of (a), calculate D for $L = 12$, $c = \dfrac{1}{4}$, $w = 200$, and $k = \dfrac{1}{500}$.

120. $1600y = 1200x - x^2$ is an equation of the orbit of a certain projectile initially fired at 200 ft/sec.
    (a) Find x for $y = 0$. (The largest value gives the range.)
    (b) Find x for $y = 200$. (This gives the horizontal distance the projectile has traveled when its vertical height is 200 ft.)

121. In chemistry, $x^2 = 4(1 - x)^2$ gives the number of moles, x, that react when 1 mole of pure ethyl alcohol is mixed with 1 mole of acetic acid. Find x.

3

RATIONAL EXPRESSIONS: FRACTIONS

I n the preceding chapter we studied some of the properties of algebraic expressions that are called polynomials. In this chapter we will consider rational algebraic expressions that are also called algebraic fractions, or simply fractions.

A **rational algebraic expression** is an expression that is obtained by adding, subtracting, multiplying, or dividing polynomials. For example, the expressions

$$t - \frac{1}{t} + 5, \quad \frac{y^2 + 9}{y + 3}, \quad \text{and} \quad \frac{x - 2}{x^2 - 5x + 6} + \frac{3x + 5}{x^2 - 9}$$

are rational expressions.

3.1 BASIC PROPERTIES OF FRACTIONS

In our initial study of polynomials we saw that the numerical value of a polynomial depends on the value substituted for the variable. **Until stated otherwise, all numbers involved in forming a rational expression shall be real numbers, and no polynomial appearing in a denominator of a fraction shall be zero.** Since the substitution of a real number for the variable results in the polynomial standing for a real number, all the properties of fractions that you learned in your study of arithmetic are equally valid in the study of rational expressions.

DEFINITION OF EQUAL RATIONAL EXPRESSIONS

If P, Q, R, and S are polynomials, and if $Q \neq 0$ and $S \neq 0$, then

$$\frac{P}{Q} = \frac{R}{S} \quad \text{if and only if} \quad P \cdot S = Q \cdot R$$

Thus $\dfrac{3}{4} = \dfrac{9}{12}$ because $3(12) = 4(9)$ and $\dfrac{2}{3} = \dfrac{4}{x}$ if and only if $2x = 12$ and $x = 6$.

Again, recognizing that a rational expression represents a real number if each polynomial stands for a real number and no division by zero occurs, we shall state the fundamental theorem of fractions in the familiar terms of real numbers. This theorem is also the fundamental theorem of rational expressions.

THE FUNDAMENTAL THEOREM OF FRACTIONS

If n, d, and k are real numbers and $d \neq 0$ and $k \neq 0$, then

$$\frac{nk}{dk} = \frac{n}{d}$$

In words, we can multiply or divide both numerator and denominator of a fraction by the same nonzero number without changing the value of the fraction.

This theorem is valid since $(nk)d = (dk)n$ by the associative and commutative axioms for multiplication.

The fundamental theorem of fractions is used to simplify a fraction—that is, to reduce a fraction to lowest terms.

The simplification of a fraction, or the reduction of a fraction to lowest terms, means the renaming of a fraction so that the numerator and denominator do not have a factor in common.

Since all common factors of the numerator and denominator must be removed, simplification of a fraction requires that the numerator and denominator be completely factored.

EXAMPLE 1 Simplify $\dfrac{35x^2y^3}{7xy}$.

Solution

$$\frac{35x^2y^3}{7xy} = \frac{5 \cdot 7 \cdot xxyyy}{7xy} = \frac{5xy^2 \cdot 7xy}{1 \cdot 7xy}$$

$$= \frac{5xy^2}{1} = 5xy^2$$

EXAMPLE 2 Simplify $\dfrac{x^2 - 9}{x^2 - 3x}$.

Solution Factoring numerator and denominator,

$$\frac{x^2 - 9}{x^2 - 3x} = \frac{(x + 3)(x - 3)}{x(x - 3)} = \frac{x + 3}{x}$$

EXAMPLE 3 Reduce $\dfrac{x^2 - 1}{x - 1}$ to lowest terms.

Solution

$$\frac{x^2 - 1}{x - 1} = \frac{(x + 1)(x - 1)}{x - 1} = \frac{(x + 1)(x - 1)}{1(x - 1)}$$

$$= \frac{x + 1}{1} = x + 1$$

EXAMPLE 4 Simplify $\dfrac{3x^2 - 75y^2}{3x^2 - 21xy + 30y^2}$.

Solution

$$\frac{3x^2 - 75y^2}{3x^2 - 21xy + 30y^2} = \frac{3(x^2 - 25y^2)}{3(x^2 - 7xy + 10y^2)}$$

$$= \frac{3(x - 5y)(x + 5y)}{3(x - 5y)(x - 2y)}$$

$$= \frac{x + 5y}{x - 2y}$$

Special attention is called to the forms $a - b$ and $b - a$, which are **additive inverses** of each other.

$$-(b - a) = -b + a = a + (-b) = a - b$$

Thus

$$-(b - a) = a - b \quad \text{and} \quad a - b = -(b - a)$$

This property is needed sometimes in the process of renaming fractions. Its use is illustrated in the following examples.

EXAMPLE 5 Simplify $\dfrac{x-1}{1-x}$.

Solution Since $1-x = -(x-1)$,

$$\frac{x-1}{1-x} = \frac{x-1}{-(x-1)} = \frac{1(x-1)}{(-1)(x-1)}$$

$$= \frac{1}{-1} = -1$$

EXAMPLE 6 Simplify $\dfrac{15-5y}{y^2-3y}$.

Solution $\dfrac{15-5y}{y^2-3y} = \dfrac{5(3-y)}{y(y-3)} = \dfrac{5[-(y-3)]}{y(y-3)}$

$$= \frac{-5(y-3)}{y(y-3)} = \frac{-5}{y}$$

In order to add or subtract certain fractions, the fractions must be renamed so that they have the same denominator, called a **common denominator.** The process of renaming a fraction by multiplying the numerator and denominator by the same number is called **raising the fraction to higher terms,** or **building up to higher terms.** This process is justified by the fundamental theorem of fractions used in its symmetric form:

$$\frac{n}{d} = \frac{nk}{dk}$$

EXAMPLE 7 Raise $\dfrac{7}{9y}$ to higher terms with denominator $18y^3$.

Solution $\dfrac{7}{9y} = \dfrac{7 \cdot k}{9y \cdot k} = \dfrac{7k}{18y^3}$

Since $18y^3 = 9y \cdot 2y^2$, the multiplier k is $2y^2$.

In other words, the multiplier can be found by dividing the new denominator by the original denominator:

$$\frac{18y^3}{9y} = 2y^2$$

Therefore $\dfrac{7}{9y} = \dfrac{7(2y^2)}{9y(2y^2)} = \dfrac{14y^2}{18y^3}$

EXAMPLE 8 Raise $\dfrac{8}{15xy^2}$ to higher terms with denominator $45x^3y^3$.

Solution Find the multiplier:

$$\frac{45x^3y^3}{15xy^2} = 3x^2y$$

Multiply the numerator and denominator by $3x^2y$:

$$\frac{8}{15xy^2} = \frac{8(3x^2y)}{15xy^2(3x^2y)} = \frac{24x^2y}{45x^3y^3}$$

Another type of problem uses the property that $x = \dfrac{x}{1}$ to build up a fraction or to change a polynomial into a rational expression with a designated denominator.

EXAMPLE 9 Express the polynomial $x - 2$ as a rational expression with denominator $x + 3$.

Solution

$$x - 2 = \frac{x - 2}{1} = \frac{(x - 2)(x + 3)}{1(x + 3)}$$
$$= \frac{x^2 + x - 6}{x + 3}$$

Some problems involve the signs of a fraction and the special property that $b - a = -(a - b)$.

EXAMPLE 10 Express $\dfrac{3}{7 - x}$ with denominator $x - 7$.

Solution

Since $7 - x = -(x - 7)$, then
$$\frac{3}{7 - x} = \frac{3}{-(x - 7)} = \frac{-3}{x - 7}$$

EXAMPLE 11 Express $\dfrac{3}{x + 2}$ in higher terms having the denominator $x^2 - 5x - 14$.

Solution

First find the multiplier:
$$\frac{x^2 - 5x - 14}{x + 2} = \frac{(x - 7)(x + 2)}{x + 2} = x - 7$$

Then
$$\frac{3}{x + 2} = \frac{3(x - 7)}{(x + 2)(x - 7)} = \frac{3x - 21}{x^2 - 5x - 14}$$

EXAMPLE 12 Find n so that $\dfrac{6}{6 - x} = \dfrac{n}{x^2 - 36}$.

Solution

$$\frac{x^2 - 36}{6 - x} = \frac{(x + 6)(x - 6)}{(-1)(x - 6)} = \frac{x + 6}{-1}$$
$$= (-1)(x + 6)$$
$$\frac{6}{6 - x} = \frac{6(-1)(x + 6)}{(-1)(x - 6)(-1)(x + 6)}$$
$$= \frac{-6(x + 6)}{x^2 - 36}$$

Thus
$$n = -6(x + 6)$$

EXERCISES 3.1

In Exercises 1–58 reduce each fraction to lowest terms. (Examples 1–6)

1. $\dfrac{6xy}{6xy}$

2. $\dfrac{7a^2b}{7ab}$

3. $\dfrac{25x^3}{75x^5}$

4. $\dfrac{75x^5}{25x^3}$

5. $-\dfrac{m}{m^3n}$

6. $\dfrac{3xy}{-6y}$

7. $\dfrac{7x + 21}{7x}$

8. $\dfrac{4y}{4y^2 - 8y}$

9. $\dfrac{x^2 - 4x}{x^2 - 2x}$

10. $\dfrac{x^2 - 4}{x^2 - 2x}$

11. $\dfrac{5y - 40}{5y}$

12. $\dfrac{6xy}{2x^2y + 6xy^2}$

13. $\dfrac{5x - 5y}{5x + 5y}$

14. $\dfrac{5x - 5y}{x^2 - xy}$

15. $\dfrac{y^2 + 5y}{25 + 5y}$

16. $\dfrac{u^2v + uv^2}{(u + v)^2}$

17. $\dfrac{4cd + 2c^2}{4cd + 8d^2}$

18. $\dfrac{3a + 3b}{9a + 9b}$

19. $\dfrac{4r + r^2}{4r + 16}$

20. $\dfrac{9t^2 - 81}{t + 3}$

21. $\dfrac{8a + 8b}{8a - 8b}$

22. $\dfrac{8a - 8b}{8b - 8a}$

23. $\dfrac{4m^2 + 4}{4m + 4}$

24. $\dfrac{t - 3}{2t^2 - 5t - 3}$

25. $\dfrac{8y^2 - 12y}{4y}$

26. $\dfrac{(r + 2s)^2}{(r + 2s)^5}$

27. $\dfrac{y + y^2}{3y + 3}$

28. $\dfrac{3a^2 + 3ab}{4b^2 + 4ab}$

29. $\dfrac{(c - d)^2}{6c - 6d}$

30. $\dfrac{2u + 2v}{6u + 6v}$

31. $\dfrac{3y + y^2}{3y + 9}$

32. $\dfrac{75n^2 - 3}{5n + 1}$

33. $\dfrac{7x - 7y}{7x + 7y}$

34. $\dfrac{7x - 7y}{7y - 7x}$

35. $\dfrac{3a^2 + 3}{3a + 3}$

36. $\dfrac{3t + 1}{4 - 36t^2}$

37. $\dfrac{3x^2 + 6x}{3x}$

38. $\dfrac{(a - b)^3}{(a - b)^2}$

39. $\dfrac{x^2 - 9}{9 - x^2}$

40. $\dfrac{3x + 15}{x^2 + 3x - 10}$

41. $\dfrac{x^2 - 1}{x^2 + 3x - 4}$

42. $\dfrac{x^2 + 2x - 8}{x^2 - 4}$

43. $\dfrac{4t^2 - 4t - 24}{6t^2 - 36t + 54}$

44. $\dfrac{z^2 - 10z + 25}{z^2 - z - 20}$

45. $\dfrac{a^2 - b^2}{(a - b)^2}$

46. $\dfrac{x^2 - 3x - 10}{x^2 - 2x - 15}$

47. $\dfrac{4a^2 - 4ab}{a^2 - b^2}$

48. $\dfrac{8t^2 - 12t - 20}{8t^2 - 28t + 20}$

49. $\dfrac{2n^3 + 4n^2 + 2n}{n^3 - n^2 - 2n}$

50. $\dfrac{x^4y^2 - 4x^3y^3}{x^3y^5 - 4x^2y^6}$

51. $\dfrac{7 - x}{x^2 - 49}$

52. $\dfrac{75n^2 - 3}{1 + 5n}$

53. $\dfrac{3n^3 - 21n^2 - 24n}{3n^3 - 3n}$

54. $\dfrac{49n^2 - 21n - 10}{49n^2 - 25}$

55. $\dfrac{6y^4 + 15y^3 - 9y^2}{9y^2 + 27y}$

56. $\dfrac{x^3 + 64}{x^2 - 16}$

57. $\dfrac{x^2 + 3x + 9}{x^3 - 27}$

58. $\dfrac{(a - b)(a - c)(b - c)}{(c - a)(c - b)(b - a)}$

Express the fractions in Exercises 59–100 in terms of the indicated denominators. (Examples 7–12)

59. $\dfrac{3x}{-8} = \dfrac{}{40x^2}$

60. $-\dfrac{2x}{5y^2} = \dfrac{}{30xy^3}$

61. $-\dfrac{3xy}{7z} = \dfrac{}{42xyz}$

62. $\dfrac{a}{b^2} = \dfrac{}{5ab^3}$

63. $\dfrac{5}{x + 1} = \dfrac{}{2x + 2}$

64. $\dfrac{x}{2x - 5} = \dfrac{}{6x - 15}$

65. $\dfrac{x - 3}{x + 4} = \dfrac{}{5x + 20}$

66. $\dfrac{2x}{x - 2} = \dfrac{}{x^2 - 2x}$

67. $\dfrac{x}{x - 5} = \dfrac{}{2x^2 - 10x}$

68. $\dfrac{2}{y} = \dfrac{}{y^2 + 2y}$

69. $\dfrac{3y}{y + 3} = \dfrac{}{y^2 - 9}$

70. $\dfrac{t + 3}{t - 2} = \dfrac{}{t^2 + t - 6}$

71. $x + 2 = \dfrac{}{x + 1}$

72. $\dfrac{1}{4x - 3} = \dfrac{}{3 - 4x}$

73. $\dfrac{-x}{7 - x} = \dfrac{}{x - 7}$

74. $\dfrac{1}{5 - 2x} = \dfrac{}{2x^2 - 5x}$

75. $\dfrac{y}{4 - y} = \dfrac{}{y^2 - 16}$

76. $\dfrac{x + 3}{x + 4} = \dfrac{}{x^2 + 6x + 8}$

77. $\dfrac{a + 1}{a - 3} = \dfrac{}{a^2 - 2a - 3}$

78. $x = \dfrac{}{6x^2 - 6x}$

79. $\dfrac{t - 2}{t + 2} = \dfrac{}{14 + 9t + t^2}$

80. $\dfrac{a - b}{a + b} = \dfrac{}{(a + b)^2}$

81. $\dfrac{2x}{2x - 1} = \dfrac{}{4x^2 - 1}$

82. $\dfrac{1 + t}{1 + 3t} = \dfrac{}{18t^2 + 6t}$

83. $\dfrac{r}{r + 4} = \dfrac{}{r^2 + r - 12}$

84. $\dfrac{u - 5}{u - 6} = \dfrac{}{u^2 - 4u - 12}$

85. $6x - 5 = \dfrac{}{x - 2}$

86. $\dfrac{1}{5 - 6y} = \dfrac{}{6y - 5}$

87. $\dfrac{-t}{1 - t} = \dfrac{}{t - 1}$

88. $\dfrac{x}{5 - x} = \dfrac{}{x^2 - 25}$

89. $\dfrac{-1}{7 - 2r} = \dfrac{}{4r^2 - 49}$

90. $\dfrac{x + 2}{x + 4} = \dfrac{}{2x^3 + 5x^2 - 12x}$

91. $\dfrac{y - 1}{y - 2} = \dfrac{}{12 - 8y + y^2}$

92. $\dfrac{6x}{x + 6} = \dfrac{}{2x^2 + 2x - 60}$

93. $x = \dfrac{}{5x + 10}$

94. $\dfrac{2x + 3y}{2x - 3y} = \dfrac{}{(2x - 3y)^2}$

95. $\dfrac{x}{3} = \dfrac{}{3x^2 - 75x}$

96. $\dfrac{3}{3y + 1} = \dfrac{}{1 - 9y^2}$

97. $\dfrac{6a^2}{5a - 1} = \dfrac{}{10a^2 + 13a - 3}$

98. $\dfrac{r + 1}{r + 2} = \dfrac{}{7r^4 - 28r^2}$

99. $\dfrac{-a}{2} = \dfrac{}{4a^2 - 10a}$

100. $\dfrac{9}{x^2 - 9} = \dfrac{}{(x - 3)(x + 3)^2}$

3.2 PRODUCTS AND QUOTIENTS OF FRACTIONS

Multiplication and division of rational expressions also follow the rules of multiplication and division of rational numbers as in arithmetic.

THEOREM: PRODUCT OF FRACTIONS

The product of two fractions is obtained by multiplying the numerators to obtain the new numerator and by multiplying the denominators to obtain the new denominator.

In symbols, if n, d, r, and s are any real numbers such that $d \neq 0$ and $s \neq 0$, then

$$\frac{n}{d} \cdot \frac{r}{s} = \frac{nr}{ds}$$

EXAMPLE 1 Express $\dfrac{7x^2}{3y} \cdot \dfrac{12y^2}{35x^3}$ as a simplified single fraction.

Solution

$$\dfrac{7x^2}{3y} \cdot \dfrac{12y^2}{35x^3} = \dfrac{7 \cdot 12x^2y^2}{3 \cdot 35x^3y}$$ (Using the commutative and associative axioms to rearrange the factors so that the numerals occur first and the letters are arranged in alphabetical order)

$$= \dfrac{4y(3 \cdot 7x^2y)}{5x(3 \cdot 7x^2y)}$$ $\left(\text{Rearranging the factors to form the pattern } \dfrac{ak}{bk}\right)$

$$= \dfrac{4y}{5x}$$ (Since $(3 \cdot 7x^2y)$ is a common factor, the fraction can be reduced)

EXAMPLE 2 Simplify $\dfrac{t^2 + 3t}{6t - 2} \cdot \dfrac{9t^2 - 1}{t^2 - 9}$.

Solution

$$\dfrac{t^2 + 3t}{6t - 2} \cdot \dfrac{9t^2 - 1}{t^2 - 9} = \dfrac{t(t + 3)}{2(3t - 1)} \cdot \dfrac{(3t - 1)(3t + 1)}{(t - 3)(t + 3)}$$ (Factoring)

$$= \dfrac{t(3t + 1)(t + 3)(3t - 1)}{2(t - 3)(t + 3)(3t - 1)}$$ (Rearranging factors)

$$= \dfrac{t(3t + 1)}{2(t - 3)}$$ (Eliminating the common factors)

The final simplified fraction in Example 2 is left in factored form.

Note that each fraction was expressed in factored form, and the terms were rearranged to simplify reducing.

EXAMPLE 3 Simplify $\dfrac{x^2 + 3x + 2}{x^2 - 4x - 12} \cdot \dfrac{x - 6}{x + 1}$.

Solution   First factor the trinomials, if possible:

$$\dfrac{x^2 + 3x + 2}{x^2 - 4x - 12} = \dfrac{(x + 2)(x + 1)}{(x + 2)(x - 6)}$$

Thus

$$\dfrac{x^2 + 3x + 2}{x^2 - 4x - 12} \cdot \dfrac{x - 6}{x + 1} = \dfrac{(x + 2)(x + 1)}{(x + 2)(x - 6)} \cdot \dfrac{(x - 6)}{(x + 1)}$$

$$= \dfrac{(x + 1)}{(x - 6)} \cdot \dfrac{(x - 6)}{(x + 1)}$$

$$= \dfrac{(x - 6)(x + 1)}{(x - 6)(x + 1)}$$

$$= 1$$

Note that the first fraction in Example 3 was simplified *before* the two fractions were multiplied. *It is usually desirable to reduce a fraction as soon as possible to avoid any cumbersome arithmetic or algebra.*

Division of rational expressions follows the rule set out by the following theorem for the division of fractions.

THEOREM: QUOTIENT OF FRACTIONS

The quotient of one fraction divided by another fraction is obtained by inverting the divisor and multiplying the resulting fractions.

In symbols, for n, d, r, and s any real numbers with $d \neq 0$, $s \neq 0$, and $r \neq 0$,

$$\frac{n}{d} \div \frac{r}{s} = \frac{n}{d} \cdot \frac{s}{r} = \frac{ns}{dr}$$

EXAMPLE 4 Simplify $\dfrac{3a^2}{5b^3} \div \dfrac{6a}{10b}$.

Solution
$$\frac{3a^2}{5b^3} \div \frac{6a}{10b} = \frac{3a^2}{5b^3} \cdot \frac{10b}{6a}$$
$$= \frac{(3a^2)(10b)}{(6a)(5b^3)} = \frac{(a)(2)}{(2)(b^2)} = \frac{a}{b^2}$$

EXAMPLE 5 Simplify $\dfrac{x+1}{x} \div \dfrac{x^2 + 3x + 2}{x^2 + x}$ by expressing the indicated quotient as a single fraction in lowest terms.

Solution
$$\frac{x+1}{x} \div \frac{x^2 + 3x + 2}{x^2 + x} = \frac{x+1}{x} \cdot \frac{x^2 + x}{x^2 + 3x + 2}$$
$$= \frac{x+1}{x} \cdot \frac{x(x+1)}{(x+2)(x+1)}$$
$$= \frac{x(x+1)(x+1)}{x(x+2)(x+1)}$$
$$= \frac{x+1}{x+2}$$

EXAMPLE 6 Simplify $\dfrac{x^2 + 4x}{x^3 + 4x^2 + 4x} \div \dfrac{5x + 20}{x^3 - 4x^2 - 12x}$ by expressing the indicated quotient as a single fraction in lowest terms.

Solution
$$\frac{x^2 + 4x}{x^3 + 4x^2 + 4x} \cdot \frac{x^3 - 4x^2 - 12x}{5x + 20}$$
$$= \frac{x(x+4)(x)(x+2)(x-6)}{x(x+2)(x+2)5(x+4)}$$
$$= \frac{x(x-6)(x)(x+2)(x+4)}{5(x+2)(x)(x+2)(x+4)}$$
$$= \frac{x(x-6)}{5(x+2)}$$

EXERCISES 3.2

In Exercises 1–50 simplify the indicated products and quotients of rational expressions.

1. $\dfrac{3x}{y-2} \cdot \dfrac{2y}{x-3}$

2. $\dfrac{3}{y} \cdot \dfrac{y^2}{6xy}$

3. $\dfrac{2}{5ab^2} \cdot \dfrac{15ab}{22}$

4. $\dfrac{4c-8}{14c} \cdot \dfrac{35c}{6c-12}$

5. $\dfrac{4abc}{-6xy^2} \cdot \dfrac{-8x^2y}{12a^2bc^2}$

6. $\dfrac{(u-v)^2}{2uv} \cdot \dfrac{8u^3v^3}{(v-u)^3}$

7. $\dfrac{x+y}{4} \cdot \dfrac{x-y}{4}$

8. $\dfrac{2}{x} \cdot \dfrac{x^2}{4}$

9. $\dfrac{5}{3x^2y} \cdot \dfrac{21y^2}{7x}$

10. $\dfrac{3a+3}{25a} \cdot \dfrac{5a}{9a+9}$

11. $\dfrac{-6xyz}{4a^2b} \cdot \dfrac{10ab^2}{15xyz^2}$

12. $\dfrac{r+s}{rs} \cdot \dfrac{r^2s^2}{(r+s)^2}$

13. $\dfrac{a^2-25}{5} \cdot \dfrac{10a}{a^2+4a-5}$

14. $\dfrac{x^2-4xy+4y^2}{9x^3} \cdot \dfrac{18x}{4x-8y}$

15. $\dfrac{x^2+6x+9}{3x^2+6x} \cdot \dfrac{x^2-4}{x^2+x-6}$

16. $\dfrac{3x^2+3}{x^2-x-6} \cdot \dfrac{x^2-5x+6}{x^2+1}$

17. $\dfrac{a^2-b^2}{3} \cdot \dfrac{3a-3b}{a^2+ab}$

18. $\dfrac{5y^2-5y}{4y-40} \cdot \dfrac{y^3}{3} \cdot \dfrac{y^2-9y-10}{2-2y}$

19. $\dfrac{2n^2-3n-2}{n^3+2n^2-3n} \cdot \dfrac{6n^2-6n}{6n^2+3n}$

20. $\dfrac{x^2-5x}{x^2-3x+2} \cdot \dfrac{x^2-x}{5x-25}$

21. $\dfrac{a^2-3a+2}{a-1} \cdot \dfrac{a+5}{a^2+3a-10}$

22. $\dfrac{-48a^3bc^2}{18ab^2} \cdot \dfrac{-9c}{24a^2c^3} \cdot 4a$

23. $\dfrac{14x^2}{9y^2} \div \dfrac{35x^2}{36y^2}$

24. $\dfrac{-5a^2b^3}{7cd^2} \div \dfrac{10a^3b^2}{21c^2d}$

25. $\dfrac{7x}{x^2-49} \div \dfrac{x^2-14x+49}{x^2+14x+49}$

26. $\dfrac{-10ab^2}{4x^2yz} \div \dfrac{-15a^2b^3}{12xy}$

27. $\dfrac{y^2-y-12}{y^2-16} \div \dfrac{y^2-9}{12y^2-36y}$

28. $\dfrac{(x-2)(x^2-6)}{8x^3+24x^2} \div \dfrac{x^4-36}{2x^9+6x^8}$

29. $\dfrac{25x^2-1}{25} \cdot \dfrac{75x}{5x^2+6x+1}$

30. $\dfrac{4a^2-a-3}{12a^2-7a-12} \div \dfrac{2a^2-a-1}{6a^2-5a-4}$

31. $\dfrac{x^2+6x+9}{x^2+4x+3} \cdot \dfrac{2x^2+2x}{x^2-9}$

32. $\dfrac{7a+14}{a^2+4a+4} \cdot \dfrac{a^2-4}{5a-10}$

33. $\dfrac{3x^2-3x}{x^2-81} \cdot \dfrac{x^2+4x-45}{3x-15}$

34. $\dfrac{10y-21}{y^2-4y+4} \cdot \dfrac{y^2+y-6}{10y^2+9y-63}$

35. $\dfrac{2x^2+13x+15}{2x^2-x-6} \cdot \dfrac{x^2-x-2}{x^2+6x+5}$

36. $\dfrac{3x^2-3x}{5x-40} \cdot \dfrac{x^2}{3} \cdot \dfrac{x^2-7x-8}{4-4x}$

37. $3xy^2 \cdot \dfrac{5x-15}{15xy-45x^2y^2}$

38. $\dfrac{10y^2-11y-6}{3y^2-6y} \cdot \dfrac{y^2-y-2}{6y^2-13y+6}$

39. $\dfrac{y^2+5y}{y^2-5y-6} \cdot \dfrac{y^2-7y+6}{y^2+4y-5} \cdot \dfrac{1}{y+1}$

40. $\dfrac{x^2+12x+36}{x^2-12x+36} \div \dfrac{x^2+5x-6}{x^2-5x-6}$

41. $\dfrac{x^2+x-2}{x^2-x-12} \div \dfrac{x^2+3x+2}{x^2-7x+12}$

42. $\dfrac{a^2+6a-16}{a^2-64} \div (a-2)$

43. $\dfrac{x^3 + 27}{x^2 + 3x} \div \dfrac{x^2 + 9x}{x^2 + 7x - 18}$

44. $\dfrac{1 + 3x - 18x^2}{6x^2 - 17x - 3} \div \dfrac{x - 3}{3x - 1}$

45. $\dfrac{x - y}{z - x} \cdot \dfrac{x - z}{z - y} \cdot \dfrac{y - z}{y - x}$

46. $\dfrac{x^2y^2 + xy - 42}{x^2y^2 - xy - 42} \cdot \dfrac{x^2y^2 - 49}{x^2y^2 - 36}$

47. $\dfrac{y^3 + 1}{y^2 + 1} \cdot \dfrac{y^4 - 1}{y + 1}$

48. $\dfrac{t^3 - 64}{t^2 + 4t + 16} \cdot \dfrac{t}{t^2 - 16}$

49. $\dfrac{x^3 + y^3}{x^2 - xy + y^2} \cdot \dfrac{x^2 - 2xy + y^2}{x^2 - y^2}$

50. $\left(\dfrac{a^3 - 8}{a^2 + 3a + 2} \div \dfrac{a^2 - 4}{a + 1} \right) \cdot \dfrac{a + 2}{2}$

3.3 DIVISION OF POLYNOMIALS: LONG AND SYNTHETIC

It is sometimes desirable to replace a single fraction by a sum or difference of two or more rational expressions. There are several applications (for example, problems in calculus) that require that a quotient of two polynomials be changed into the sum of fractions in which the numerator of each has a degree smaller than the degree of the denominator.

The method used to replace a quotient of two polynomials by a sum or difference is similar to the long-division algorithm used in arithmetic. (An algorithm is a method or pattern of performing a calculation.) For example, the division of 865 by 23 is done as follows:

$$
\begin{array}{r}
37 \\
23\overline{)865} \\
69 \\
\hline
175 \\
161 \\
\hline
14
\end{array}
$$

Thus

$$\frac{865}{23} = 37 + \frac{14}{23} = 37\,\frac{14}{23}$$

Check

23(37) + 14 = 851 + 14 = 865

In arithmetic, the division process is stopped when the remainder is less than the divisor.

In algebra, the division process is stopped when the degree of the remainder polynomial is less than the degree of the divisor polynomial.

EXAMPLE 1 Express $\dfrac{x^3 - 5x^2 + 4x + 7}{x^2 + 2}$ as a sum of rational expressions with the degree of the polynomial in the numerator of any fraction smaller than the degree of the polynomial in its denominator.

Solution

1. Divide x^3 by x^2 to obtain the partial quotient, x:

2. Subtract the product of x and $x^2 + 2$ from the dividend:

$$
\begin{array}{r}
x\ -\ 5 \\
x^2 + 2\overline{)\,x^3 - 5x^2 + 4x + 7} \\
\underline{x^3\qquad\ \ + 2x} \\
-\,5x^2 + 2x + 7
\end{array}
$$

3. Repeat the process:
   Divide $-5x^2$ by x^2 to obtain -5.
4. Subtract the product of -5 and $x^2 + 2$ from the remainder polynomial, $-5x^2 + 2x + 7$:

$$
\begin{array}{r}
-\,5x^2\qquad\ \ -\ 10 \\
\hline
2x + 17
\end{array}
$$

5. Since the degree of the remainder, $2x + 17$, is less than the degree of the divisor, $x^2 + 2$, the process is stopped at this point, and the quotient is expressed as the following sum:

$$\frac{x^3 - 5x^2 + 4x + 7}{x^2 + 2} = x - 5 + \frac{2x + 17}{x^2 + 2}$$

Check

A quick check can be made numerically by assigning a value to x. For instance, let $x = 2$. Then

$$x^3 - 5x^2 + 4x + 7 = 2^3 - 5(2^2) + 4(2) + 7 = 3$$

and

$$x^2 + 2 = 2^2 + 2 = 6$$

For $x = 2$,

$$\frac{x^3 - 5x^2 + 4x + 7}{x^2 + 2} = \frac{3}{6} = \frac{1}{2}$$

Also, for $x = 2$,

$$x - 5 + \frac{2x + 17}{x^2 + 2} = 2 - 5 + \frac{4 + 17}{4 + 2}$$

$$= -3 + \frac{21}{6} = \frac{-18 + 21}{6} = \frac{3}{6} = \frac{1}{2}$$

Since

$$\frac{x^3 - 5x^2 + 4x + 7}{x^2 + 2} = x - 5 + \frac{2x + 17}{x^2 + 2} \quad \text{for} \quad x = 2$$

there is assurance that the problem has been worked correctly.

EXAMPLE 2 Divide $\dfrac{x^3 + 1}{x + 1}$ by using the long-division algorithm.

Solution The polynomial $x^3 + 1$ can also be considered as

$$x^3 + 0 \cdot x^2 + 0 \cdot x + 1$$

so places must be provided for the missing x^2 and x terms:

$$
\begin{array}{r}
x^2 - x + 1 \\
x + 1 \overline{)x^3 \qquad\qquad + 1} \\
\underline{x^3 + x^2} \\
- x^2 \qquad + 1 \\
\underline{- x^2 - x} \\
+ x + 1 \\
\underline{x + 1}
\end{array}
$$

Thus $\dfrac{x^3 + 1}{x + 1} = x^2 - x + 1$

In Example 2, the remainder is zero. Thus $x + 1$ is a factor of $x^3 + 1$.

Note that

$$\frac{x^3 + 1}{x + 1} = \frac{(x + 1)(x^2 - x + 1)}{(x + 1)} \qquad \text{(Factoring the sum of cubes)}$$

$$= x^2 - x + 1$$

A quotient of a polynomial, $P(x)$, divided by a polynomial of the form $D(x) - x - a$, is frequently encountered in problems, so a shorter and less tedious method than long division has been devised for this type of problem. This process is called **synthetic division.**

To illustrate, let us first divide $x^4 + 3x^2 + 2x + 1$ by $x - 2$, using the long-division algorithm:

$$
\begin{array}{r}
x^3 + 2x^2 + 7x + 16 \\
x - 2 \overline{)x^4 \qquad\quad + 3x^2 + 2x + 1} \\
\underline{x^4 - 2x^3} \\
+ 2x^3 + 3x^2 \\
\underline{+ 2x^3 - 4x^2} \\
+ 7x^2 + 2x \\
\underline{+ 7x^2 - 14x} \\
+ 16x + 1 \\
\underline{+ 16x - 32} \\
+ 33 \quad \text{(Remainder)}
\end{array}
$$

Notice that both polynomials were arranged in descending powers of x, and that a space was left for the missing power of x whose coefficient was zero, $(0x^3)$.

If $P(x)$ is the dividend polynomial and $Q(x)$ the quotient polynomial, the coefficients of the terms of $P(x)$ are 1, 0, 3, 2, 1, and the coefficients of the terms of $Q(x)$ are 1, 2, 7, 16. The remainder is 33. Thus

$$\frac{x^4 + 3x^2 + 2x + 1}{x - 2} = x^3 + 2x^2 + 7x + 16 + \frac{33}{x - 2}, \qquad x \neq 2$$

and

$$x^4 + 3x^2 + 2x + 1 = (x - 2)(x^3 + 2x^2 + 7x + 16) + 33$$

even if $\qquad\qquad x = 2$

Therefore

$$P(x) = (x - 2)Q(x) + 33$$

Rewriting the division problem, but this time omitting the variable and using only the numerical coefficients,

```
           1    2    7    16
1 −2) 1    0    3    2     1
       (1) −2
           +2   (3)
          (+2) −4
                +7   (2)
               (+7) −14
                     +16   (1)
                    (+16) −32
                          +33
```

Notice that the numerals in parentheses are repetitions of the numerals directly above them. Omitting these repetitions, the problem may be condensed as follows:

```
           1    2    7    16
1 −2) 1    0    3    2     1
          −2   −4  −14   −32
      1    2    7    16  |33
```

Except for the last term (the remainder 33), the bottom row is identical to the top row, and therefore the top row may be omitted. The coefficient 1 of the divisor is not needed for the subtractions involved, so it may be omitted. Moreover, the subtractions may be changed to additions by using the definition of subtraction—that is, $a - b = a + (-b)$. Replacing -2 by $+2$, the division may be expressed as follows:

```
2| 1  0  3   2   1      or   1  0  3   2   1|2
      2  4  14  32               2  4  14  32
   1  2  7  16 |33            1  2  7  16 |33
```

A detailed discussion of the application of synthetic division to this problem follows.

Divide $x^4 + 3x^2 + 2x + 1$ by $x - 2$.

1. Arrange the terms in the dividend and divisor in descending powers of the variable.

$$x^4 + 0x^3 + 3x^2 + 2x + 1$$
$$x - 2$$

2. Write the coefficients of the terms of the dividend in this order, and be sure to use 0 for every missing power of the variable. On

$$+1 \quad 0 \quad +3 \quad +2 \quad +1 \quad |+2$$

the right, write *a* from the divisor $(x - a)$.

3. Bring down the first term to the third row ($+1$ in the example) and multiply by *a* ($+2$ in the example). Write this product under the second coefficient and add. Multiply this sum by *a* and repeat the process until the last column has been totaled.

```
+1    0   +3   +2   +1  |+2
     +2   +4  +14  +32
+1↗ +2↗ +7↗ +16↗ +33
```

4. The last entry represents the remainder, and the other numbers are the coefficients of the quotient. Since the dividend is a polynomial of degree *n*, and the divisor is a first-degree polynomial, the degree of the quotient is $n - 1$,

$$\left(\frac{ax^n}{x} = ax^{n-1}\right)$$

$$1(x^3) + 2(x^2) + 7(x) + 16 \underline{\lvert +33}$$

$$x^3 + 2x^2 + 7x + 16$$
Remainder: 33

Therefore

$$(x^4 + 3x^2 + 2x + 1) \div (x - 2) = x^3 + 2x^2 + 7x + 16 + \frac{33}{x - 2}$$

Note that the synthetic divisor may be placed to the right or to the left.

EXAMPLE 3 Use synthetic division to find the quotient.

$$\frac{x^5 - 1}{x - 1}$$

Solution

$$x^5 - 1 = x^5 + 0x^4 + 0x^3 + 0x^2 + 0x - 1; \ a = 1$$

```
+1|  +1    0    0    0    0   -1
          +1   +1   +1   +1   +1
     +1   +1   +1   +1   +1  |0
```

Quotient: $x^4 + x^3 + x^2 + x + 1$, with no remainder.

The result can be checked by multiplying $(x^4 + x^3 + x^2 + x + 1)(x - 1)$, or by giving *x* a value *other* than 1. (Why?)

EXAMPLE 4   Use synthetic division to find the quotient.

$$\frac{x^3 + 3x^2 + 4}{x + 2}$$

Solution

$$
\begin{array}{rrrr|r}
+1 & +3 & 0 & +4 & \underline{-2} \\
 & -2 & -2 & +4 & \\
\hline
+1 & +1 & -2 & \underline{|+8} &
\end{array}
$$

Note that $x + 2 =$
$x - (-2)$.
Therefore
$a = -2$.

Quotient: $x^2 + x - 2$; remainder: $+8$.
The answer may be written

$$x^2 + x - 2 + \frac{8}{x + 2}$$

Check   If $P(x) = x^3 + 3x^2 + 4$ and $Q(x) = x^2 + x - 2$, then
$P(x) = (x + 2) \cdot Q(x) + 8$
(The dividend is equal to the product of the divisor and the quotient, plus the remainder.)

$$(x + 2) \cdot Q(x) = (x + 2)(x^2 + x - 2) = x^3 + 3x^2 - 4$$
$$(x + 2) \cdot Q(x) + 8 = x^3 + 3x^2 - 4 + 8 = x^3 + 3x^2 + 4 = P(x)$$

EXERCISES 3.3

In Exercises 1–10 use the long-division algorithm to divide, and check as indicated. (Examples 1 and 2)

1. $\dfrac{2x^3 - 10x^2 + 3x - 14}{x - 5}$
   Check using $x = 2$.

2. $\dfrac{t^2 - 10t + 8 + t^3}{t + 4}$
   Check using $t = 1$.

3. $\dfrac{x^4 - 1}{x - 1}$
   Check using $x = 3$.

4. $\dfrac{4x^2 + 7}{2x - 1}$
   Check using $x = 1$.

5. $\dfrac{2x^5 + 4x^2 - 6x^3 - 5}{x^2 - 3}$
   Check using $x = 2$.

6. $\dfrac{y^4 + 64}{y^2 - 4y + 8}$
   Check using $y = 2$.

7. $\dfrac{25a^3 + ab^2 + b^3}{5a + 2b}$
   Check using $a = 1, b = -2$.

8. $\dfrac{x^6 + x^4 + x^2 + 1}{x^2 + 1}$
   Check using $x = 2$.

9. $\dfrac{x^2 - y^2 - 6y - 9}{x - y - 3}$
   Check using $x = 4, y = 2$.

10. $\dfrac{x^2 - y^2 + 4x - 6y - 5}{x + y + 5}$
    Check using $x = 3, y = 2$.

In Exercises 11–18 use the long-division algorithm to divide. Check either by multiplication or by selecting a suitable value for x. (Examples 1 and 2)

11. $\dfrac{2x^3 + 5x^2 - 3x + 7}{x + 4}$

12. $\dfrac{5x^4 + x^3 - 4x^2 + 2x - 3}{x + 2}$

13. $\dfrac{2x^3 - 13x^2 + 13x + 10}{x - 5}$

14. $\dfrac{6x^3 - 25x^2 + 3x + 4}{2x - 1}$

15. $\dfrac{4x^3 + 10x^2 - 16x - 14}{2x + 7}$

16. $\dfrac{x^4 - 7x^2 + 9}{x^2 - x - 3}$

17. $\dfrac{5x^4 - 30x^2 + 2x - 1}{x^2 - 6}$

18. $\dfrac{x^3 + 3x^2 + 2x + 7}{x + 3}$

In Exercises 19–40 use synthetic division to determine the quotient and remainder. Check by using $P(x) = (x - a) \cdot Q(x) + R$. (Examples 3 and 4)

19. $\dfrac{2x^3 - 3x^2 + 4x - 5}{x - 2}$

20. $\dfrac{2x^4 - 4x^3 + x^2 - 1}{x - 1}$

21. $\dfrac{x^3 - 5x^2 + 2x + 3}{x + 1}$

22. $\dfrac{x^3 - x^2 - 11x + 15}{x - 3}$

23. $\dfrac{x^4 + x^3 - x^2 - x - 2}{x + 2}$

24. $\dfrac{x^4 - x^3 + x^2 - x + 2}{x + 2}$

25. $\dfrac{x^4 + x^2 - 6}{x + 3}$

26. $\dfrac{2x^4 - 17x^2 - 4}{x + 3}$

27. $\dfrac{2x^4 - 200x + 6}{x - 5}$

28. $\dfrac{y^3 - 11y + 6}{y - 3}$

29. $\dfrac{5y^4 + 12y^3 - 30y^2 - 32}{y + 4}$

30. $\dfrac{x^3 - 4x^2 + 8x - 32}{x - 4}$

31. $\dfrac{x^5 + 1}{x + 1}$

32. $\dfrac{x^5 + 32}{x + 2}$

33. $\dfrac{5x^5 + 2x^4 - 50x^3 - x + 1}{x - 3}$

34. $\dfrac{y^6 - 64}{y - 2}$

35. $\dfrac{4x^3 - 10x^2 + x - 1}{x - \frac{1}{2}}$

36. $\dfrac{8x^3 - 6x^2 - 3x + 1}{x - \frac{1}{4}}$

37. $\dfrac{8x^3 - 10x^2 + 7}{4x + 3}$ (*Hint:* Use synthetic division to divide by $x + \dfrac{3}{4}$. Then divide the result by 4, but be careful about the remainder—it is *not* divided!)

38. $\dfrac{8x^3 - 10x^2 + 9}{4x + 3}$

39. $\dfrac{3x^3 - 7x^2 - x + 1}{3x - 1}$

40. $\dfrac{5x^4 + 8x^3 - 4x^2 + 10x - 6}{5x - 2}$

41. Determine p so that the remainder is 0 when $3x^2 + 2x - p$ is divided by $x - 4$.

42. In Exercise 41 $P(x) = 3x^2 + 2x - p$. Replace p by your answer for Exercise 41 and evaluate $P(4)$.

43. Let $P(x) = x^3 - 3x + 6$. Use synthetic division to find the remainder when $P(x)$ is divided by $x - 2$. Compare the remainder R with $P(2)$.

44. Let $P(x) = x^5 - 3x^2 + x - 4$. Find the remainder R when $P(x)$ is divided by $x + 3$. Compare R with $P(-3)$.

3.4 REMAINDER AND FACTOR THEOREMS (Optional)

Let $P(x)$ represent any polynomial. Then $P(2)$ is the value of the polynomial when x is replaced by 2; $P(0)$ is the value of the polynomial when x is replaced by zero; and in general, $P(a)$ is the value of the polynomial when x is replaced by a.

To illustrate, let

$$P(x) = 5x^2 + 3x + 2$$

Then

$$P(2) = 5(2)^2 + 3(2) + 2 = 28$$

and

$$P(0) = 5(0)^2 + 3(0) + 2 = 2$$

Also

$$P(a) = 5a^2 + 3a + 2$$

Let $P(x) = 5x^2 + 3x + 2$; divide $P(x)$ by $x + 3$. By synthetic division,

$$
\begin{array}{rrr|r}
5 & 3 & 2 & \underline{-3} \\
 & -15 & +36 & \\
\hline
5 & -12 & \underline{38} & \text{(Remainder)}
\end{array}
$$

Therefore

$$\frac{5x^2 + 3x + 2}{x + 3} = 5x - 12 + \frac{38}{x + 3}$$

Also

$$P(-3) = 5(-3)^2 + 3(-3) + 2 = 38$$

Let $Q(x) = 2x^3 + 7x - 1$; divide $Q(x)$ by $x + 2$:

$$
\begin{array}{rrrr|r}
2 & 0 & 7 & -1 & \underline{-2} \\
 & -4 & +8 & -30 & \\
\hline
2 & -4 & +15 & \underline{-31} & \text{(Remainder)}
\end{array}
$$

Thus

$$\frac{2x^3 + 7x - 1}{x + 2} = 2x^2 - 4x + 15 + \frac{-31}{x + 2}$$

Also

$$Q(-2) = 2(-2)^3 + 7(-2) - 1 = -31$$

Note that the remainder, after dividing $P(x)$ by $x + 3$, is $P(-3)$, and that the remainder, after dividing $Q(x)$ by $x + 2$, is $Q(-2)$. This is not just coincidence but can be proved by the following theorem.

3.4 REMAINDER AND FACTOR THEOREMS

THE REMAINDER THEOREM

If a is any real number and if a polynomial $P(x)$ is divided by a divisor of the form $(x - a)$, then the remainder R is equal to $P(a)$.

This theorem may be proved as follows:

Let $P(x)$ be any polynomial. Let $Q(x)$ represent the quotient and R the remainder when $P(x)$ is divided by $x - a$.

$$\frac{P(x)}{x - a} = Q(x) + \frac{R}{x - a}, \quad x \neq a \tag{1}$$

or

$$P(x) = Q(x) \cdot (x - a) + R \tag{2}$$

Statement (2) is true for all values of x; in particular, it is true for $x = a$. Then

$$
\begin{aligned}
P(a) &= Q(a) \cdot (a - a) + R \\
&= Q(a) \cdot 0 + R \\
&= 0 + R \\
&= R
\end{aligned}
$$

We can now confirm the results of the two examples just worked by applying the remainder theorem.

EXAMPLE 1   Find the remainder when $5x^2 + 3x + 2$ is divided by $x + 3$.

Solution

$$P(x) = 5x^2 + 3x + 2$$
$$(x + 3) = [x - (-3)]$$

Therefore

$$a = -3$$
$$P(-3) = 5(-3)^2 + 3(-3) + 2 = 38$$

Therefore the remainder is 38.

EXAMPLE 2   Find the remainder when $2x^3 + 7x - 1$ is divided by $x + 2$.

Solution

$$Q(x) = 2x^3 + 7x - 1, \quad a = -2$$
$$Q(-2) = 2(-2)^3 + 7(-2) - 1 = -31$$

Therefore the remainder is -31.

Examples 1 and 2 were checked by synthetic division and are shown to be correct.

Another useful theorem results as a direct consequence of the remainder theorem.

THE FACTOR THEOREM

Let $P(x)$ be a polynomial and let a be any real number. Then $(x - a)$ is a **factor** of $P(x)$ if and only if $P(a) = 0$.

Note the words "if and only if" in the factor theorem. These words indicate that *two* statements must be proved in order for the theorem to be valid:

(a) If $(x - a)$ is a factor of $P(x)$, then $P(a) = 0$.

The **converse** of statement (a) must also be proved.

(b) If $P(a) = 0$, then $(x - a)$ is a factor of $P(x)$.

Proof (a) *Assume* $(x - a)$ is a factor of $P(x)$ and *prove* that $P(a) = 0$.

If $(x - a)$ is a factor of $P(x)$, then there exists a polynomial $Q(x)$ such that

$$P(x) = Q(x) \cdot (x - a)$$

and also

$$P(a) = Q(a) \cdot (a - a) = Q(a) \cdot 0 = 0$$

therefore

$$P(a) = 0$$

(b) *Assume* that for polynomial $P(x)$, $P(a) = 0$ and *prove* that $(x - a)$ is a factor of $P(x)$.

By the remainder theorem,

$$
\begin{aligned}
P(x) &= Q(x) \cdot (x - a) + R \\
&= Q(x) \cdot (x - a) + P(a) \quad &\text{(Since } R = P(a)) \\
&= Q(x) \cdot (x - a) + 0 \quad &(P(a) = 0 \text{ by assumption}) \\
&= Q(x) \cdot (x - a)
\end{aligned}
$$

Since $P(x)$ is shown to be the product of two polynomials, each of these polynomials is a factor of $P(x)$; in particular, $(x - a)$ is a factor of $P(x)$. This proves the factor theorem.

EXAMPLE 3 Show that $x + 2$ is a factor of the polynomial
$$P(x) = x^4 + 5x^3 + 8x^2 + 5x + 2$$

Solution
$$
\begin{aligned}
P(-2) &= (-2)^4 + 5(-2)^3 + 8(-2)^2 + 5(-2) + 2 \\
&= 16 - 40 + 32 - 10 + 2 \\
&= 0
\end{aligned}
$$

Since $P(-2) = 0$, by the factor theorem, $x + 2$ is a factor of
$$x^4 + 5x^3 + 8x^2 + 5x + 2$$

EXAMPLE 4 Determine if $x - 6$ is a factor of
$$2x^5 - x^4 - 16x^3 + 5x^2 - 1832x + 12$$

Solution   Use synthetic division to find $P(6)$:

$$
\begin{array}{r|rrrrrr}
6 & 2 & -1 & -16 & 5 & -1832 & 12 \\
  &   & 12 & 66 & 300 & 1830 & -12 \\
\hline
  & 2 & 11 & 50 & 305 & -2 & 0
\end{array}
$$

$P(6) = R = 0$. Thus $x - 6$ is a factor.

EXAMPLE 5 Determine if $x + 5$ is a factor of
$$x^3 + 8x^2 + 16x + 10$$

Solution   Use synthetic division to find $P(-5)$:

$$\begin{array}{r|rrrr} -5 & 1 & 8 & 16 & 10 \\ & & -5 & -15 & -5 \\ \hline & 1 & 3 & 1 & 5 \end{array}$$

$P(-5) = R = 5$. Since $R \neq 0$, $x + 5$ is *not* a factor.

EXAMPLE 6   Find k so that $x - 3$ is a factor of $kx^3 - 6x^2 + 2kx - 12$

Solution   By the factor theorem, $x - 3$ is a factor of

$$P(x) = kx^3 - 6x^2 + 2kx - 12 \text{ if } P(3) = 0$$
$$P(3) = k(3)^3 - 6(3)^2 + 2k(3) - 12$$
$$= 27k - 54 + 6k - 12 = 33k - 66$$

Let

$$P(3) = 33k - 66 = 0$$

Then

$$k = 2$$

Therefore, when $k = 2$, $x - 3$ is a factor of $kx^3 - 6x^2 + 2kx - 12$.

EXERCISES 3.4 (Optional)

In Exercises 1–10 use the remainder theorem to find the remainder for each indicated division. (Examples 1 and 2)

1. $\dfrac{4x^3 - 3x^2 + x - 4}{x - 3}$

2. $\dfrac{3x^3 + 2x^2 - 4x + 5}{x - 1}$

3. $\dfrac{x^6 + 3x^2 - x + 2}{x + 2}$

4. $\dfrac{x^7 + 3x^3 + x - 2}{x - 2}$

5. $\dfrac{x^9 - 1}{x - 1}$

6. $\dfrac{x^9 - 1}{x + 1}$

7. $\dfrac{3x^{12} + x^3 - x + 2}{x + 1}$

8. $\dfrac{5x^{20} - 7x^{15} + 3x - 1}{x - 1}$

9. $\dfrac{x^5 - 3x^3 - 8}{x - 2}$

10. $\dfrac{x^6 - 4x^5 + x^4 - x^2 + 2}{x + 2}$

In Exercises 11–20 use the factor theorem to determine if $Q(x)$ is a factor of $P(x)$. (Examples 3, 4, and 5)

11. $P(x) = x^3 + 3x^2 + 2x - 24$
$Q(x) = x - 2$

12. $P(x) = 2x^3 - 11x^2 + 12x + 9$
$Q(x) = x - 3$

13. $P(x) = x^4 - x^2 + 2x + 6$
$Q(x) = x + 2$

14. $P(x) = 2x^3 - 7x^2 - 21x + 54$
$Q(x) = x + 3$

15. $P(x) = x^4 - 12x^2 - 5x + 12$
$Q(x) = x + 3$

16. $P(x) = 2x^3 - 3x^2 - 2x + 6$
$Q(x) = x + 1$

17. $P(x) = x^3 + 3x^2 + 3x - 7$
$Q(x) = x - 1$

18. $P(x) = x^3 - 3x^2 - 9x - 5$
$Q(x) = x - 5$

19. $P(x) = x^8 + 1$
$Q(x) = x + 1$

20. $P(x) = x^6 - 1$
$Q(x) = x - 1$

For Exercises 21–26 let $P(x) = x^4 - 6x^2 - 7x - 6$. Find the indicated value, $P(a)$, for each exercise by using (a) the remainder theorem, and (b) synthetic division.

21. $P(1)$

22. $P(-1)$

23. $P(2)$

24. $P(-2)$

25. $P(3)$

26. $P(-3)$

27. Use the results of Exercises 21–26 to factor $P(x)$ over the integers.

For Exercises 28–33 let $P(x) = x^3 - 26x + 5$. Find the indicated value, $P(a)$, for each exercise by using (a) the remainder theorem, and (b) synthetic division.

28. $P(1)$

29. $P(-1)$

30. $P(5)$

31. $P(-5)$

32. $P(2)$

33. $P(-2)$

34. Use the results of Exercises 28–33 to factor $P(x)$ over the integers.

35. For $P(x) = 2x^3 - 3x^2 - 3x + 2$, (a) find $P(1)$, $P(-1)$, $P(2)$, $P(-2)$, $P\left(\dfrac{1}{2}\right)$, $P\left(-\dfrac{1}{2}\right)$, and (b) factor $P(x)$ as a product of linear factors.

36. For $P(x) = 3x^3 - 23x^2 + 13x + 7$, (a) find $P(1)$, $P(-1)$, $P(7)$, $P(-7)$, $P\left(\dfrac{1}{3}\right)$, $P\left(-\dfrac{1}{3}\right)$, and (b) factor $P(x)$ as a product of linear factors.

37. If $P(x) = x^3 + 3x^2 - 2x + k$, find k so that (a) $P(2) = 6$, (b) $P(2) = -4$, and (c) $x - 2$ is a factor of $P(x)$.

38. If $P(x) = x^4 + 2x^3 + x + k$, find k so that (a) $P(3) = 35$, (b) $P(3) = -12$, and (c) $x - 3$ is a factor of $P(x)$.

39. Find k so that $x + 2$ is a factor of $x^4 + 2x^3 + x + k$.

40. Find k so that $x - 3$ is a factor of $2x^3 - 3x^2 + x + k$.

41. Find k so that $x - 2$ is a factor of $x^3 - 4kx^2 + kx + 6$.

42. Find k so that $x - 3$ is a factor of $2x^3 + kx^2 - 5x + 4k$.

43. Find p so that if $Q(x) = 2x^3 - x^2p - 2xp + 5$, then $Q(-1) = 0$.

44. Find m so that if $Q(x) = x^4 + mx^3 + 3x^2 - m$, then $Q(2) = 0$.

45. Find p so that when $2x^3 + x^2p - 2xp + 5$ is divided by $x + 2$ the remainder is -11.

46. Find r so that when $x^3 - rx^2 + 2x + r$ is divided by $x - 2$ the remainder is 3.

47. Find the remainder when $2x^4 - 3x^3 + 5x - 1$ is divided by $x - 4$ by (a) long division, (b) synthetic division, and (c) the remainder theorem.

48. Find the remainder when $3x^5 - 46x^3 + x - 9$ is divided by $x + 4$ by (a) long division, (b) synthetic division, and (c) the remainder theorem.

3.5 SUMS OF FRACTIONS

In this chapter we have used the fundamental theorem of fractions to reduce rational expressions, to raise them to higher terms, to multiply rational expressions, and to divide them. The fundamental theorem of fractions is also the basis for finding sums of rational expressions, and the procedure for performing these operations is based on the theorems for finding sums of quotients of real numbers.

Let n, m, and d be any real numbers with $d \neq 0$.

THE SUM OF QUOTIENTS THEOREM

$$\frac{n}{d} + \frac{m}{d} = \frac{n + m}{d}$$

Examination of this theorem for the quotients of real numbers reveals that fractions cannot be combined into a single fraction by the operations of addition and subtraction unless the denominators of the fractions being combined are the same. The fundamental theorem of fractions provides a technique for renaming fractions so that they will have a common denominator. A common denominator can always be found by multiplying the denominators of the fractions that are to be added or subtracted. A procedure for finding a common denominator that simplifies the arithmetic calculations involved and sometimes (but not always!) avoids final simplification is called finding the **least common denominator,** or **L.C.D.** The objective is to find the smallest number that contains *each* of the original denominators as a factor. Thus to find the least common denominator, the original denominators must be factored into prime factors. The L.C.D. is the product of all the primes that occur in each factorization, with each prime taken the greatest number of times it occurs in any denominator.

For example, to find the L.C.D. of $\frac{5}{12}$ and $\frac{7}{18}$, factor 12 and 18:

$12 = 2 \cdot 2 \cdot 3$ and $18 = 2 \cdot 3 \cdot 3$

Then the L.C.D. is $2 \cdot 2 \cdot 3 \cdot 3 = 36$.

Now the sum $\frac{5}{12} + \frac{7}{18}$ is obtained as follows:

$$\frac{5}{12} + \frac{7}{18} = \frac{5 \cdot 3}{12 \cdot 3} + \frac{7 \cdot 2}{18 \cdot 2} = \frac{15 + 14}{36} = \frac{29}{36}$$

The procedure for finding sums of rational expressions is similar, only the objective of finding the smallest number for the common denominator is changed to finding the polynomial of the smallest degree for the common denominator.

EXAMPLE 1 Express $\frac{3}{5x} + \frac{7}{5x}$ as a single fraction.

Solution   The denominators of the two fractions are the same; therefore the addition of quotients theorem can be applied directly:

$$\frac{3}{5x} + \frac{7}{5x} = \frac{3 + 7}{5x} = \frac{10}{5x}$$

But

$$\frac{10}{5x} = \frac{5 \cdot 2}{5 \cdot x} = \frac{2}{x}$$

Therefore, in lowest terms,

$$\frac{3}{5x} + \frac{7}{5x} = \frac{2}{x}$$

EXAMPLE 2 Express $\frac{7}{6x^2} + \frac{2}{9x}$ as a single fraction.

Solution   The two fractions have unlike denominators.

1. Find the L.C.D. by factoring $6x^2$ and $9x$:

$$6x^2 = 2 \cdot 3 \cdot x \cdot x$$

and

$$9x = 3 \cdot 3 \cdot x$$

Thus the L.C.D. is $2 \cdot 3 \cdot 3 \cdot x \cdot x = 18x^2$.

2. Rename the fractions by using the fundamental theorem of fractions:

$$\frac{7}{6x^2} = \frac{7 \cdot 3}{6x^2 \cdot 3} = \frac{21}{18x^2}$$

and

$$\frac{2}{9x} = \frac{2 \cdot 2x}{9x \cdot 2x} = \frac{4x}{18x^2}$$

3. Apply the theorem for the addition of quotients:

$$\frac{7}{6x^2} + \frac{2}{9x} = \frac{21}{18x^2} + \frac{4x}{18x^2} = \frac{21 + 4x}{18x^2}$$

$$= \frac{4x + 21}{18x^2} \qquad \text{(Conventional form)}$$

EXAMPLE 3   Express $\dfrac{3}{y} + \dfrac{2}{3} - \dfrac{y}{y + 3}$ as a single fraction.

Solution

1. The L.C.D. is $3y(y + 3)$ since $(y + 3)$ cannot be factored further. Thus *all* factors must appear in the L.C.D.

2. Rename the fractions so that each has the denominator $3y(y + 3)$:

$$\frac{3}{y} = \frac{3 \cdot 3(y + 3)}{y \cdot 3(y + 3)} = \frac{9y + 27}{3y(y + 3)}$$

$$\frac{2}{3} = \frac{2 \cdot y(y + 3)}{3 \cdot y(y + 3)} = \frac{2y^2 + 6y}{3y(y + 3)}$$

$$\frac{y}{y + 3} = \frac{y \cdot 3y}{(y + 3) \cdot 3y} = \frac{3y^2}{3y(y + 3)}$$

3. Apply the theorem for the addition and subtraction of quotients and combine like terms:

$$\frac{9y + 27}{3y(y + 3)} + \frac{2y^2 + 6y}{3y(y + 3)} - \frac{3y^2}{3y(y + 3)}$$

$$= \frac{(9y + 27) + (2y^2 + 6y) - (3y^2)}{3y(y + 3)}$$

$$= \frac{-y^2 + 15y + 27}{3y(y + 3)}$$

In subtracting fractions, it is especially important to remember that the bar, the horizontal line separating the numerator and denominator, is a grouping symbol indicating that the numerator is to be considered as a single number and that the denominator is to be considered as a single number.

In Example 4 the parentheses are used to enclose the numerator of each fraction in order to emphasize this fact.

EXAMPLE 4   Simplify $\dfrac{5}{x^2 - 6x + 9} - \dfrac{4}{x^2 - 9}$.

Solution

1. $x^2 - 6x + 9 = (x - 3)(x - 3)$
   $x^2 - 9 = (x - 3)(x + 3)$
   Thus the L.C.D. $= (x - 3)^2(x + 3)$.

2. $\dfrac{5}{(x - 3)^2} = \dfrac{5(x + 3)}{(x - 3)^2(x + 3)} = \dfrac{5x + 15}{(x - 3)^2(x + 3)}$

   $\dfrac{4}{x^2 - 9} = \dfrac{4(x - 3)}{(x - 3)(x + 3)(x - 3)} = \dfrac{4x - 12}{(x - 3)^2(x + 3)}$

3. $\dfrac{5}{(x-3)^2} - \dfrac{4}{x^2-9} = \dfrac{(5x+15)}{(x-3)^2(x+3)} - \dfrac{(4x-12)}{(x-3)^2(x+3)}$

$$= \dfrac{(5x+15) - (4x-12)}{(x-3)^2(x+3)}$$

$$= \dfrac{x+27}{(x-3)^2(x+3)}$$

EXAMPLE 5 Simplify $\dfrac{2}{x^2-4x+4} + \dfrac{3x}{4-x^2}$.

Solution

1. $x^2 - 4x + 4 = (x-2)(x-2)$

   $4 - x^2 = -(x^2-4) = -(x-2)(x+2)$

   It is advisable at this time to rewrite the second fraction.

   $$\dfrac{3x}{4-x^2} = \dfrac{3x}{-(x^2-4)} = \dfrac{-3x}{x^2-4}$$

   Thus the L.C.D. $= (x-2)(x-2)(x+2)$

   $\qquad\qquad\quad = (x-2)^2(x+2)$

2. $\dfrac{2}{x^2-4x+4} = \dfrac{2(x+2)}{(x-2)(x-2)(x+2)} = \dfrac{2x+4}{(x-2)^2(x+2)}$

   $\dfrac{-3x}{x^2-4} = \dfrac{-3x(x-2)}{(x-2)(x-2)(x+2)} = \dfrac{-3x^2+6x}{(x-2)^2(x+2)}$

3. $\dfrac{2}{x^2-4x+4} + \dfrac{3x}{4-x^2} = \dfrac{2}{x^2-4x+4} + \dfrac{-3x}{x^2-4}$

   $$= \dfrac{(2x+4) + (-3x^2+6x)}{(x-2)^2(x+2)}$$

   $$= \dfrac{-3x^2+8x+4}{(x-2)^2(x+2)}$$

EXAMPLE 6 Simplify $\dfrac{3}{x-2} - \dfrac{2}{3x} - \dfrac{18}{3x^2-6x}$.

Solution

1. $\left.\begin{array}{l} x - 2 = x - 2 \\ 3x = 3x \end{array}\right\}$ These expressions are already in factored form.

   $3x^2 - 6x = 3x(x-2)$

   Thus the L.C.D. $= 3x(x-2)$.

2. $\dfrac{3}{x-2} - \dfrac{2}{3x} - \dfrac{18}{3x^2-6x}$

   $$= \dfrac{3(3x)}{3x(x-2)} - \dfrac{2(x-2)}{3x(x-2)} - \dfrac{18}{3x(x-2)}$$

   $$= \dfrac{3(3x) - 2(x-2) - 18}{3x(x-2)}$$

   $$= \dfrac{9x - 2x + 4 - 18}{3x(x-2)}$$

   $$= \dfrac{7x - 14}{3x(x-2)}$$

   $$= \dfrac{7(x-2)}{3x(x-2)} = \dfrac{7}{3x}$$

EXERCISES 3.5

In Exercises 1–10 find the L.C.D. for each indicated group of denominators.

1. $8, x^2, 12x$

2. x^2y, xy^2, xy

3. $x^2 - 1, (x - 1)^2$

4. $x^2 - 1, x + 1, x$

5. $2, x, x + 2$

6. $9 - x^2, x + 3, 3x$

7. $x^2 - 4x - 12, 2x^2 - 9x - 12$

8. $7x^2y^4 - 28x^2y^2, 8x, x + 2$

9. $x^3 - 1, x^2 - 1, x - 1$

10. $x^3 + 1, (x + 1)^2, x^2 - 1$

Simplify Exercises 11–62 and reduce the answer to lowest terms.

11. $\dfrac{2x}{9y^2} - \dfrac{3x}{8y^2} + \dfrac{5x}{18y^2}$

12. $\dfrac{5x - 2}{x} - \dfrac{2x - 5}{x}$

13. $\dfrac{4}{9} + \dfrac{5}{12} - \dfrac{2}{15}$

14. $\dfrac{5}{14} + \dfrac{3}{28} - \dfrac{1}{21}$

15. $\dfrac{1}{3} + \dfrac{1}{4} + \dfrac{1}{5} + \dfrac{1}{6}$

16. $\dfrac{2}{3y} + \dfrac{5}{12y} + \dfrac{1}{18y}$

17. $\dfrac{1}{2x} + \dfrac{1}{6x} - \dfrac{1}{3x}$

18. $\dfrac{2}{x} + \dfrac{3}{x^2}$

19. $\dfrac{1}{2x} + \dfrac{4}{3x^2} - \dfrac{1}{6x^3}$

20. $\dfrac{3}{x} + \dfrac{2}{x + 1}$

21. $\dfrac{1}{b - 1} - \dfrac{1}{b}$

22. $\dfrac{x}{x + 2} + \dfrac{1}{x} + \dfrac{1}{2}$

23. $\dfrac{4}{y + 1} - \dfrac{3}{y + 2}$

24. $\dfrac{2}{a + 3} + \dfrac{5}{a} - \dfrac{1}{3}$

25. $\dfrac{3}{5y} + \dfrac{2}{15y^2} - \dfrac{1}{10y^3}$

26. $\dfrac{1}{a} + \dfrac{2}{a + 1}$

27. $\dfrac{x}{y + 1} - \dfrac{x}{y}$

28. $\dfrac{m}{m + 4} + \dfrac{1}{m} + \dfrac{3}{4}$

29. $\dfrac{2}{x - 1} - \dfrac{x}{x - 2}$

30. $\dfrac{3}{a} + \dfrac{2}{a + 5} - \dfrac{1}{5}$

31. $\dfrac{a}{b} + \dfrac{b}{a}$

32. $\dfrac{5x}{(3x - 1)^2} + \dfrac{4}{3x - 1}$

33. $\dfrac{y}{(2y - 3)^2} + \dfrac{5}{2y - 3} - \dfrac{3}{2y + 3}$

34. $2x + \dfrac{3}{x}$

35. $2x - \dfrac{x^2}{x + 1}$

36. $2 - \dfrac{x - 1}{x + 1}$

37. $\dfrac{x}{x^2 + 5x + 4} + \dfrac{3}{x^2 + 4x + 3}$

38. $\dfrac{x + 2}{3x^2 + 5x + 2} - \dfrac{x + 3}{3x^2 - 16x - 12}$

39. $\dfrac{2y}{y^2 - 5y + 6} + \dfrac{3}{2 + 3y - 2y^2}$

40. $\dfrac{2}{x - 2} + \dfrac{3}{2 - x}$

41. $\dfrac{a - 2}{3a + 3} + \dfrac{a - 3}{2a + 2}$

42. $\dfrac{x}{5x - 5} - \dfrac{2}{3 - 3x}$

43. $\dfrac{x^2 + 3}{24 - 2x - x^2} + \dfrac{2x + 1}{2x - 8}$

44. $\dfrac{2}{c - 5} + \dfrac{3}{c + 5} - 1$

45. $\dfrac{3}{y^2 - 5y} + \dfrac{2}{y^2 + 5y} - \dfrac{4}{y^2 - 25}$

46. $\dfrac{1}{x^2 - 49} + \dfrac{1}{(x + 7)^2} - \dfrac{1}{(x - 7)^2}$

47. $1 + \dfrac{2}{x} - \dfrac{x - 1}{x^2 - x}$

48. $2a - 5 + \dfrac{25}{2a + 5}$

49. $\dfrac{x}{x + y} + \dfrac{y}{x - y} - \dfrac{2xy}{x^2 - y^2}$

50. $\dfrac{7y}{2y^2 + 5y - 3} - \dfrac{10y}{3y^2 + 8y - 3}$

51. $\dfrac{a}{b - a} + \dfrac{b}{a - b}$

52. $\dfrac{x + 7}{4 - 6x} + \dfrac{3x^2 + 14}{9x^2 - 4}$

53. $\dfrac{y + 9}{4y^2 - 5y - 6} - \dfrac{y + 7}{5y^2 - 11y + 2}$

54. $\dfrac{5}{2x + 8} - \dfrac{7}{3x - 12} + \dfrac{20}{x^2 - 16}$

55. $\dfrac{1}{a^2 - 1} - \dfrac{1}{a^2 + 2a + 1}$

56. $\dfrac{2}{2x + x^2} - \dfrac{3}{2x - x^2} - \dfrac{4}{x^2 - 4}$

57. $\dfrac{t^2 - 2t - 21}{t^2 - 4t - 5} + \dfrac{t - 6}{5 - t}$

58. $\dfrac{x^2 - 2x + 4}{x - 2} - \dfrac{x^2 + 2x + 4}{x + 2} - \dfrac{8x}{x^2 - 4}$

59. $\dfrac{4a^2}{x^2 - a^2} - \dfrac{x - a}{x + a} - \dfrac{x + a}{x - a}$

60. $\dfrac{x + a}{x^3 - a^3} - \dfrac{1}{x^2 - a^2}$

61. $\dfrac{x + 6}{x^3 + 5x^2 + 4x + 20} - \dfrac{x - 2}{x^3 - 3x^2 + 4x - 12}$

62. $\dfrac{1}{x^3 + x^2 - 5x - 5} - \dfrac{3 - x}{x^3 - x^2 - 5x + 5}$

3.6 COMPLEX FRACTIONS

A **complex fraction** is a fraction that contains a fraction in the numerator or in the denominator or in both numerator and denominator.

There are two basic methods for simplifying a complex fraction. The method that usually involves the least number of calculations is based on the fundamental theorem of fractions. The numerator and denominator are both multiplied by the least common denominator of all the fractions that appear in the numerator and denominator.

The other method for simplifying a complex fraction considers the complex fraction as an indicated quotient. First the numerator and denominator are expressed as single fractions, then the resulting fractions are divided.

EXAMPLE 1 Simplify

$$\dfrac{3 + \dfrac{1}{3}}{2 - \dfrac{3}{5}}$$

Solution   Multiplying numerator and denominator by 15, the L.C.D.,

$$\frac{\left(3 + \dfrac{1}{3}\right)(15)}{\left(2 - \dfrac{3}{5}\right)(15)} = \frac{45 + 5}{30 - 9} = \frac{50}{21}$$

Alternate Solution   First rewriting the numerator and the denominator each as single fractions,

$$\frac{3 + \dfrac{1}{3}}{2 - \dfrac{3}{5}} = \frac{\dfrac{9}{3} + \dfrac{1}{3}}{\dfrac{10}{5} - \dfrac{3}{5}} = \frac{\dfrac{10}{3}}{\dfrac{7}{5}} = \frac{10}{3} \cdot \frac{5}{7} = \frac{50}{21}$$

EXAMPLE 2   Simplify

$$\frac{\dfrac{1}{x} + \dfrac{1}{2}}{\dfrac{1}{4}}$$

Solution   Multiplying numerator and denominator by $4x$, the L.C.D.,

$$\frac{\dfrac{1}{x} + \dfrac{1}{2}}{\dfrac{1}{4}} \cdot \frac{4x}{4x} = \frac{\left(\dfrac{1}{x} + \dfrac{1}{2}\right)4x}{\dfrac{1}{4} \cdot 4x}$$

$$= \frac{\dfrac{4x}{x} + \dfrac{4x}{2}}{\dfrac{4x}{4}} = \frac{4 + 2x}{x}$$

Alternate Solution

$$\frac{\dfrac{1}{x} + \dfrac{1}{2}}{\dfrac{1}{4}} = \frac{\dfrac{2 + x}{2x}}{\dfrac{1}{4}} = \frac{2 + x}{2x} \div \frac{1}{4}$$

$$= \frac{2 + x}{2x} \cdot \frac{4}{1}$$

$$= \frac{2(2 + x)}{x} = \frac{4 + 2x}{x}$$

A complex fraction may contain a complex fraction in either its numerator or denominator or in both. If this is the case, the complex fraction in the numerator or denominator is simplified first by either of the two methods described above, and then the resulting fraction is simplified.

EXAMPLE 3 Simplify

$$\dfrac{\dfrac{x}{x-y}-\dfrac{y}{x+y}}{\dfrac{x}{x+y}-\dfrac{y}{y-x}}$$

Solution   The denominators within the fraction are $x - y$ and $x + y$ in the numerator and $x + y$ and $y - x$ in the denominator.

Since $y - x = -(x - y)$, it is useful to rename

$$\dfrac{y}{y-x}\quad\text{as}\quad\dfrac{-y}{x-y}$$

$$\dfrac{\dfrac{x}{x-y}-\dfrac{y}{x+y}}{\dfrac{x}{x+y}-\dfrac{y}{y-x}}=\dfrac{\dfrac{x}{x-y}-\dfrac{y}{x+y}}{\dfrac{x}{x+y}-\dfrac{-y}{x-y}}$$

$$=\dfrac{(x-y)(x+y)\left(\dfrac{x}{x-y}-\dfrac{y}{x+y}\right)}{(x-y)(x+y)\left(\dfrac{x}{x+y}+\dfrac{y}{x-y}\right)}$$

$$=\dfrac{x(x+y)-y(x-y)}{x(x-y)+y(x+y)}$$

$$=\dfrac{x^2+xy-xy+y^2}{x^2-xy+xy+y^2}$$

$$=\dfrac{x^2+y^2}{x^2+y^2}=1$$

EXAMPLE 4 Simplify

$$\cfrac{1}{1+\cfrac{1}{1+\cfrac{1}{2}}}$$

Solution   First simplify the complex fraction that occurs in the denominator:

$$\dfrac{1}{1+\dfrac{1}{2}}=\dfrac{1\cdot 2}{\left(1+\dfrac{1}{2}\right)2}=\dfrac{2}{2+1}=\dfrac{2}{3}$$

Thus

$$\dfrac{1}{1+\left(\dfrac{1}{1+\dfrac{1}{2}}\right)}=\dfrac{1}{1+\dfrac{2}{3}}$$

$$=\dfrac{1\cdot 3}{\left(1+\dfrac{2}{3}\right)3}$$

$$=\dfrac{3}{3+2}=\dfrac{3}{5}$$

Alternate Solution

$$\cfrac{1}{1+\cfrac{1}{1+\cfrac{1}{2}}} = \cfrac{1}{1+\cfrac{1}{\cfrac{3}{2}}}$$

$$= \cfrac{1}{1+\cfrac{2}{3}} = \cfrac{1}{\cfrac{5}{3}} = \cfrac{3}{5}$$

EXERCISES 3.6

In Exercises 1–40 simplify and write as a single fraction in lowest terms.

1. $\dfrac{2+\dfrac{3}{4}}{3+\dfrac{7}{8}}$

2. $\dfrac{1+\dfrac{1}{3}}{1+\dfrac{1}{9}}$

3. $\dfrac{3-\dfrac{5}{7}}{5+\dfrac{3}{14}}$

4. $\dfrac{1-\dfrac{1}{5}}{1-\dfrac{1}{25}}$

5. $\dfrac{\dfrac{5}{2}}{\dfrac{3}{4}+\dfrac{7}{8}}$

6. $\dfrac{1}{\dfrac{2}{3}+\dfrac{3}{4}}$

7. $\dfrac{1}{\dfrac{1}{4}-\dfrac{1}{5}}$

8. $\dfrac{\dfrac{2}{3}}{\dfrac{1}{4}+\dfrac{2}{5}}$

9. $\dfrac{\dfrac{1}{2}+\dfrac{1}{3}}{1-\left(\dfrac{1}{2}\right)\left(\dfrac{1}{3}\right)}$

10. $\dfrac{\dfrac{1}{x}+\dfrac{1}{2}}{\dfrac{x}{2}-\dfrac{2}{x}}$

11. $\dfrac{\dfrac{1}{4}-\dfrac{1}{3}}{1+\left(\dfrac{1}{4}\right)\left(\dfrac{1}{3}\right)}$

12. $\dfrac{x-\dfrac{y}{5}}{x+\dfrac{y}{5}}$

13. $\dfrac{\dfrac{x}{y}+\dfrac{y}{5}}{\dfrac{xy}{10}}$

14. $\dfrac{a+\dfrac{3}{a}}{a^2-\dfrac{9}{a^2}}$

15. $\dfrac{\dfrac{y}{3} - \dfrac{3}{y}}{\dfrac{1}{y} + \dfrac{1}{3}}$

16. $\dfrac{3 - \dfrac{1}{y}}{9 - \dfrac{1}{y^2}}$

17. $\dfrac{y - \dfrac{9}{y}}{y - 7 + \dfrac{12}{y}}$

18. $\dfrac{\dfrac{1}{3} + \dfrac{1}{n}}{1 - \dfrac{1}{3n}}$

19. $\dfrac{x + 3 - \dfrac{10}{x}}{x - \dfrac{25}{x}}$

20. $\dfrac{x + 6}{\dfrac{1}{x} + \dfrac{1}{6}}$

21. $\dfrac{1}{\dfrac{1}{x} + \dfrac{1}{y}}$

22. $\dfrac{2x}{\dfrac{x}{30} + \dfrac{x}{50}}$

23. $\dfrac{1}{\dfrac{2}{a} + \dfrac{3}{b}}$

24. $\dfrac{\dfrac{x}{a} - \dfrac{x}{b}}{\dfrac{x}{ab}}$

25. $\dfrac{1 + \dfrac{x}{y}}{\dfrac{1}{x} + \dfrac{1}{y}}$

26. $\dfrac{1 - \dfrac{1}{x + 1}}{1 + \dfrac{1}{x - 1}}$

27. $\dfrac{1}{1 + \dfrac{3}{x}}$

28. $\dfrac{1 + \dfrac{5}{x - 5}}{1 - \dfrac{5}{x + 5}}$

29. $\dfrac{3 - \dfrac{2}{\dfrac{1}{5} + \dfrac{1}{3}}}{5 - \dfrac{2}{\dfrac{1}{5} + \dfrac{1}{3}}}$

30. $\dfrac{\dfrac{1}{a} - \dfrac{2}{a^2} - \dfrac{3}{a^3}}{1 - \dfrac{9}{a^2}}$

31. $\dfrac{2 + \dfrac{1}{\dfrac{1}{4} + \dfrac{1}{3}}}{4 - \dfrac{1}{\dfrac{1}{4} + \dfrac{1}{3}}}$

32. $\dfrac{\dfrac{a}{6b} - \dfrac{3b}{2a}}{\dfrac{a}{2b} + \dfrac{3b}{2a} - 2}$

33. $\dfrac{1}{1 - \dfrac{1}{1 - \dfrac{1}{x}}}$

34. $\dfrac{1}{y - \dfrac{1}{y + \dfrac{1}{y}}}$

35. $\dfrac{1}{1 + \dfrac{1}{2 + \dfrac{1}{3}}}$

38. $\dfrac{1}{2 - \dfrac{1}{2 + \dfrac{1}{2 - \dfrac{1}{2}}}}$

36. $\dfrac{1}{1 + \dfrac{1}{1 + \dfrac{1}{4}}}$

39. $\dfrac{1}{x + \dfrac{1}{x - \dfrac{1}{x}}}$

37. $\dfrac{\dfrac{1}{x - 2} - \dfrac{1}{x + 2}}{1 + \dfrac{1}{x^2 - 4}}$

40. $\dfrac{1}{x - 1 + \dfrac{1}{1 + \dfrac{x}{4 - x}}}$

3.7 FRACTIONAL EQUATIONS

A **fractional equation** is an equation whose terms are rational expressions.

For example, $\dfrac{1}{x} = 3$ and $\dfrac{2}{x + 2} + \dfrac{3}{x - 3} = 15$ are fractional equations.

To transform a fractional equation into an equivalent equation whose terms are integral expressions, both sides of the equation must be multiplied by the least common denominator of all the fractions involved in the equation. Since this may require multiplication by an expression involving the variable, *an equivalent equation is obtained only when the variable is restricted to those numbers that do not make the multiplier zero.*

For example, to solve $\dfrac{1}{x} = 3$, both sides of the equation are multiplied by x, and x is restricted so that $x \neq 0$. Then, if $x \neq 0$,

$$\frac{1}{x}(x) = 3(x)$$
$$1 = 3x$$
$$3x = 1$$
$$x = \frac{1}{3}$$

To solve $\dfrac{x}{x-3} = \dfrac{3}{x-3}$, both sides are multiplied by $x - 3$, and x is restricted, so that $x - 3 \neq 0$; that is, $x \neq 3$.

Thus, if $x \neq 3$,

$$\frac{x}{x-3}(x-3) = \frac{3}{x-3}(x-3) \quad \text{and} \quad x = 3$$

The solution set of

$$\frac{x}{x-3} = \frac{3}{x-3}$$

is the set of real numbers, x, such that $x \neq 3$ and $x = 3$. There is no value for x that makes both of these statements true, so there is no solution. In other words, the solution set is the empty set, $\varnothing$.

The equivalence theorem for multiplication, introduced earlier, is restated below for convenience.

EQUIVALENCE THEOREM FOR MULTIPLICATION

If each side of an equation is multiplied by the same *nonzero* number, the resulting products are equal, and the two equations are equivalent. In symbols, if A, B, and C are any real numbers, then $A = B$ if and only if $AC = BC$ and $C \neq 0$.

It is important to be aware that the "if and only if" statement works two ways: (1) If $A = B$ and $C \neq 0$, then $AC = BC$; and (2) if $AC = BC$ and $C \neq 0$, then $A = B$.

This theorem is used in solving fractional equations, and care must be taken to exclude any values of a variable that might yield a zero value for C. The values for which $C \neq 0$ are called **restricted values of the variable.**

EXAMPLE 1   State the restricted values of y and solve:

$$\frac{2}{y-5} + \frac{1}{y+5} = \frac{11}{y^2 - 25}$$

Solution   First write all denominators in factored form:

$$\frac{2}{y-5} + \frac{1}{y+5} = \frac{11}{(y-5)(y+5)}$$

The L.C.D. is $(y-5)(y+5)$. Since this product is 0 when $y - 5 = 0$ or $y + 5 = 0$, $y \neq 5$ and $y \neq -5$. Multiplying both sides by the L.C.D. $(y-5)(y+5)$,

$$\left(\frac{2}{y-5}+\frac{1}{y+5}\right)(y-5)(y+5)=\frac{11(y-5)(y+5)}{(y-5)(y+5)}$$

$$2(y+5)+(y-5)=11$$
$$3y+5=11$$
$$3y=6$$
$$y=2$$

Since the common solution to $y \neq 5, y \neq -5$, and $y = 2$ is 2, {2} is the solution set of the original equation.

Check

$$\frac{2}{y-5}+\frac{1}{y+5}=\frac{2}{2-5}+\frac{1}{2+5}=\frac{-2}{3}+\frac{1}{7}=\frac{-11}{21}$$

$$\frac{11}{y^2-25}=\frac{11}{(2)^2-25}=\frac{11}{4-25}=\frac{-11}{21}$$

EXAMPLE 2 State the restricted values of the variable and solve:

$$\frac{x+1}{x+2}-\frac{x+1}{x-3}=\frac{5}{x^2-x-6}$$

Solution The L.C.D. is $(x+2)(x-3)$.
Since $x + 2 \neq 0$ and $x - 3 \neq 0$, $x \neq -2$ and $x \neq 3$,

$$\left(\frac{x+1}{x+2}\right)(x+2)(x-3)-\left(\frac{x+1}{x-3}\right)(x+2)(x-3)$$
$$=\frac{5}{(x+2)(x-3)}(x+2)(x-3)$$

$$(x+1)(x-3)-(x+1)(x+2)=5$$
$$(x^2-2x-3)-(x^2+3x+2)=5$$
$$-5x-5=5$$
$$-5x=10$$
$$x=-2$$

Thus $x \neq -2$ and $x \neq 3$ and $x = -2$. Since there is no common solution, the solution set is the empty set, $\varnothing$.

EXAMPLE 3 State the restricted values of the variable and solve:

$$\frac{x}{(x-4)^2}+\frac{2}{x-4}=\frac{3x-8}{(x-4)^2}$$

Solution
$x \neq 4$, because $x - 4 \neq 0$.

$$(x-4)^2\frac{x}{(x-4)^2}+(x-4)^2\frac{2}{x-4}=\frac{3x-8}{(x-4)^2}(x-4)^2$$
$$x+2(x-4)=3x-8$$
$$3x-8=3x-8$$
$$3x=3x$$
$$x=x$$

Since $x = x$ is true for all real numbers, the solution set is the set of all real numbers except 4; $R, x \neq 4$.

135

EXAMPLE 4 State the restricted values and solve:

$$1 + \frac{2}{x - 1} = \frac{2}{x(x - 1)}$$

Solution $x \neq 1$ and $x \neq 0$. (Why?)

$$1 + \frac{2}{x - 1} = \frac{2}{x(x - 1)}$$

$$\frac{x(x - 1)}{x(x - 1)} + \frac{2x}{x(x - 1)} = \frac{2}{x(x - 1)}$$

Multiplying each side of the equation by $x(x - 1)$,

$$x^2 - x + 2x = 2$$

This is a quadratic equation, so we must find an equivalent equation with one side equal to 0 and apply the zero product theorem to find the solution set.

$$x^2 + x - 2 = 0$$
$$(x + 2)(x - 1) = 0$$
$$x + 2 = 0 \quad \text{or} \quad x - 1 = 0$$
$$x = -2 \quad \text{or} \quad x = 1$$

Since $x \neq 1$, $\{-2\}$ is the solution set.

EXAMPLE 5 State the restricted values and solve

$$\frac{a}{x} - \frac{b}{a} = \frac{1}{3x} \text{ for } x$$

Solution

L.C.D. $= 3ax$; $x \neq 0$.

$$\frac{a}{x} - \frac{b}{a} = \frac{1}{3x}$$

$$3ax \left(\frac{a}{x}\right) - 3ax \left(\frac{b}{a}\right) = 3ax \left(\frac{1}{3x}\right)$$

$$3a^2 - 3bx = a$$
$$-3bx = a - 3a^2$$
$$x = \frac{a - 3a^2}{-3b}$$
$$x = \frac{a(3a - 1)}{3b}$$

EXERCISES 3.7

In Exercises 1–34 state the restricted values of the variable and solve. Check each solution.

1. $\dfrac{1}{2x} - \dfrac{2}{3x} = \dfrac{1}{24}$

2. $\dfrac{2}{5x} - \dfrac{1}{2x} = \dfrac{x+2}{10x}$

3. $\dfrac{1}{a} + \dfrac{1}{2a} + \dfrac{1}{3a} = \dfrac{1}{a+5}$

4. $\dfrac{2}{y} + \dfrac{1}{2y} + \dfrac{4}{3y} = \dfrac{3}{y+2}$

5. $\dfrac{2}{3x-7} = \dfrac{5}{x+2}$

6. $\dfrac{6}{y} - \dfrac{1}{y-2} = \dfrac{3}{y^2-2y}$

7. $\dfrac{3}{y} + \dfrac{5}{1-y} + \dfrac{2y+3}{y^2-1} = 0$

8. $\dfrac{x}{5} - \dfrac{x}{6} = 3$

9. $\dfrac{5}{x} = 20$

10. $\dfrac{3}{2x-1} = \dfrac{7}{3x+1}$

11. $\dfrac{20+y}{10y+5} = \dfrac{2}{5}$

12. $\dfrac{3x+1}{25} - \dfrac{1}{10} = \dfrac{3x-1}{30}$

13. $\dfrac{5}{x} = 0$

14. $\dfrac{x+1}{x-5} - \dfrac{x+3}{x+2} = \dfrac{2}{x^2-3x-10}$

15. $\dfrac{x-6}{x+6} - \dfrac{x-2}{x+1} = \dfrac{15}{x^2+7x+6}$

16. $\dfrac{4t}{9t+18} + \dfrac{t+2}{3t-6} = \dfrac{7}{9}$

17. $\dfrac{y+12}{y^2-16} + \dfrac{1}{4-y} = \dfrac{1}{4+y}$

18. $\dfrac{8y-1}{5} = \dfrac{16y+3}{10} - \dfrac{2y-5}{5y-1}$

19. $\dfrac{5+x}{x} = 0$

20. $\dfrac{9y+2}{12} + \dfrac{17}{9} = \dfrac{21y-8}{18} - \dfrac{2y-3}{3}$

21. $\dfrac{3}{2x-6} + \dfrac{1}{4x+2} = \dfrac{2x-3}{2x^2-5x-3}$

22. $\dfrac{44}{2x^2-9x-5} - \dfrac{6}{2x+1} = \dfrac{4}{x-5}$

23. $\dfrac{5}{t} + \dfrac{1}{2t} - \dfrac{2}{3t} = 29$

24. $\dfrac{2x-29}{x^2+7x-8} - \dfrac{5}{x+8} = \dfrac{3}{x-1}$

25. $\dfrac{5t^2}{t^2-t-20} = \dfrac{3t+2}{t+4} + \dfrac{2t+3}{t-5}$

26. $\dfrac{y-8}{y-3} - \dfrac{y+8}{y+3} = \dfrac{y}{9-y^2}$

27. $\dfrac{2}{x+1} + \dfrac{1}{3x+3} = \dfrac{1}{6}$

28. $\dfrac{x}{x+2} - \dfrac{3}{x-2} = \dfrac{x^2-8}{x^2-4}$

29. $\dfrac{5}{5x+3} + \dfrac{2}{1-2x} = \dfrac{4-6x}{10x^2+x-3}$

30. $\dfrac{1}{2x-3} + \dfrac{x}{4x^2-9} = \dfrac{1}{8x+12}$

31. $1 + \dfrac{2}{x-1} = \dfrac{20}{x(x-1)}$

32. $2 + \dfrac{3}{x-2} - \dfrac{3}{x(x-2)} = 0$

33. $\dfrac{x}{x+3} - \dfrac{2}{x+1} = 0$

34. $\dfrac{x}{x-2} - \dfrac{30}{(x+4)(x-2)} = \dfrac{5}{x+4}$

In Exercises 35–38 state the restricted values and solve for the indicated variable. (Example 5)

35. $\dfrac{a-2x}{b-x} = \dfrac{3}{2}$,　　for x

36. $\dfrac{x}{a} - \dfrac{a}{b} = \dfrac{b}{c}$,　　for x

37. $\dfrac{3}{x + a} - \dfrac{2}{x - a} = \dfrac{1}{x}$,　　for x

38. $\dfrac{a - 2}{b} + \dfrac{3}{2b} = \dfrac{2}{x}$,　　for x

39. The total resistance R in an electrical circuit with two resistances that have values S and T that are connected in parallel can be expressed in the given ratios:

$$\frac{R}{S} = \frac{T}{S + T}$$

Solve for S.

40. Solve Exercise 39 for T.

41. The relationship between the total resistance R and two other resistances with values S and T as discussed in Exercise 39 are often written in the form

$$\frac{1}{R} = \frac{1}{S} + \frac{1}{T}$$

(a) Solve for R.

(b) Show that the formula given in Exercise 39 is equivalent to the formula given in Exercise 41.

42. In anthropology it is held that the following proportional relationship exists between the cephalic index C, the width of a head W, and the length of the head L:

$$\frac{W}{C} = \frac{L}{100}$$

Solve for C.

43. The efficiency E of an engine is related to the heat input I and the heat output O by the following formula:

$$E = \frac{I - O}{I}$$

Solve for O.

44. Solve Exercise 43 for I.

45. In acoustics the relationships between the frequencies of two sources of sound waves, F and f, and the speed V of sound in air and the speed v of a moving source may be expressed by the formula

$$F = \frac{fV}{V + v}$$

Solve for V.

46. Solve Exercise 45 for v.

47. In optometry a formula for corrective lenses is

$$P = \frac{100(p + q)}{pq}$$

where P is the power of the lens in diopters, p is the distance of an object from the lens, and q is the image distance from the lens. Solve the formula for p.

48. Solve Exercise 47 for q.

3.8 APPLICATIONS: RATIO, PROPORTION, AND VARIATION

A fraction is the **ratio** of two numbers, the numerator and the denominator. For example, the fraction $\dfrac{3}{4}$ is also a ratio of 3 to 4, and the fraction $\dfrac{x}{x+3}$ is a ratio of x to $(x + 3)$. An **equation** which states that two ratios are equal is called a **proportion.** Thus a proportion has the form

$$\frac{a}{b} = \frac{c}{d}$$

A proportion is a simple fractional equation.

If two numbers are in the ratio $\dfrac{a}{b}$, then the numbers may be represented as ax and bx $(x \neq 0)$ by applying the fundamental theorem of fractions:

$$\frac{ax}{bx} = \frac{a}{b}$$

EXAMPLE 1   A sum of \$350 is to be divided between two partners in the ratio $\dfrac{3}{4}$. How much does each partner receive?

Solution   Let $x =$ the share of one partner. Then $350 - x =$ the share of the other partner. Then

$$\frac{x}{350 - x} = \frac{3}{4}$$
$$4x = 3(350 - x)$$
$$7x = 3(350)$$
$$x = \$150$$
$$350 - x = \$200$$

Therefore, one partner receives \$150 and the other partner receives \$200.

EXAMPLE 2 The ratio of women students to men students at a certain college is $\frac{7}{9}$. If there are 2135 women students, how many men students are there?

Solution Let x = the number of men students. Then

$$\frac{7}{9} = \frac{2135}{x}$$
$$7x = 2135 \cdot 9$$
$$x = 305 \cdot 9$$
$$x = 2745$$

Thus there are 2745 men students.

The terms a, b, c, and d of the proportion $\frac{a}{b} = \frac{c}{d}$ also have special names. In the order in which they are stated above, a is the first term, b the second, c the third, and d the fourth. The first and fourth terms are called the **extremes** of a proportion, and the second and third terms are called the **means** of the proportion.

THEOREM

In a proportion the product of the means is equal to the product of the extremes. In symbols, if $\frac{a}{b} = \frac{c}{d}$, then $bc = ad$.

EXAMPLE 3 If 3, 4, and 5 are the first three terms, in that order, of a proportion, find the fourth term.

Solution Let d = the fourth term. Then

$$\frac{3}{4} = \frac{5}{d}$$
$$3d = 20$$
$$d = \frac{20}{3} = 6\frac{2}{3}$$

Thus the fourth term is $6\frac{2}{3}$.

Special applications in various branches of science are found for the proportion $\frac{y}{x} = k$, where k is constant and x and y are variable. This equation is equivalent to $y = kx$ and, if $k \neq 0$, is defined as follows.

DEFINITION—Direct Variation

Two variables are said to **vary directly,** or to be **directly proportional,** if they are related by an equation of the form $y = kx$, where k is a nonzero constant, and x and y are variable. The number k is called the **constant of proportionality,** or the **constant of variation.**

The problem illustrated in Example 4 is from physics.

EXAMPLE 4 The work W done by a constant force k is directly proportional to the distance s the object is moved. If it requires 60 ft-lb of work to move an object 4 ft, how much work is required to move the same object 20 ft?

Solution From the problem, k is the constant of proportionality and W is directly proportional to s. From the definition,

$$W = ks \quad \text{and} \quad \frac{W}{s} = k, \qquad \text{a constant}$$

From the given information we can solve for k:

$$\frac{60}{4} = k \quad \text{and} \quad k = 15$$

Thus $W = 15s$, and when $s = 20$, $W = (15)(20) = 300$ ft-lb. That is, 300 ft-lb of work is required.

Another important relationship between variables is **inverse variation.** Keep in mind that the constant of variation is never zero.

DEFINITION—Inverse Variation

If one variable is equal to a constant times the reciprocal of the other variable, then the variables are said to **vary inversely,** or they are said to be **inversely proportional** to each other.

This relationship is expressed by the form

$$y = \frac{k}{x}, \qquad xy \neq 0$$

where x and y are the variables, and k is the constant of variation.

EXAMPLE 5 Under constant temperature, the volume V of a gas varies inversely as the pressure p. If the volume of a certain gas is 30 cu ft under 4 lb of pressure, what is its volume at the same temperature when the pressure is 10 lb?

Solution Since $V = \dfrac{k}{p}$, $Vp = k$, a constant.

Thus

$$V \cdot 10 = 30 \cdot 4 = 120 = k$$

and

$$V = \frac{120}{10} = 12$$

Thus the volume is 12 cu ft.

Last, but by no means least, is the relationship between three variables involving variation.

DEFINITION—Joint Variation

Three variables are said to **vary jointly** if they are related by an equation having the form

$$z = kxy, \qquad k \neq 0$$

The variables are z, x, and y, and the constant of variation is k.

EXAMPLE 6 By Newton's law of gravitation, the force F between two particles of mass kept at a constant distance apart varies jointly as their masses m and n.
(a) If $m = 5$, $n = 2$, and $F = 1$, find the constant of variation.
(b) Find F if $m = 4$ and $n = 3$.

Solution

(a) The formula is $F = kmn$. For $F = 1$, $m = 5$, $n = 2$,

$$1 = k(5)(2)$$

$$k = \frac{1}{10}$$

The constant of variation is $\frac{1}{10}$.

(b) Using $F = \frac{1}{10} mn$ for $m = 4$ and $n = 3$,

$$F = \frac{1}{10}(4)(3)$$

$$F = \frac{6}{5}$$

EXERCISES 3.8

1. An inheritance of $56,000 is divided between two heirs in a ratio of 5 to 2. How much money does each heir receive?
2. After contesting the will, the heirs in Exercise 1 were awarded the $56,000 in a ratio of 3 to 2. How much will each heir receive now?
3. By weight, the ratio of oxygen to hydrogen in pure distilled water is 8 to 1. How many grams of each element are there in 100 g of water? In 250 g of water?
4. The ratio of sodium to chlorine in common table salt is 35 to 23. Find the amount of each element in a salt compound weighing 290 lb.
5. A blueprint has a scale 1 in. = 4 ft. What are the dimensions of a rectangular living room that measures $4\frac{1}{2}$ by 5 in. on the blueprint?

6. On a certain map, $1\frac{1}{2}$ in. represents 50 mi. If two cities are $6\frac{1}{4}$ in. apart on the map, find the number of miles from one city to the other.

7. A copy machine has the capacity to reduce printed material in a ratio of 5 to 3. If the dimensions of a printed advertisement in rectangular format are 25 cm by 18 cm, what are the dimensions after reduction by the machine?

8. A piece of wire 50 ft long is cut so that the parts are in a ratio of 4 to 3. How long is each part?

9. On a given day, the exchange rate of DM (German marks) to U.S. dollars is in a ratio of 2.25 to 1. A camera lens costing DM 787.50 is purchased in Frankfurt, Germany, and billed through an American credit card company to a customer in San Francisco. Assuming no other charges, what is the cost of this lens in dollars?

10. Assuming the same exchange rate as in Exercise 9, find the DM equivalence in cost of a pair of blue denims that cost $28.00.

In Exercises 11–20 write each statement as an algebraic equation. Use the definitions for variations, and let k be the constant of proportionality.

11. F varies directly as d.

12. F varies directly as the square of s.

13. u varies inversely as v.

14. v is inversely proportional to u.

15. P varies inversely as the cube of t.

16. S varies jointly as X, Y, and Z.

17. y varies jointly as x and the square of z.

18. The volume V of a cone varies jointly as its height h and the square of its radius r.

19. The area A of a triangle varies jointly as its base b and the altitude on that base a.

20. The distance d that an object travels at a uniform speed k is directly proportional to the number of seconds t that it travels.

In Exercises 21–36 use the definitions for variations and Examples 4, 5, and 6 to solve the problems.

21. y varies directly as the square of x. If $y = 45$ when $x = 3$, find y when $x = 7$.

22. y varies directly as the sum of a and b. If $y = 35$ when $a = 3$ and $b = 2$, find y when $a = 9$ and $b = -3$.

23. The distance that an object travels at a uniform speed is directly proportional to the time it travels. If an object travels a distance of 850 km in 7 hr, how far does it travel at the same speed in 12 hr?

24. Water pressure varies directly as the depth beneath the surface of the water. If the pressure is 1250 lb/sq ft at a depth of 20 ft, find the pressure at a depth of 1 mi (1 mi = 5280 ft).

25. The force required to stretch a spring varies directly as the elongation of the spring. If a force of 30 lb is required to stretch a certain spring 2 in., what force will be required to stretch the spring 5 in.?

26. The time required to complete a certain job varies inversely as the number of men working on the job. If 12 men can paint an office building in 5 days, in how many days could 15 men paint the building?

27. For a fixed volume the area of the base of a cylinder varies inversely as the height of the cylinder. If the area of the base of a tin can is 12 sq in. when the height of the can is 8 in., what should be the height of a can with a base of 15 sq in. to produce the same volume?

28. Under constant tension the frequency (vibrations per second) of a string varies inversely as the length of the string. A frequency of 440 vibrations per second (concert A) is obtained by a certain string 32 in. long. Using the same type of string, what length is needed for a frequency of 256 vibrations per second (middle C)?

29. The intensity of illumination of an object varies inversely as the square of the distance of the object from the source of light. If the intensity is 8 lumens at a distance of 2 m from the light source, what is the intensity at a distance of 5 m from the same source?

30. The gravitational attraction between two bodies varies inversely as the square of the distance between the bodies. If a force F of 25 force units results from two bodies that are 6 units apart, find the gravitational force F generated by the two bodies if they are 10 units apart.

31. w varies jointly as x and y. If $w = 12$ when $x = 18$ and $y = 2$, find w when $x = 30$ and $y = 8$.

32. y varies jointly as the quotient of x and the square of z. If $y = \dfrac{3}{4}$ when $x = 3$ and $z = 4$, find y when $x = 8$ and $z = 5$.

33. The pressure p of a gas varies jointly as its absolute temperature t and its density d. If the pressure is 20 lb when the temperature is 220 degrees and the density is 1.4, find the pressure when the density is 0.80 and the temperature is 280 degrees.

34. Use the information from Exercise 33 to find the absolute temperature of a gas with density 1.54 if the pressure is 30 lb.

35. If the potential energy e of an object varies jointly as its mass m and its height h above sea level, and it is known that $e = 200$ when $m = 20$ and $h = 15$, find the potential energy of an object with $m = 35$ and $h = 12$.

36. The weight of a body is inversely proportional to the square of its distance from the center of the earth. If we assume the radius of the earth to be 4000 mi, what will a person weigh 80 mi above the earth if she weighs 120 lb on the earth's surface?

In Exercises 37–40 refer to the theorem on proportions and Example 3.

37. Find the fourth term of a proportion if its first three terms are 7, 14, and 20, in that order.

38. If the first three terms of a proportion are 2, 4, and 6, respectively, find the fourth term.

39. Find the means of a proportion if the product of the extremes is 36 and if the ratio of one mean to another is 4 to 1.

40. The first term of a proportion is 6 and the second term is 8. The fourth term is the sum of the second and the third terms. Find the third and fourth terms.

DIAGNOSTIC TEST

For Problems 1–10 determine if each statement is true or false. If the statement is false, correct it.

1. $\dfrac{3x + 6}{3x} = 6$

2. $\dfrac{x^2 - y^2}{x - y} = x - y$

3. $\dfrac{x^2 - x - 6}{2x^2 - 5x - 3} \div \dfrac{x^2 + x - 2}{2x^2 - x - 1} = 1$

4. If $x^3 + 4x^2 - 2$ is divided by $x - 1$, the remainder is -12.

5. For the problem $\dfrac{2}{x + 3} - \dfrac{1}{x} + \dfrac{x}{3}$ the common denominator of the three fractions is $x + 3$.

6. If $\dfrac{x}{x - 3} = \dfrac{3}{x - 3}$, then $x = 3$.

7. If $\dfrac{x + 5}{x + 2} = \dfrac{5}{2}$, then $x = 0$.

8. If the first three terms of a proportion are 2, 3, and 4 in that order, then the fourth term is 5.

9. If $\dfrac{x + a}{x - b} = \dfrac{x - b}{x - a}$, then $x = \dfrac{a^2 + b^2}{2b}$.

10. The complex fraction $\dfrac{\dfrac{1}{9} + \dfrac{3}{x^2}}{\dfrac{x}{3} - \dfrac{2}{x}}$ can be simplified by multiplying its numerator and denominator by $9x^2$.

For Problems 11–20 fill in the missing blanks.

11. $\dfrac{x^2 - 16}{4 - x} = $ _____

12. $x = \dfrac{}{4x^2 - 4x}$

13. If $3x^2 + 4x - p$ is divided by $x - 2$, then $p = $ _____ if the remainder is 6.

14. If $\dfrac{2}{x^2 - 6x + 9} + \dfrac{3x}{9 - x^2}$ is written as a single fraction, the denominator of this fraction is _____ , and the numerator is

_____ .

15. The restricted values of x for the equation $\dfrac{5}{5x + 3} + \dfrac{2}{x(1 - 2x)} =$

$\dfrac{4 - 6x}{10x^2 + x - 3}$ are _____ .

16. In the proportion $\dfrac{3}{x} = \dfrac{y}{4}$, x and y are called the _____

of the proportion, and 3 and 4 are called the _____ .

17. If $y = \dfrac{k}{x}$, $xy \neq 0$, and k is constant, then x and y are said to be

_____ to each other.

18. If y varies directly as the square of x, and $y = 20$ when $x = 2$, the constant of proportionality equals _____ .

19. The value of $\dfrac{x - y}{x + y}$ when $x = \dfrac{2}{3}$ and $y = \dfrac{1}{2}$ is _____ .

20. If $\dfrac{1}{x} + \dfrac{2}{3} = \dfrac{3 + 2x}{3x}$, then the solution set of this equation is

_____ .

REVIEW EXERCISES

Simplify Exercises 1–22 and write each answer as a single fraction reduced to lowest terms. (Refer to Sections 3.1, 3.2, 3.5, and 3.6.)

1. $\dfrac{15x^3y^2}{27xy^3z}$

2. $\dfrac{8x + 24}{8x}$

3. $\dfrac{x - 7}{7 - x}$

4. $\dfrac{9x^2 - 36y^2}{x^2 - 4xy + 4y^2}$

5. $\dfrac{2x^2 + 5x - 150}{2x^2 + 17x - 30}$

6. $\dfrac{x^2 - 2x + 1}{15x^2} \cdot \dfrac{20x^5}{1 - x^2}$

7. $\dfrac{x^3 + 5x^2 + 6x}{x^2 - x} \cdot \dfrac{x^2 - 3x + 2}{(x^2 - 4)(x + 3)}$

8. $\dfrac{5xy + 15y}{15y} \div \dfrac{x^2 + 4x + 3}{x^2 - 1}$

9. $\dfrac{a^2 + 8a + 12}{a^2 + 11a + 30} \cdot \dfrac{20 - a - a^2}{16 - 8a + a^2}$

10. $\left(\dfrac{a + b}{a^2 + b^2} \cdot \dfrac{a}{a - b}\right) \div \dfrac{(a + b)^2}{a^4 - b^4}$

REVIEW EXERCISES

11. $\dfrac{5}{x+3} - \dfrac{x}{x+3}$

12. $\dfrac{x}{x+3} + \dfrac{5x^2}{x^2-9}$

13. $\dfrac{x}{x^2-6x+5} + \dfrac{3}{x-1}$

14. $\dfrac{2}{x-1} - \dfrac{4}{x} + \dfrac{2}{x+1}$

15. $\dfrac{1}{x} + \dfrac{1}{2} + \dfrac{5}{x+2}$

16. $\dfrac{3}{x^2+5x+4} - \dfrac{2}{x^2+4x+3}$

17. $\dfrac{1}{(x+2)^2} - \dfrac{2}{x^2-4} + \dfrac{1}{(x-2)^2}$

18. $\left(x + \dfrac{3}{y}\right) \div \left(x - \dfrac{3}{y}\right)$

19. $\dfrac{\dfrac{1}{x} - \dfrac{1}{3}}{x - \dfrac{9}{x}}$

20. $\left(\dfrac{3x-1}{3x} - \dfrac{3x}{3x+1}\right) \div \left(\dfrac{5x-2}{5x} - \dfrac{5x}{5x+2}\right)$

21. $3x - \dfrac{x}{4x - \dfrac{5x}{2}}$

22. $4x + \dfrac{6}{2x - \dfrac{6x}{3}}$

In Exercises 23–27 use the long division algorithm to divide. Check by multiplication or by selecting a suitable value to substitute for the variable. (Refer to Section 3.3.)

23. $\dfrac{x^3 - 4x^2 + 3x - 1}{x - 3}$

24. $\dfrac{x^4 + x^2 - 2}{x^2 + 3}$

25. $\dfrac{39 + 20x^2 + 15x^3}{5x + 10}$

26. $\dfrac{a^3 - 1}{a - 1}$

27. $\dfrac{7x^3 - 3x + x^2 + 3x^3 - 5x - 10}{2x + 1}$

In Exercises 28–32 use synthetic division to determine the quotient and the remainder. Check by using $P(x) = (x - a) \cdot Q(x) + R$. (Refer to Section 3.3.)

28. $\dfrac{2x^5 - 30x^3 + 9x^2 + 5x - 3}{x + 4}$

29. $\dfrac{3x^4 + 2x^2 - x + 4}{x - 4}$

30. $\dfrac{x^4 + 81}{x + 3}$

31. $\dfrac{4x^3 - 9x^2 - x + 9}{4x + 3}$

32. $\dfrac{3x^5 - 2x^4 + x^2 - x + 9}{x - 2}$

In Exercises 33–38 refer to Section 3.4.

33. Use the remainder theorem to check your answers to Exercises 28–32.

34. Use the factor theorem to determine if $x - 3$ is a factor of $2x^3 - 7x^2 - 21x + 54$.

35. Let $P(x) = x^4 - x^3 - 7x^2 + x + 6$ and evaluate $P(1)$, $P(-1)$, $P(2)$, $P(-2)$, $P(3)$, and $P(-3)$.

36. Use the results of Exercise 35 to factor $P(x)$ over the integers.

37. If $P(x) = 4x^3 - 3x^2 + 2x + k$, find k so that:
(a) $P(2) = 4$, (b) $P(3) = 12$, and (c) $P(-1) = 0$.

38. Find m so that when $2x^3 + mx^2 + x + m$ is divided by $x - 1$, the remainder is 4.

Solve and check Exercises 39–50 and state the restricted values of the variable. (Refer to Section 3.7.)

39. $\dfrac{5}{x - 6} + \dfrac{3}{x + 6} = \dfrac{8}{x^2 - 36}$

40. $\dfrac{x + 3}{x + 2} = \dfrac{3}{2}$

41. $\dfrac{2}{x + 3} - \dfrac{2}{x - 3} = \dfrac{1}{3 - x}$

42. $\dfrac{2x + 3}{x - 1} = \dfrac{2x - 5}{x + 3}$

43. $\dfrac{5}{3x + 1} - \dfrac{x + 2}{4x} + \dfrac{1}{4} = 0$

44. $\dfrac{x}{x + 4} - \dfrac{4}{x - 4} = \dfrac{x^2 + 16}{x^2 - 16}$

45. $\dfrac{1}{1 - x} + \dfrac{x}{x - 1} = 1$

46. $x + \dfrac{5x}{x - 3} = \dfrac{15}{x - 3}$

47. Solve $\dfrac{x - a}{x + b} = \dfrac{x + b}{x - a}$ for x.

48. Solve $\dfrac{p - x}{x} = c + \dfrac{1}{x}$ for x.

49. Solve $\dfrac{x + b}{x + c} = \dfrac{x + c}{x + b}$ for x.

50. Solve $\dfrac{1}{a} + \dfrac{1}{b} = \dfrac{1}{c} - \dfrac{1}{b}$ for b.

In Exercises 51–60 refer to Section 3.8.

51. Divide 30 into two parts whose ratio is 5 to 1.

52. Separate 180 into two parts in a ratio of 2 to 7.

53. If 8 cans of sodapop cost $1.00, what is the cost of 36 cans?

54. 3, 5, and 9 are the first, second, and third terms of a proportion in that order. Find the fourth term.

55. For what value of x is the proportion $\dfrac{x + 2}{x + 9} = \dfrac{5}{12}$ true?

56. The force applied to stretch an elastic spring varies directly as the amount of elongation. If a force of 18 lb determines an elongation of 2.5 in., what is the constant of variation?

57. Find the amount of force it takes for the spring in Exercise 56 to stretch 6 in.

58. If F varies inversely as the square of d and if $F = 6$ when $d = 2$, find the value of F when $d = 6$.

59. The area of a triangle varies jointly as the length of the base b and the altitude h upon that base. If $A = 6$ when $b = 3$ and $h = 4$, find the area of a triangle with base 12 units and altitude 7 units.

60. The length L of a pendulum of a clock varies directly as the square of the period P of the pendulum. (The period is the time it takes for the end of the pendulum to make a movement from one extreme position to the other and then return.) If a pendulum

18 in. long has a period of $1\frac{1}{2}$ sec, how much should the pendulum be shortened so that the new period will be exactly 1 sec?

61. Given the algebraic expression

$$\left(\frac{x-2}{x^2+1}\right) \div \left(\frac{k-3}{x+3}\right)$$

(a) For what value(s) of x will the expression be undefined?

(b) For what value(s) of k will the expression be undefined?

(c) For what value(s) of x will the expression equal 0?

Why?

EXPONENTS AND RADICALS

In Chapter 2 we defined the expression x^n for any real number x and any natural number n. The five theorems of exponents relied on this definition are restated below for reference.

THE EXPONENT THEOREMS

Theorem 1 $x^n x^m = x^{n+m}$ Product of two powers

Theorem 2 $(x^n)^m = x^{nm}$ Power of a power

Theorem 3 $\dfrac{x^n}{x^m} = x^{n-m}$ if $n > m$ and $x \neq 0$

$\dfrac{x^n}{x^m} = \dfrac{1}{x^{m-n}}$ if $m > n$ and $x \neq 0$ Quotient of two powers

$\dfrac{x^n}{x^m} = 1$ if $m = n$ and $x \neq 0$

Theorem 4 $(xy)^n = x^n y^n$ Power of a product

Theorem 5 $\left(\dfrac{x}{y}\right)^n = \dfrac{x^n}{y^n}$ if $y \neq 0$ Power of a quotient

What happens if *n* is not a natural number? The possibility that *n* is either 0 or a negative integer, and the need for dealing with this possibility, can easily be seen by looking at Theorem 3, the quotient of two powers. If a practical meaning is to be established for x^0 and x^{-n}, where *x* is any *nonzero* real number and *n* is a natural number, the meaning must be consistent with the "rules" or theorems for exponents which are natural numbers. A similar need arises for the possibility that the exponent is a rational number, or indeed *any* real number.

4.1 INTEGRAL EXPONENTS

We first establish a meaning for the expressions x^0 and x^{-n}, where $x \neq 0$ and *n* is a natural number. We require the first theorem of exponents to remain valid when zero is used as an exponent; therefore

$$x^n x^0 = x^{n+0} = x^n = x^n \cdot 1$$

But there is only one multiplicative identity. Therefore, if x^0 is to be defined, it must be defined as the number 1.

DEFINITION OF x^0

$x^0 = 1$ if *x* is a real number and $x \neq 0$. (0^0 is not defined.)

As examples, $5^0 = 1$, $(-3)^0 = 1$, and $(x + 2)^0 = 1$ if $x \neq -2$.

It can be shown that all the exponent theorems remain valid when x^0 is defined as 1.

The first theorem of exponents must also remain valid when an exponent is a negative integer, so we examine the following statement for *n*, a positive integer and $x \neq 0$, to arrive at a definition for x^{-n}.

$$x^n x^{-n} = x^{n+(-n)} = x^0 = 1 = \frac{x^n}{x^n} = x^n \cdot \frac{1}{x^n}$$

Since each nonzero real number has exactly one reciprocal, if x^{-n} is to be defined, x^{-n} must be defined as $1/x^n$, the reciprocal of x^n.

DEFINITION OF x^{-n}

$$x^{-n} = \frac{1}{x^n}$$

if x is any real number and $x \neq 0$.

Noting that

$$\frac{1}{x^n} = \frac{1}{\underbrace{x \cdot x \cdots x}_{n \text{ factors}}} = \underbrace{\frac{1}{x} \cdot \frac{1}{x} \cdots \frac{1}{x}}_{n \text{ factors}} = \left(\frac{1}{x}\right)^n$$

the following theorem can be stated:

THEOREM

$$\frac{1}{x^n} = \left(\frac{1}{x}\right)^n \quad \text{if} \quad x \neq 0$$

As examples,

$$5^{-1} = \frac{1}{5}$$

$$2^{-3} = \frac{1}{2^3} = \frac{1}{8}$$

$$\left(\frac{1}{3}\right)^{-2} = \left(\frac{1}{\frac{1}{3}}\right)^2 = 3^2 = 9$$

$$x^{-4} = \frac{1}{x^4}$$

$$\left(-\frac{2}{5}\right)^{-3} = \left(-\frac{5}{2}\right)^3 = -\frac{125}{8} \quad \text{since} \quad \frac{1}{-\frac{2}{5}} = -\frac{5}{2}$$

$$\frac{1}{3^{-4}} = \left(\frac{1}{3}\right)^{-4} = 3^4 = 81$$

It can be shown that all five theorems of exponents remain valid when the exponent is a negative integer. These theorems are illustrated in the following examples. Assume all variables are nonzero.

EXAMPLE I Using Theorem 1,

$$2^{-3} \cdot 2^{-2} = 2^{(-3)+(-2)} = 2^{-5} = \frac{1}{2^5} = \frac{1}{32}$$

$$2^4 \cdot 2^{-7} = 2^{4+(-7)} = 2^{-3} = \frac{1}{2^3} = \frac{1}{8}$$

$$x^3 \cdot x^{-5} = x^{3+(-5)} = x^{-2} = \frac{1}{x^2}$$

EXAMPLE 2 Using Theorem 2,

$$(2^{-3})^2 = 2^{-6} = \frac{1}{2^6} = \frac{1}{64}$$

$$(2^{-2})^{-4} = 2^{(-2)(-4)} = 2^8 = 256$$

$$(x^3)^{-2} = x^{3(-2)} = x^{-6} = \frac{1}{x^6}$$

EXAMPLE 3 Using Theorem 3,

$$\frac{3^4}{3^7} = 3^{4-7} = 3^{-3} = \frac{1}{3^3} = \frac{1}{27}$$

$$\frac{x^4}{x^6} = x^{4-6} = x^{-2} = \frac{1}{x^2}$$

$$\frac{x^{-2}}{x^{-5}} = x^{-2-(-5)} = x^{-2+5} = x^3$$

EXAMPLE 4 Using Theorem 4,

$$(2x^{-2})^{-3} = 2^{-3}(x^{-2})^{-3} = 2^{-3}x^6 = \frac{x^6}{8}$$

EXAMPLE 5 Using Theorem 5,

$$\left(\frac{x^{-1}}{5}\right)^{-4} = \frac{(x^{-1})^{-4}}{5^{-4}} = \frac{x^4}{5^{-4}} = 5^4 x^4 = 625x^4$$

$$\left(\frac{2^{-3}}{y^{-3}}\right)^{-2} = \frac{(2^{-3})^{-2}}{(y^{-3})^{-2}} = \frac{2^6}{y^6}$$

EXAMPLE 6 Simplify $(x^{-2}y^{-2})^{-2}$.

Solution

$$(x^{-2}y^{-2})^{-2} = (x^{-2})^{-2}(y^{-2})^{-2} \qquad \text{(Theorem 4)}$$
$$= x^4 y^4 \qquad\qquad\qquad \text{(Theorem 2)}$$

EXAMPLE 7 Simplify $(2^{-3} \cdot 2^{-5})^{-1}$.

Solution

$$(2^{-3} \cdot 2^{-5})^{-1} = (2^{-3-5})^{-1} = (2^{-8})^{-1} \qquad \text{(Theorem 1)}$$
$$= 2^8 \qquad\qquad\qquad \text{(Theorem 2)}$$
$$= 256 \qquad\qquad \text{(Definition)}$$

EXAMPLE 8 Simplify $(2^{-3} + 2^{-5})^{-1}$.

Solution   There is no exponent theorem that applies to a power of a sum (or difference), so the definition must be used.

$$(2^{-3} + 2^{-5})^{-1} = \left(\frac{1}{2^3} + \frac{1}{2^5}\right)^{-1}$$
$$= \left(\frac{2^2}{2^5} + \frac{1}{2^5}\right)^{-1} \qquad \text{(2^5 is the L.C.D. for the}$$
$$\qquad\qquad\qquad\qquad\qquad \text{two fractions)}$$
$$= \left(\frac{4}{32} + \frac{1}{32}\right)^{-1}$$
$$= \left(\frac{5}{32}\right)^{-1}$$
$$= \frac{32}{5} \qquad\qquad \left(\text{Definition } x^{-1} = \frac{1}{x}\right)$$

EXAMPLE 9 Simplify $\dfrac{2^{-3} + 2^{-5}}{2^{-7}}$.

Solution

$$\frac{2^{-3} + 2^{-5}}{2^{-7}} = \left(\frac{2^{-3} + 2^{-5}}{2^{-7}}\right)\left(\frac{2^7}{2^7}\right)$$

$$= \frac{2^7(2^{-3} + 2^{-5})}{(2^7)(2^{-7})}$$

$$= \frac{(2^7 2^{-3}) + (2^7 2^{-5})}{2^7 2^{-7}} \qquad \text{(Distributive axiom)}$$

$$= \frac{2^4 + 2^2}{2^0} \qquad \text{(Theorem 1)}$$

$$= \frac{16 + 4}{1} = 20 \qquad \text{(Definition)}$$

The preceding example illustrates the use of the distributive axiom to distribute the operation of multiplication over a sum. Theorem 1 for exponents is also used in the simplification. It is important to note the differences illustrated by the last four examples. It is especially important to remember that the exponent theorems apply to products and quotients and *not* to sums and differences.

EXAMPLE 10 Does $(x + 2)^2 = x^2 + 4$? Why?

Solution By the definition of squaring,
$$(x + 2)^2 = (x + 2)(x + 2) = x^2 + 4x + 4$$
If
$$(x + 2)^2 = x^2 + 4$$
then
$$x^2 + 4x + 4 = x^2 + 4$$
and
$$4x = 0 \quad \text{and} \quad x = 0$$
Thus $(x + 2)^2 = x^2 + 4$ only if $x = 0$.

EXAMPLE 11 Simplify $\dfrac{2x^{-1} + x^{-2}}{x^{-2} - x}$.

Solution Apply the definition of x^{-n} and write

$$\frac{2 \cdot \dfrac{1}{x} + \dfrac{1}{x^2}}{\dfrac{1}{x^2} - x} = \left(\frac{\dfrac{2}{x} + \dfrac{1}{x^2}}{\dfrac{1}{x^2} - x}\right)\left(\frac{x^2}{x^2}\right)$$

$$= \frac{2x + 1}{1 - x^3}$$

EXERCISES 4.1

Use the definitions and theorems to express each of the following in simplest form without zero or negative exponents. Assume all variables to be nonzero.

1. 2^{-1}

2. 5^{-3}

3. 10^{-1}

4. 6^{-2}

5. $\left(\dfrac{1}{2}\right)^{-4}$

6. $\left(\dfrac{1}{10}\right)^{-5}$

7. $\left(\dfrac{1}{5}\right)^{-3}$

8. $\left(\dfrac{1}{10}\right)^{-6}$

9. $\dfrac{1}{3^{-2}}$

10. $\dfrac{1}{5^{-4}}$

11. $\dfrac{1}{2^{-3}}$

12. $\dfrac{1}{10^{-5}}$

13. 10^{-3}

14. $\left(\dfrac{3}{5}\right)^{-1}$

15. 10^{-4}

16. $\left(\dfrac{4}{3}\right)^{-2}$

17. $\left(-\dfrac{10}{3}\right)^{-4}$

18. $3^{5} \cdot 3^{-2}$

19. $\left(-\dfrac{3}{16}\right)^{-1}$

20. $5^{4} \cdot 5^{-7}$

21. $2^{5} \cdot 2^{-9}$

22. $10^{6} \cdot 10^{-8}$

23. $10^{-4} \cdot 10^{2}$

24. $10^{-5} \cdot 10^{-4}$

25. $10^{-2} \cdot 10^{-3}$

26. $5^{4} \cdot 5^{-4}$

27. $10^{6} \cdot 10^{-6}$

28. $6^{-5} \cdot 6^{5}$

29. $\dfrac{10^{2}}{10^{-4}}$

30. $\dfrac{10^{-3}}{10^{2}}$

31. $\dfrac{10^{6}}{10^{-3}}$

32. $\dfrac{10^{-5}}{10^{-7}}$

33. $\dfrac{10^{-4}}{10^{\;4}}$

34. $\dfrac{10^{-5}}{10^{-2}}$

35. $(10^{-2})^{-1}$

36. $\dfrac{10^{-6}}{10^{-6}}$

37. $(2^{-3})^{2}$

38. $(3^{-2} \cdot 5)^{-2}$

39. $(2^{-1} \cdot 3^{-2})^{2}$

40. $(-3^{-1})^{2}$

41. $(-2^{-2})^{-1}$

42. $\left(\dfrac{5^{-2}}{2^{-2}}\right)^{-1}$

43. $(-5^{-1})^{-1}$

44. $\left(\dfrac{5}{8^{-1}}\right)^{-2}$

45. $\dfrac{2^{-5} \cdot 3^{4}}{2^{-4} \cdot 3^{-1}}$

46. $\left(\dfrac{2^{-3}}{5}\right)^{-2}$

47. $\dfrac{3^{-2} \cdot 5^{-1}}{3^2 \cdot 5^{-2}}$

48. $\left(\dfrac{5^{-3}}{3^{-4}}\right)^{-1}$

49. $10^{-3}(10^4 \cdot 10^{-1})$

50. $5^{-2}(5^3 + 5^2)$

51. $5^4(5^{-3} \cdot 5^{-1})$

52. $10^8(10^{-6} + 10^{-8})$

53. $4^7(4^{-7} - 4^{-5})$

54. $\left(\dfrac{2^{-2}}{5^0}\right)^{-3}$

55. $3^{-5}(6^5 - 3^6)$

56. $\left(\dfrac{4^0}{4^{-3}}\right)^{-1}$

57. $(2^{-1} \cdot 5^{-1})^{-1}$

58. $(2^{-1} + 5^{-1})^{-1}$

59. $(2^{-2} \cdot 5^{-2})^{-2}$

60. $(2^{-2} + 5^{-2})^{-2}$

61. $10^{-2}(2^{-1} \cdot 5^{-1})$

62. $10^{-2}(2^{-1} + 5^{-1})$

63. $\dfrac{2^{-2} \cdot 5^{-2}}{10^{-2}}$

64. $\dfrac{2^{-2} + 5^{-2}}{10^{-2}}$

65. $(2x)^{-3}$

66. $2x^{-3}$

67. $-3(x^2)^{-4}$

68. $(-3 \cdot x^2)^{-4}$

69. $(4^{-1} \cdot x)^{-2}$

70. $\left(\dfrac{5}{x}\right)^{-3}$

71. $(10^{-3} \cdot x^3)^{-2}$

72. $\left(\dfrac{x}{4}\right)^{-2}$

73. $(2^{-1} \cdot x^{-1})^{-2}$

74. $(2^{-1} + x^{-1})^{-2}$

75. $(x^{-2} + 5^{-2})^{-1}$

76. $(x^{-2} \cdot 5^{-2})^{-1}$

77. $\dfrac{y}{y^{-1}}$

78. $\dfrac{1}{2x^{-2}}$

79. $\dfrac{x^{-3}}{y^{-2}}$

80. $\dfrac{1}{4x^{-3}}$

81. $(xx^{-4} \cdot x^3)^{10}$

82. $(x^{-2}y)^{-1}$

83. $\left(\dfrac{x^{-3}}{y^{-2}}\right)\left(\dfrac{x^{-5}}{x^7}\right)^0$

84. $(y^{-5} - y^2)^{-3}(y^{-5} - y^2)^3$

85. $(x^{-1} - y^{-1})(x - y)^{-1}$

86. $\dfrac{x^{-1} + y^{-1}}{x^{-1} - y^{-1}}$

87. $\dfrac{a^4b - a^{-1}b^{-4}}{a^5b^{-5}}$

88. $\dfrac{a + b^{-1}}{a^{-1} + b}$

89. $(x^2 + 2)^{-2}$

90. $(2 - x^{-2})^{-2}$

91. $(-2x^{-2}y^{-2})y^{-3}$

92. $(-2x^{-2}y^{-2})^{-3}$

93. $(x^{-2} + y^{-2})^n(x^{-2} + y^{-2})^{-n}$

94. $x^{-n}(x^n - x^{n-1})$

95. $(2^n + 2^{-n})(2^n - 2^{-n})$

96. $(2^n + 2^n)^{-n}$

97. $(x^{-n} + y^{-n})^0(x^{-n}y^n)^{-1}$

98. $\dfrac{(x^{n+1}x^{2n-1})^{-2}}{x^{4n}}$

99. $\dfrac{2^{-3n} \cdot 3^{-2n}}{2^{2n} \cdot 3^{-n}}$

100. $\dfrac{x^{-n}y^{2n}}{x^{2n}y^{-n}}$

4.2 SQUARE ROOTS, CUBE ROOTS, AND nTH ROOTS

DEFINITION

The number a is a square root of b if and only if $a^2 = b$.

This definition implies that the operations of squaring and extracting square roots are inverses of each other, just as addition and subtraction are inverse operations and multiplication and division are inverse operations.

For example,

3 is a square root of 9 since $3^2 = 9$

-3 is a square root of 9 since $(-3)^2 = 9$

$\dfrac{5}{8}$ is a square root of $\dfrac{25}{64}$ since $\left(\dfrac{5}{8}\right)^2 = \dfrac{25}{64}$

$-\dfrac{5}{8}$ is a square root of $\dfrac{25}{64}$ since $\left(-\dfrac{5}{8}\right)^2 = \dfrac{25}{64}$

and

0 is a square root of 0 since $0^2 = 0$

Examination of these examples and similar ones reveals the following properties:

1. If x is a real number, then $x^2 \geq 0$. (In other words, the square of a real number is never negative.)
2. Every positive real number has two square roots, a positive real number and its negative. Thus if a is a positive real number ($a > 0$), then the square roots of a^2 are a and $-a$.

It is useful to denote a square root by the symbol $\sqrt{x}$. However, the symbolic expression $\sqrt{x}$ must represent exactly one number. Suppose, for example, that $\sqrt{9} = 3$ and $\sqrt{9} = -3$. Then, by the transitive axiom of the equal relation, it would follow that $3 = -3$, and this is impossible. Therefore $\sqrt{9}$ and, in general, $\sqrt{x}$ must represent exactly one number in order to avoid contradictions. Mathematicians agree that $\sqrt{9}$ shall be used to designate 3, the positive square root of 9. The negative square root, -3, is indicated by $-\sqrt{9}$.

In general, the following definition is made:

DEFINITION OF $\sqrt{x}$

If $x > 0$, then $\sqrt{x}$ is the unique positive real number such that $(\sqrt{x})^2 = x$.

Note: If $x < 0$, then $\sqrt{x}$ is not a real number.

The following useful theorem aids in the solution of problems involving square roots.

THEOREM

$\sqrt{x^2} = x$ if and only if $x \geq 0$.

Alternatively, this theorem can be stated in terms of the absolute value of x as follows:

THEOREM

$\sqrt{x^2} = |x|$ for every real number x.

The positive real number $\sqrt{x}$ is called the **principal square root** of x. Its negative is also a square root and is designated by the symbol $-\sqrt{x}$.

Thus $\sqrt{25} = 5$, and 5 is the principal square root of 25. Also, $-\sqrt{25} = -5$, and -5 is the other square root of 25.

Thus there is exactly one symbolic representation for each of the square roots of a positive real number.

Although every positive real number x has two real square roots—namely, $\sqrt{x}$ and $-\sqrt{x}$, it does not follow that every positive rational number has two *rational* roots.

Rational numbers such as 49, 81, and $\frac{9}{16}$ are called **perfect squares** because their square roots are also rational.

On the other hand, the number 2 is not a perfect square, and its square roots, $\sqrt{2}$ and $-\sqrt{2}$, are not rational but irrational.

Q, the set of rational numbers, was defined as the set of quotients of integers, p/q, where $q \neq 0$. It may be shown that $\sqrt{2}$ cannot be expressed as the quotient of two integers, and thus $\sqrt{2}$ is irrational.

The terminating decimals and the nonterminating repeating decimals represent the rational numbers. Therefore it is convenient to think of the irrational numbers as the set of nonterminating, nonrepeating decimals and the real numbers as the union of these two sets—that is, the set of all decimals.

The table of squares and square roots on the inside cover of this book is useful in providing a first approximation to the square root of a number. The square root values in the table are not exact but are approximations to the nearest thousandth.

Pocket calculators are increasingly being used to find roots of numbers, but the user must be aware that approximations and round-offs may yield answers that are different from those in tables. However, these differences are minimal and should be examined in view of the degree of accuracy desired in a specific problem. You should become thoroughly familiar with your calculator and use it not only to obtain numerical answers, but also to check problems.

When we introduced the set of real numbers in Chapter 1, a visual representation was given in the form of the number line. The number line can be used not only to represent the integers and the rational numbers, but also the irrational numbers. The following axiom, called the axiom of completeness, describes this property of the set of real numbers.

THE AXIOM OF COMPLETENESS

Each point on the number line corresponds to exactly one real number, and each real number corresponds to exactly one point on the number line.

A point on the number line that corresponds to an irrational square root can be found by using the theorem of Pythagoras.

THE THEOREM OF PYTHAGORAS

The square of the length of the hypotenuse of a right triangle is equal to the sum of the squares of the lengths of the legs of the right triangle.

The hypotenuse of a right triangle, the longest side, is opposite the right angle. The legs of a right triangle are the other two sides—that is, the sides that form the right angle.

Thus if *c* represents the length of the hypotenuse in Figure 4.1 and if *a* and *b* represent the lengths of the legs, then

$$c^2 = a^2 + b^2$$

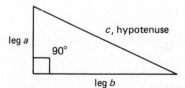

FIG. 4.1

Therefore a length of $\sqrt{2}$ can be represented by the diagonal of a square (see Figure 4.2). Since

$$c^2 = 1^2 + 1^2$$
$$c^2 = 2$$

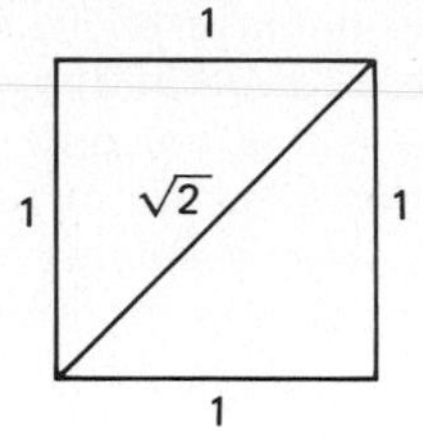

FIG. 4.2

Thus $c = \sqrt{2}$

A length of $\sqrt{3}$ can be represented as the diagonal of a rectangle whose sides are 1 and $\sqrt{2}$ (see Figure 4.3). Since

$$c^2 = (\sqrt{2})^2 + 1^2$$
$$c^2 = 2 + 1 = 3$$

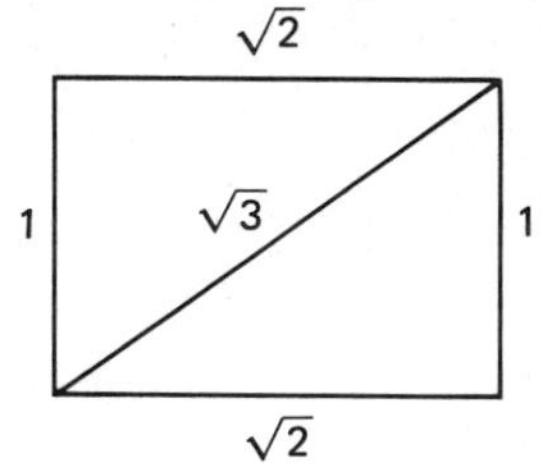

FIG. 4.3

Thus $c = \sqrt{3}$

In a similar way, the numbers $\sqrt{4} = 2$, $\sqrt{5}$, $\sqrt{6}$, and so on, can be represented as the lengths of the diagonals of rectangles. Now using circles with centers at the origin and with radii equal successively to these diagonals, the square roots can be located on the number line. This is illustrated in Figure 4.4.

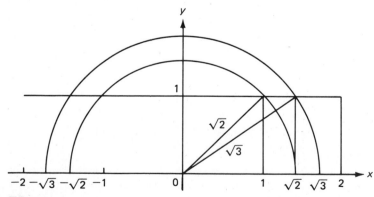

FIG. 4.4

160

HISTORICAL NOTE

Although there are many stories about the life of Pythagoras (ca. 580–501 B.C.) little is known for certain. He was probably born on the island of Samos, probably traveled to Egypt and Babylonia, and is known to have settled in Crotona on the Italian coast. In Crotona he founded a brotherhood composed of some 300 wealthy young aristocrats. This group, known as the Pythagoreans, became the prototype of all the secret societies of Europe and America. Their motto, "Number rules the universe," expressed the combination of mathematics and mysticism in which they believed. Shakespeare refers to the Pythagorean belief in immortality and transmigration of the soul in _The Merchant of Venice:_

> Thou almost mak'st me waver in my faith,
> To hold opinion with Pythagoras,
> That souls of animals infuse themselves
> Into the trunks of men.

The name of Pythagoras is most famous in connection with the relationship of the squares of the sides of a right triangle. While Pythagoras did not discover this property (it was already known to the Babylonians), he may have offered the first proof of this statement.

EXAMPLE 1 Simplify $\sqrt{64} - \sqrt{(-4)^2}$.

Solution

$$\sqrt{64} - \sqrt{(-4)^2} = \sqrt{64} - \sqrt{16}$$
$$= 8 - 4 = 4$$

EXAMPLE 2 Approximate $\sqrt{12} - \sqrt{20}$ to the nearest hundredth by using a table or calculator.

Solution   From the table in this book,
$$\sqrt{12} = 3.464$$
and
$$\sqrt{20} = 4.472$$
$$\sqrt{12} - \sqrt{20} = 3.464 - 4.472$$
$$= -1.008$$
$$= -1.01 \text{ to the nearest hundredth}$$

EXAMPLE 3 Find the hypotenuse c of a right triangle if the lengths of its legs a and b are such that $a = 12$ and $b = 35$.

Solution   Using the theorem of Pythagoras,
$$c^2 = a^2 + b^2$$
$$c^2 = (12)^2 + (35)^2$$
$$c^2 = 144 + 1225 = 1369$$
$$c = \sqrt{1369} = 37$$

EXAMPLE 4 If a, b, and c are the sides of a right triangle whose hypotenuse is c, find b if $a = 16$ and $c = 18$.

Solution   Using
$$a^2 + b^2 = c^2$$
$$(16)^2 + b^2 = (18)^2$$
$$b^2 = (18)^2 - (16)^2 = (18 + 16)(18 - 16)$$
$$b^2 = 68$$
$$b = 8.25 \text{ to the nearest hundredth}$$

EXAMPLE 5   Solve:
(a) $x = \sqrt{9}$
(b) $x^2 = 9$

Solution
(a) Since $\sqrt{9} = 3$, the positive square root of 9,
$$x = \sqrt{9} \quad \text{means} \quad x = 3$$
The solution set is {3}.
(b) Since $(+3)^2 = 9$ and $(-3)^2 = 9$, the solution set of $x^2 = 9$ is {3, -3}.
An alternate solution to (b) is as follows:
$$x^2 = 9$$
$$x^2 - 9 = 0$$
$$(x + 3)(x - 3) = 0$$
$$x + 3 = 0 \quad \text{or} \quad x - 3 = 0$$
$$x = -3 \quad \text{or} \qquad x = 3$$
and the solution set is {3, -3}.

Many mathematical problems are concerned with roots higher than square roots. Cube roots, fourth roots, fifth roots, and higher roots of numbers are often needed for the solution of problems. These roots are referred to as **nth roots,** where n is any integer greater than or equal to 2. We defined the square root of a number b by saying that a is a square root of b if and only if $a^2 = b$, and we can now extend this definition to include nth roots.

DEFINITION

The number a is an nth root of b if and only if $a^n = b$.

EXAMPLES

2 is a cube root of 8 because $2^3 = 8$.
-3 is a cube root of -27 because $(-3)^3 = -27$.
5 is a 4th root of 625 because $5^4 = 625$.

If n is an *even* number, the following properties of the nth roots of numbers may be stated.

1. Every positive real number has exactly two real _n_th roots for _n_ even: one positive and one negative. For example, 5 and −5 are 4th roots of 625.
2. Negative real numbers do _not_ have real _n_th roots for _n_ even. For example, there is no number _x_ such that $x^4 = -625$.

   If _n_ is an _odd_ number, the following properties of _n_th roots of numbers may be stated.

1. Every real number has exactly one real _n_th root for _n_ odd.
2. The real _n_th root of a positive number is positive for _n_ odd.
3. The real _n_th root of a negative number is negative for _n_ odd. For example, 2 is the real 5th root of 32, and −2 is the real 5th root of −32.

   The symbol $\sqrt[n]{x}$ (read "the principal _n_th root of _x_") is used to indicate exactly one _n_th root of the real number _x_.

DEFINITION OF $\sqrt[n]{x}$, x POSITIVE

$\sqrt[n]{x}$ is the unique positive real number such that $(\sqrt[n]{x})^n = x$.

DEFINITION OF $\sqrt[n]{-x}$, x POSITIVE, n ODD

$$\sqrt[n]{-x} = -\sqrt[n]{x}$$

Note that $\sqrt[n]{-x}$ is not defined for _n_ even and −_x_ negative.

The symbol $\sqrt[n]{x}$ is called a **radical,** the natural number _n_ is called the **index,** and the real number _x_ is called the **radicand.**

THEOREM

If _x_ is a **positive** real number and _n_ is a natural number, then

$$\sqrt[n]{x^n} = x$$

EXAMPLE 6 Simplify, if possible:

(a) $\sqrt[3]{8}$
(b) $\sqrt[3]{-8}$
(c) $\sqrt[4]{81}$
(d) $\sqrt[4]{-81}$
(e) $\sqrt[5]{243}$
(f) $\sqrt[5]{-243}$

Solution

(a) $\sqrt[3]{8} = \sqrt[3]{2^3} = 2$
(b) $\sqrt[3]{-8} = \sqrt[3]{(-2)^3} = -2$
(c) $\sqrt[4]{81} = \sqrt[4]{3^4} = 3$ since 3 is positive
(d) $\sqrt[4]{-81}$ is not a real number
(e) $\sqrt[5]{243} = \sqrt[5]{3^5} = 3$
(f) $\sqrt[5]{-243} = \sqrt[5]{(-3)^5} = -3$

EXAMPLE 7 Simplify $\dfrac{(\sqrt[3]{25})^3}{\sqrt[3]{64}}$.

Solution

$$\frac{(\sqrt[3]{25})^3}{\sqrt[3]{64}} = \frac{25}{\sqrt[3]{4^3}} = \frac{25}{4}$$

EXAMPLE 8 Approximate $2\sqrt[3]{15} - \sqrt[3]{6}$, correct to the nearest hundredth.

Solution Using a table,

$$\sqrt[3]{15} = 2.466$$
$$2\sqrt[3]{15} = 4.932$$
$$\underline{\sqrt[3]{6} = 1.817}$$
$$2\sqrt[3]{15} - \sqrt[3]{6} = 3.115$$
$$= 3.12 \text{ to the nearest hundredth}$$

EXAMPLE 9 Solve for all real values of x:
$$x^5 = 32$$

Solution By the definition of nth root, $x^5 = 32$ if and only if $x = \sqrt[5]{32}$. By trial and error or by use of a calculator, $\sqrt[5]{32} = 2$, since $2^5 = 2 \cdot 2 \cdot 2 \cdot 2 \cdot 2 = 32$. So $x = 2$.

EXERCISES 4.2

Simplify Exercises 1–10. (Examples 1 and 2)

1. $\sqrt{16}$

2. $-\sqrt{36}$

3. $\sqrt{(-3)^2}$

4. $\sqrt{\dfrac{4}{25}}$

5. $\sqrt{400}$

6. $\sqrt{5^2 - 4^2}$

7. $\sqrt{5^2} - \sqrt{4^2}$

8. $\sqrt{(-5)^2 + (12)^2}$

9. $\sqrt{(-5)^2} + \sqrt{(12)^2}$

10. $\dfrac{3\sqrt{64 + 36}}{3(\sqrt{64} + \sqrt{36})}$

In Exercises 11–22 a and b represent the lengths of the legs of a right triangle and c represents the length of the hypotenuse of the triangle. Evaluate the length of the missing side to the nearest hundredth. (Examples 3 and 4)

11. $a = 12, b = 5$

12. $b = 5, c = 13$

13. $a = 4, b = 5$

14. $b = 5, c = 15$

15. $a = 1, b = 1$

16. $c = 10, a = b$

17. $a = \sqrt{3}, b = \sqrt{5}$

18. $b = \sqrt{10}, c = \sqrt{30}$

19. $a = 1, c = \sqrt{3}$

20. $a = 1, b = 2$

21. $a = 2t, c = t^2 + 1, t > 1$

22. $b = t^2 - 9, c = t^2 + 9, t > 3$

Solve Exercises 23–30 over the set of real numbers (*that is, for solutions that are real numbers*). (*Example 5*)

23. $x = \sqrt{25}$

24. $x^2 = 25$

25. $x = -\sqrt{36}$

26. $x = \sqrt{36}$

27. $x^2 = 36$

28. $x^2 = -36$

29. $x^2 = -49$

30. $x = -\sqrt{49}$

Simplify Exercises 31–50 using the table inside the cover of this book or a calculator, if necessary. (*Examples 6, 7, and 8*)

31. $\sqrt[3]{64}$

32. $\sqrt[3]{-27}$

33. $\sqrt[3]{-125}$

34. $\sqrt[3]{729}$

35. $\sqrt[4]{16}$

36. $\sqrt[4]{625}$

37. $\sqrt[3]{(-8)^3}$

38. $\sqrt[4]{(-6)^4}$

39. $\sqrt[4]{(-7)^4}$

40. $\sqrt[4]{(7)^4}$

41. $(\sqrt[3]{15})^3$

42. $(\sqrt[4]{25x^2})^4$

43. $\sqrt[5]{32}$

44. $\sqrt[5]{-32}$

45. $(\sqrt[3]{-3x^3y^6})^3$

46. $(\sqrt[9]{428})^9$

47. $\sqrt[7]{(-41)^7}$

48. $\sqrt[7]{-1}$

49. $\sqrt[8]{1}$

50. $\sqrt[9]{1}$

Solve for all real values of x in Exercises 51–60. (*Example 9*)

51. $x^3 = 64$

52. $x^3 = -64$

53. $x = \sqrt[3]{64}$

54. $x = \sqrt[3]{-64}$

55. $x^4 = 16$

56. $x = \sqrt[4]{16}$

57. $x^4 = -16$

58. $x = -\sqrt[4]{16}$

59. $x^5 = 10^5$

60. $x^5 = (-10)^5$

4.3 RATIONAL EXPONENTS

It is useful to assign a meaning to expressions such as $9^{1/2}$, $8^{1/3}$, $x^{3/4}$, and $x^{-2/3}$. To see if it is possible to define a power having a fractional exponent and still have the five exponent theorems remain valid, the second theorem of exponents, $(x^m)^n = x^{mn}$, is examined first. To give a meaning to such expressions and still hold the second theorem valid, consider the following:

$$(9^{1/2})^2 = 9^{1/2 \cdot 2} = 9^1 = 9$$

But

$$(9^{1/2})^2 = 9^{1/2} \cdot 9^{1/2} = 9$$

and

$$\sqrt{9}\sqrt{9} = 9$$

Thus a possibility is $9^{1/2} = \sqrt{9} = 3$.

Similarly,

$$(8^{1/3})^3 = 8^{1/3 \cdot 3} = 8^1 = 8$$
$$(8^{1/3})(8^{1/3})(8^{1/3}) = 8$$
$$\sqrt[3]{8}\sqrt[3]{8}\sqrt[3]{8} = 8$$

This suggests the possibility $8^{1/3} = \sqrt[3]{8} = 2$.

Accordingly, if x is a nonnegative real number and n is a natural number, then $x^{1/n}$ is defined as $\sqrt[n]{x}$, the principal real root of x.

DEFINITIONS OF $x^{1/n}$ AND $(-x)^{1/n}$

x is a positive real number, n is a natural number.

$$x^{1/n} = \sqrt[n]{x}$$
$$(-x)^{1/n} = -\sqrt[n]{x} \text{ if } n \text{ is odd } (1, 3, 5, 7, \ldots)$$
$$(-x)^{1/n} \text{ is undefined if } n \text{ is even } (2, 4, 6, 8, \ldots)$$
$$0^{1/n} = 0$$

For example,

$$(64)^{1/3} = \sqrt[3]{64} = 4$$
$$(-64)^{1/3} = -\sqrt[3]{64} = -4$$
$$(64)^{1/2} = \sqrt{64} = 8$$
$$(-64)^{1/2} = \sqrt{-64}$$

Note that the last expression, $\sqrt{-64}$, is not defined in the real numbers because the radicand, -64, is negative and the index is 2, an even number.

A similar approach is used for defining $x^{m/n}$ and $x^{-m/n}$, with the same restrictions on x and n as in the above definitions to ensure that the expression is defined in the set of real numbers.

Thus

$x^{m/n}$ is defined as $(x^m)^{1/n} = \sqrt[n]{x^m}$

and

$x^{-m/n}$ is defined as $\dfrac{1}{x^{m/n}}$

Note that $x^{m/n} = (x^{1/n})^m = (\sqrt[n]{x})^m$.

It can be shown that the five exponent theorems remain valid with these definitions. The results on exponents developed so far are summarized in Table 4.1 with the following assumptions: x and y are real numbers, $x \neq 0$ and $y \neq 0$; n and m are natural numbers; and a and b are rational numbers.

TABLE 4.1   EXPONENT DEFINITIONS AND THEOREMS

DEFINITIONS	*THEOREMS*
1. $x^n = xx \cdots x$ (n factors)	1. $x^a x^b = x^{a+b}$
2. $x^1 = x$	2. $(x^a)^b = x^{ab}$
3. $x^0 = 1$	3. $\dfrac{x^a}{x^b} = x^{a-b}$
4. $x^{-n} = \dfrac{1}{x^n}$	4. $(xy)^a = x^a y^a$
5. $x^{1/n} = \sqrt[n]{x}$ ($x > 0$ if n even)	5. $\left(\dfrac{x}{y}\right)^a = \dfrac{x^a}{y^a}$
6. $x^{m/n} = \sqrt[n]{x^m}$ ($x > 0$ if n even)	
7. $x^{-m/n} = \dfrac{1}{\sqrt[n]{x^m}}$ ($x > 0$ if n even)	
8. $0^a = 0$ if $a \neq 0$	

EXAMPLE 1   Simplify $(16)^{3/4}$.

Solution   Factoring the base, $16 = 2^4$, then
$$(16)^{3/4} = (2^4)^{3/4} = 2^{4 \cdot 3/4} = 2^3 = 8$$

EXAMPLE 2   Simplify $(64)^{-2/3}$.

Solution   Factoring the base, $64 = 2^6$, then
$$(64)^{-2/3} = (2^6)^{-2/3} = 2^{6(-2/3)}$$
$$= 2^{-4} = \frac{1}{2^4} = \frac{1}{16}$$

EXAMPLE 3   Simplify $\left(\dfrac{1}{16}\right)^{-1/2}$.

Solution
$$\left(\frac{1}{16}\right)^{-1/2} = (16)^{1/2} = \sqrt{16} = 4$$

EXAMPLE 4 Simplify $[(-6)^2]^{1/2}$.

Solution

$$[(-6)^2]^{1/2} = \sqrt{(-6)^2} = \sqrt{36} = 6$$

Caution: It is important to note that the theorems do not apply whenever the base is negative and an exponent indicates a root whose index is even. Thus

$$[(-6)^2]^{1/2} \neq (-6)^1$$

EXAMPLE 5 Simplify $(10^{-2/3} \cdot 10^{5/6})^3$.

Solution

$$
\begin{aligned}
(10^{-2/3} \cdot 10^{5/6})^3 &= (10^{-4/6} \cdot 10^{5/6})^3 \\
&= (10^{1/6})^3 = 10^{3/6} \\
&= 10^{1/2} = \sqrt{10}
\end{aligned}
$$

EXAMPLE 6 Simplify $\left(\dfrac{2^{3/4}x^{-3/4}}{2^{1/2}}\right)^{-4}$.

Solution

$$
\begin{aligned}
\left(\frac{2^{3/4}x^{-3/4}}{2^{1/2}}\right)^{-4} &= (2^{3/4-2/4}x^{-3/4})^{-4} \\
&= (2^{1/4})^{-4}(x^{-3/4})^{-4} \\
&= 2^{-1}x^3 = \frac{x^3}{2}
\end{aligned}
$$

With the increased use of calculators, the writing of fractional exponents in decimal form takes on importance.

EXAMPLE 7 Simplify each of the following:
(a) $5^{0.5} \cdot 5^{3.5}$
(b) $8^{0.4} \cdot 8^{0.6}$
(c) $2^{-1.3} \cdot 2^{-0.7}$
(d) $(4^6)^{-0.5}$

Solution

(a) $5^{0.5} \cdot 5^{3.5} = 5^{0.5+3.5} = 5^4 = 625$
(b) $8^{0.4} \cdot 8^{0.6} = 8^{0.4+0.6} = 8^1 = 8$

(c) $2^{-1.3} \cdot 2^{-0.7} = 2^{-1.3+(-0.7)} = 2^{-2} = \dfrac{1}{2^2} = \dfrac{1}{4}$

(d) $(4^6)^{-0.5} = 4^{6(-0.5)} = 4^{-3} = \dfrac{1}{4^3} = \dfrac{1}{64}$

EXAMPLE 8 Use a calculator to find $2^{0.37}$ correct to the nearest hundredth.

Solution The problem involves the use of the y^x key on your calculator, with $y = 2$ and $x = 0.37$. The answer, correct to the nearest hundredth, of $2^{0.37} = 1.29$.

This answer could not have been found by using the tables in this book.

EXERCISES 4.3

In Exercises 1–40 express each in simplified radical form. No exponents should appear in the answer. (Examples 1–5 and 7)

1. $4^{1/2}$

2. $16^{1/2}$

3. $25^{1/2}$
4. $27^{1/3}$
5. $16^{1/4}$
6. $81^{1/4}$
7. $243^{1/5}$
8. $216^{1/3}$
9. $-4^{1/2}$
10. $32^{1/5}$

11. $\left(\dfrac{27}{125}\right)^{1/3}$

12. $\left(\dfrac{81}{256}\right)^{1/4}$

13. $64^{2/3}$
14. $64^{3/2}$
15. $8^{-2/3}$
16. $25^{-3/2}$
17. $100{,}000^{-3/5}$

18. $\left(\dfrac{1}{27}\right)^{-2/3}$

19. $\left(-\dfrac{1}{125}\right)^{-2/3}$

20. $\left(\dfrac{216}{343}\right)^{2/3}$

21. $\left(\dfrac{1}{16}\right)^{3/4}$

22. $\left(\dfrac{1}{16}\right)^{-3/4}$

23. $(-8)^{2/3}$
24. $-8^{2/3}$
25. $(-8)^{-2/3}$
26. $-8^{-2/3}$
27. $(5^{3/2})^4$
28. $(8^{3/2})^{-2/3}$
29. $(10^{-3/5})^{-5/6}$
30. $(4^2)^{1/6}$

31. $\left(\dfrac{1}{14}\right)^{-1/4}$

32. $8^{1/2} \cdot 8^{1/6}$

33. $5^{1/2} \cdot 5^{-1/6}$
34. $10^{1.2} \cdot 10^{3.8}$
35. $10^{-1.4} \cdot 10^{1.9}$
36. $10^{-0.4} \cdot 10^{-0.6}$
37. $6^{3.5} \cdot 6^{-1.5}$

38. $16^{-0.08} \cdot 16^{-0.17}$

39. $(2 \cdot 2^{1/2})^{1/2}$

40. $(2 \cdot 2^{1/3})^{1/2}$

In Exercises 41–60, simplify. Assume all variables to be positive numbers. Leave no fractional, negative, or zero exponents in your answers. (Examples 5 and 6)

41. $8x^{-2/3}$
42. $(8x)^{-2/3}$
43. $(4x^{2/3})^{-3}$
44. $4(x^{2/3})^{-3}$
45. $(9x^{-4}y^2)^{1/2}$
46. $(10x^{1/6}y^{5/6})^{-6}$
47. $(x^{1/2} + y^{1/2})^2$

48. $(x^{1/2} + 3^{1/2})(x^{1/2} - 3^{1/2})$

49. $x^{1/2}(2x^{1/2} + 5x^{-1/2})$
50. $(x^{1/2} - x^{-1/2})^2$

51. $(5 - x^{1/2})(5 + x^{1/2})$
52. $(x^{1/2} - y^{1/2})(x^{1/2} + y^{1/2})$
53. $x^{-1/2}(x^{-1/2} + x^{1/2})$
54. $x^{1/4}(x^{3/4} + x^{-3/4} + x^{-1/4})$
55. $\{[(81)(81)^{1/2}]^{1/3}\}^{1/2}$
56. $7^{1/2} \cdot 7^{1/3} \cdot 7^{1/6}$
57. $(2^3 \cdot 2^{1/2})^{1/2}$

58. $\dfrac{(216)^{1/4}}{6^{1/12}}$

59. $2^{1/2} \cdot 4^{1/3} \cdot 32^{1/6}$
60. $(3^2 + 4^2)^{1/2}$

4.4 RADICALS: SIMPLIFICATION AND PRODUCTS

A radical of the form $\sqrt{M}$, where M is a monomial with an integer for its coefficient, is said to be simplified if no perfect squares are factors of the radicand M.

Similarly, a radical of the form $\sqrt[3]{M}$ is said to be simplified if no perfect cubes are factors of the radicand.

Renaming a number denoted by a radical so that any resulting radical is simplified is called **reducing the radicand.**

To reduce the radicand of a square root radical, the product of square roots theorem is used.

THE PRODUCT OF SQUARE ROOTS THEOREM

If r and s are nonnegative real numbers, then
$$\sqrt{r}\sqrt{s} = \sqrt{rs} \quad \text{and} \quad \sqrt{rs} = \sqrt{r}\sqrt{s}$$

Since $r \geq 0$ and $s \geq 0$, then by using exponents,
$$\sqrt{r}\sqrt{s} = r^{1/2}s^{1/2} = (rs)^{1/2} = \sqrt{rs}$$

EXAMPLE 1 Simplify $\sqrt{3}\sqrt{12}$.

Solution
$$\sqrt{3}\sqrt{12} = \sqrt{3 \cdot 12} = \sqrt{36} = 6$$

The following theorem is useful for simplifying square root radicals:

THEOREM

If x and y are any nonnegative real numbers,
$$\sqrt{x^2y} = \sqrt{x^2}\sqrt{y} = x\sqrt{y}$$

EXAMPLE 2 Simplify $\sqrt{75}$.

Solution   The basic idea is to factor the radicand to find the perfect square factors:
$$75 = 3 \cdot 5^2$$
$$\sqrt{75} = \sqrt{5^2 \cdot 3} = \sqrt{5^2}\sqrt{3} = 5\sqrt{3}$$

Alternate Solution
$$\sqrt{75} = (5^2 \cdot 3)^{1/2} = (5^2)^{1/2}(3^{1/2}) = 5\sqrt{3}$$

EXAMPLE 3   Simplify $\sqrt{6x}\sqrt{12xy^3}$ where $x \geq 0$ and $y \geq 0$.

Solution

$$
\begin{aligned}
\sqrt{6x}\sqrt{12xy^3} &= \sqrt{6x(12xy^3)} \\
&= \sqrt{(2 \cdot 3)(2 \cdot 2 \cdot 3)x^2y^2y} \\
&= \sqrt{2^2 3^2 x^2 y^2}\sqrt{2y} \\
&= \sqrt{(6xy)^2}\sqrt{2y} \\
&= 6xy\sqrt{2y}
\end{aligned}
$$

Alternate Solution

$$
\begin{aligned}
(6x)^{1/2}(12xy^3)^{1/2} &= (72x^2y^3)^{1/2} \\
&= (2^2 \cdot 3^2 \cdot 2x^2y^2y)^{1/2} \\
&= (6^2x^2y^2)^{1/2}(2y)^{1/2} \\
&= 6xy\sqrt{2y}
\end{aligned}
$$

Reducing the radicand of a cube root, a fourth root, or, in general, an nth root can be accomplished by the following theorem, which takes into account whether n is odd or even.

THEOREM—The Product of nth Roots

1. If a and b are any real numbers and n is an *odd* integer greater than or equal to 3, then
$$\sqrt[n]{a}\sqrt[n]{b} = \sqrt[n]{ab} \quad \text{and} \quad \sqrt[n]{ab} = \sqrt[n]{a}\sqrt[n]{b}$$
Alternately,
$$a^{1/n}b^{1/n} = (ab)^{1/n}$$

2. If a and b are any *nonnegative* numbers and n is an *even* integer greater than or equal to 2, then
$$\sqrt[n]{a}\sqrt[n]{b} = \sqrt[n]{ab} \quad \text{and} \quad \sqrt[n]{ab} = \sqrt[n]{a}\sqrt[n]{b}$$
Alternately,
$$a^{1/n}b^{1/n} = (ab)^{1/n}$$

EXAMPLE 4   Simplify $\sqrt[3]{20}\sqrt[3]{50}$.

Solution

$$
\begin{aligned}
\sqrt[3]{20}\sqrt[3]{50} &= \sqrt[3]{20(50)} = \sqrt[3]{1000} \\
&= \sqrt[3]{10^3} = 10
\end{aligned}
$$

Alternate Solution

$$
\begin{aligned}
20 &= 2 \cdot 2 \cdot 5; \quad\quad 50 = 2 \cdot 5 \cdot 5 \\
20 \cdot 50 &= 2 \cdot 2 \cdot 2 \cdot 5 \cdot 5 \cdot 5 = 2^3 \cdot 5^3 \\
\sqrt[3]{20}\sqrt[3]{50} &= \sqrt[3]{(20)(50)} = \sqrt[3]{2^3 \cdot 5^3} \\
&= (2)(5) = 10
\end{aligned}
$$

EXAMPLE 5   Simplify $\sqrt[5]{972}$.

Solution

$$
\begin{aligned}
972 &= 2 \cdot 2 \cdot 3 \cdot 3 \cdot 3 \cdot 3 \cdot 3 = 2^2 3^5 \\
\sqrt[5]{972} &= \sqrt[5]{2^2 3^5} = \sqrt[5]{3^5}\sqrt[5]{2^2} = 3\sqrt[5]{4}
\end{aligned}
$$

Alternate Solution

$$\sqrt[5]{972} = (3^5 2^2)^{1/5} = (3^{5/5})(2^{2/5}) = 3\sqrt[5]{4}$$

EXAMPLE 6 Simplify $\sqrt[3]{-81x^6}$.

Solution

$$\sqrt[3]{-81x^6} = -\sqrt[3]{81x^6}$$
$$= -\sqrt[3]{3^4x^6}$$
$$= -\sqrt[3]{3^3 \cdot 3 \cdot (x^2)^3}$$
$$= -\sqrt[3]{3^3(x^2)^3}\sqrt[3]{3}$$
$$= -3x^2\sqrt[3]{3}$$

Alternate Solution

$$\sqrt[3]{-81x^6} = -(3^4x^6)^{1/3}$$
$$= -(3^{4/3})(x^{6/3})$$
$$= -(3^{1+1/3})(x^2)$$
$$= -(3 \cdot 3^{1/3})(x^2)$$
$$= -3x^2\sqrt[3]{3}$$

EXAMPLE 7 Simplify $(\sqrt{2x} + 3)(\sqrt{2x} - 3)$.

Solution Recognize this product as the form $(A + B)(A - B)$, which is the difference of two squares $A^2 - B^2$.

If $A = \sqrt{2x}$, then $A^2 = (\sqrt{2x})^2 = 2x$.

If $B = 3$, then $B^2 = 9$.

Thus

$$(\sqrt{2x} + 3)(\sqrt{2x} - 3) = 2x - 9$$

Notice that the same result is achieved by simply expanding $(\sqrt{2x} + 3)(\sqrt{2x} - 3)$.

EXERCISES 4.4

Simplify Exercises 1–46. Assume all variables to be positive. (Examples 1–6)

1. $\sqrt{5}\sqrt{20}$
2. $\sqrt{50}\sqrt{2}$
3. $\sqrt{3x}\sqrt{12x}$
4. $\sqrt{y}\sqrt{y^3}$
5. $\sqrt{12}$
6. $\sqrt{125}$
7. $\sqrt{405}$
8. $\sqrt{9x}$
9. $\sqrt{24x^2}$
10. $\sqrt{40y^3}$
11. $\sqrt{x}\sqrt{x^3}$
12. $\sqrt{y^3}\sqrt{y^4}$
13. $\sqrt{6x}\sqrt{6x^2}$
14. $\sqrt{40x^5}$
15. $\sqrt{32x}$
16. $\sqrt{1452}$

17. $\sqrt{343x^3}$
18. $5\sqrt{72}$
19. $\sqrt{x^3}\sqrt{x^5}$
20. $\sqrt{2x}\sqrt{8x^3}$
21. $\sqrt{x^3y^5}$
22. $\sqrt{54y^7}$
23. $\sqrt{108y}$
24. $\sqrt{184}$
25. $\sqrt{112x^3}$
26. $4\sqrt{363}$
27. $\sqrt[3]{-128}$
28. $\sqrt[3]{500}$
29. $\sqrt[3]{625}$
30. $\sqrt[3]{5}\sqrt[3]{25}$
31. $\sqrt[3]{12}\sqrt[3]{18}$
32. $\sqrt[3]{98x^2}\sqrt[3]{28x}$

33. $\sqrt[3]{24x^2}$

34. $\sqrt[3]{256y}$

35. $\sqrt[3]{56y^5}$

36. $\sqrt[3]{x^2}\sqrt[3]{x^4}$

37. $\sqrt[3]{25y^3}\sqrt[3]{25y^6}$

38. $5y^2\sqrt[3]{80y^4}$

39. $\sqrt[3]{49x}\sqrt[3]{7x^2}$

40. $-5\sqrt[3]{-625}$

41. $\sqrt[4]{96}$

42. $\sqrt[6]{192}$

43. $\sqrt[5]{128x^7y^6}$

44. $\sqrt[5]{-1024a^8}$

45. $2x\sqrt{54x^4}$

46. $10xy\sqrt{800x^2y}$

In Exercises 47–56, simplify. Assume all variables to be positive. (Example 7)

47. $(\sqrt{5} + \sqrt{2})(\sqrt{5} - \sqrt{2})$

48. $(4 - \sqrt{3})(4 + \sqrt{3})$

49. $(\sqrt{6} - 2)(\sqrt{6} + 2)$

50. $(1 - \sqrt{2})(1 + \sqrt{2})$

51. $(\sqrt{8} - \sqrt{2})(\sqrt{8} + \sqrt{2})$

52. $(\sqrt{6} - \sqrt{2})(\sqrt{6} + \sqrt{2})$

53. $\dfrac{\sqrt{3} + 1}{2} \cdot \dfrac{\sqrt{3} - 1}{2}$

54. $(\sqrt{x} - \sqrt{3})(\sqrt{x} + \sqrt{3})$

55. $(2\sqrt{x} + y)(2\sqrt{x} - y)$

56. $(\sqrt{3x} - \sqrt{2x})(\sqrt{3x} + \sqrt{2x})$

4.5 RADICALS: RATIONALIZING

In order to rename an expression that has a radical in a numerator or denominator, or a fraction in a radicand, the quotient of square roots theorem is used for expressions involving square roots. This is called **rationalizing** the denominator, or the numerator, as the case may be. We start with the following theorem.

THE QUOTIENT OF SQUARE ROOTS THEOREM

If r and s are positive real numbers, then

$$\sqrt{\frac{r}{s}} = \frac{\sqrt{r}}{\sqrt{s}} \quad \text{and} \quad \frac{\sqrt{r}}{\sqrt{s}} = \sqrt{\frac{r}{s}}$$

Since r and s are positive,

$$\sqrt{\frac{r}{s}} = \left(\frac{r}{s}\right)^{1/2} = \frac{r^{1/2}}{s^{1/2}} = \frac{\sqrt{r}}{\sqrt{s}}$$

EXAMPLE 1 Simplify $\sqrt{\dfrac{2}{3}}$ by writing the fraction with a rational denominator.

Solution

$$\sqrt{\frac{2}{3}} = \frac{\sqrt{2}}{\sqrt{3}} \qquad \text{(By the Quotient of Square Roots Theorem)}$$

$$= \frac{\sqrt{2}\sqrt{3}}{\sqrt{3}\sqrt{3}} \qquad \text{(By the Fundamental Theorem of Fractions—the numerator and denominator are multiplied by the number that causes the radicand of the denominator to become a perfect square)}$$

Thus

$$\sqrt{\frac{2}{3}} = \frac{\sqrt{6}}{3}$$

EXAMPLE 2 Rationalize the denominator of $\dfrac{2}{\sqrt{27}}$.

Solution

$$\frac{2}{\sqrt{27}} = \frac{2}{3\sqrt{3}} \qquad \text{(Reducing the radicand)}$$

$$= \frac{2\sqrt{3}}{3\sqrt{3}\sqrt{3}} \qquad \text{(Multiplying numerator and denominator by } \sqrt{3} \text{ to clear the denominator of radicals)}$$

$$= \frac{2\sqrt{3}}{9} \qquad \text{(Simplifying)}$$

EXAMPLE 3 Rationalize the numerator of $\dfrac{\sqrt{x+3}}{2}$.

Solution

$$\frac{\sqrt{x+3}}{2} \cdot \frac{\sqrt{x+3}}{\sqrt{x+3}} = \frac{x+3}{2\sqrt{x+3}}$$

If the denominator is a sum or difference of terms involving a radical, then the difference of squares theorem provides a technique for rationalizing denominators involving a square root. In other words, a fraction having an irrational denominator may be renamed as a fraction with a rational denominator. The basic idea involved is

$$X^2 - Y^2 = (X + Y)(X - Y)$$
$$a - b = (\sqrt{a} + \sqrt{b})(\sqrt{a} - \sqrt{b})$$

If the denominator has the form $\sqrt{a} + \sqrt{b}$, then multiplication by $\sqrt{a} - \sqrt{b}$ (called its **conjugate**) will produce the rational number $a - b$. This assumes, of course, that a and b are positive rational numbers.

Similarly, multiplying $\sqrt{a} - \sqrt{b}$ by its conjugate, $\sqrt{a} + \sqrt{b}$, will produce the rational number $a - b$.

EXAMPLE 4 Rationalize the denominator of $\dfrac{1}{\sqrt{5}-1}$.

Solution

$$\frac{1}{\sqrt{5}-1} \cdot \frac{\sqrt{5}+1}{\sqrt{5}+1} = \frac{\sqrt{5}+1}{5-1}$$
$$= \frac{\sqrt{5}+1}{4}$$

EXAMPLE 5 Rationalize the denominator of $\dfrac{2}{6+\sqrt{2}}$.

Solution

$$\frac{2}{6+\sqrt{2}} \cdot \frac{6-\sqrt{2}}{6-\sqrt{2}} = \frac{2(6-\sqrt{2})}{36-2}$$
$$= \frac{2(6-\sqrt{2})}{34}$$
$$= \frac{6-\sqrt{2}}{17}$$

EXAMPLE 6 Express with a rational denominator $\dfrac{3}{\sqrt{5}-\sqrt{2}}$.

Solution

$$\frac{3}{\sqrt{5}-\sqrt{2}} \cdot \frac{\sqrt{5}+\sqrt{2}}{\sqrt{5}+\sqrt{2}} = \frac{3(\sqrt{5}+\sqrt{2})}{5-2}$$
$$= \frac{3(\sqrt{5}+\sqrt{2})}{3}$$
$$= \sqrt{5}+\sqrt{2}$$

The technique of multiplying by a conjugate also works for rationalizing the numerator of a fraction.

EXAMPLE 7 Rationalize the numerator of $\dfrac{\sqrt{x}+3}{4}$.

Solution

$$\frac{\sqrt{x}+3}{4} \cdot \frac{\sqrt{x}-3}{\sqrt{x}-3} = \frac{x-9}{4(\sqrt{x}-3)}$$

EXERCISES 4.5

In Exercises 1–40 rationalize the denominator and write in simplest radical form. Assume all variables to be positive.

1. $\sqrt{\dfrac{1}{2}}$

2. $\sqrt{\dfrac{1}{6}}$

3. $\sqrt{\dfrac{5}{12}}$

4. $\dfrac{1}{\sqrt{5}}$

5. $\dfrac{1}{\sqrt{20}}$

6. $\sqrt{\dfrac{2}{25}}$

7. $\sqrt{\dfrac{3}{50}}$

8. $\dfrac{8}{\sqrt{12}}$

9. $\sqrt{\dfrac{3}{8}}$

10. $\dfrac{1}{\sqrt{10}}$

11. $\dfrac{1}{\sqrt{18}}$

12. $\sqrt{\dfrac{3}{12}}$

13. $\sqrt{\dfrac{3}{32}}$

14. $\dfrac{9}{\sqrt{45}}$

15. $\dfrac{4}{2\sqrt{200}}$

16. $\dfrac{21\sqrt{8y^3}}{\sqrt{49y}}$

17. $\sqrt{\dfrac{1}{3x}}$

18. $\sqrt{\dfrac{125}{y^3}}$

19. $\dfrac{4}{\sqrt{32x}}$

20. $\sqrt{\dfrac{7}{8x}}\,\sqrt{\dfrac{2x^3}{49}}$

21. $\dfrac{14\sqrt{9}}{3\sqrt{7}}$

22. $\dfrac{\sqrt{64x^4y^6}}{\sqrt{128x^6y^6}}$

23. $\sqrt{\dfrac{2}{x}}$

24. $\sqrt{\dfrac{8}{y}}$

25. $\dfrac{6}{\sqrt{72x^3}}$

26. $3\sqrt{\dfrac{1}{3}}\cdot 5\sqrt{\dfrac{3}{5}}$

27. $\dfrac{\sqrt{3}}{\sqrt{6x}}$

28. $\dfrac{1}{\sqrt{2}-1}$

29. $\dfrac{\sqrt{2x}}{\sqrt{10x}}$

30. $\dfrac{1}{1+\sqrt{3}}$

31. $\dfrac{\sqrt{3}}{\sqrt{7}-\sqrt{3}}$

32. $\dfrac{4}{\sqrt{7}+\sqrt{5}}$

33. $\dfrac{4}{\sqrt{7}+\sqrt{3}}$

34. $\dfrac{\sqrt{5}}{2-\sqrt{5}}$

35. $\dfrac{5}{\sqrt{11}+1}$

36. $\dfrac{2}{\sqrt{x}-1}$

37. $\dfrac{x-y}{\sqrt{x}-\sqrt{y}}$

38. $\dfrac{x-y}{\sqrt{x}+\sqrt{y}}$

39. $\dfrac{1}{\sqrt{4x}-\sqrt{x}}$

40. $\dfrac{x-2y}{\sqrt{x}+\sqrt{2y}}$

In Exercises 41–50 rationalize the numerator. Assume all variables to be positive and write the answer in simplest form.

41. $\dfrac{\sqrt{3}}{2}$

42. $\dfrac{\sqrt{2}}{2}$

43. $\dfrac{\sqrt{3}}{3}$

44. $\dfrac{3\sqrt{2}}{2}$

45. $\dfrac{\sqrt{x-1}}{x}$

46. $\dfrac{\sqrt{4-3x}}{2x}$

47. $\dfrac{\sqrt{2}-\sqrt{3}}{3}$

48. $\dfrac{\sqrt{3}+1}{2}$

49. $\dfrac{\sqrt{x}-\sqrt{y}}{x-y}$

50. $\dfrac{\sqrt{2x}+\sqrt{y}}{2x-y}$

4.6 RADICALS: SUMS AND DIFFERENCES

It is often possible to express a sum or difference of two radical terms as a single term. For example, since $\sqrt{5}$ is a real number, so are $3\sqrt{5}$ and $4\sqrt{5}$ and $3\sqrt{5}+4\sqrt{5}$ by the closure axioms for real numbers. Now using the distributive axiom,

$$3\sqrt{5}+4\sqrt{5}=(3+4)\sqrt{5}=7\sqrt{5}$$

Radicals that have the same radicands *and* the same indices are called **like radicals.** Radicals that are not like are called **unlike.** For example,

$3\sqrt{5x}$ and $4\sqrt{5x}$ contain like radicals

$\sqrt{5x}$ and $\sqrt[3]{5x}$ are unlike radicals

$\sqrt{5x}$ and $\sqrt{3x}$ are unlike radicals

Sometimes it is necessary to reduce each radical term to simplified radical form in order to identify like radicals.

EXAMPLE 1 Simplify $\sqrt{50x}+\sqrt{18x}$ if possible.

Solution

$$\sqrt{50x}+\sqrt{18x}=5\sqrt{2x}+3\sqrt{2x} \qquad \text{(Simplifying each radical)}$$

$$=(5+3)\sqrt{2x} \qquad \text{(Using the distributive axiom)}$$

$$=8\sqrt{2x}$$

EXAMPLE 2   Simplify $\sqrt{48} + \sqrt{18} - \sqrt{12}$.

Solution

$$\sqrt{48} + \sqrt{18} - \sqrt{12} = 4\sqrt{3} + 3\sqrt{2} - 2\sqrt{3} \qquad \text{(Simplifying)}$$

$$= (4\sqrt{3} - 2\sqrt{3}) + 3\sqrt{2} \qquad \text{(Collecting like radicals)}$$

$$= (4 - 2)\sqrt{3} + 3\sqrt{2} \qquad \text{(Using distributive axiom)}$$

$$= 2\sqrt{3} + 3\sqrt{2} \qquad \text{(Simplified form)}$$

EXAMPLE 3   Simplify $\sqrt{40} - \sqrt{\dfrac{2}{5}}$.

Solution

$$\sqrt{40} - \sqrt{\frac{2}{5}} = \sqrt{4 \cdot 10} - \sqrt{\frac{2}{5} \cdot \frac{5}{5}}$$

$$= 2\sqrt{10} - \frac{\sqrt{10}}{5}$$

$$= \left(2 - \frac{1}{5}\right)\sqrt{10}$$

$$= \frac{9\sqrt{10}}{5}$$

EXAMPLE 4   Simplify $(\sqrt{3x} + 2)(2\sqrt{3x} - 1)$.

Solution

$$(\sqrt{3x} + 2)(2\sqrt{3x} - 1)$$
$$= (\sqrt{3x})(2\sqrt{3x}) + (\sqrt{3x})(-1) + 2(2\sqrt{3x}) + 2(-1)$$
$$= 6x - \sqrt{3x} + 4\sqrt{3x} - 2$$
$$= 6x + 3\sqrt{3x} - 2$$

EXAMPLE 5   Simplify $(6 - \sqrt{2})^2$.

Solution

$$(6 - \sqrt{2})^2 = 6^2 - 2(6)(\sqrt{2}) + (\sqrt{2})^2$$
$$= 36 - 12\sqrt{2} + 2$$
$$= 38 - 12\sqrt{2}$$

EXAMPLE 6   Evaluate $x^2 - 2x + 3$ when $x = 2 + \sqrt{3}$.

Solution   If $x = 2 + \sqrt{3}$, then
$$x^2 = (2 + \sqrt{3})^2$$
$$= 4 + 4\sqrt{3} + (\sqrt{3})^2$$
$$= 4 + 4\sqrt{3} + 3$$
$$= 7 + 4\sqrt{3}$$

Therefore
$$x^2 - 2x + 3 = 7 + 4\sqrt{3} - 2(2 + \sqrt{3}) + 3$$
$$= 7 + 4\sqrt{3} - 4 - 2\sqrt{3} + 3$$
$$= 6 + 2\sqrt{3}$$

EXERCISES 4.6

Write Exercises 1–40 in simplest radical form. Assume all variables to be positive. (Examples 1–5)

1. $6\sqrt{5} + 2\sqrt{5}$

2. $7\sqrt{2} - 4\sqrt{2}$

3. $5\sqrt{6} + \sqrt{6}$

4. $9\sqrt{10} - 8\sqrt{10}$

5. $3\sqrt{5} + 5\sqrt{3}$

6. $5\sqrt{3} - 4\sqrt{3} + 2\sqrt{3}$

7. $6\sqrt{14x} - 4\sqrt{14x}$

8. $2\sqrt{2x} + 3\sqrt{3x}$

9. $\sqrt{27} + \sqrt{3}$

10. $\sqrt{40} - \sqrt{10}$

11. $3\sqrt{24} + 2\sqrt{54}$

12. $\sqrt{32x} + \sqrt{98x}$

13. $\sqrt{52x} - \sqrt{13x}$

14. $\sqrt{63x^2} + \sqrt{28x^2}$

15. $4\sqrt{9x^2} - 3\sqrt{4x^2}$

16. $\sqrt{4y^2} + 4\sqrt{y^2}$

17. $\sqrt{\dfrac{2}{5}} + \sqrt{\dfrac{1}{10}}$

18. $\sqrt{18} - \sqrt{\dfrac{1}{18}}$

19. $\sqrt{20} + \sqrt{12} + \sqrt{45}$

20. $\sqrt{16x} - \sqrt{4y} - \sqrt{9y}$

21. $\sqrt{\dfrac{42}{25}} + \sqrt{2\dfrac{5}{8}}$

22. $\sqrt{3x} - \sqrt{18x} + \sqrt{12x}$

23. $\sqrt{54} + \sqrt{32} - \sqrt{24}$

24. $\sqrt{25x} - \sqrt{36y} - \sqrt{x}$

25. $\sqrt{\dfrac{7}{8}} + \sqrt{\dfrac{5}{6}} - \sqrt{\dfrac{7}{2}}$

26. $\sqrt{\dfrac{2}{x}} + \sqrt{\dfrac{1}{2x}} + \sqrt{2x}$

27. $\dfrac{2}{\sqrt{12}} + 2\sqrt{81} + \dfrac{\sqrt{27}}{\sqrt{3}}$

28. $(\sqrt{5} - \sqrt{3})^2$

29. $(2\sqrt{5} + 1)^2$

30. $(\sqrt{10x} - \sqrt{5x})^2$

31. $\sqrt{2}(\sqrt{8} - \sqrt{18} - \sqrt{12})$

32. $\sqrt{6}(\sqrt{12} + \sqrt{3} - \sqrt{2})$

33. $(\sqrt{x} - 3)(2\sqrt{x} + 4)$

34. $(\sqrt{2x} - \sqrt{3})(\sqrt{2x} + \sqrt{5})$

35. $\sqrt{224} + 4\sqrt{\dfrac{1}{2}} - \dfrac{1}{2}\sqrt{50}$

36. $(\sqrt{3} + \sqrt{2})^2$

37. $(2\sqrt{3} - 3\sqrt{2})^2$

38. $(\sqrt{6x} + \sqrt{3x})^2$

39. $\sqrt{3}(\sqrt{6} + \sqrt{12} + \sqrt{24})$

40. $\sqrt{5}(\sqrt{10} - \sqrt{15} - \sqrt{40})$

Write Exercises 41–44 in simplest radical form. (Example 6)

41. Find the value of $x^2 - 10x + 23$ for $x = 5 + \sqrt{2}$.

42. Find the value of $x^2 + x - 1$ for $x = \sqrt{5} - 1$.

43. Find the value of $x^2 - x + 2$ for $x = \sqrt{2} + \sqrt{3}$.

44. Find the value of $x^2 + 2x - 2$ for $x = \sqrt{3} - 1$.

4.7 FURTHER SIMPLIFICATION OF RADICALS

If a radical has the form $\sqrt[mn]{x^m}$, where m and n are natural numbers, then by changing to exponential notation,

$$\sqrt[mn]{x^m} = (x^m)^{1/mn} = x^{1/n} = \sqrt[n]{x}, \qquad x \geq 0$$

This process is called **lowering the index,** and $\sqrt[n]{x}$ is the simplified form of $\sqrt[mn]{x^m}$.

EXAMPLE 1   Simplify $\sqrt[6]{8x^3}$, $x \geq 0$.

Solution

$$\sqrt[6]{8x^3} = (2x)^{3(1/6)} = (2x)^{1/2} = \sqrt{2x}$$

EXAMPLE 2   Simplify $\sqrt[6]{25y^4}$, $y \geq 0$.

Solution

$$\sqrt[6]{25y^4} = \sqrt[6]{(5y^2)^2} = (5y^2)^{2(1/6)}$$
$$= (5y^2)^{1/3} = \sqrt[3]{5y^2}$$

EXAMPLE 3   Simplify $\sqrt[4]{36}$.

Solution

$$\sqrt[4]{36} = (6^2)^{1/4} = 6^{1/2} = \sqrt{6}$$

A product or quotient of radicals is considered simplified when it is written as an expression containing at most one radical in the numerator and no radical in the denominator. Certain products and quotients of radicals are most easily simplified by rewriting the expression in exponential form and by applying the theorems on exponents.

EXAMPLE 4   Simplify $\sqrt{2} \cdot \sqrt[3]{2}$.

Solution

$$\sqrt{2} \cdot \sqrt[3]{2} = 2^{1/2}2^{1/3} = 2^{1/2+1/3} = 2^{5/6}$$
$$= \sqrt[6]{2^5} = \sqrt[6]{32}$$

EXAMPLE 5   Simplify $\dfrac{\sqrt[3]{16}}{\sqrt[6]{4}}$.

Solution

$$\frac{\sqrt[3]{16}}{\sqrt[6]{4}} = \frac{(2^4)^{1/3}}{(2^2)^{1/6}} = \frac{2^{4/3}}{2^{1/3}} = 2^{4/3-1/3} = 2^1 = 2$$

EXAMPLE 6 Simplify $\dfrac{\sqrt[12]{3}}{\sqrt[4]{27}}$.

Solution

$$\frac{\sqrt[12]{3}}{\sqrt[4]{27}} = \frac{3^{1/12}}{(3^3)^{1/4}} = \frac{3^{1/12}}{3^{3/4}} \cdot \frac{3^{1/4}}{3^{1/4}}$$

$$= \frac{3^{1/12+3/12}}{3}$$

$$= \frac{3^{4/12}}{3} = \frac{\sqrt[3]{3}}{3}$$

EXAMPLE 7 Simplify $\sqrt[3]{7\sqrt{7}}$.

Solution

$$\sqrt[3]{7\sqrt{7}} = (7 \cdot 7^{1/2})^{1/3}$$

$$= (7^{3/2})^{1/3}$$

$$= 7^{1/2} = \sqrt{7}$$

EXERCISES 4.7

Simplify each of the following. Assume all variables to be positive.

1. $\sqrt[6]{125}$

2. $\sqrt[6]{100}$

3. $\sqrt[4]{49}$

4. $\sqrt[4]{121x^2}$

5. $\sqrt[8]{81t^4}$

6. $\sqrt[6]{16t^4}$

7. $\sqrt[4]{400}$

8. $\sqrt[4]{324a^2}$

9. $\sqrt[4]{9x^2}$

10. $\sqrt[6]{27y^3}$

11. $\sqrt[9]{216x^3}$

12. $\sqrt[4]{900y^6}$

13. $\sqrt[4]{256t^2}$

14. $\sqrt[6]{512y^2}$

15. $\sqrt[8]{64r^6}$

16. $\sqrt[3]{\sqrt[4]{27n^3}}$

17. $\sqrt[4]{3}\sqrt[12]{3}$

18. $\sqrt{8x^3}\sqrt[3]{4x^2}$

19. $\sqrt{7}\sqrt[3]{49}$

20. $\sqrt[6]{25}\sqrt[3]{25}$

21. $\sqrt{\sqrt[3]{4c^2}}$

22. $\sqrt[3]{25}\sqrt{5}$

23. $\sqrt{6}\sqrt[4]{6}$

24. $\sqrt[3]{4y^2}\sqrt[6]{4y^2}$

25. $\sqrt{5}\sqrt[6]{5}$

26. $\sqrt[12]{10}\sqrt[4]{10}$

27. $\dfrac{\sqrt[3]{81}}{\sqrt[6]{9}}$

28. $\dfrac{\sqrt{7}}{\sqrt[4]{7}}$

29. $\dfrac{\sqrt[4]{8}}{\sqrt{8}}$

30. $\dfrac{\sqrt[6]{81x^4}}{\sqrt[3]{3x}}$

31. $\dfrac{\sqrt{10}}{\sqrt[3]{5}}$

32. $\sqrt[3]{5}\sqrt{5}$

33. $\dfrac{\sqrt[3]{5}}{\sqrt[6]{5}}$

34. $\dfrac{\sqrt[3]{100x^2}}{\sqrt[6]{10x}}$

35. $\dfrac{\sqrt[12]{6}}{\sqrt[4]{216}}$

36. $\dfrac{\sqrt{15}}{\sqrt[3]{15}}$

37. $\dfrac{\sqrt{10}}{\sqrt[3]{2}\,\sqrt[6]{2}}$

38. $\sqrt[3]{6x}\,\sqrt{6x}$

39. $\sqrt{2y}\,\sqrt[3]{2y}$

40. $\sqrt{2}\,\sqrt[4]{4}\,\sqrt[6]{32}$

41. $\sqrt{2}\,\sqrt[3]{2}$

42. $\sqrt{6}\,\sqrt[3]{6}\,\sqrt[6]{6}$

43. $\sqrt[3]{9}\,\sqrt[4]{3}\,\sqrt[12]{3}$

44. $\sqrt{\sqrt[3]{\sqrt[4]{2^8}}}$

45. $\sqrt{81\sqrt[3]{81\sqrt{81}}}$

46. $6\sqrt{\dfrac{1}{3}} + \sqrt{18} - \sqrt[4]{9}$

47. $\sqrt[6]{x}\,\sqrt[10]{x}\,\sqrt[15]{x}$

48. $\sqrt{\sqrt[3]{\sqrt[4]{5^{12}}}}$

49. $\sqrt[3]{25\sqrt{25\sqrt[3]{25}}}$

50. $\sqrt[6]{x^3} + \sqrt[4]{16x^2} - \sqrt{9x}$

51. $\dfrac{6}{\sqrt{12}} + \sqrt[4]{729} + \sqrt[6]{27}$

52. $\dfrac{\sqrt[3]{4}\,\sqrt[6]{4}}{\sqrt[4]{4}}$

53. $\sqrt[3]{4}(\sqrt[6]{4} - \sqrt[12]{16})$

54. $\sqrt{\sqrt[3]{\sqrt[4]{x^{60}}}}$

55. $\sqrt[6]{8x^3} + \sqrt[4]{400x^2} + \sqrt{50x}$

56. $\dfrac{\sqrt[4]{7}\,\sqrt[12]{7}}{\sqrt{49}}$

57. $\sqrt[6]{10y}(\sqrt{10y} + \sqrt[3]{10y})$

58. $\sqrt[a]{\sqrt[b]{\sqrt[c]{x^{abc}}}}$

4.8 SCIENTIFIC NOTATION

Many scientific measurements require the use of extremely large or extremely small numbers. For example, the speed of light is about 30,000,000,000 cm/sec; the number of molecules in 1 cu cm of gas at 0 degrees Celsius is about 30,000,000,000,000,000,000; the time for an electronic computer to do a certain arithmetical operation is 0.000 000 0024 sec; and the mass of the hydrogen atom is about 0.000 000 000 000 000 000 000 001 672 g.

A very convenient and effective system for expressing such numbers is called **scientific notation.**

4.8 SCIENTIFIC NOTATION

A number is expressed in scientific notation when it is written as a product of a decimal fraction between 1 and 10 and an integral power of 10. In symbols, a number written in scientific notation has the form

$$N \times 10^k$$

where N is a number between 1 and 10 in decimal form and k is an integer.

EXAMPLE 1

Ordinary Notation	Scientific Notation
3.14	3.14×10^0
20.5	2.05×10^1
608	6.08×10^2
5,000,000	5×10^6
0.14	1.4×10^{-1}
0.025	2.5×10^{-2}
0.000 000 167	1.67×10^{-7}

It may be noted that the exponent on 10 for the scientific notation indicates the number of places to move the decimal point to obtain the ordinary notation (to the right if the exponent is positive, to the left if the exponent is negative, and no change if the exponent is 0).

EXAMPLE 2 Express 30,000,000,000 cm/sec, the speed of light, in scientific notation.

Solution 3.00×10^{10}, since $\overset{+10}{\overrightarrow{30,000,000,000}}$ involves moving the decimal point ten places in the positive direction.

EXAMPLE 3 Express 3.1×10^{-5} in., the diameter of an average red blood corpuscle, in ordinary notation.

Solution 0.000 031, since 3.1×10^{-5} means moving the decimal point five places in the negative direction:

$$\overset{-5}{\overleftarrow{00003.1}}$$

Scientific notation may also be used to simplify a problem in arithmetic.

EXAMPLE 4 Calculate using scientific notation. Express the answer in ordinary notation:

$$\frac{(2.5 \times 10^{-3})(4.2 \times 10^5)}{1.4 \times 10^{-1}}$$

Solution First multiply each number in scientific notation by $10^k \times 10^{-k}$ to eliminate the decimal point:

$$\frac{(2.5 \times 10^{-3})(4.2 \times 10^5)}{1.4 \times 10^{-1}}$$

$$= \frac{(25 \times 10^{-4})(42 \times 10^4)}{14 \times 10^{-2}}$$

$$= \frac{(25)(42)}{14} \cdot \frac{(10^{-4})(10^4)}{10^{-2}}$$

$$= (25)(3) \times 10^{-4+4-(-2)}$$

$$= 75 \times 10^2$$

$$= 7500$$

EXAMPLE 5 Evaluate

$$\sqrt{(3.0 \times 10^{-3})^2 - 4(4.8 \times 10^{-4})(0.3 \times 10^{-2})}$$

Solution

$$\sqrt{(3.0 \times 10^{-3})^2 - 4(4.8 \times 10^{-4})(0.3 \times 10^{-2})}$$

$$= \sqrt{(9 \times 10^{-6}) - (5.76)(10^{-6})}$$

$$= \sqrt{(9.00 - 5.76)10^{-6}}$$

$$= \sqrt{3.24 \times 10^{-6}}$$

$$= \sqrt{(324)(10^{-8})}$$

$$= 18 \times 10^{-4}$$

$$= 0.0018$$

EXERCISES 4.8

In Exercises 1–20 write each number in scientific notation. (Examples 1, 2, and 3)

1. 92,900,000 mi; distance from the earth to the sun
2. 6,600,000,000,000,000,000,000 tons; weight of the earth
3. 11,400,000; population of Tokyo
4. 0.000 000 095 cm; wavelength of certain X rays
5. 0.000 000 0024 sec; time for an electronic computer to do an addition
6. 2,210,000,000; heartbeats per normal lifetime
7. 0.00061 atm; a gas pressure
8. 120,000; seating capacity of a football stadium
9. 0.002 205 lb; weight of 1 g
10. 0.000 000 0667; constant of gravitation
11. 2,000,000,000 light-years; probable diameter of the universe

12. 5,870,000,000,000 mi; the distance light travels in a year, called a light-year

13. 300,000,000,000 dollars; national debt

14. 0.000 000 015 cm; radius of an atom

15. 0.000 011 ft per 0 degrees Celsius; expansion of steel pipe

16. 0.000 005 cm; size of a certain virus

17. 0.03 mm/sec; rate of certain plant growth

18. 0.001 5625 sq mi; area of 1 acre

19. 603,000,000,000,000,000,000,000; Avogadro's number

20. 3,500,000,000; approximate world population

In Exercises 21–40 write each number in ordinary notation. (Example 1)

21. 2.3×10^3; pounds of pollution per car per year

22. -4.60×10^2 degrees Fahrenheit; absolute zero

23. 1.80×10^{-5}; ionization constant of acetic acid

24. 8.64×10^5 mi; diameter of the sun

25. 3.03×10^{-8} cm; grating space in calcite crystals

26. 1.745×10^{-2}; number of radians in 1 degree

27. 1.6667×10^{-1} in; width of an em space (printing industry)

28. 4.80×10^{-10} absolute electrostatic units; electronic charge

29. 1.87×10^9 dollars; a congressional appropriation

30. 6.3×10^{18} electrons per second; for 1 amp of current

31. 4.9×10^{10} dollars; federal investment in public water supplies

32. 9.11×10^{-28} g; mass of an electron

33. 8.31×10^7 ergs per degree-mole; molar gas constant

34. 3.4×10^4 cm/sec; velocity of sound

35. 1×10^{-8} cm; equals 1 angstrom, unit used to measure wavelengths

36. 3.937×10^{-1}; number of inches in 1 cm

37. 2.5×10^{-2} sec; shutter speed of a motion picture camera

38. 7.1×10^8 years; half-life of uranium-235

39. 5×10^{-7} cm; thickness of an oil film

40. 1.256×10^8 cu yd; amount of earth in Fort Peck Dam, largest dam in the world

Calculate Exercises 41–50, using scientific notation. Express the answer in ordinary notation. (Examples 4 and 5)

41. $(3.6 \times 10^{-5})(1.1 \times 10^4)$

42. $(4.3 \times 10^{-1})(4.7 \times 10^{-3})$

43. $\sqrt[3]{\dfrac{1.5 \times 10^{-12}}{1.2 \times 10^{-17}}}$

44. $\dfrac{(2.4 \times 10^{-5})(1.5 \times 10^4)}{1.8 \times 10^3}$

45. $\dfrac{(1.4 \times 10^{-3})^2}{2.8 \times 10^{-4}}$

46. $(5.4 \times 10^{-3})(2.0 \times 10^5)$

47. $\sqrt{(1.25 \times 10^{-2})(8.0 \times 10^{-3})}$

48. $\dfrac{6.9 \times 10^{-8}}{2.3 \times 10^{-6}}$

49. $\dfrac{(3.75 \times 10^{-6})(2.00 \times 10^9)}{2.5 \times 10^{-2}}$

50. $\dfrac{(1.2 \times 10^{-2})^3}{3.6 \times 10^{-5}}$

51. Find the number of radians in 60 degrees if 1 degree = 1.745×10^{-2} radians.

52. Find the number of em spaces across the width of a page 7 in. wide if 1 em space = 1.6667×10^{-1} in.

53. How long does it take a spaceship traveling 2.8×10^4 mph to travel the 2.48×10^5 mi from the earth to the moon?

54. (Nuclear physics) Find the force F with which a helium nucleus and a neon nucleus repel each other when separated by a distance of 4×10^{-9} m using

$$F = \frac{khn}{d^2}$$

where
$$k = 9 \times 10^9$$
$$h = 3.2 \times 10^{-19}$$
$$n = 1.6 \times 10^{-18}$$
$$d = 4 \times 10^{-9}$$

DIAGNOSTIC TEST

Determine if each statement in Problems 1–20 is true or false. If the statement is false, correct it. (You may assume that all variables stand for positive numbers.)

1. $3x^{-2} = \dfrac{1}{3x^2}$

2. $(x^{-1} + 3)^2 = x^{-2} + 9$

3. $\dfrac{x - y^{-1}}{x^{-1} - y} = -\dfrac{x}{y}$

4. $\sqrt{64 + 36} = \sqrt{64} + \sqrt{36} = 8 + 6 = 14$

5. $\left(\dfrac{1}{8}\right)^{-2/3} = 4$

6. $(3^{0.5})(3^{-1.5}) = -3$

7. $(x^{1/2} + 3^{1/2})(x^{1/2} - 3^{1/2}) = x - 3$

8. $(2x + \sqrt{5})^2 = 4x^2 + 5$

9. $\sqrt{128x^3}\sqrt{2xy^3} = 16x^2y\sqrt{y}$

10. $\dfrac{2}{x - \sqrt{3}} = \dfrac{2(x + \sqrt{3})}{x - 3}$

11. $\dfrac{5 + \sqrt{3}}{2} = \dfrac{2}{2(5 - \sqrt{3})} = \dfrac{1}{5 - \sqrt{3}}$

12. If $x = -2 + \sqrt{6}$, then $x^2 + 4x - 2 = 0$.

13. $\sqrt[3]{8}\sqrt{8} = 2\sqrt{2}$

14. In scientific notation, $120{,}000 = 1.2 \times 10^5$.

15. In scientific notation, $0.00034 = 3.4 \times 10^{-4}$.

16. $1.42 \times 10^{-3} = 0.142$

17. $3.415 \times 10^2 = 341.5$

18. If the legs of a right triangle are 2 units and 5 units in length, respectively, then the length of the hypotenuse is $\sqrt{7}$ units.

19. If the length of the hypotenuse of a right triangle is 10 units, and one leg is 4 units long, then the other leg is $2\sqrt{21}$ units.

20. Every real number has at least one real square root.

REVIEW EXERCISES

Assume all variables to be positive.

1. If a and b are the legs of a right triangle and c is the hypotenuse, find the missing side. (Section 4.2)

(a) $a = \sqrt{7}, b = \sqrt{5}$ (c) $a - b, c = 14$

(b) $a = 24, c = 26$

2. Approximate the following to the nearest hundredth. (Section 4.2)

(a) $\sqrt{73}$ (c) $\sqrt[3]{25}$

(b) $\sqrt{5 \times 10^{-4}}$ (d) $\sqrt[3]{4 \times 10^{-5}}$

3. Simplify. (Section 4.2)

(a) $\sqrt[3]{-81}$ (d) $\sqrt[7]{(0.0013)^7}$

(b) $\sqrt{(-7)^2}$ (e) $\sqrt[5]{32x^5y^5}$

(c) $\sqrt[3]{\dfrac{8}{27}}$

4. Write in simplest radical form. (Section 4.2)

(a) $\sqrt{192}$ (d) $\sqrt{60} + \sqrt{1500}$

(b) $2\sqrt{45}$ (e) $\sqrt[4]{32x^5y^3}$

(c) $\sqrt[3]{144}$

5. Rationalize the denominator and write in simplest radical form. (Section 4.5)

(a) $\dfrac{3}{\sqrt{5}}$ (d) $\dfrac{\sqrt{24} - \sqrt{75}}{\sqrt{3}}$

(b) $\dfrac{\sqrt{3}}{\sqrt{5}}$ (e) $\dfrac{\sqrt{3}}{1 - \sqrt{2}}$

(c) $\dfrac{24\sqrt{60}}{8\sqrt{5}}$ (f) $\dfrac{3}{\sqrt{x} + 2}$

6. Simplify. (Section 4.4)

(a) $\sqrt{10}\sqrt{15}$ (d) $(2\sqrt{8} + 8\sqrt{2})^2$

(b) $\sqrt{5}(\sqrt{10} - \sqrt{5} - \sqrt{45})$ (e) $\sqrt{6} - \dfrac{1}{2}\sqrt{54}$

(c) $\sqrt{252} + 14\sqrt{\dfrac{1}{7}} - \dfrac{1}{2}\sqrt{72}$ (f) $\dfrac{2}{\sqrt{120}} + \sqrt{27} + \sqrt{30}$

7. Simplify. (Section 4.4)

(a) $(a^2b^3)(a^3b^5)^2$

(c) $\left(\dfrac{1}{3} a^2b\right)\left(\dfrac{1}{4} a^2b\right)^3$

(b) $\dfrac{a^3b^4}{5m^2n^2} \cdot \dfrac{-36a^5b^3}{10m^6n}$

(d) $-(y)^2(-y)^2$

8. Simplify by writing equivalent statements containing only positive exponents. (Section 4.1)

(a) m^{-4}

(d) $\dfrac{a^{-2} - b^{-2}}{a^{-1} - b^{-1}}$

(b) $\dfrac{1}{p^{-2}}$

(e) $(x^{-1} + y^{-1})^{-2}$

(c) $\dfrac{x^0 y^{-1} z^2}{x^{-2} y^0 z}$

9. Write each of the following in radical form. (Section 4.3)

(a) $x^{1/4}$

(d) $(xy)^{-2/3}$

(b) $2a^{1/2}$

(e) $3x^{0.5}$

(c) $x^{3/4}$

10. Write each of the following in exponential form. (Section 4.3)

(a) $\sqrt[3]{y}$

(c) $x\sqrt{y}$

(b) $\sqrt[6]{x^3y^5}$

(d) $\sqrt[3]{(xy)^2}$

11. Simplify. (Section 4.3)

(a) $2(4)^{-1/2}$

(c) $16^{-3/4}$

(b) $2^3 \cdot 2^{-1} + 2^0$

(d) $5^0 - 36^{-1/2} + 32^{2/5}$

12. Find the value of $x^2 - 8x + 10$ for the following. (Section 4.6)

(a) $x = 4 + \sqrt{6}$

(c) $x = 4 + 2\sqrt{6}$

(b) $x = 4 - \sqrt{6}$

13. Find the value of $\sqrt{b^2 - 4ac}$ for each of the following. (Section 4.6)

(a) $a = 3, b = -5, c = -2$

(c) $a = 4, b = 4, c = -5$

(b) $a = 1, b = -7, c = 1$

14. Find the value of $100(1 - R^{-2/5})$ for the following. (Section 4.8)

(a) $R = \dfrac{3125}{243}$

(b) $R = 32 \times 10^5$

15. Evaluate, using scientific notation; express the answer in ordinary notation. (Section 4.8)

(a) $\sqrt{(2.4 \times 10^9)(1.5 \times 10^{-4})^3}$

(b) $\dfrac{(1.60 \times 10^{-5})(2.50 \times 10^{-3})^2}{1.25 \times 10^{-15}}$

16. Simplify. (Section 4.7)

(a) $\sqrt[4]{64x^6} + 3x\sqrt[6]{8x^3}$

(c) $\dfrac{\sqrt[6]{15xy}\,\sqrt[3]{15xy}}{\sqrt[4]{25y^2}}$

(b) $\sqrt[6]{125} + \sqrt[3]{5}\sqrt{5}$

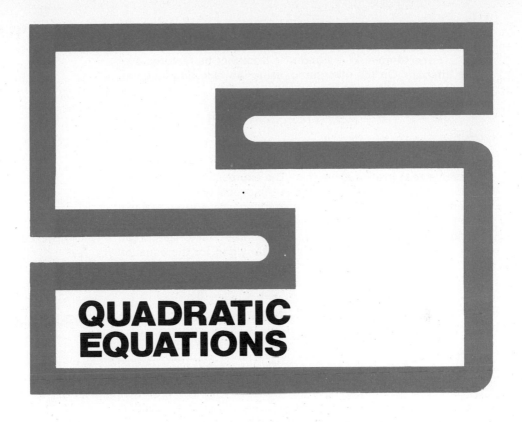

QUADRATIC EQUATIONS

The solution of first-degree equations in one variable has been treated previously. If a polynomial can be expressed as a product of factors of the form $(x + c)$, c constant, then the corresponding polynomial equation can be reduced to the solution of first-degree equations. So far factoring has been restricted to factors with integers for coefficients. Not all polynomials can be factored in this way.

For example, $x^2 - 2$ and $x^2 + 1$ cannot be factored over the integers. On the other hand, using the factoring theorem,
$$x^2 - a^2 = (x - a)(x + a)$$
one could write
$$x^2 - 2 = x^2 - (\sqrt{2})^2 = (x - \sqrt{2})(x + \sqrt{2})$$
and
$$x^2 + 1 = x^2 - (-1) = x^2 - (\sqrt{-1})^2 = (x - \sqrt{-1})(x + \sqrt{-1})$$
The radical $\sqrt{2}$ has been defined, and it designates a **positive real irrational number.** The expression $\sqrt{-1}$ has not been defined so far. If $\sqrt{-1}$ has the property $(\sqrt{-1})^2 = -1$, then $\sqrt{-1}$ cannot be a real number since the square of a real number is either positive or zero. By defining $\sqrt{-1}$ to be a new kind of number, called an *imaginary number,* factorization can be extended, and solutions can be provided for the equation $x^2 + 1 = 0$:
$$x^2 + 1 = (x - \sqrt{-1})(x + \sqrt{-1}) = 0$$
$$x - \sqrt{-1} = 0 \quad \text{or} \quad x + \sqrt{-1} = 0$$
$$x = \sqrt{-1} \quad \text{or} \quad x = -\sqrt{-1}$$

DEFINITION—Imaginary Numbers

Numbers having the form $a + b\sqrt{-1}$, where a and b are real numbers with $b \neq 0$, are called **imaginary numbers.**

These numbers are needed if every polynomial equation is to have a solution, and they will be discussed in this chapter.

It is now possible to provide a more general solution method for quadratic polynomial equations having the form
$$ax^2 + bx + c = 0$$
where $a \neq 0$. This equation is called a quadratic equation. Its solution and some of its many applications are the major theme of this chapter.

HISTORICAL NOTE

The Italian mathematician Girolamo Cardano (1501–1576) was the first to use the square root of a negative number in a computation, in his *Ars Magna* of 1545. In 1572 the Italian mathematician Raffael Bombelli (ca. 1530–after 1572) introduced a consistent theory of imaginary numbers in his *Algebra*. The French mathematician René Descartes (1596–1650) classified numbers as "real" and "imaginary" and discussed complex numbers as the solutions of equations in his *La Géométrie* of 1637. The letter i to designate $\sqrt{-1}$ was introduced in 1748 by the Swiss mathematician Leonard Euler (1707–1783). The name "complex number" was introduced in 1832 by the great German mathematician Karl Friedrich Gauss (1777–1855).

5.1 COMPLEX NUMBERS: DEFINITION, SUMS, AND DIFFERENCES

SQUARE ROOTS OF NEGATIVE NUMBERS

Up to this point a meaning has not been assigned to an *even* root of a *negative real number,* for example, $\sqrt{-1}$, $\sqrt{-9}$, and $\sqrt[4]{-16}$. In the discussions of the properties of the set of real numbers, it was observed that some real numbers are not rational but irrational, such as $\sqrt{2}$, $\sqrt{3}$, $\sqrt[3]{5}$, $\sqrt{5}$, and $\sqrt[3]{-5}$. The set of rational numbers was established as being closed under the operations of addition, subtraction, multiplication, division (by a nonzero rational), and raising to a power. However, it is not closed with respect to the root-extraction operation. For example, $\sqrt{2}$ is not a rational number. To have a solution for $x^2 = 2$, it is necessary to include the irrational numbers $\sqrt{2}$ and $-\sqrt{2}$ in the number system since $(\sqrt{2})^2 = \sqrt{2} \cdot \sqrt{2} = 2$ and $(-\sqrt{2})^2 = (-\sqrt{2})(-\sqrt{2}) = 2$.

Thus the set of rationals was extended to the set of real numbers for two basic reasons:

1. To help close the number system with respect to the root-extraction operation.
2. To establish a one-to-one correspondence between the real numbers and the points of the number line (the completeness axiom).

Although the real number system has the property that exactly one real number can be assigned to each point on the number line, still the set of real numbers is not closed with respect to the operation of root extraction. **There is no real number whose square is a negative real number.**

In particular, there is no real number x such that $x^2 = -1$. To obtain closure, a number is *invented* with the property that its square is -1. This number is assigned the symbolic name $\sqrt{-1}$. Thus $(\sqrt{-1})^2 = \sqrt{-1}\sqrt{-1} = -1$. It is convenient to designate this new number by the letter i.

DEFINITION—The Imaginary Unit, i

$$i = \sqrt{-1} \quad \text{and} \quad i^2 = -1$$

191

Now it is still necessary to include numbers whose squares are the other negative real numbers. In other words, a meaning must be established for expressions such as $\sqrt{-4}$, $\sqrt{-5}$, and $\sqrt{-\frac{4}{9}}$. Therefore the following definition is stated:

DEFINITION

$\sqrt{-a} = i\sqrt{a}$ if a is a positive real number.

The number named by the symbol $\sqrt{-a}$ is called the **principal square root** of $-a$. The definition states that the principal square root of a negative real number can be expressed as the product of the new number i and a positive real number.

For example,

$$\sqrt{-4} = i\sqrt{4} = i \cdot 2 = 2i$$
$$\sqrt{-5} = i\sqrt{5} = \sqrt{5}i$$

Numbers of the form bi, where b is a real number, are called **pure imaginary numbers.** Numbers of the form $a + bi$, where a and b are real numbers and $b \neq 0$, are called **imaginary numbers.**

For example, $5i$, $-\frac{2}{3}i$, $\frac{1}{2} + 6i$, and $-7 - i\sqrt{2}$ are imaginary numbers, but of these, only $5i$ and $-\frac{2}{3}i$ are pure imaginary numbers.

Note that $i\sqrt{5} = \sqrt{5}i$ and that $i\sqrt{2} = \sqrt{2}i$. When the multiplier of i is a radical, it is conventional to write $i\sqrt{5}$ rather than $\sqrt{5}i$ to avoid the accidental error of writing $\sqrt{5i}$, since $\sqrt{5i} \neq \sqrt{5}i$.

We are now ready to define a set of numbers that includes all the real numbers and all the imaginary numbers. This set is called the set of **complex numbers.**

DEFINITION—Complex Numbers

The set of numbers of the form $a + bi$, where a and b are real numbers, and i is the imaginary unit, is called the set of **complex numbers.**

A complex number, $a + bi$, is a **real number** if $b = 0$. For example, $3 + 0i = 3$, and 3 is a real number. If $b \neq 0$, the number is imaginary, as shown above.

Every complex number can be expressed in the **standard form** $a + bi$, where a and b are real numbers and i is the imaginary unit such that $i^2 = -1$.

The real number a is called the **real part** of $a + bi$.

The real number b is called the **imaginary part** of $a + bi$.

5.1 COMPLEX NUMBERS: DEFINITION, SUMS, AND DIFFERENCES

Two complex numbers are said to be equal if and only if their real parts are equal and their imaginary parts are equal.

DEFINITION—Equality of Two Complex Numbers
$$a + bi = c + di \quad \text{if and only if} \quad a = c \quad \text{and} \quad b = d$$

EXAMPLE 1 Determine the real numbers x and y for which $x + yi = 3 - 5i$ is true.

Solution Since

$a + bi = c + di$ if and only if $a = c$ and $b = d$
$x + yi = 3 - 5i$ if and only if $x = 3$ and $y = -5$

EXAMPLE 2 Determine the real numbers x and y for which $(x - 2y) + 2xi = 8i$.

Solution

$a + bi = c + di$ if and only if $a = c$ and $b = d$
$(x - 2y) + 2xi = 0 + 8i$ if and only if
$x - 2y = 0 \quad$ and $\quad 2x = 8$
$x - 2y \quad$ and $\quad x = 4$

If $x = 4$, then $y = 2$.

EXAMPLE 3 Express each of the following in terms of the imaginary unit i:
(a) $\sqrt{-36}$
(b) $\sqrt{-5}$
(c) $\sqrt{-4x^2}$ ($x > 0$)

Solution
(a) $\sqrt{-36} = \sqrt{+36}\sqrt{-1} = 6i$
(b) $\sqrt{-5} - \sqrt{5}\sqrt{-1} = \sqrt{5}i = i\sqrt{5}$, preferred form
(c) $\sqrt{-4x^2} = \sqrt{4x^2}\sqrt{-1} = 2xi$

EXAMPLE 4 Write each of the following in standard form, $a + bi$:
(a) $3 + \sqrt{-4}$
(b) $5 - \sqrt{-9}$
(c) $\sqrt{-12}$
(d) 6

Solution
(a) $3 + \sqrt{-4} = 3 + 2i$
(b) $5 - \sqrt{-9} = 5 - 3i$
(c) $\sqrt{-12} = \sqrt{4}\sqrt{3}\sqrt{-1} = 2\sqrt{3}i = 0 + 2i\sqrt{3}$, preferred form
(d) $6 = 6 + 0i$

SUMS AND DIFFERENCES OF COMPLEX NUMBERS

The sum and difference of two complex numbers are defined as follows:

DEFINITION OF SUM

$$(a + bi) + (c + di) = (a + c) + (b + d)i$$

DEFINITION OF DIFFERENCE

$$(a + bi) - (c + di) = (a - c) + (b - d)i$$

The above definitions state that two complex numbers are added or subtracted in the usual way where the imaginary unit i is treated as a literal constant. The addition properties for the set of complex numbers are similar to those for the set of real numbers, as can be seen in the following summary.

1. Closure
   $(a + bi) + (c + di)$ is a complex number.
   $(a + bi) - (c + di)$ is a complex number.
2. Commutativity
   $(a + bi) + (c + di) = (c + di) + (a + bi)$
3. Associativity
   $[(a + bi) + (c + di)] + (e + fi)$
   $$= (a + bi) + [(c + di) + (e + fi)]$$
4. Identity. There is exactly one complex number, $0 = 0 + 0 \cdot i$, so that
   $(a + bi) + 0 = a + bi$ for any $a + bi$.
5. Inverse. For each $a + bi$, there is exactly one complex number, $-(a + bi)$, called the additive inverse of $a + bi$, so that
   $(a + bi) + [-(a + bi)] = 0$
Moreover,
   $-(a + bi) = -a + (-bi)$

EXAMPLE 5 Add $(3 - 4i) + (5 + 2i)$.

Solution
$$(3 - 4i) + (5 + 2i) = (3 + 5) + (-4 + 2)i = 8 - 2i$$

EXAMPLE 6 Subtract $(-2 - i) - (7 + 6i)$.

Solution
$$(-2 - i) - (7 + 6i) = (-2 -7) + (-1 - 6)i = -9 - 7i$$

EXAMPLE 7 Simplify $(5 + 2\sqrt{-9}) - (4 - \sqrt{-25})$.

Solution First, express each complex number in the standard form, $a + bi$:
$$(5 + 2\sqrt{-9}) - (4 - \sqrt{-25})$$
$$= (5 + 2 \cdot 3i) - (4 - 5i)$$
$$= (5 + 6i) - (4 - 5i)$$
$$= (5 + 6i) + (-4 + 5i)$$
$$= (5 - 4) + (6 + 5)i$$
$$= 1 + 11i$$

EXERCISES 5.1

Express Exercises 1–20 in terms of the imaginary unit i. Assume x and y to be positive real numbers. (Example 3)

1. $\sqrt{-4}$

2. $\sqrt{-81}$

3. $\sqrt{-x^2}$

4. $\sqrt{-\dfrac{1}{4}}$

5. $\sqrt{-2}$

6. $\sqrt{-8}$

7. $\sqrt{-16}$

8. $\sqrt{-25y^2}$

9. $\sqrt{-\dfrac{36}{49}}$

10. $5\sqrt{-9}$

11. $-2\sqrt{-100}$

12. $\sqrt{-98}$

13. $4\sqrt{-27}$

14. $-4\sqrt{-27}$

15. $5\sqrt{-64}$

16. $-5\sqrt{-64}$

17. $2\sqrt{-18}$

18. $-2\sqrt{-18}$

19. $3\sqrt{-49x^2}$

20. $-5\sqrt{-121y^4}$

Write the numbers in Exercises 21–34 in standard form, a + bi. (Example 4)

21. $1 - \sqrt{-36}$

22. $9 + \sqrt{-50}$

23. $4 - 2\sqrt{-72}$

24. -6

25. $-2\sqrt{-25}$

26. i^2

27. $7 + 2\sqrt{-9}$

28. $8 - \sqrt{-12}$

29. 5

30. 0

31. $\sqrt{-4}$

32. $\sqrt{-49} + \sqrt{-36}$

33. $\sqrt{-16} + \sqrt{-4}$

34. $\sqrt{-25} - \sqrt{-9}$

Determine the real numbers x and y for which each of the equations in Exercises 35–40 is true. (Examples 1 and 2)

35. $x + yi = 4 + 2i$

36. $3x + yi = 12 - 5i$

37. $2x + 3yi = 6i$

38. $5x - 2yi = 10$

39. $(x - y) + 2xi = 4 + 6i$

40. $2x + (x + y)i = 10$

In Exercises 41–56 perform the indicated operations and express the results in standard form. (Examples 5, 6, and 7)

41. $(2 + 3i) + (5 - i)$

42. $(3 + 4i) + (2 + 3i)$

43. $(4 - 5i) - (6 - 2i)$

44. $(5 - 2i) - (4 + 3i)$

45. $(6 - 4i) + (-2 + 2i)$

46. $(-6 - i) + (3 - 4i)$

47. $(2 + 5i) - (4 - i)$

48. $(3 - 2i) - (7 + 3i)$

49. $(5 + \sqrt{-9}) + (2 - \sqrt{-25})$

50. $(5 - \sqrt{-36}) - (8 + 2\sqrt{-49})$

51. $(8 - \sqrt{-8}) - (9 - \sqrt{-18})$

52. $(\sqrt{45} + \sqrt{-24}) + (\sqrt{20} - \sqrt{-54})$

53. $(4 - 2i) - (5 + 6i) - (3 - 8i)$

54. $(4 - \sqrt{-9}) + (1 + \sqrt{-16}) + (3 - \sqrt{-1})$

55. $\sqrt{-25} + \sqrt{-36} - \sqrt{-49}$

56. $(6 - \sqrt{-9}) - (9 - 2\sqrt{-16}) + (4 + 3\sqrt{-25})$

57. Under what conditions will $a + bi = a - bi$?

5.2 COMPLEX NUMBERS: PRODUCTS AND QUOTIENTS

The product of two complex numbers is obtained by treating i as a literal constant, multiplying the two numbers as if they were binomials, and finally replacing i^2 by -1 as shown below.

$$(a + bi)(c + di) = (a + bi)c + (a + bi)di$$
$$= (ac + bci) + (adi + bdi^2)$$
$$= ac + (bc + ad)i + bd(-1)$$
$$= (ac - bd) + (bc + ad)i$$

This leads to the formal definition of the product of two complex numbers.

DEFINITION—Product of Two Complex Numbers

$$(a + bi)(c + di) = (ac - bd) + (bc + ad)i$$

With this definition, the following properties can now be established for the multiplication of two complex numbers:

1. Closure

   $(a + bi)(c + di)$ is a complex number.

2. Commutativity

   $(a + bi)(c + di) = (c + di)(a + bi)$

3. Associativity

   $[(a + bi)(c + di)](e + fi) = (a + bi)[(c + di)(e + fi)]$

4. Identity. For any complex number $a + bi$ there is exactly one number, $1 = 1 + 0i$, so that $(a + bi) \cdot 1 = a + bi$.

5. Inverse. For each $a + bi \neq 0$ there is exactly one complex number, $1/(a + bi)$, called the *reciprocal* of $a + bi$, so that $(a + bi)[1/(a + bi)] = 1$.

6. Distributive Property

   $(a + bi)[(c + di) + (e + fi)] = (a + bi)(c + di) + (a + bi)(e + fi)$

Although the definition is necessary to show how these properties are logically derived, it is too difficult to remember to use directly in computations. Instead, the product is obtained by multiplying in the usual way, as for real numbers, treating i as a literal constant *but* replacing i^2 by -1.

EXAMPLE 1   Multiply $(3 + 2i)(2 - 5i)$.

Solution

$$(3 + 2i)(2 - 5i) = 6 - 15i + 4i - 10i^2$$
$$= 6 - 11i - 10i^2$$
$$= 6 - 11i - 10(-1)$$
$$= 16 - 11i$$

Note that direct application of the definition gets the same result:

$$(3 + 2i)(2 - 5i) = [(3)(2) - (2)(-5)] + [(2)(2) + (3)(-5)]i$$
$$= [6 + 10] + [4 - 15]i$$
$$= 16 - 11i$$

When a problem is stated with the square root of a negative number, always change it to standard form, $a + bi$, before performing any operations or calculations.

EXAMPLE 2   Simplify $(4 + \sqrt{-9})(3 - \sqrt{-25})$.

Solution   Before performing the multiplication, rewrite each number in terms of the imaginary unit i.

$$(4 + \sqrt{-9})(3 - \sqrt{-25})$$
$$= (4 + \sqrt{9}\sqrt{-1})(3 - \sqrt{25}\sqrt{-1})$$
$$= (4 + 3i)(3 - 5i)$$
$$= 12 - 20i + 9i - 15i^2$$
$$= 12 - 11i - 15(-1)$$
$$= 27 - 11i$$

EXAMPLE 3   Multiply $\sqrt{-16}\sqrt{-25}$.

Solution

$$\sqrt{-16}\sqrt{-25} = (4i)(5i) = 20i^2 = -20$$

Note that $\sqrt{-16}\sqrt{-25} \neq \sqrt{(-16)(-25)}$ since $\sqrt{(-16)(-25)} = \sqrt{400} = 20$. The theorem $\sqrt{a}\sqrt{b} = \sqrt{ab}$ is not valid when a and b are negative numbers.

Powers of i, in order to be consistent with the theorems on exponents as defined for the real numbers, follow the pattern shown below:

$i^1 = i$
$i^2 = -1$ (by definition)
$i^3 = i^2 \cdot i = -i$
$i^4 = i^2 \cdot i^2 = (-1)(-1) = 1$

Thus $i^1 = i$, $i^2 = -1$, $i^3 = -i$, and $i^4 = 1$. This pattern repeats itself, so that $i^5 = i^4 \cdot i = i$, $i^6 = i^4 \cdot i^2 = -1$, and so forth.

EXAMPLE 4   Simplify i^7, i^{21}, i^{40}, and i^{14}.

Solution

$$i^7 = i^4 \cdot i^3 = (1)(i^3) = -i$$
$$i^{21} = (i^4)^5 \cdot i^1 = (1)(i^1) = i$$
$$i^{40} = (i^4)^{10} = 1$$
$$i^{14} = (i^4)^3 \cdot i^2 = (1)(-1) = -1$$

The quotient of two complex numbers is defined so that division retains its meaning as the inverse operation of multiplication; thus

$$\frac{a + bi}{c + di} = (a + bi)\left(\frac{1}{c + di}\right)$$

Therefore it is necessary to investigate the reciprocal of a complex number. To do this, the concept of the conjugate of a complex number is introduced.

DEFINITION

The **conjugate** of $a + bi$ is $a - bi$.
The **conjugate** of $a - bi$ is $a + bi$.

THE PRODUCT OF CONJUGATES THEOREM

$$(a + bi)(a - bi) = a^2 + b^2, \quad \text{a real number}$$

For example, $(3 + 4i)(3 - 4i) = 9 + 16 = 25$.
Similarly, $(-2 + 5i)(-2 - 5i) = 4 + 25 = 29$.

To change a number of the form $\dfrac{1}{c + di}$ to standard form, $a + bi$, it is necessary to eliminate the i term from the denominator. Since $(c + di)(c - di)$ (the product of a complex number and its conjugate) is a real number, we multiply the numerator and the denominator of $\dfrac{1}{c + di}$ by $c - di$.

Thus

$$\frac{1}{c + di} \cdot \frac{c - di}{c - di} = \frac{c - di}{c^2 + d^2}$$

by applying the product of conjugates theorem. Therefore in standard form,

$$\frac{1}{c + di} = \frac{c}{c^2 + d^2} + \frac{-d}{c^2 + d^2}i$$

EXAMPLE 5   Express $\dfrac{1}{3 + 5i}$ in the $a + bi$ form.

Solution Multiplying the numerator and denominator by $3 - 5i$, the conjugate of $3 + 5i$,

$$\frac{1}{3 + 5i} = \frac{3 - 5i}{(3 + 5i)(3 - 5i)}$$

$$= \frac{3 - 5i}{9 - 25i^2} = \frac{3 - 5i}{9 + 25}$$

$$= \frac{3}{34} + \frac{-5}{34}i$$

EXAMPLE 6 Write $\dfrac{2 + \sqrt{-9}}{3 - \sqrt{-4}}$ in standard form.

Solution First rewriting in terms of i,

$$\frac{2 + \sqrt{-9}}{3 - \sqrt{-4}} = \frac{2 + 3i}{3 - 2i}$$

$$= \frac{(2 + 3i)(3 + 2i)}{(3 - 2i)(3 + 2i)} \qquad \text{(Multiplying numerator and}$$
$$\text{denominator by } 3 + 2i,$$
$$\text{the conjugate of } 3 - 2i)$$

$$= \frac{6 + 4i + 9i + 6i^2}{9 - 4i^2}$$

$$= \frac{6 + 13i + 6(-1)}{9 - 4(-1)}$$

$$= \frac{13i}{13}$$

$$= i, \qquad \text{or } 0 + i \text{ in standard form}$$

EXAMPLE 7 Show that $1 + i\sqrt{6}$ and $1 - i\sqrt{6}$ are solutions of the equation $x^2 - 2x + 7 = 0$.

Solution Substitute each value for x into the equation to see if the equation is true.

$$\text{If} \quad x = 1 + i\sqrt{6}$$
$$x^2 - 2x + 7 = (1 + i\sqrt{6})^2 - 2(1 + i\sqrt{6}) + 7$$
$$= 1 + 2i\sqrt{6} - 6 - 2 - 2i\sqrt{6} + 7$$
$$= (1 - 6 - 2 + 7) + (2i\sqrt{6} - 2i\sqrt{6})$$
$$= 0 + 0i = 0$$
$$\text{If} \quad x = 1 - i\sqrt{6}$$
$$x^2 - 2x + 7 = (1 - i\sqrt{6})^2 - 2(1 - i\sqrt{6}) + 7$$
$$= 1 - 2i\sqrt{6} - 6 - 2 + 2i\sqrt{6} + 7$$
$$= (1 - 6 - 2 + 7) + (-2i\sqrt{6} + 2i\sqrt{6})$$
$$= 0 + 0i = 0$$

Therefore $1 + i\sqrt{6}$ and $1 - i\sqrt{6}$ are solutions of $x^2 - 2x + 7 = 0$.

Using more advanced methods, it can be shown that the set of complex numbers is closed with respect to the operation of root extraction. Thus the set of complex numbers is closed with respect to all six operations of elementary algebra. It can also be shown that every polynomial in one variable with coefficients from the set of real numbers or complex numbers has a solution in the set of complex numbers. This statement is called the Fundamental Theorem of Algebra. The German mathematician Karl Friedrich Gauss (1777–1855) gave the first satisfactory proof of this theorem when he was only twenty-one years old.

EXERCISES 5.2

In Exercises 1–32, simplify. (Examples 1, 2, and 3)

1. $\sqrt{-5}\sqrt{-20}$
2. $\sqrt{-8}\sqrt{-9}$
3. $\sqrt{-12}\sqrt{3}$
4. $\sqrt{2}\sqrt{-3}$
5. $\sqrt{-2}\sqrt{-3}\sqrt{-6}$
6. $2i(4 + 3i)$
7. $3i(5 - 2i)$
8. $\sqrt{-2}(\sqrt{18} - \sqrt{-18})$
9. $(3 + \sqrt{-4})(2 - \sqrt{-9})$
10. $(4 + 5i)(3 - 2i)$
11. $(4 + 2i)(4 - 2i)$
12. $(\sqrt{-2} + \sqrt{-3})(\sqrt{-2} - \sqrt{-3})$
13. $(3 - i)(1 + 2i)$
14. $5i(-3i)(1 - 6i)$
15. $(3 + 2i)^2$
16. $2i(3 + 4i)(5 - 6i)$

17. $(2 + i)[(3 + 2i) + (4 - 5i)]$
18. $4(3 + 2i) - 5i(3 + 2i)$
19. $i(i - 1)(i - 2)$
20. $(1 + 2i)(1 - 2i)(1 - 3i)$
21. $(1 - 3i)(1 - 2i)(1 + 2i)$
22. $(5 - 4i)^2$
23. $(2 + i)^2 - 4(2 + i) + 5$
24. $(2 - i)^2 - 4(2 - i) + 5$
25. $(5 + \sqrt{-36})(5 - \sqrt{-36})$
26. $(3 - 2i)[(2 - 4i) + (5 + 2i)]$
27. $\sqrt{-5}(\sqrt{80} + \sqrt{-80})$
28. $6(5 - 4i) - 2i(5 - 4i)$
29. $(2i + 7)(2i - 7)$
30. $(\sqrt{-2} + \sqrt{-8})^2$
31. $(i + 1)(4 + 3i)(4 - 3i)$
32. $(4 - 3i)(i + 1)(4 + 3i)$

Use Example 4 to simplify Exercises 33–40.

33. i^5
34. i^6
35. i^7
36. i^8

37. i^{15}
38. i^{22}
39. $i^3 + i^9$
40. $i^3 + i^5 + i^7$

Find the conjugate of each of the complex numbers in Exercises 41–50.

41. $5 + 7i$
42. $6 - 4i$
43. $2 + i$
44. $4 - 3i$
45. i

46. $-i$
47. 3
48. $i - 1$
49. 0
50. -2

Express each of the quotients in Exercises 51–68 in standard form.
(Examples 5 and 6)

51. $\dfrac{1}{3 + 2i}$

52. $\dfrac{1}{3i - 5}$

53. $\dfrac{2}{3 - 2i}$

54. $\dfrac{4}{2i - 3}$

55. $\dfrac{2i}{2i + 3}$

56. $\dfrac{2i + 3}{2i}$

57. $\dfrac{1 + i}{1 - i}$

58. $\dfrac{3 + 4i}{2i}$

59. $\dfrac{2i + 3}{3 + 2i}$

60. $\dfrac{1 + 2i}{1 - 2i}$

61. $\dfrac{\sqrt{2} + i}{\sqrt{3} - 2i}$

62. $\dfrac{2 + 3\sqrt{-3}}{3 + 2\sqrt{-2}}$

63. $\dfrac{3i}{2 + 4i}$

64. $\dfrac{1 - \sqrt{-3}}{1 + \sqrt{-3}}$

65. $\dfrac{\sqrt{-3} + 3\sqrt{-1}}{\sqrt{-3} - 3\sqrt{-1}}$

66. $\dfrac{\sqrt{2} + 3i}{1 + i\sqrt{2}}$

67. $\dfrac{\sqrt{-2} + 2\sqrt{-1}}{\sqrt{-3} - 3\sqrt{-1}}$

68. $\dfrac{a + bi}{a - bi}$ (*a* and *b* are real numbers)

Simplify Exercises 69–70, expressing the results in standard form.

69. $\dfrac{2 + i}{1 + 2i} + \dfrac{2 - i}{1 - 2i}$

70. $\dfrac{2 + 3i}{3 + i} - \dfrac{1 + i}{1 + 2i}$

For Exercises 71–74, use Example 7.

71. Show that $3 + i$ and $3 - i$ are solutions of $x^2 - 6x + 10 = 0$.
72. Show that $2i$ is a solution of $x^2 + ix + 6 = 0$.
73. Show that $2 + i\sqrt{5}$ and $2 - i\sqrt{5}$ are solutions of
$x^2 - 4x + 9 = 0$.
74. Show that $1 + i$ is a solution of $x^2 - 2ix - 2 = 0$.

5.3 QUADRATIC EQUATIONS: SOLUTION BY FACTORING

A **quadratic equation** is an equation that can be expressed in the form

$$ax^2 + bx + c = 0, \qquad a \neq 0$$

Previously, the coefficients a, b, and c of the quadratic polynomial

$$ax^2 + bx + c, \quad a \neq 0$$

were restricted to belonging to the set of integers. The factorization of such polynomials was also restricted to polynomial factors with integral coefficients. These polynomials will now be extended so that a, b, and c can be any complex numbers, and polynomial factors can also have complex coefficients. Now expressions such as $x^2 + 4$, which were not factorable over the set of integers, can be factored over the set of complex numbers:

$$x^2 + 4 = x^2 - (-4) = (x + \sqrt{-4})(x - \sqrt{-4})$$
$$= (x + 2i)(x - 2i) \quad (i^2 = -1)$$

The solution of quadratic equations by factoring involves four basic steps:

1. Find an equivalent quadratic equation whose right side is zero.
2. Factor the quadratic polynomial.
3. Set each factor equal to zero.
4. Solve for the variable.

This method of solution is based on the zero-product theorem which states that if the product of two complex numbers is zero, then at least one of those numbers must equal zero.

THE ZERO-PRODUCT THEOREM

Let r and s be any complex numbers. If $rs = 0$, then $r = 0$ or $s = 0$.

It is important to remember in solving an equation by the factoring method that the right side of the equation must be zero before the zero-product theorem can be applied and each factor on the left side of the equation can be equated to zero.

EXAMPLE 1 Solve for x: $x^2 - x = 6$

Solution Write an equivalent equation whose right side is zero.

$$x^2 - x = 6$$
$$x^2 - x - 6 = 0$$
$$(x - 3)(x + 2) = 0 \qquad \text{(Factor the quadratic polynomial)}$$
$$x - 3 = 0 \quad \text{or} \quad x + 2 = 0 \qquad \text{(Apply the zero-product theorem)}$$
$$x = 3 \quad \text{or} \qquad x = -2 \qquad \text{(Solve the two resulting equations)}$$

Check

When $x = 3$, $x^2 - x = (3)^2 - 3 = 9 - 3 = 6$.

When $x = -2$, $x^2 - x = (-2)^2 - (-2) = 4 + 2 = 6$.

Since both values check, the solution set is $\{3, -2\}$.

EXAMPLE 2 Solve: $(x - 2)(x - 5) = 40$

Solution

$$(x - 2)(x - 5) = 40$$
$$x^2 - 7x + 10 = 40$$
$$x^2 - 7x - 30 = 0$$
$$(x + 3)(x - 10) = 0$$
$$x + 3 = 0 \quad \text{or} \quad x - 10 = 0$$
$$x = -3 \quad \text{or} \quad x = 10$$

The solution set is $\{-3, 10\}$.

Check

For $x = -3$

$$(x - 2)(x - 5) = (-3 - 2)(-3 - 5)$$
$$= (-5)(-8) = 40.$$

For $x = 10$

$$(x - 2)(x - 5) = (10 - 2)(10 - 5)$$
$$= (8)(5) = 40.$$

A special case of the quadratic equation $ax^2 + bx + c = 0$ occurs when $b = 0$. Then

$$ax^2 + c = 0$$

Since $a \neq 0$, we can divide by a:

$$x^2 + \frac{c}{a} = 0$$

$$x^2 - \left(-\frac{c}{a}\right) = 0$$

$$\left(x + \sqrt{-\frac{c}{a}}\right)\left(x - \sqrt{-\frac{c}{a}}\right) = 0$$

$$x = -\sqrt{-\frac{c}{a}} \quad \text{or} \quad x = +\sqrt{-\frac{c}{a}}$$

The term $\sqrt{-\dfrac{c}{a}}$ designates a real number if either c or a (but not both) is a negative number, and an imaginary number if a and c are both positive or both negative numbers.

A useful shortcut to this problem is:

$$ax^2 + c = 0$$
$$ax^2 = -c$$
$$x^2 = -\frac{c}{a}$$
$$x = \pm\sqrt{-\frac{c}{a}} \quad \left(\text{read } x = +\sqrt{-\frac{c}{a}} \quad \text{or} \quad x = -\sqrt{-\frac{c}{a}}\right)$$

EXAMPLE 3 Solve for x: $3x^2 - 2 = 0$

Solution In this problem $b = 0$.

$$3x^2 - 2 = 0$$
$$3x^2 = 2$$
$$x^2 = \frac{2}{3}$$
$$x = \pm \sqrt{\frac{2}{3}} = \pm \frac{\sqrt{6}}{3}$$

The solution set is $\left\{ -\dfrac{\sqrt{6}}{3}, \dfrac{\sqrt{6}}{3} \right\}$.

Check

For $x = +\sqrt{\dfrac{2}{3}}$,

$$3x^2 - 2 = 3 \left(\sqrt{\frac{2}{3}} \right)^2 - 2$$
$$= 3 \left(\frac{2}{3} \right) - 2 = 2 - 2 = 0$$

For $x = -\sqrt{\dfrac{2}{3}}$,

$$3x^2 - 2 = 3 \left(-\sqrt{\frac{2}{3}} \right)^2 - 2$$
$$= 3 \left(\frac{2}{3} \right) - 2 = 2 - 2 = 0$$

It is important to remember that the expression $\sqrt{a^2}$ designates exactly one number, the principal square root of a^2. However, the quadratic equation $x^2 = a^2$ has *two* solutions, and both these solutions must be found—that is,

$$x = +\sqrt{a^2} \quad \text{and} \quad x = -\sqrt{a^2}$$

Another special case occurs when $c = 0$. The equation reads

$$ax^2 + bx = 0$$

Now there is a common factor, x, for the left member of the equation:

$$ax^2 + bx = 0$$
$$x(ax + b) = 0$$
$$x = 0 \quad \text{or} \quad ax + b = 0 \quad \text{(Zero-product theorem)}$$
$$ax = -b$$
$$x = -\frac{b}{a}$$

The solution set is $\left\{ 0, -\dfrac{b}{a} \right\}$.

EXAMPLE 4 Solve by factoring: $3x^2 + 2x = 0$

Solution
$$3x^2 + 2x = 0$$
$$x(3x + 2) = 0$$
$$x = 0 \quad \text{or} \quad 3x + 2 = 0$$
$$3x = -2$$
$$x = -\frac{2}{3}$$

Therefore the solution set is $\left\{0, -\frac{2}{3}\right\}$.

Check

If $x = 0$, $3(0)^2 + 2(0) = 0 + 0 = 0$.

If $x = -\frac{2}{3}$, $3x^2 + 2x = 3\left(-\frac{2}{3}\right)^2 + 2\left(-\frac{2}{3}\right)$

$$= 3\left(\frac{4}{9}\right) - \frac{4}{3} = \frac{4}{3} - \frac{4}{3} = 0$$

An important point to stress is that the solution of these equations depends on the zero-product theorem, and that it is very possible, as shown in Example 4, that a solution for the equation may be $x = 0$. For this reason, the student must be cautioned: **never divide by the variable.** Division by the variable causes possible division by zero (an undefined operation), or the loss of a possible solution.

EXAMPLE 5 Solve $4x^2 = 12x$.

Solution
$$4x^2 = 12x$$
$$4x^2 - 12x = 0$$
$$4x(x - 3) = 0$$
$$4x = 0 \quad \text{or} \quad x - 3 = 0$$

Since $4 \neq 0$, $x = 0$ or $x = 3$.

Check

For $x = 0$, $4x^2 = 4(0)^2 = 0$ and
$12x = 12(0) = 0$.

For $x = 3$, $4(3)^2 = 4(9) = 36$ and
$12x = 12(3) = 36$.

A quadratic equation may have only one number in its solution set. Such a solution is called a **double root,** and it implies that the roots are equal.

In solving a quadratic equation, it is always necessary to account for *two* solutions. By considering a double root as a solution counted twice, it can then be stated that *a quadratic equation, an equation of degree 2, always has two solutions in the set of complex numbers.*

EXAMPLE 6 Solve $x^2 - 6x + 9 = 0$.

Solution

$$x^2 - 6x + 9 = 0$$
$$(x - 3)(x - 3) = 0$$
$$x - 3 = 0 \quad \text{or} \quad x - 3 = 0$$
$$x = 3 \quad \text{or} \quad x = 3$$

The solution set is {3}, and 3 is called a double root.

EXAMPLE 7 Solve: $\dfrac{x + 2}{x - 1} - \dfrac{x - 2}{3x} = \dfrac{3}{x - 1}$

Solution

$$\frac{x + 2}{x - 1} - \frac{x - 2}{3x} - \frac{3}{x - 1} = 0$$

Simplifying the left side of the equation to write it as a single fraction:

$$\frac{2x^2 - 2}{3x(x - 1)} = 0$$

Reducing this fraction,

$$\frac{2(x + 1)(x - 1)}{3x(x - 1)} = \frac{2(x + 1)}{3x} = 0 \quad \text{if } x \neq 1$$

A fraction $\dfrac{A}{B} = 0$ if and only if $A = 0$ and $B \neq 0$. Thus this fraction is equal to zero if and only if its denominator, $3x$, is not equal to zero, but its numerator, $2(x + 1)$, is equal to zero.

If $2(x + 1) = 0$, then $x + 1 = 0$ and $x = -1$.
If $3x \neq 0$, then $x \neq 0$.
Therefore the solution is $x = -1$, and $\{-1\}$ is the solution set. The check is left for the student.

The zero-product theorem can be extended to apply to more than two factors whose product is zero. For example, if $a \cdot b \cdot c \cdot d = 0$, then $a = 0$ or $b = 0$ or $c = 0$ or $d = 0$. This enables us to find solutions for equations of degree greater than 2.

EXAMPLE 8 Solve for x: $x^3 - x^2 = 6x$

Solution

$$x^3 - x^2 = 6x$$
$$x^3 - x^2 - 6x = 0 \qquad \text{(Set right side equal to zero in}$$
$$x(x^2 - x - 6) = 0 \qquad \text{preparation for the zero-product}$$
$$x(x + 2)(x - 3) = 0 \qquad \text{theorem)}$$
$$x = 0 \quad \text{or} \quad x + 2 = 0 \quad \text{or} \quad x - 3 = 0 \qquad \text{(Apply zero-}$$
$$x = 0 \quad \text{or} \qquad x = -2 \quad \text{or} \quad x = 3 \qquad \text{product}$$
$$\text{theorem)}$$

Therefore the solution set is {0, -2, 3}.

EXERCISES 5.3

In Exercises 1–50 find all values of x that are solutions for the given equations.

1. $(x + 5)(x - 1) = 0$
2. $(x - 3)(x + 2) = 0$
3. $5x(x - 2) = 0$
4. $2x(x + 3) = 0$
5. $x^2 + x - 6 = 0$
6. $x^2 + 5x + 6 = 0$
7. $x^2 = 3x$
8. $4x^2 = 25$
9. $x^2 + 9 = 0$
10. $4x^2 = 25x$
11. $x^2 + 9x = 0$
12. $4x^2 + 25 = 0$
13. $x^2 + 3x = -2$
14. $3x^2 + x = 2$
15. $(x + 3)(x - 2) = 0$
16. $(x - 2)(x - 1) = 0$
17. $(x + 3)(x - 2) = 14$
18. $(x - 2)(x - 1) = 6$
19. $(x - 4)(x - 3) = 42$
20. $5x^2 = 20x$
21. $10x = 4x^2$
22. $2x^2 + 1 = 0$
23. $2x^2 - 1 = x$
24. $x^2 + 4x + 4 = 0$
25. $x^2 - 10x + 25 = 0$

26. $(x - 3)^2 = 1$
27. $(x - 2)^2 = 1$
28. $(x - a)^2 = n^2$
29. $(ax + b)^2 = c^2 \qquad (a \neq 0)$
30. $(ax + b)^2 = 0 \qquad (a \neq 0)$
31. $8x^2 - 28x = 60$
32. $x^2 - 2x = 63$
33. $x^3 - 6x^2 + 5x = 0$
34. $x^3 - 5x^2 + 4x = 0$
35. $(x + 3)(x + 4)(x - 1)(x + 1) = 0$
36. $x(x - 5)(2x + 1)(3x + 2) = 0$
37. $x^4 - 13x^2 + 40 = 4$
38. $x^4 - 16 = 0$
39. $\dfrac{4}{3x + 12} + \dfrac{x}{9} = \dfrac{1}{x + 4}$
40. $\dfrac{x - 4}{3} = \dfrac{3}{x + 4}$
41. $\dfrac{3}{x + 2} + \dfrac{x + 8}{x(x + 2)} = 1$
42. $2 - \dfrac{1}{x^2} + \dfrac{1}{x} = 0$
43. $\dfrac{3}{x - 2} - \dfrac{x + 1}{x} = \dfrac{2}{x(x - 2)}$
44. $\dfrac{x}{x + 5} + \dfrac{2}{x - 5} = \dfrac{4x}{x^2 - 25}$
45. $\dfrac{3}{x - 1} - \dfrac{3}{x + 1} = \dfrac{2}{5}$
46. $\dfrac{x}{x + 1} + \dfrac{3}{x + 3} + \dfrac{2}{(x + 1)(x + 3)} = 0$
47. $x + 3 = \dfrac{15}{2x - 1}$
48. $\dfrac{x}{a} - \dfrac{a}{x} = 0 \quad (a > 0)$
49. $x + \dfrac{12}{x} = 7$
50. $1 + \dfrac{a - b}{x} = \dfrac{ab}{x^2}$

5.4 QUADRATIC EQUATIONS: COMPLETING THE SQUARE

Not all quadratic polynomials can be factored easily, and the solution of quadratic equations could be very time-consuming and cumbersome if the factors had to be determined by the trial-and-error method. In order to find the solution more rapidly, other methods are available. One such method is called **completing the square.** The aim of this method is to make the left side of the equation a perfect square trinomial which is equal to a constant—that is, $(x + a)^2 = k$.

The equation $(x + a)^2 = k$ is equivalent to the equation
$$(x + a)^2 - k = 0$$
whose left side can now be factored by the difference of two squares theorem:
$$(x + a)^2 - (\sqrt{k})^2 = 0$$
$$(x + a - \sqrt{k})(x + a + \sqrt{k}) = 0$$
$$x + a - \sqrt{k} = 0 \quad \text{or} \quad x + a + \sqrt{k} = 0$$
$$x = -a + \sqrt{k} \quad \text{or} \quad x = -a - \sqrt{k}$$

In practice, it is convenient to write the solution in the shorter form indicated below:
$$(x + a)^2 = k$$
$$x + a = \pm\sqrt{k}$$
$$x = -a + \sqrt{k} \quad \text{or} \quad x = -a - \sqrt{k}$$

EXAMPLE 1 Solve for x: $(x - 3)^2 = 13$

Solution Applying the short method suggested above,
$$(x - 3)^2 = 13$$
$$x - 3 = \pm\sqrt{13}$$
$$x = 3 + \sqrt{13} \quad \text{or} \quad x = 3 - \sqrt{13}$$
In practical applications it is useful to have an approximation for these roots, so by using either the table inside the cover of this book or a calculator, we find $\sqrt{13}$ approximately equal to 3.61. Therefore $x = 3 + 3.61 = 6.61$, or $x = 3 - 3.61 = -0.61$, correct to two decimal places.

The method of **completing the square** is based on the formula for a perfect square trinomial—that is,
$$x^2 + 2nx + n^2 = (x + n)^2$$

5.4 QUADRATIC EQUATIONS: COMPLETING THE SQUARE

Note that the constant term n^2 is the square of one-half the coefficient of x.

Example 2 gives the steps for completing the square when the quadratic trinomial $ax^2 + bx + c$ is such that a, the coefficient of x^2, equals 1 ($a = 1$).

EXAMPLE 2 Solve $x^2 + 6x + 4 = 0$ by completing the square.

Solution

1. Subtract the constant from both sides of the equation: $x^2 + 6x = -4$
2. Add to each side the square of one-half the coefficient of x: $x^2 + 6x + 9 = -4 + 9$
3. Write the left side as the square of a binomial and simplify the right side: $(x + 3)^2 = 5$
4. Take the square root of each side: $x + 3 = \pm\sqrt{5}$
5. Solve for x: $x = -3 + \sqrt{5}$ or $x = -3 - \sqrt{5}$

Check If $x = -3 + \sqrt{5}$,

$$x^2 + 6x + 4 = 0$$
$$(-3 + \sqrt{5})^2 + 6(-3 + \sqrt{5}) + 4 - 0$$
$$9 - 6\sqrt{5} + 5 - 18 + 6\sqrt{5} + 4 = 0$$
$$0 = 0$$

If $x = -3 - \sqrt{5}$,

$$x^2 + 6x + 4 = 0$$
$$(-3 - \sqrt{5})^2 + 6(-3 - \sqrt{5}) + 4 = 0$$
$$9 + 6\sqrt{5} + 5 - 18 - 6\sqrt{5} + 4 = 0$$
$$0 = 0$$

Therefore the solution set is $\{-3 + \sqrt{5}, -3 - \sqrt{5}\}$.

In the quadratic trinomial $ax^2 + bx + c$, if $a \neq 1$, then it is convenient to divide by a first.

EXAMPLE 3 Solve $3x^2 + 6x + 1 = 0$ by completing the square.

Solution

1. Divide by 3: $3x^2 + 6x + 1 = 0$

$$x^2 + 2x + \frac{1}{3} = 0$$

2. Subtract $\frac{1}{3}$ from each side of the equation: $x^2 + 2x = -\frac{1}{3}$

3. Complete the square and add this term to each side: $x^2 + 2x + 1 = -\frac{1}{3} + 1$

4. Write the left side as the square of a binomial and simplify the right side:

$$(x + 1)^2 = \frac{2}{3}$$

5. Take the square root of each side of the equation:

$$x + 1 = \pm\sqrt{\frac{2}{3}} = \pm\frac{\sqrt{6}}{3}$$

6. Solve for x:

$$x = -1 + \frac{\sqrt{6}}{3} \quad \text{or}$$

$$x = -1 - \frac{\sqrt{6}}{3}$$

Check   For $x = -1 + \dfrac{\sqrt{6}}{3}$,

$$3x^2 + 6x + 1 = 3\left(-1 + \frac{\sqrt{6}}{3}\right)^2 + 6\left(-1 + \frac{\sqrt{6}}{3}\right) + 1$$

$$= 3\left(1 - \frac{2\sqrt{6}}{3} + \frac{6}{9}\right) - 6 + 2\sqrt{6} + 1$$

$$= 3 - 2\sqrt{6} + 2 - 5 + 2\sqrt{6}$$

$$= 5 - 5 - 2\sqrt{6} + 2\sqrt{6}$$

$$= 0$$

For $x = -1 - \dfrac{\sqrt{6}}{3}$,

$$3x^2 + 6x + 1 = 3\left(-1 - \frac{\sqrt{6}}{3}\right)^2 + 6\left(-1 - \frac{\sqrt{6}}{3}\right) + 1$$

$$= 3\left(1 + \frac{2\sqrt{6}}{3} + \frac{6}{9}\right) - 6 - 2\sqrt{6} + 1$$

$$= 3 + 2\sqrt{6} + 2 - 5 - 2\sqrt{6}$$

$$= 0$$

Therefore the solution set is $\left\{-1 - \dfrac{\sqrt{6}}{3}, -1 + \dfrac{\sqrt{6}}{3}\right\}$.

EXAMPLE 4   Solve $x^2 + x + 1 = 0$ by completing the square.

Solution

$$x^2 + x + 1 = 0$$

$$x^2 + x = -1$$

$$x^2 + x + \frac{1}{4} = -1 + \frac{1}{4}$$

$$\left(x + \frac{1}{2}\right)^2 = -\frac{3}{4}$$

$$x + \frac{1}{2} = \pm\sqrt{-\frac{3}{4}} = \pm\frac{i\sqrt{3}}{2}$$

$$x = \frac{-1 \pm i\sqrt{3}}{2}$$

Note that both roots are imaginary.

The solution set is $\left\{\dfrac{-1 + i\sqrt{3}}{2}, \dfrac{-1 - i\sqrt{3}}{2}\right\}$.

Check For $x = \dfrac{-1 + i\sqrt{3}}{2}$,

$$x^2 + x + 1 = \left(\dfrac{-1 + i\sqrt{3}}{2}\right)^2 + \left(\dfrac{-1 + i\sqrt{3}}{2}\right) + 1$$

$$= \dfrac{1 - 2i\sqrt{3} - 3}{4} + \dfrac{-1 + i\sqrt{3}}{2} + 1$$

$$= \dfrac{-2 - 2i\sqrt{3}}{4} + \dfrac{-1 + i\sqrt{3}}{2} + 1$$

$$= \dfrac{-1 - i\sqrt{3}}{2} + \dfrac{-1 + i\sqrt{3}}{2} + 1$$

$$= \dfrac{-2}{2} + 1 = 0$$

The check is similar for $x = \dfrac{-1 - i\sqrt{3}}{2}$.

The method of solution by completing the square may also be used for solving literal equations.

EXAMPLE 5 Solve for y: $y^2 - xy - x^2 = 0$

Solution Since we are solving for y, x is held constant, and by comparing with the form $ay^2 + by + c = 0$, $a = 1$, $b = -x$, and $c = -x^2$. Thus

$$y^2 - xy = x^2$$

$$y^2 - xy + \left(\dfrac{x}{2}\right)^2 = x^2 + \left(\dfrac{x}{2}\right)^2$$

$$\left(y - \dfrac{x}{2}\right)^2 = \dfrac{5x^2}{4}$$

$$y - \dfrac{x}{2} = \pm \dfrac{x\sqrt{5}}{2}$$

$$y = \dfrac{x + x\sqrt{5}}{2} \quad \text{or} \quad y = \dfrac{x - x\sqrt{5}}{2}$$

and the solution set is $\left\{\dfrac{x + x\sqrt{5}}{2}, \dfrac{x - x\sqrt{5}}{2}\right\}$.

EXERCISES 5.4

Use Example 1 to solve the equations in Exercises 1–6.

1. $(x - 5)^2 = 2$

2. $(x + 3)^2 = 4$

3. $(x - 5)^2 = -2$

4. $(x + 3)^2 = -4$

5. $(2x - 3)^2 = 7$

6. $(3x + 2)^2 = -5$

Solve each quadratic equation in Exercises 7–42 by completing the square. (Examples 2, 3, and 4)

7. $x^2 - 4x + 1 = 0$

8. $x^2 - 2x + 3 = 0$

9. $x^2 - 2x - 3 = 0$

10. $y^2 - 8y + 15 = 0$

11. $x^2 - 10x + 20 = 0$

12. $x^2 - 6x - 3 = 0$

13. $x^2 - 6x + 3 = 0$

14. $x^2 + 6x - 3 = 0$

15. $x^2 + 6x + 3 = 0$

16. $y^2 = 2(5y - 4)$

17. $y^2 - 8y + 17 = 0$

18. $x^2 = 2x + 19$

19. $x(x - 1) = 1$

20. $t + 2 = 3t^2$

21. $t^2 + 2 = 2t$

22. $4x - x^2 = 3$

23. $y^2 + 4y = 0$

24. $4x(3 - x) = 7$

25. $y(2y - 1) = 1$

26. $u^2 + 3 = u$

27. $4z = z^2 + 9$

28. $z + 10 = 2z^2$

29. $4u = 15 - 4u^2$

30. $t^2 + 8t + 20 = 0$

31. $2(6 - t^2) = 5t$

32. $3y^2 = 7y + 6$

33. $x(3 - x) = 4$

34. $x + 2 = \dfrac{3}{x}$

35. $y + \dfrac{5}{y} = 2$

36. $\dfrac{10}{x^2} = \dfrac{20}{(10 - x)^2}$

37. $\dfrac{1}{t} + \dfrac{3 - t}{1 - t} = 0$

38. $x^2 - 2\pi x + 1 = 0$

39. $x - 10 + \dfrac{95}{x} = 0$

40. $\dfrac{x}{2} + \dfrac{2}{3} = \dfrac{x^2}{6}$

41. $\dfrac{y + 3}{y - 2} = \dfrac{13}{y}$

42. $\dfrac{t}{2} = \dfrac{1 - t}{2 - t}$

In Exercises 43–50 solve for the indicated letter. (Example 5)

43. $x^2 - 6xy + 5y^2 = 0$, for x

44. $x^2 + xy + y^2 = 0$, for y

45. $x^2 - xy - y^2 = 0$, for x

46. $x^2 - xy - 2y^2 = 0$, for x

47. $4x^2 + 25y^2 = 100$, for y

48. $4x^2 + 25y^2 = 100$, for x

49. $ax^2 + bx + c = 0$, for x

50. $x^2 - 2bx + c = 0$, for x

5.5 THE QUADRATIC FORMULA

In the preceding section it was seen that any quadratic equation of the form
$$ax^2 + bx + c = 0, \qquad a \neq 0$$
can be solved by completing the square. The process is some-

times lengthy, so the method is applied to the general equation and a formula is developed:

$$ax^2 + bx + c = 0$$

1. Divide each side by a: $x^2 + \dfrac{b}{a}x + \dfrac{c}{a} = 0$

2. Subtract $\dfrac{c}{a}$ from each side:

$$x^2 + \frac{b}{a}x = -\frac{c}{a}$$

3. Complete the square: $x^2 + \dfrac{b}{a}x + \left(\dfrac{b}{2a}\right)^2 = -\dfrac{c}{a} + \left(\dfrac{b}{2a}\right)^2$

4. Write the left side as the square of a binomial and simplify the right side:

$$\left(x + \frac{b}{2a}\right)^2 = -\frac{c}{a} + \frac{b^2}{4a^2}$$

$$= \frac{b^2 - 4ac}{4a^2}$$

5. Take the square root of each side:

$$x + \frac{b}{2a} = \pm\frac{\sqrt{b^2 - 4ac}}{2a}$$

6. Solve for x:

$$x = \frac{-b + \sqrt{b^2 - 4ac}}{2a}$$

or

$$x = \frac{-b - \sqrt{b^2 - 4ac}}{2a}$$

If $b^2 - 4ac \geq 0$, then $\sqrt{b^2 - 4ac}$ is a real number, and x designates a real number. If $b^2 - 4ac < 0$, then $\sqrt{b^2 - 4ac}$ represents an imaginary number, and the solutions of the equation are imaginary.

If $b^2 - 4ac = 0$, then the equation has a double root— namely, $-\dfrac{b}{2a}$. The roots are also said to be equal in this case.

Because the expression $b^2 - 4ac$ determines whether the roots are real, imaginary, or equal, $b^2 - 4ac$ is called the **discriminant** of the quadratic equation.

Since a, b, and c were chosen completely arbitrarily except $a \neq 0$, the set of equations

$$\left\{x = \frac{-b + \sqrt{b^2 - 4ac}}{2a}, \quad x = \frac{-b - \sqrt{b^2 - 4ac}}{2a}\right\}$$

can be used as a formula for the solution of a quadratic equation.

THE QUADRATIC FORMULA

The quadratic equation $ax^2 + bx + c = 0$ $(a \neq 0)$ has the solutions

$$x = \frac{-b + \sqrt{b^2 - 4ac}}{2a}, \quad x = \frac{-b - \sqrt{b^2 - 4ac}}{2a}$$

DEFINITION—Discriminant

$b^2 - 4ac$ is the **discriminant** of the quadratic equation
$$ax^2 + bx + c = 0, \qquad a \neq 0$$
If $b^2 - 4ac$ is positive, the solutions are real.
If $b^2 - 4ac$ is negative, the solutions are imaginary.
If $b^2 - 4ac = 0$, the solutions are real and equal—that is, the equation has a double root which is a real number.

EXAMPLE 1   Solve $2x^2 + 3x + 1 = 0$ by using the quadratic formula.

Solution   Comparing $2x^2 + 3x + 1 = 0$ with $ax^2 + bx + c = 0$, it is seen that $a = 2$, $b = 3$, and $c = 1$. Therefore

$$x = \frac{-b \pm \sqrt{b^2 - 4ac}}{2a}$$

yields

$$x = \frac{-3 + \sqrt{9 - 4(2)(1)}}{4} = \frac{-3 + 1}{4}$$

$$= \frac{-2}{4} = -\frac{1}{2}$$

or

$$x = \frac{-3 - \sqrt{9 - 4(2)(1)}}{4} = \frac{-3 - 1}{4}$$

$$= \frac{-4}{4} = -1$$

Check

If $x = -\dfrac{1}{2}$,

$$2x^2 + 3x + 1 = 0$$

$$2\left(-\frac{1}{2}\right)^2 + 3\left(-\frac{1}{2}\right) + 1 = 0$$

$$\frac{1}{2} - \frac{3}{2} + 1 = 0$$

$$0 = 0$$

If $x = -1$,

$$2x^2 + 3x + 1 = 0$$
$$2(-1)^2 + 3(-1) + 1 = 0$$
$$2 - 3 + 1 = 0$$
$$0 = 0$$

Therefore $\left\{-\dfrac{1}{2}, -1\right\}$ is the solution set.

Note in this example that the discriminant $b^2 - 4ac = 9 - 4(2)(1) = 9 - 8 = 1$, a positive number, indicating that the roots are real numbers. Moreover, the discriminant is a perfect square since $1^2 = 1$. *Whenever the discriminant is a perfect square, the roots are rational numbers.*

EXAMPLE 2 Solve $2x^2 - 3x = x^2 - 1$ by using the quadratic formula.

Solution   First an equivalent equation of the form $ax^2 + bx + c = 0$ must be found:

$$2x^2 - 3x = x^2 - 1$$
$$x^2 - 3x + 1 = 0$$
$$a = 1, b = -3, \quad c = 1$$
$$x = \frac{-(-3) \pm \sqrt{(-3)^2 - 4(1)(1)}}{2}$$
$$x = \frac{3 + \sqrt{5}}{2} \quad \text{or} \quad x = \frac{3 - \sqrt{5}}{2}$$

The solution set is $\left\{ \dfrac{3 + \sqrt{5}}{2}, \dfrac{3 - \sqrt{5}}{2} \right\}$.

Check   For $x = \dfrac{3 + \sqrt{5}}{2}$,

$$2x^2 - 3x = 2 \left(\frac{3 + \sqrt{5}}{2} \right)^2 - 3 \left(\frac{3 + \sqrt{5}}{2} \right)$$
$$= \frac{2(14 + 6\sqrt{5})}{4} - \frac{3(3 + \sqrt{5})}{2}$$
$$= \frac{14 + 6\sqrt{5} - 9 - 3\sqrt{5}}{2} = \frac{5 + 3\sqrt{5}}{2}$$
$$x^2 - 1 = \left(\frac{3 + \sqrt{5}}{2} \right)^2 - 1 = \frac{14 + 6\sqrt{5}}{4} - \frac{4}{4}$$
$$= \frac{10 + 6\sqrt{5}}{4}$$
$$= \frac{5 + 3\sqrt{5}}{2}$$

The check for $x = \dfrac{3 - \sqrt{5}}{2}$ is similar.

Note that the discriminant $b^2 - 4ac = (-3)^2 - 4(1)(1) = 9 - 4 = 5$ a positive number, indicating that the roots are real. Also, since 5 is not a perfect square, the roots are irrational.

The solutions $\dfrac{3 + \sqrt{5}}{2}$ and $\dfrac{3 - \sqrt{5}}{2}$ are said to be in simplified exact form. If an approximation to a solution is desired, a table of square roots or a calculator may be used, and the answer can be expressed in decimal form for any specified accuracy.

Thus, expressing $\dfrac{3 + \sqrt{5}}{2}$ correct to the nearest hundredth,

$$\frac{3 + \sqrt{5}}{2} = \frac{3 + 2.236}{2} = \frac{5.236}{2} = 2.618$$

Therefore $\dfrac{3 + \sqrt{5}}{2} = 2.62$ correct to the nearest hundredth.

EXAMPLE 3 Solve for x: $\dfrac{x^2}{4} - \dfrac{x}{2} + 1 = 0$

Solution Although this problem can be worked with $a = \dfrac{1}{4}$, $b = -\dfrac{1}{2}$, and $c = 1$, it is easier to solve by first finding an equivalent equation with integers for coefficients. Multiplying each term by 4, the desired equivalent equation is obtained:

$$x^2 - 2x + 4 = 0$$

then $a = 1, \quad b = -2, \quad c = 4$

$$x = \frac{-(-2) \pm \sqrt{(-2)^2 - 4(1)(4)}}{2(1)}$$

$$= \frac{2 \pm \sqrt{4 - 16}}{2}$$

$$= \frac{2 \pm \sqrt{-12}}{2}$$

$$= \frac{2 \pm 2i\sqrt{3}}{2}$$

$$= 1 \pm i\sqrt{3}$$

The solution set is $\{1 + i\sqrt{3}, \ 1 - i\sqrt{3}\}$.

Check For $x = 1 + i\sqrt{3}$,

$$\frac{x^2}{4} - \frac{x}{2} + 1 = \frac{(1 + i\sqrt{3})^2}{4} - \frac{1 + i\sqrt{3}}{2} + 1$$

$$= \frac{-2 + 2i\sqrt{3}}{4} - \frac{1 + i\sqrt{3}}{2} + 1$$

$$= \frac{-1 + i\sqrt{3} - 1 - i\sqrt{3}}{2} + 1$$

$$= -1 + 1 = 0$$

The check is similar for $x = 1 - i\sqrt{3}$.

Note that the discriminant $b^2 - 4ac = (-2)^2 - 4(1)(4) = 4 - 16 = -12$, a negative number, indicating that the roots are imaginary.

EXAMPLE 4 Solve $9x^2 + 30x + 25 = 0$.

Solution $a = 9, b = 30, \quad c = 25$

$$x = \frac{-30 \pm \sqrt{(30)^2 - 4(9)(25)}}{2(9)}$$

$$= \frac{-30 \pm \sqrt{900 - 900}}{2(9)}$$

$$= \frac{-30}{18}$$

$$= -\frac{5}{3}$$

The solution set is $\left\{-\dfrac{5}{3}\right\}$.

Check For $x = -\dfrac{5}{3}$,

$$9x^2 + 30x + 25 = 9\left(-\frac{5}{3}\right)^2 + 30\left(-\frac{5}{3}\right) + 25$$

$$= 9\left(\frac{25}{9}\right) + 10(-5) + 25$$

$$= 25 - 50 + 25$$

$$= 0$$

In this case, $-\dfrac{5}{3}$ is a double root. Note that the discriminant $b^2 - 4ac = 0$:

$$b^2 - 4ac = (30)^2 - 4(9)(25) = 900 - 900 = 0$$

Note also that $9x^2 + 30x + 25 = (3x + 5)^2$, a perfect square trinomial.

Since the quadratic formula may be used for solving all quadratic equations, we use it for solving literal quadratic equations as in the following example.

EXAMPLE 5 Solve for x: $5x^2 + 6xy - 4y^2 = 0$

Solution Since we are solving for x, we hold y constant. Comparing with $ax^2 + bx + c = 0$, $a = 5$, $b = 6y$, and $c = -4y^2$.

$$x = \frac{-6y \pm \sqrt{36y^2 - 4(5)(-4y^2)}}{2(5)}$$

$$= \frac{-6y \pm \sqrt{116y^2}}{10} = \frac{-6y \pm 2y\sqrt{29}}{10}$$

And in simplified form,

$$x = \frac{-3y \pm y\sqrt{29}}{5}$$

The solution set is $\left\{ \dfrac{-3y + y\sqrt{29}}{5}, \dfrac{-3y - y\sqrt{29}}{5} \right\}$.

EXERCISES 5.5

(a) Write the equations in Exercises 1–10 as equivalent equations in the form $ax^2 + bx + c = 0$ and state the value of a, b, and c in each equation.

(b) Find the value of the discriminant $b^2 - 4ac$ and state whether the roots of the equation are real and unequal, imaginary, or real and equal.

1. $2x^2 = 3x + 1$

2. $4x^2 = 12x - 9$

3. $\dfrac{5}{x} = x$

4. $(x + 2)(x + 3) = 6$

5. $4x - 8 = x^2$

6. $4x^2 - 2 = 2x$

7. $(x + 1)^2 = x + 3$

8. $9 = 6x + x^2$

9. $(5 - x)(4 + x) = 40$

10. $2x + \dfrac{1}{8x} = 1$

Solve each of the equations in Exercises 11–38 by using the quadratic formula. Check each solution. (Examples 1–4)

11. $3x^2 + 5x + 2 = 0$

12. $3x^2 - 5x - 2 = 0$
13. $x^2 - 3x + 1 = 0$
14. $x^2 + 3x - 1 = 0$
15. $x^2 - 3x + 3 = 0$

16. $x^2 + x + 1 = 0$

17. $4x^2 + 7x + 3 = 0$

18. $4x^2 + 7x + 4 = 0$

19. $4x^2 + 7x - 2 = 0$
20. $4x^2 - 4x + 1 = 0$

21. $x^2 - 4x - 4 = 0$

22. $x^2 + 2x + 4 = 0$

23. $y^2 - y + \dfrac{1}{5} = 0$

24. $\dfrac{y^2 + 1}{6} = \dfrac{y}{2}$

25. $y^2 - 2y - \dfrac{3}{4} = 0$

26. $3y^2 = 2(y - 1)$
27. $y + 2 = 3y(y + 1)$
28. $t^2 + 16 = 0$
29. $5t = 3t^2$

30. $4x - 36x^2 = \dfrac{1}{9}$

31. $80t - 16t^2 = 80$

32. $\dfrac{1}{2}d^2 - \dfrac{30}{8}d + \dfrac{36}{8} = 0$

33. $3(y + 1) = 5 - 2y^2$
34. $9 = 6t + t^2$

35. $\dfrac{t - 1}{t} - \dfrac{t - 1}{6} = \dfrac{1}{6}$

36. $25x - 100x^2 = 0$

37. $25u^2 = 40u - 16$

38. $\dfrac{1}{2}d^2 = 6d + 9$

Solve Exercises 39–48 for the variable indicated. (Example 5)

39. $x^2 + 2px + q = 0$, for x
40. $y^2 - ny - n^2 = 0$, for y
41. $3x^2 + 8xy - 3y^2 = 0$, for x
42. $y^2 - 4y + 2x = 6$, for y

43. $s = vt + \dfrac{1}{2}gt^2$, for t

44. $s = (k - s)^2$, for s
45. $cx^2 + 2bx + a = 0$, for x
46. $y^2 - 4ny + 2n^2 = 0$, for y
47. $4x^2 - 6xy + y^2 = 0$, for x

48. $\dfrac{W}{L} = \dfrac{L - W}{W}$, for W

5.6 EQUATIONS IN QUADRATIC FORM

We have now seen that equations of the form $ax^2 + bx + c = 0$ ($a \neq 0$) can be solved by three methods: factoring, completing the square, or the quadratic formula. There are also many other equations that are not quadratic but that may be expressed in **quadratic form** and can then be solved by quadratic methods.

If the equation is a polynomial equation, the degree of the polynomial determines the number of complex roots of the equation. For example, a quadratic, or second-degree, equation has exactly two roots; a cubic, or third-degree, equation has exactly three roots; and a seventh-degree polynomial equation has seven roots. Some of these roots may be multiple roots, as we have seen in previous problems, and the roots may be real or imaginary. Thus we can say that a polynomial equation of degree n has exactly n roots in the set of complex numbers.

EXAMPLE 1 Find the solution set of the equation
$$x^4 - 11x^2 + 28 = 0$$

Solution This equation is of the *fourth* degree and *four* roots must be found. By means of a substitution, it can be expressed in quadratic form. Let $u = x^2$; then $u^2 = x^4$, and the equation may now be written
$$u^2 - 11u + 28 = 0$$
Factoring,
$$(u - 4)(u \quad 7) = 0$$
$$u - 4 = 0 \quad \text{or} \quad u - 7 = 0$$
$$u = 4 \quad \text{or} \quad u = 7$$
If $u = 4$,
$$x^2 = 4 \quad \text{and} \quad x = +2 \quad \text{or} \quad x = -2$$
If $u = 7$,
$$x^2 = 7 \quad \text{and} \quad x = +\sqrt{7} \quad \text{or} \quad x = -\sqrt{7}$$
Therefore the solution set is $\{2, -2, \sqrt{7}, -\sqrt{7}\}$.

It is important to recognize that an equation of the form $ax^p + bx^r + c = 0$ may be written in quadratic form if $p = 2r$. If $u = x^r$, then $u^2 = x^{2r} = x^p$ since $p = 2r$. This will result in the equation $au^2 + bu + c = 0$, which may be solved by the methods of solution of quadratic equations.

EXAMPLE 2 Find the solution set of the equation
$$y^{-4} - 8y^{-2} + 15 = 0$$
Solution This is *not* a polynomial equation, and the number of roots cannot be determined readily. Again, a substitution can be made to obtain an equation in quadratic form. Let $u = y^{-2}$; then $u^2 = y^{-4}$, so that
$$u^2 - 8u + 15 = 0$$
This equation can be solved by factoring:
$$(u - 3)(u - 5) = 0$$
$$u = 3 \quad \text{or} \quad u = 5$$

But $u = y^{-2} = \dfrac{1}{y^2}$; therefore

$$\dfrac{1}{y^2} = 3 \qquad\qquad \text{or} \quad \dfrac{1}{y^2} = 5$$

$$y^2 = \dfrac{1}{3} \qquad\qquad \text{or} \quad y^2 = \dfrac{1}{5}$$

$$y = \pm\sqrt{\dfrac{1}{3}} = \pm\dfrac{\sqrt{3}}{3} \quad \text{or} \quad y = \pm\sqrt{\dfrac{1}{5}} = \pm\dfrac{\sqrt{5}}{5}$$

Thus the solution set is $\left\{ \pm\dfrac{\sqrt{3}}{3},\ \pm\dfrac{\sqrt{5}}{5} \right\}$.

EXAMPLE 3 Solve for x: $x^{2/3} - x^{1/3} = 72$

Solution Since $\dfrac{2}{3} = 2\left(\dfrac{1}{3}\right)$ let $u = x^{1/3}$ and $u^2 = x^{2/3}$.

Then
$$x^{2/3} - x^{1/3} - 72 = 0$$
$$u^2 - u - 72 = 0$$
$$(u - 9)(u + 8) = 0$$
$$u = 9 \qquad \text{or} \qquad u = -8$$
$$x^{1/3} = 9 \qquad \text{or} \qquad x^{1/3} = -8$$
$$(x^{1/3})^3 = 9^3 \qquad \text{or} \quad (x^{1/3})^3 = (-8)^3$$
$$x = 729 \quad \text{or} \qquad x = -512$$
The solution set is $\{729, -512\}$.

EXAMPLE 4 Solve for x: $x^6 = 27 - 26x^3$

Solution
$$x^6 + 26x^3 - 27 = 0$$
Let $u = x^3$; then $u^2 = x^6$ and
$$u^2 + 26u - 27 = 0$$
$$(u + 27)(u - 1) = 0$$
$$u + 27 = 0 \ \text{or} \ u - 1 = 0$$
$$x^3 + 27 = 0 \ \text{or} \ x^3 - 1 = 0$$
$$(x + 3)(x^2 - 3x + 9) = 0 \ \text{or} \ (x - 1)(x^2 + x + 1) = 0$$
$$x + 3 = 0 \ \text{ or } \ x^2 - 3x + 9 = 0 \ \text{ or } \ x - 1 = 0 \ \text{ or } \ x^2 + x + 1 = 0$$
$$x = -3 \ \text{ or } \ x = \dfrac{3 \pm 3i\sqrt{3}}{2} \ \text{ or } \ x = 1 \ \text{ or } \ x = \dfrac{-1 \pm i\sqrt{3}}{2}$$

The solution set is
$$\left\{ -3,\ 1,\ \dfrac{3 + 3i\sqrt{3}}{2},\ \dfrac{3 - 3i\sqrt{3}}{2},\ \dfrac{-1 + i\sqrt{3}}{2},\ \dfrac{-1 - i\sqrt{3}}{2} \right\}$$

EXAMPLE 5 Solve for x: $(x^2 + 5x)^2 + 5(x^2 + 5x) = 6$

Solution Note that the expression $x^2 + 5x$ appears raised to the second power and multiplied by 5. Use the substitution
$$u = x^2 + 5x \quad \text{and} \quad u^2 = (x^2 + 5x)^2$$
to obtain an equation in u which is quadratic.

$$(x^2 + 5x)^2 + 5(x^2 + 5x) - 6 = 0$$
$$u^2 + 5u - 6 = 0$$
$$(u - 1)(u + 6) = 0$$
$$u - 1 = 0 \quad \text{or} \quad u + 6 = 0$$
$$x^2 + 5x - 1 = 0 \quad \text{or} \quad x^2 + 5x + 6 = 0$$

Using the quadratic formula to solve
$$x^2 + 5x - 1 = 0$$
$$x = \frac{-5 \pm \sqrt{29}}{2}$$

Factoring to solve $x^2 + 5x + 6 = 0$
$$(x + 3)(x + 2) = 0$$
$$x + 3 = 0 \quad \text{or} \quad x + 2 = 0$$
$$x = -3 \quad \text{or} \quad x = -2$$

The solution set is
$$\left\{ \frac{-5 + \sqrt{29}}{2}, \frac{-5 - \sqrt{29}}{2}, -3, -2 \right\}$$

EXERCISES 5.6

Find the solution set of each equation in Exercises 1–26 over the set of complex numbers. Check each solution.

1. $x^4 - 14x^2 + 45 = 0$
2. $x^4 + x^2 = 12$
3. $x^4 + x^2 = 6$
4. $x^6 - 7x^3 - 8 = 0$
5. $x^6 + 64 = 16x^3$
6. $y^{-2} - y^{-1} = 20$
7. $10x^{-2} + 3x^{-1} = 1$

8. $4x^{-4} - 17x^{-2} + 4 = 0$

9. $9x^{-4} + 5x^{-2} = 4$

10. $\dfrac{6}{x^2} + \dfrac{1}{x^4} = 7$

11. $y^{1/2} - 7y^{1/4} + 12 = 0$

12. $3z^{1/2} + 2z^{1/4} - 5 = 0$
13. $a^{2/3} + 2a^{1/3} - 8 = 0$
14. $x^{2/3} = 20 + x^{1/3}$
15. $(p - 3)^4 - 13(p - 3)^2 + 36 = 0$
16. $(x + 2)^4 + 24 = 10(x + 2)^2$
17. $(x^2 + 8x)^2 = 5(x^2 + 8x) + 36$
18. $(y^2 - 4y)^2 - 36 = 9(y^2 - 4y)$

19. $\left(\dfrac{x}{x - 1} \right)^2 = \dfrac{9x}{2x - 2} + \dfrac{5}{2}$

20. $16 \left(\dfrac{x}{x + 1} \right)^4 + 9 = 25 \left(\dfrac{x}{x + 1} \right)^2$

21. $2 \left(x + \dfrac{1}{x} \right)^2 = \left(x + \dfrac{1}{x} \right) + 10$

22. $(x^2 - 5x) + 12 = 8(x^2 - 5x)^{1/2}$

23. $(x^2 - 6x + 1)^{2/3} - (x^2 - 6x + 1)^{1/3} = 6$
24. $(x^2 - 6x)^{1/2} - 3(x^2 - 6x)^{1/4} + 2 = 0$
25. $y^4 - 12y^2 + 32 = 0$
26. $28x^{-2/3} - 3x^{-1/3} = 1$

5.7 RADICAL EQUATIONS

An equation containing a variable in a radicand is called a **radical equation.**

Some examples of radical equations are

$$\sqrt{x} = 4$$
$$\sqrt{x + 2} = 3$$
$$5\sqrt{x + 2} - \sqrt{x + 1} = 5$$

Solutions of radical equations can often be found by applying the following theorem.

THEOREM

The solution set of the open equation $A = B$ is a **subset** of the solution set of $A^n = B^n$, where n is a natural number.

If two numbers are equal, then their squares are equal, their cubes are equal, and, in general, their nth powers are equal. However, if the nth powers of two numbers are equal, the numbers may not necessarily be equal. For example, $(-5)^2 = (+5)^2$, but $-5 \neq +5$.

EXAMPLE 1 Compare the solution sets of the equations $x = 3$ and $x^2 = 9$.

Solution   The solution set of $x = 3$ is $\{3\}$. The solution set of $x^2 = 9$ is $\{3, -3\}$. The solution set of $x = 3$ is a subset of the solution set of $x^2 = 9$. In other words, the solution set of $x = 3$ is contained in the solution set of $x^2 = 9$.

EXAMPLE 2 Solve $\sqrt{x^2 - 9} = 4$

Solution

$$\sqrt{x^2 - 9} = 4$$
$$x^2 - 9 = 16 \qquad \text{(Squaring each side)}$$
$$x^2 - 25 = 0$$
$$(x - 5)(x + 5) = 0$$
$$x = 5 \quad \text{or} \quad x = -5$$

The solution set of $x^2 - 9 = 16$ is $\{5, -5\}$. The solution set of $\sqrt{x^2 - 9} = 4$ is a subset of $\{5, -5\}$. Therefore each member of $\{5, -5\}$

must be checked in the original equation to determine if it is a solution or not.

Check

$$\sqrt{x^2 - 9} = 4 \text{ for } x = 5,$$
$$\sqrt{(5)^2 - 9} = \sqrt{25 - 9} = \sqrt{16} = 4$$

Therefore 5 is a solution.

Check

$$\sqrt{x^2 - 9} = 4 \text{ for } x = -5,$$
$$\sqrt{(-5)^2 - 9} = \sqrt{25 - 9} = \sqrt{16} = 4$$

Therefore -5 is a solution.

The solution set of $\sqrt{x^2 - 9} = 4$ is $\{5, -5\}$.

EXAMPLE 3 Solve $\sqrt{x - 1} = 3 - x$.

Solution

$$(\sqrt{x - 1})^2 = (3 - x)^2 \qquad \text{(Squaring both sides)}$$
$$x - 1 = 9 - 6x + x^2$$
$$x^2 - 6x + 9 = x - 1$$
$$x^2 - 7x + 10 = 0$$
$$(x - 2)(x - 5) = 0$$
$$x = 2 \quad \text{or} \quad x = 5$$

Check

$$\sqrt{x - 1} = 3 - x \text{ for } x = 2,$$
$$\sqrt{2 - 1} = 3 - 2$$
$$1 = 1$$

Thus 2 is a solution.

Check

$$\sqrt{x - 1} = 3 - x \text{ for } x = 5,$$
$$\sqrt{5 - 1} = 3 - 5$$
$$2 \neq -2$$

Thus 5 is not a solution. The solution set is $\{2\}$.

When a number is a root (solution) of the transformed equation, $A^n = B^n$, and it is not a root of the original equation, $A = B$, it is called an **extraneous root.** Actually this term is misleading, since it is not a root at all.

Some equations involving radicals may require several applications of raising each side to an nth power. If an equation contains more than one radical, it is usually desirable to first find an equivalent equation containing a single radical on one side of the equation. This process is called **isolating a radical.**

EXAMPLE 4 Solve $\sqrt{x} - \sqrt{x-5} = 1$.

Solution

$$\sqrt{x} - \sqrt{x-5} = 1$$
$$\sqrt{x} - 1 = \sqrt{x-5} \qquad \text{(Separating the radicals)}$$
$$(\sqrt{x} - 1)^2 = (\sqrt{x-5})^2 \qquad \text{(Squaring both sides)}$$
$$x - 2\sqrt{x} + 1 = x - 5$$
$$-2\sqrt{x} + 1 = -5$$
$$-2\sqrt{x} = -6 \qquad \left\{ \text{(Isolating the radical)} \right.$$
$$\sqrt{x} = 3$$
$$(\sqrt{x})^2 = 3^2 \qquad \text{(Squaring both sides)}$$
$$x = 9$$

Check

$$\sqrt{x} - \sqrt{x-5} = 1$$
$$\sqrt{9} - \sqrt{9-5} = 1$$
$$3 - 2 = 1$$
$$1 = 1$$

The solution set is {9}.

EXAMPLE 5 Solve $\sqrt{2x - \sqrt{x-5}} = 4$.

Solution

$$(\sqrt{2x - \sqrt{x-5}})^2 = (4)^2 \qquad \text{(Squaring both sides)}$$
$$2x - \sqrt{x-5} = 16$$
$$2x - 16 = \sqrt{x-5} \qquad \text{(Isolating the radical)}$$
$$4x^2 - 64x + 256 = x - 5 \qquad \text{(Squaring both sides again)}$$
$$4x^2 - 65x + 261 = 0$$
$$(4x - 29)(x - 9) = 0$$
$$x = \frac{29}{4} \quad \text{or} \quad x = 9$$

Check For $x = \frac{29}{4}$,

$$\sqrt{2x - \sqrt{x-5}} = \sqrt{\frac{29}{2} - \sqrt{\frac{29}{4} - \frac{20}{4}}}$$
$$= \sqrt{\frac{29}{2} - \frac{3}{2}}$$
$$= \sqrt{13}$$

Since $\sqrt{13} \neq 4$, $\frac{29}{4}$ is not a solution.

Check For $x = 9$,

$$\sqrt{2x - \sqrt{x-5}} = \sqrt{18 - \sqrt{9-5}}$$
$$= \sqrt{18 - 2} = \sqrt{16} = 4$$

Therefore 9 is a solution. The solution set is {9}.

It should be noted that checking is part of the solution process for a radical equation. In other words, **checking is mandatory.**

In fact, some radical equations may not have any solution, as the next example illustrates.

EXAMPLE 6 Solve $\sqrt{x - 3} - \sqrt{x} = 3$.

Solution

$$\sqrt{x - 3} = \sqrt{x} + 3$$
$$(\sqrt{x - 3})^2 = (\sqrt{x} + 3)^2$$
$$x - 3 = x + 6\sqrt{x} + 9$$
$$-12 = 6\sqrt{x}$$
$$\sqrt{x} = -2$$
$$x = 4$$

Check For $x = 4$,
$$\sqrt{x - 3} - \sqrt{x} = \sqrt{4 - 3} - \sqrt{4}$$
$$= 1 - 2 = -1$$

Since $-1 \neq 3$, 4 is not a solution. The solution set is $\varnothing$, the empty set.

Note that $\sqrt{x} \geq 0$ for x, a nonnegative real number; $\sqrt{x} = -2$ has no real solution.

EXERCISES 5.7

Find the solution set of each equation in Exercises 1–33. Check each equation to verify the solutions.

1. $\sqrt{2x + 3} = 5$
2. $\sqrt{2x} + 3 = 5$
3. $x + 2 + \sqrt{x + 8} = 0$
4. $x + 2 - \sqrt{x + 8} = 0$
5. $\sqrt{x + 8} = 3$
6. $\sqrt{x} + 8 = 3$
7. $\sqrt{x} + \sqrt{2} = \sqrt{x + 2}$
8. $\sqrt{2t + 13} = t + 7$
9. $2y = 9 - \sqrt{8y + 9}$
10. $2y = 9 + \sqrt{8y + 9}$
11. $\sqrt{2x - 3} + \sqrt{x + 2} = 0$
12. $r - \sqrt{r - 5} = 5$
13. $1 + \sqrt{5x^2 - 5x - 1} = 2x$
14. $\sqrt{2x - 3} - \sqrt{x + 2} = 0$
15. $\sqrt{3 - x} + \sqrt{2 + x} = 3$
16. $\sqrt{3 - x} - \sqrt{2 + x} = 3$
17. $2\sqrt{2x + 5} = 1 + \sqrt{8x + 1}$

18. $\sqrt{x - 2} + \sqrt{x} - 2 = 0$
19. $\sqrt{x + 5} = x + 5$
20. $x - 1 + \sqrt{x - 1} = 0$
21. $\sqrt{2y + 45} - \sqrt{54 - y} = 3$
22. $\sqrt{5y + 1} - \sqrt{3y - 5} = 2$
23. $\sqrt{y^2 - \sqrt{2y - 2}} + y = 2$
24. $\sqrt{2x} + \sqrt{x + 4} = 2$
25. $\sqrt{y^2 + 3y + 2} - \sqrt{y + 1} = 0$
26. $\sqrt{2x} + \sqrt{2x + 8} = 2$
27. $x - 1 - \sqrt{x - 1} = 0$
28. $\sqrt{2x - 5} + \sqrt{3x + 1} = 3$
29. $\sqrt{7t + 4} - \sqrt{3t + 40} = 2$
30. $\sqrt{y + 6} - \sqrt{y + 2} = \sqrt{8y - 1}$
31. $\sqrt{x} - \sqrt{x + 3} = 3$
32. $\sqrt{8x + 25} - \sqrt{2x + 5} = \sqrt{2x + 8}$
33. $\sqrt{x} - \sqrt{x + 3} = \sqrt{2x + 7}$

5.8 POLYNOMIAL EQUATIONS: SOLUTION BY FACTORING

The techniques for solving polynomial equations of degree 2 (quadratic equations) have been shown in this chapter. Polynomial equations that can be solved like quadratic equations by making an appropriate substitution were presented in Section 5.6. It was also stated that a polynomial equation of degree n has exactly n roots in the set of complex numbers. In this section we examine polynomial equations that can be solved by factoring, the use of the factor theorem (see Chapter 3, Section 4), and the zero-product theorem.

If $P(x) = (x - a)(x - b)(x - c) \ldots (x - n) = 0$, then by the zero-product theorem,

$$x - a = 0 \quad \text{or} \quad x - b = 0 \quad \text{or} \quad x - c = 0 \ldots \quad \text{or} \quad x - n = 0$$

EXAMPLE 1 Solve $(x - 1)(x - 2)(x + 5) = 0$.

Solution By the zero-product theorem,
$$x - 1 = 0 \quad \text{or} \quad x - 2 = 0 \quad \text{or} \quad x + 5 = 0$$
Thus $x = 1$ or $x = 2$ or $x = -5$, and the solution set is $\{1, 2, -5\}$.

EXAMPLE 2 Solve $x^3 - 3x^2 - 4x + 12 = 0$.

Solution Factoring by grouping,
$$
\begin{aligned}
(x^3 - 3x^2) - (4x - 12) &= x^2(x - 3) - 4(x - 3) \\
&= (x^2 - 4)(x - 3) \\
&= (x + 2)(x - 2)(x - 3)
\end{aligned}
$$
Thus $x^3 - 3x^2 - 4x + 12 = (x + 2)(x - 2)(x - 3) = 0$ if and only if $x + 2 = 0$ or $x - 2 = 0$ or $x - 3 = 0$. The solution set is $\{-2, 2, 3\}$.

Note in Example 2 that the number of solutions in the solution set is 3, the same as the degree of the polynomial $x^3 - 3x^2 - 4x + 12$ (the degree of a polynomial in one variable is the greatest exponent that occurs on the variable). In general, *a polynomial of degree n has n first-degree factors, and therefore n roots must be accounted for when solving a polynomial equation.*

It is also convenient to notice that if all the coefficients of the polynomial are real numbers and the equation has one imaginary root, then the conjugate of that imaginary root is also a root of the equation.

EXAMPLE 3   Solve $x^3 - 8 = 0$.

Solution   Factoring $x^3 - 8$,

$$x^3 - 8 = (x - 2)(x^2 + 2x + 4) = 0$$
$$x - 2 = 0 \quad \text{or} \quad x^2 + 2x + 4 = 0$$
$$x = 2 \quad \text{or} \quad x = -1 \pm i\sqrt{3}$$

The solution set is $\{2, -1 + i\sqrt{3}, -1 - i\sqrt{3}\}$.

EXAMPLE 4   Solve $x^4 - 4x^3 + 4x^2 - (x^2 - 4x + 4) = 0$.

Solution   Factoring,

$$x^2(x^2 - 4x + 4) - (x^2 - 4x + 4) = 0$$
$$(x^2 - 4x + 4)(x^2 - 1) = 0$$
$$x^2 - 4x + 4 = 0 \quad \text{or} \quad x^2 - 1 = 0$$
$$(x - 2)^2 = 0 \quad \text{or} \quad (x + 1)(x - 1) = 0$$
$$x = 2 \quad \text{or} \quad x = 2 \quad \text{or} \quad x = -1 \quad \text{or} \quad x = 1$$

The solution set is $\{2, -1, 1\}$, where 2 is called a **double root** or a root of **multiplicity 2.** With the root 2 counted twice, it may then be said that the number of roots (4) is equal to the degree of the polynomial (4).

EXAMPLE 5   Solve $x^3 - 5x - 12 = 0$.

Solution   Using the factor theorem to find r so that $P(r) = 0$:

If $x = 1$, then $P(1) = 1 - 5 - 12 \neq 0$
If $x = 2$, then $P(2) = 8 - 10 - 12 \neq 0$
If $x = 3$, then $P(3) = 27 - 15 - 12 = 0$

Thus $x - 3$ is a factor.

Now by using synthetic division to find the quotient polynomial,

$$
\begin{array}{rrrr|l}
1 & 0 & -5 & -12 & 3 \\
  & 3 & 9 & 12 & \\
\hline
1 & 3 & 4 & 0 &
\end{array}
$$

Thus

$$x^3 - 5x - 12 = (x - 3)(x^2 + 3x + 4) = 0$$

and

$$x - 3 = 0 \quad \text{or} \quad x^2 + 3x + 4 = 0$$
$$x = 3 \quad \text{or} \quad x = \frac{-3 \pm i\sqrt{7}}{2}$$

The solution set is

$$\left\{3, \frac{-3 + i\sqrt{7}}{2}, \frac{-3 - i\sqrt{7}}{2}\right\}$$

Although the statement that every polynomial equation of degree n has exactly n roots is always true under the right conditions, finding these roots is not always possible by the methods of factoring shown in this section. If a polynomial cannot be readily factored, the solutions for the polynomial equations must be found by techniques beyond the scope of this text, and often these solutions may only be approximations.

EXERCISES 5.8

In Exercises 1–30 solve for x.

1. $x^3 + 3x^2 - x - 3 = 0$
2. $x^3 - 26x + 5 = 0$
3. $9x^4 - 13x^2 + 4 = 0$
4. $3x^4 - x^3 - 27x^2 + 9x = 0$
5. $x^5 - 9x^3 + 8x^2 - 72 = 0$
6. $2x^3 + 7x^2 - 9 = 0$
7. $3x^4 + 2x^3 - 9x^2 + 4 = 0$
8. $x^3 + 64 = 0$
9. $x^3 - 1 = 0$
10. $x^4 - 4x^3 - 6x^2 + 4x + 5 = 0$
11. $x^3 - 3x^2 = 16$
12. $x^4 - 14x^2 + 49 = 0$
13. $x^4 - x^3 - 6x - 36 = 0$
14. $x^3 + x^2 - 12 = 0$
15. $x^3 - 19x + 30 = 0$
16. $x^3 - 48x - 7 = 0$
17. $x^3 + 5x^2 - 2x - 10 = 0$
18. $x^5 + x^4 - 25x^3 - 25x^2 = 0$
19. $4x^4 - 15x^2 - 4 = 0$
20. $x^3 - 7x^2 + 36 = 0$
21. $x^5 - x^3 - 27x^2 + 27 = 0$
22. $8x^3 - 125 = 0$
23. $x^3 + 6x^2 + 4x - 5 = 0$
24. $27x^3 + 1 = 0$
25. $x^4 - 6x^3 + 5x^2 + 24x - 36 = 0$
26. $x^3 + 6x^2 + 12x + 8 = 0$
27. $x^4 - 4x^3 - 2x^2 + 12x + 9 = 0$
28. $x^5 + 2x^2 - x - 2 = 0$
29. $x^4 - 4x^3 + 6x^2 - 4x + 1 = 0$
30. $x^4 - 4x^2 + 12x - 9 = 0$

5.9 APPLICATIONS

There are many practical applications that involve solving a quadratic equation. When finding the solutions to verbal problems that involve the application of quadratic equations, it is especially important to check each root of the equation in the statement of the problem to see if the necessary conditions are met. Often the equation will have two roots, but only one may apply to a given problem. For example, lengths of sides of rectangles, triangles, and the like are always positive numbers; ages of individuals are positive numbers; digits in a numeral cannot be fractions; and the number of people present at a certain gathering cannot be fractional or negative.

EXAMPLE 1   If the legs of a right triangle measure 5 in. and 12 in., respectively, what is the length of the hypotenuse?

Solution   The theorem of Pythagoras states that if a and b are the measures of the legs of a right triangle and c represents the measure of the hypotenuse, then
$$a^2 + b^2 = c^2$$
Let $a = 5$, $b = 12$; then
$$5^2 + 12^2 = c^2$$
$$25 + 144 = c^2$$
$$169 = c^2$$
$$\pm 13 = c$$
Since length is a positive number, the condition $c > 0$ is implied. Thus the common solution of $c = \pm 13$ and $c > 0$ is $c = 13$.

EXAMPLE 2   If the hypotenuse of a right triangle is 25 in. long and one leg measures 24 in., how long is the other leg?

Solution   Applying the theorem of Pythagoras as in the preceding example, let $c = 25$, $a = 24$, and b designate the other leg:
$$a^2 + b^2 = c^2$$
$$(24)^2 + b^2 = (25)^2$$
$$b^2 = (25)^2 - (24)^2$$
$$b^2 = (25 - 24)(25 + 24)$$
$$b^2 = 49$$
$$b = \pm 7$$
Again we disregard the solution $b = -7$. Therefore the other leg is 7 in. long.

EXAMPLE 3   A square flower bed has a 3-ft walk surrounding it. If the walk were to be replaced and planted with flowers, the new flower bed would have 4 times the area of the original bed. What is the length of one side of the original bed?

Solution   Let x = length of a side of the original flower bed.

Then

$$x^2 = \text{area of original bed}$$
$$x + 6 = \text{length of a side of the new flower bed}$$

and

$$(x + 6)^2 = \text{area of the new flower bed}$$

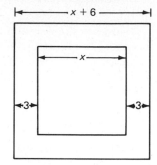

$$(x + 6)^2 = 4x^2$$
$$x^2 + 12x + 36 = 4x^2$$
$$3x^2 - 12x - 36 = 0$$
$$3(x^2 - 4x - 12) = 0$$
$$x^2 - 4x - 12 = 0$$
$$(x + 2)(x - 6) = 0$$
$$x + 2 = 0 \quad \text{or} \quad x - 6 = 0$$
$$x = -2 \quad \text{or} \quad x = 6$$

Again we discard the negative answer, and the length of the original side is 6 ft.

Check

$$\text{Area of original bed} = x^2 = 6^2$$
$$= 36 \text{ sq ft}$$
$$\text{Area of new bed} = (x + 6)^2 = (6 + 6)^2$$
$$= 12^2 = 144 \text{ sq ft}$$
$$\text{Area of new bed} = 4 \text{ times area of old bed}$$
$$144 = 4 \times 36$$
$$144 = 144$$

EXAMPLE 4   A plane flies 300 mi with a tail wind of 10 mph and returns against a wind of 20 mph. What is the speed of the plane in still air if the total flying time is $4\frac{1}{2}$ hr?

Solution   Let x represent the speed of the plane in still air.

Then

$$x + 10 = \text{speed with the tail wind}$$

and

$$x - 20 = \text{speed against the wind}$$

Formula	r	t	=	d
With wind	$x + 10$	$\dfrac{300}{x + 10}$		300
Against wind	$x - 20$	$\dfrac{300}{x - 20}$		300

Equation   Time going + time returning = total time = $4\frac{1}{2}$ hr = $\frac{9}{2}$ hr.

$$\frac{300}{x + 10} + \frac{300}{x - 20} = \frac{9}{2}$$

Multiplying both sides of this equation by the L.C.D.,

$$2(x + 10)(x - 20)$$

yields

$$600(x - 20) + 600(x + 10) = 9(x + 10)(x - 20)$$

Dividing both sides by 3,

$$200(x - 20) + 200(x + 10) = 3(x + 10)(x - 20)$$
$$200x - 4000 + 200x + 2000 = 3x^2 - 30x - 600$$
$$3x^2 - 430x + 1400 = 0$$
$$(3x - 10)(x - 140) = 0$$
$$3x - 10 = 0 \quad \text{or} \quad x - 140 = 0$$
$$x = 3\frac{1}{3} \quad \text{or} \quad x = 140$$

Since speed is always a positive number, $x \neq 3\frac{1}{3}$ because

$$x - 20 = 3\frac{1}{3} - 20$$

which is a negative number.

Therefore the speed of the plane in still air is 140 mph.

EXAMPLE 5   Working alone, a carpenter can make a set of cabinets in 3 hr less time than his helper can. Working together, they can make the set of cabinets in 6 hr. Find the time (correct to the nearest minute) that each requires to make the set alone.

Solution
Let

x = time it takes helper alone

Then

$x - 3$ = time it takes carpenter alone

Formulas

$tr = w$, and w of first + w of second = w of both = 1
where t = time, r = rate, and w = amount of work done.

	Working Alone				Working Together (whole job)		
	t	$\cdot$ r	= w		t	$\cdot$ r	= w
Carpenter	$x - 3$	$\dfrac{1}{x - 3}$	1		6	$\dfrac{1}{x - 3}$	$\dfrac{6}{x - 3}$
Helper	x	$\dfrac{1}{x}$	1		6	$\dfrac{1}{x}$	$\dfrac{6}{x}$

Equation   Work of helper + work of carpenter = whole job

$$\frac{6}{x} + \frac{6}{x-3} = 1, \quad \text{with restrictions } x > 0 \text{ and } x - 3 > 0$$

$$6(x-3) + 6x = x(x-3)$$
$$12x - 18 = x^2 - 3x$$
$$x^2 - 15x + 18 = 0$$

Since $x^2 - 15x + 18$ is not readily factorable, and since $\dfrac{b}{a} = -15$

is not an even integer, the method using the quadratic formula is recommended for solving the equation.

Solution of equation

$$ax^2 + bx + c = 0 \quad \text{if and only if} \quad x = \frac{-b \pm \sqrt{b^2 - 4ac}}{2a}$$

If $x^2 - 15x + 18 = 0$, then $a = 1$, $b = -15$, $c = 18$.

Thus

$$x = \frac{-(-15) \pm \sqrt{(-15)^2 - 4(18)}}{2} = \frac{15 \pm \sqrt{225 - 72}}{2}$$

$$x = \frac{15 + \sqrt{153}}{2} \quad \text{or} \quad x = \frac{15 - \sqrt{153}}{2}$$

Approximating (by using a calculator or the tables), $\sqrt{153} = 12.37$. Thus to the nearest hundredth,

$$x = \frac{15 + 12.37}{2} \quad \text{or} \quad x = \frac{15 - 12.37}{2}$$

$$x = \frac{27.37}{2} \quad \text{or} \quad x = \frac{2.63}{2}$$

$$x = 13.69 \quad \text{or} \quad x = 1.32$$

and

$$x - 3 = 10.69 \quad \text{or} \quad x - 3 = -1.68$$

(This solution is rejected since $x - 3 > 0$.)

Thus $x = 13.69$ and $x - 3 = 10.69$, correct to the nearest hundredth. Since there are 60 min in an hour,

$$0.69 \text{ hr} = 0.69(60) \text{ min} = 41.4 \text{ min}$$

Therefore correct to the nearest minute,

$$x = 13 \text{ hr } 41 \text{ min, time of helper alone}$$
$$x - 3 = 10 \text{ hr } 41 \text{ min, time of carpenter alone}$$

EXERCISES 5.9

1. The hypotenuse of a right triangle is 3 units and the legs are equal in length. Find the length of a leg of the triangle.
2. The length of a rectangle exceeds 3 times its width by 1 in. The area is 52 sq in. Find the dimensions of the rectangle.
3. The hypotenuse of a right triangle is 13 in. If one leg is 7 in. longer than the other, how long are the legs of the triangle?
4. If 4 times a number is added to 3 times its square, the sum is 95. Find the number.

5. The sum of two numbers is 7 and the difference of their recipro-
cals is $\frac{1}{12}$. Find the numbers.

6. A concrete walk of uniform width extends around a rectangular
lawn having dimensions of 20 ft by 80 ft. Find the width of the
walk if the area of the walk is 864 sq ft.

7. A rectangular piece of sheet metal is twice as long as it is wide.
From each of its four corners a square piece 2 in. on a side is cut
out. The flaps are then turned up to form an uncovered metal
box. If the volume of this box is 320 cu in., find the dimensions of
the original piece of sheet metal.

8. It takes John 3 hr longer to do a certain job than it does his
brother Bob. For 3 hr they worked together; then John left and
Bob finished the job in 1 hr. How many hours would it have
taken Bob to do the whole job by himself?

9. One man can do a job in 8 days less time than another man, but
he charges $50 a day, whereas the slower man charges $20 a
day. When the two men work together, they take 3 days to complete the job. Which would cost the less, to have the faster man
do the job alone, to have the slower man do the job alone, or to
have both work together? State the cost for each case.

10. One of two inlets can fill a swimming pool in 6 hr. The time for
the other inlet to fill the pool is 2 hr longer than the two inlets
together. Find the time it takes for the two inlets together to fill
the pool, correct to the nearest minute.

11. One inlet pipe takes 12 min longer than another inlet pipe to fill a
certain tank. An outlet pipe can empty the tank in 45 min. When
all three pipes are open, it takes 15 min to fill the tank. Find the
time it takes to fill the tank if only the larger inlet pipe is open.

12. A private plane flew from San Francisco to Lake Tahoe, a distance of 180 mi, with a tail wind and then returned against the
same wind. If the total flying time was $2\frac{1}{2}$ hr and if the speed
of the plane in still air was 150 mph, find the speed of the wind.

13. A jet plane flying against a head wind of 20 mph takes 20 min
longer to fly a distance of 2610 mi than a plane with the same
still air speed flying in the opposite direction. Find the still air
speed of the plane.

14. A boat that travels 12 mph in still water takes 2 hr less time to go
45 mi downstream than to return the same distance upstream.
Find the rate of the current.

15. A baseball diamond has the shape of a square with each side
90 ft long. The pitcher's mound is 60.5 ft from home plate on the
line joining home plate to second base. Find the distance from
the pitcher's mound to second base.

16. A 36-in. length of copper tubing is bent to form a right triangle
having a 15-in. hypotenuse. Find the lengths of the other two
sides of the triangle.

17. A wire is stretched from the top of a 4-ft fence to the top of a 20-ft vertical pole. If the fence and the pole are 30 ft apart, find the length of the wire.

18. A fisherman trolled upstream in a motorboat to a spot 6 mi from his campsite and then returned to camp. If the rate of the current was $1\frac{1}{2}$ mph and if the round trip took 3 hr, find the rate of the motorboat in still water.

19. The span s of a circular arch is related to its height h and its radius r by the formula

$$s^2 = 8rh - 4h^2$$

Find the height of a circular arch whose span is 80 ft and whose radius is 50 ft. (Assume the circular arch is less than a semicircle.)

20. A circular hole has a radius of 5 in. How much larger should the radius be so that a new circular hole will have a cross-sectional area twice as large as the original area? (Area of circle, $A = \pi r^2$.)

21. For a certain electric motor having a mechanical output of 25,000 watts, a resistance in the armature of 0.04 ohms, and a line voltage of 110 volts, the armature current I in amps is given by

$$25,000 = 110\,I - 0.04\,I^2$$

Find the current I.

22. If I is the intensity of illumination in lumens, c is the candlepower of the source of light, and s is the distance in feet from the source, then

$$I = \frac{c}{s^2}$$

A 20-candlepower light is 3 ft to the left of a 45-candlepower light. Find the distance x from the 20-candlepower source on the line joining the two sources so that the illumination is the same from each source. Use

$$\frac{20}{x^2} = \frac{45}{(3 - x)^2}$$

23. A fisherman in a boat on a small lake sees some plants growing in the water with their roots at the bottom of the lake. To find the depth of the lake, he pushes a plant extending 8 in. out of the

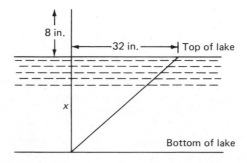

water so that the plant is completely submerged with its tip just touching the water. He measures the distance from the point where the plant first emerged from the water to the new position of its tip and finds this to be 32 in. Find the depth of the lake.

DIAGNOSTIC TEST

Complete the statements in Problems 1–10 by filling in the blanks.

1. If a and b are real numbers, and $i = \sqrt{-1}$, then $a + bi$ and $a - bi$ are called _____ .

2. The product $(a + bi)(a - bi)$ is always a _____ number.

3. If $3x - 5yi = 12 + 10i$, then $x =$ _____ and $y =$ _____ .

4. If the equation $x^2 + 2x + 2 = 0$ has a root $x = -1 + i$, then it also has a root $x =$ _____ .

5. For the equation $3x^2 + 2x - 5 = 0$, the discriminant equals _____ .

6. The value of the discriminant in Problem 5 tells us that the roots of the given equation are _____ and _____ .

7. The equation $x^{2/3} - 3x^{1/3} - 4 = 0$ may be solved by means of a quadratic equation by using the substitution $u =$ _____ .

8. If $x^2 + 4x + 6 = k$, the value of k for which the roots of the equation are real and equal is $k =$ _____ .

9. $(3 - \sqrt{-2})^2$ may be written in standard form, $a + bi$, as _____ .

10. If $x^2 - 6x = 20$, and we wish to solve this equation by completing the square, then the right side of the equation is _____ after the square has been completed on the left side.

Select the best answer for each of the following problems. If the best answer is "none of these" write a correct answer.

11. The solution set for the equation $x^2 + 1 = 0$ is
    (a) $\{1\}$ (d) $\{i, -i\}$
    (b) $\{-1\}$ (e) none of these
    (c) $\{1, -1\}$

12. The solution set for the equation $x + \sqrt{x - 1} = 3$ is
    (a) $\{2\}$ (d) $\varnothing$
    (b) $\{5\}$ (e) none of these
    (c) $\{2, 5\}$

13. $(3 + 2\sqrt{-3})(2 - 3\sqrt{-3})$ equals
    (a) $24 - 5i\sqrt{3}$ (d) 9
    (b) $-12 - 5i\sqrt{3}$ (e) none of these
    (c) $12 - 5i\sqrt{3}$

14. $(1 + i)^3$ equals

(a) $3 + 3i$ (d) $-2 + 2i$

(b) $1 - i$ (e) none of these

(c) $1 + i^3$

15. If $\sqrt{5x - 4} = \sqrt{x} + 3$, then

(a) $5x - 4 = x + 9$ (d) $(5x - 4)^2 = (x + 3)^2$

(b) $5x - 4 = x + 3\sqrt{x} + 9$ (e) none of these

(c) $5x - 4 = x + 6\sqrt{x} + 9$

16. The equation $x^4 - 11x^2 + 24 = 0$ has

(a) 4 real roots

(b) 4 imaginary roots

(c) 3 real roots and 1 imaginary root

(d) 2 real roots and 2 imaginary roots

(e) none of these

17. The roots of the equation $3x^2 + 5x - 8 = 0$ are

(a) real, rational, and unequal

(b) real, rational, and equal

(c) real and irrational

(d) imaginary

(e) none of these

18. The conjugate of 3 is

(a) -3 (d) 3

(b) $3 - i$ (e) none of these

(c) $3 + i$

REVIEW EXERCISES

In Exercises 1–6 write each of the complex numbers in the standard form, a + bi. (Section 5.1)

1. $2 + \sqrt{-16}$ **4.** $(2 + 3i)(3 - 4i)$

2. $(3 + 5i) - (2 - 3i)$ **5.** $2i(3 + 4i)^2$

3. $\dfrac{4 + \sqrt{-12}}{2}$ **6.** $\dfrac{1 - i}{2 + 3i}$

7. For what real numbers x and y is the equation $3x + 2yi = 5$ true? (Section 5.1)

Solve each equation in Exercises 8–13 by the factoring method and check each solution. (Section 5.3)

8. $x^2 - 6x - 40 = 0$ **11.** $x^2 + 9 = 0$

9. $12x^2 + 32x + 5 = 0$ **12.** $x^3 + 3x^2 - x - 3 = 0$

10. $3x^2 - 4x = 0$ **13.** $x^4 - 1 = 0$

Solve each equation in Exercises 14–17 by completing the square and check each solution. (Section 5.4)

14. $x^2 + 8x + 15 = 0$

15. $2x - x^2 - 3 = 0$

16. $3x^2 = 12x + 3$

17. $4x^2 - 8x = 23$

Solve each equation in Exercises 18–25 by the quadratic formula and check each solution. (Section 5.5)

18. $2x^2 + 5x - 1 = 0$

19. $6x^2 - 7x - 20 = 0$

20. $2x^2 - x = 2$

21. $3x^2 + 8 = 16x$

22. $(x - 3)(x + 2) = 1$

23. $4 - 3x - 5x^2 = 0$

24. $8x^2 + 3x = 0$

25. $3x^2 - 7 = 0$

Solve each equation in Exercises 26–30 by any method you wish and check each solution. (Sections 5.3–5.5)

26. $\frac{3}{4}x^2 = \frac{7}{8}$

27. $x^2 + 4x + 8 = 0$

28. $(2x + 3)(x - 2) = 4$

29. $3x^2 + 5x = 1$

30. $(2x + 5)^2 = 9$

In Exercises 31–35 use the discriminant $b^2 - 4ac$, and the fact that the roots of the corresponding quadratic equation $ax^2 + bx + c = 0$ ($a \neq 0$) are equal if the discriminant equals zero to find the value of k for which the roots of the equations given are equal. (Section 5.5)

31. $x^2 + 6x + k = 0$

32. $x^2 + 2kx + 4 = 0$

33. $kx^2 + 5x + 5 = 0$

34. $kx^2 + 8x + k = 0$

35. $x^2 + 4x + 5 = k$

Solve each equation in Exercises 36 and 37 for y. (Section 5.4)

36. $10x^2 = y^2 - 3xy$

37. $y^2 + 2x^2y = x^4$

Find the solution set for each equation in Exercises 38–43. (Sections 5.6–5.8)

38. $3x^4 + 5x^2 - 78 = 0$

39. $(y + 2)^{1/2} - (y + 2)^{1/4} = 6$

40. $x^{2/3} - 6x^{1/3} + 5 = 0$

41. $5y^{-2} + 9y^{-1} - 2 = 0$

42. $17 - \sqrt{x - 3} = 10 + \sqrt{32 + x}$

43. $\sqrt{y - 2} + \sqrt{2y - 2} = \sqrt{3y + 20}$

44. John can mow his lawn in 20 min less time with his power mower than with his hand mower. One day his power mower broke down 15 min after he started mowing, and he had to complete the job with his hand mower. It took him 25 min to finish mowing by hand. How long does it take John to do the complete job with the power mower?

45. A pilot left a Chicago airport and flew 200 mi south to a town *T* with a tail wind of 20 mph. From T he flew back to Chicago against a head wind of 30 mph. If his total flying time was $2\frac{1}{3}$ hr, what was the average speed of the plane in still air?

46. A farmer has 90 ft of fencing he can use to enclose a rectangular piece of land. Find the dimensions of the rectangle if the area is to be 450 sq ft.

47. One leg of a right triangle is 9 in. longer than the other leg. The hypotenuse is 45 in. long. Find the lengths of the legs of the triangle.

48. The height H of a projectile at the end of t sec is given by

$$H = cvt - \frac{1}{2} gt^2$$

Solve for t.

49. The total surface area T of a right circular cylinder of radius r and height h is given by

$$T = 2\pi r(r + h)$$

Solve for r.

50. The ancient Greeks considered the most beautiful rectangle to be one in which the ratio of the length to the width was equal to the ratio of width to the length minus the width.

$$\frac{L}{W} = \frac{W}{L - W}$$

Find the exact numerical value of $\frac{L}{W}$, the "Golden Ratio." Compare $\frac{L}{W}$ with $\frac{W}{L}$.

INEQUALITIES AND ABSOLUTE VALUES IN ONE VARIABLE

I n Chapter 1, Section 2, inequalities and their graphs on the number line were introduced. In this chapter we briefly review these concepts and expand them to find solution sets for first- and second-degree inequalities in one variable. Similarly, we will expand on the concept of absolute value to find solutions for absolute value equalities and then tie inequalities and absolute value together for solutions of absolute value inequalities in one variable.

6.1 INEQUALITIES AND THE NUMBER LINE

The trichotomy axiom states: If *a* and *b* are real numbers, then exactly one of the following statements is true:

$a < b$ or $a = b$ or $a > b$

On the number line drawn horizontally with its positive direction to the right, it is seen that $a < b$ if and only if the point whose coordinate is *a* is to the left of the point whose coordinate is *b*, and $a > b$ if and only if the point whose coordinate is *a* is located to the right of the point whose coordinate is *b*.

From Figures 6.1 and 6.2 it may be observed that $a < b$ if and only if $b > a$.

FIG. 6.1   **FIG. 6.2**

A formal definition of the "less than" relation between two real numbers is stated below.

DEFINITION

If *a* and *b* are real numbers, then $a < b$ (*a* is less than *b*) if and only if there exists a positive real number *p* such that $a + p = b$.

$a > b$ if and only if $b < a$

For example, $3 < 5$ because $3 + 2 = 5$, and $5 > 3$ because $3 < 5$.

The relations "less than" and "greater than" ($<$ and $>$) are called **order relations.** A statement involving an order relation is called an **inequality.** Examples of inequalities are

$3 < 5$, $2x + 1 < 4$, $y \geq 2x + 1$

The inequalities $a < b$ and $c < d$ are called **inequalities of the same order** or **inequalities having the same sense,** while the inequalities $a < b$ and $c > d$ are called **inequalities of opposite order** or **inequalities having the opposite sense.**

Since all numbers to the right of 0 on the number line are said to be positive, it follows that "$a > 0$" is another way of writing "*a* is a positive number." Similarly, "$a < 0$" is another way of writing "*a* is a negative number." Note that no minus sign is necessary to indicate the negative nature of *a*.

The number line is often a useful graphic aid in the solution of inequalities. For example, the set $\{x|x > 3\}$ is graphed as shown in Figure 6.3.

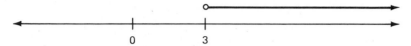

FIG. 6.3

The circle above the numeral 3 indicates that 3 is excluded from the solution set; the solution set is indicated by the half-line starting at 3 (but not including 3), and all values greater than 3 as shown by the direction of the line.

The sex $\{x|x \leq -2\}$ is graphed as shown in Figure 6.4.

FIG. 6.4

This time a solid dot over the -2 coordinate indicates that -2 is included in the solution set, as well as all points to the left of -2 since *x* is less than or equal to -2.

The statement $x \leq a$ means "*x* is less than *a* or *x* equals *a*," and the statement $x \geq a$ means "*x* is greater than *a* or *x* equals *a*."

The statement $a < x$ and $x < b$ may be expressed as $a < x < b$, read "*x* is between *a* and *b*," whenever $a < b$.

It is often convenient to graph "intersections" and "unions" of inequalities.

The terms "intersection" and "union" are briefly defined as follows. If *A* and *B* are sets, then the **intersection** of the two sets (symbolized $A \cap B$) is the set that contains all elements in set *A* *and* in set *B*. These are the elements that the two sets have in common. For example, if

$A = \{1, 2, 3, 4\}$ and $B = \{2, 4, 6, 8, 10\}$

then $A \cap B = \{2, 4\}$, or the common elements in set *A* and in set *B*.

The **union** of two sets *A* and *B* (symbolized $A \cup B$) is the set that contains all elements found in either *A* *or* *B*, that is, *all* the elements found in either set. For example, the union of sets *A* and *B*, as defined above, is $A \cup B = \{1, 2, 3, 4, 6, 8, 10\}$.

The words *and* and *or* are very important and must be

241

used correctly. Remember, intersection, ∩, corresponds to the word *and*, whereas union, ∪, corresponds to the word *or*.

The following summary of definitions is given for quick reference.

DEFINITIONS

$$a < x < b \text{ means } a < x \text{ and } x < b \text{ for } a < b$$

Unions and Intersections

$$\{x \mid a < x\} \cap \{x \mid x < b\} = \{x \mid a < x \text{ and } x < b\} = \{x \mid a < x < b\}$$
$$\{x \mid x < a\} \cup \{x \mid x > b\} = \{x \mid x < a \text{ or } x > b\}$$
$$\{x \mid x > a\} \cap \{x \mid x > b\} = \{x \mid x > a\} \text{ if } a > b$$
$$\{x \mid x < a\} \cap \{x \mid x < b\} = \{x \mid x < a\} \text{ if } a < b$$

EXAMPLE 1 The graph of $\{x \mid -2 < x < 1\}$ is illustrated in Figure 6.5.

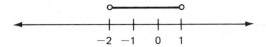

FIG. 6.5

In words, the graph illustrates $x > -2$ *and* $x < 1$. It also represents $\{x \mid x > -2\} \cap \{x \mid x < 1\}$.

EXAMPLE 2 The graph of $\{x \mid x < -2 \text{ or } x \geq 1\}$ is illustrated in Figure 6.6. The circle over -2 indicates that -2 is not included in the set, and the solid dot over 1 indicates that 1 is included in the set. In set symbols Figure 6.6 is the graph of $\{x \mid x < -2\} \cup \{x \mid x \geq 1\}$.

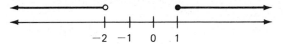

FIG. 6.6

EXAMPLE 3 Graph on a number line and write a verbal statement to describe the set $\{x \mid -2 \leq x < -1\} \cup \{x \mid x > 0\}$.

Solution   The union of the two stated sets is needed; thus *all* the numbers in both sets will be in the union set. In words, "x is greater than or equal to -2 and x is less than -1" or "x is greater than zero" (Figure 6.7).

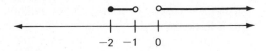

FIG. 6.7

EXAMPLE 4 Graph on a number line and write a verbal statement to describe the set $\{x \mid -4 < x \le 1\} \cap \{x \mid -1 < x \le 2\}$.

Solution This time the intersection of the two sets is needed, so the final set will contain the numbers which are in both sets. From Figure 6.8, it is clear that the intersection is

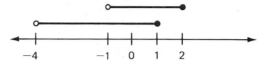

FIG. 6.8

the set $\{x \mid -1 < x \le 1\}$ (see Figure 6.9). In words, "x is greater than -1" *and* "x is less than or equal to 1."

FIG. 6.9

EXERCISES 6.1

Graph Exercises 1–20 on a number line and write a verbal statement describing the set.

1. $\{x \mid x < 3\}$
2. $\{x \mid x > -2\}$
3. $\{x \mid x \ge 3\}$
4. $\{x \mid x \le -2\}$
5. $\{x \mid 2 > x\}$
6. $\{x \mid -5 \ge x\}$
7. $\{x \mid x < 0\}$
8. $\{x \mid x \ge 0\}$

9. $\{x \mid x > -4\} \cap \{x \mid x < 4\}$
10. $\{x \mid x \ge -4\} \cup \{x \mid x \le 4\}$
11. $\{x \mid -2 \le x\} \cup \{x \mid x \ge 3\}$
12. $\{x \mid -2 \ge x\} \cup \{x \mid x \ge 3\}$
13. $\{x \mid x > -2\} \cap \{x \mid x \le 3\}$
14. $\{x \mid 1 < x < 5\}$
15. $\{x \mid -2 \le x < 3\}$
16. $\{x \mid 0 < x < 4\}$

17. $\{x \mid -2 < x \le 0\} \cup \{x \mid 2 < x < 3\}$
18. $\{x \mid 0 < x < 1\} \cup \{x \mid 1 \le x < 3\}$
19. $\{x \mid -1 < x < 2\} \cap \{x \mid 1 \le x < 3\}$
20. $\left\{x \mid -2 \le x \le \dfrac{1}{2}\right\} \cap \{x \mid 0 < x \le 15\}$

For Exercises 21–40 write a symbolic statement and graph each set on a number line.

21. *x* is greater than 2.
22. *x* is less than or equal to 1.
23. *x* is less than negative 4.
24. *x* is greater than negative 1.

25. 1 is greater than *x*.

26. Negative 2 is greater than or equal to *x*.

27. 1 is less than or equal to *x*.

28. Negative 3 is less than or equal to *x*.

29. Negative 5 is less than *x* and *x* is less than 5.

30. Negative 5 is less than or equal to *x* and *x* is less than 5.

31. *x* is less than or equal to negative 1 and *x* is less than 2.

32. Negative 5 is greater than *x* and *x* is greater than 5.

33. *x* is greater than negative 5 or *x* is less than 5.

34. *x* is less than negative 1 or *x* is less than or equal to 2.

35. 2 is less than or equal to *x* and *x* is less than 4.

36. Negative 3 is less than *x* and *x* is less than or equal to 0.

37. Negative 1 is less than *x* and *x* is less than or equal to 1, or *x* is greater than 3.

38. Negative 2 is less than or equal to *x* and *x* is negative, or *x* is positive and less than 2.

39. *x* is greater than or equal to negative 2 and *x* is less than or equal to negative 1, and *x* is greater than or equal to negative $\frac{3}{2}$ and *x* is less than 5.

40. *x* is a negative number greater than -3, and *x* is greater than or equal to 2.

6.2 FIRST-DEGREE INEQUALITIES IN ONE VARIABLE

A **solution of an inequality** is a number that makes the inequality true when its variable is replaced by this number.

The **solution set** of an inequality is the set of all solutions of the inequality.

Two inequalities are equivalent if and only if they have the same solution set.

Solving an inequality is similar to solving an equation, but there are two important exceptions. One has to do with exchanging sides, and the other has to do with multiplying (or dividing) by a negative number.

THEOREM ON EXCHANGING SIDES

If one inequality is obtained from another one by exchanging sides and by changing the order symbol, then these two inequalities are equivalent. In symbols,

$$a < b \quad \text{if and only if} \quad b > a$$

For example,

$3 < 8$ and $8 > 3$ are equivalent.

$x \geq 9$ and $9 \leq x$ are equivalent.

$12 - x > 3x$ and $3x < 12 - x$ are equivalent.

ADDITION THEOREM

If the same number is added to each side of an inequality, then the resulting inequality having the same sense is equivalent to the original one.

In symbols, for any real numbers a, b, and c,

$$\text{if} \quad a < b, \quad \text{then} \quad a + c < b + c$$

For example,

$3 < 8$ and $3 + 4 < 8 + 4$ are equivalent.

$x - 4 > 5$ and $x - 4 + 4 > 5 + 4$ (that is, $x > 9$) are equivalent.

Also $x + 2 < 3$ and $x + 2 - 2 < 3 - 2$ (that is, $x < 1$) are equivalent.

MULTIPLICATION BY A POSITIVE NUMBER THEOREM

If each side of an inequality is multiplied by the same positive number, then the resulting inequality having the same sense is equivalent to the original one. In symbols,

$$\text{if} \quad a < b \quad \text{and} \quad c > 0, \quad \text{then} \quad ac < bc$$

For example,

$5 > 2$ and $5(4) > 2(4)$ are equivalent.

$\dfrac{x}{4} \leq 6$ and $x \leq 24$ are equivalent.

MULTIPLICATION BY A NEGATIVE NUMBER THEOREM

If each side of an inequality is multiplied by the same *negative* number and if the order symbol is changed, then the resulting inequality, *having the opposite sense,* is equivalent to the original one. In symbols,

$$\text{if} \quad a < b \quad \text{and} \quad c < 0, \quad \text{then} \quad ac > bc$$

For example,
$-2 < 8$ and $-2(-3) > 8(-3)$ (that is, $6 > -24$) are equivalent.
$-2x \geq 8$ and $x \leq -4$ are equivalent.

EXAMPLE 1   Solve the inequality $2x + 3 < x + 5$ and graph the solution set (see Figure 6.10).

Solution
$$2x + 3 < x + 5$$
$$2x < x + 2 \qquad \text{(Addition theorem—adding } -3)$$
$$x < 2 \qquad \text{(Addition theorem—adding } -x)$$
Solution set $= \{x \mid x < 2\}$.

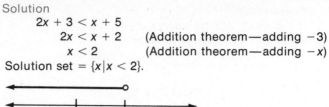

FIG. 6.10

A quick test for the correctness of the answer may be made by selecting a number in the solution set to see if the statement is true for that value, and selecting a number that is *not* in the solution set to see if the statement is false for that value. In Example 1, for instance, 0 is in the proposed solution set. When $x = 0$, $2(0) + 3 = 3$, and $0 + 5 = 5$, and $3 < 5$, so the statement is true when $x = 0$. The number 3 is not in the solution set. $2(3) + 3 = 9$, $3 + 5 = 8$, and the statement $9 < 8$ is false.

EXAMPLE 2   Solve for x and graph the solution set
$$2 - x \leq 4 + 3x$$
Solution
$$2 - x \leq 4 + 3x$$
$$2 - 4x \leq 4 \qquad \text{(Addition theorem—adding } -3x)$$
$$-4x \leq 2 \qquad \text{(Addition theorem—adding } -2)$$
$$x \geq -\frac{1}{2} \qquad \text{(Multiplication theorem—note that mul-}$$
$$\text{tiplication by } -\frac{1}{4} \text{ reversed the order}$$
$$\text{of the inequality)}$$
Solution set: $\left\{ x \mid x \geq -\dfrac{1}{2} \right\}$ (see Figure 6.11).

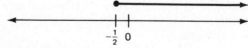

FIG. 6.11

To test the solution set, let $x = 0$ (in the solution set). Then $2 - 0 = 2, 4 + 3(0) = 4$, and $2 < 4$, a true statement. Now try $x = -1$ (*not* in the solution set) to show that the statement is false for that value.

EXAMPLE 3 Solve $\dfrac{x}{x + 1} > 2$ and graph the solution set.

Solution   Since the inequality has a variable in the denominator, it is necessary to consider two cases, $x + 1 > 0$ and $x + 1 < 0$. The solution set of the inequality is the union of the solution sets for the two cases.

Case 1 $x + 1 > 0$ and thus $x > -1$. Since $x + 1 > 0$, multiplying each side of the inequality by $x + 1$ does *not* change the order.

$$x > 2x + 2$$
$$-x > 2 \qquad \text{(Subtracting } 2x \text{ from both sides)}$$
$$x < -2 \qquad \text{(Multiplication by } -1 \text{ reversed the order)}$$

Since x was restricted to $x > -1$, the solution set is $\{x \mid x > -1\} \cap \{x \mid x < -2\} = \varnothing$.

In other words, there are no values of x such that x is greater than -1 *and* (at the same time) less than -2.

Case 2 $x + 1 < 0$ and thus $x < -1$. Since $x + 1 < 0$, multiplying each side of the inequality by $x + 1$ *does change the order.*

$$x < 2x + 2$$
$$-x < 2$$
$$x > -2$$

Since x was restricted to $x < -1$, the solution set is

$$\{x \mid x < -1\} \cap \{x \mid x > -2\} = \{x \mid -2 < x < -1\}$$

The union of the solution sets for the two cases is

$$\{x \mid -2 < x < -1\} \cup \varnothing$$
$$= \{x \mid -2 < x < -1\}$$

The union of the two cases is the set of solutions from case 1 *or* from case 2. Thus since the solution set from case 1 was the empty set, only the solutions from case 2 are valid. Figure 6.12 is a graph of the solution set.

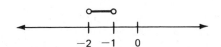

FIG. 6.12

$-2 \quad -1 \quad 0$

Note that the third possibility, $x + 1 = 0$, was not considered since this would have made $\dfrac{x}{x + 1} = -\dfrac{1}{0}$, and division by zero is not defined.

A verbal statement of the solution set $\{x \mid -2 < x < -1\}$ is "all real numbers x such that x is greater than -2 *and* x is less than -1"; or "x is any real number between -2 and -1."

EXERCISES 6.2

Solve Exercises 1–50 and graph the solution sets.

1. $2x - 4 > 0$
2. $2x + 4 > 0$
3. $3x - 2 > 1 + 2x$
4. $3x + 5 < x + 7$
5. $2x + 1 < x - 5$
6. $6x + 10 < 2 + 4x$
7. $4 - 2x \leq 0$
8. $8 - 2x \geq 0$
9. $15 - 3x \geq 0$
10. $15 - 3x \leq 0$
11. $1 < 2x + 3$
12. $1 \leq 2x - 3$
13. $2x - 3(x + 1) > 0$
14. $2x - 3(x + 1) \leq 0$
15. $5 - x \leq 3 - (x - 2)$
16. $4x + 2 \leq 4x$
17. $x + 4 - 3x \leq 2x + 4$
18. $3x - (x + 4) \leq 4 - 2x$
19. $-2(x + 3) \geq 4(2x + 1)$
20. $2(x + 3) < -4(2x + 1)$
21. $33 - 9x \leq 2x$

22. $x - 12 \geq 0$
23. $12 - x \geq 0$
24. $\dfrac{x + 3}{2} < 0$
25. $\dfrac{x + 3}{5} \geq 0$
26. $2x - \dfrac{1}{3} \leq 0$
27. $\dfrac{x - 6x}{2} < -20$
28. $\dfrac{x - 6}{2} < -20$
29. $\dfrac{6 - x}{2} < -20$
30. $2(x + 3) \geq 8(2 - x)$
31. $\dfrac{x}{3} < 0$
32. $\dfrac{x}{3} > 0$
33. $x > 4 + x$
34. $x < 4 + x$
35. $x - 2 < 2 - x$
36. $x - 2 < 2 + x$
37. $-3(2 - 3x) < 15 - (x + 1)$
38. $15 - (x + 2) \leq 0$
39. $-2(x + 4) \geq -2(x - 4)$
40. $3 \leq \dfrac{x + 2}{5}$
41. $\dfrac{3}{x} \leq 2$
42. $\dfrac{x + 5}{x} \geq 6$

43. $\dfrac{1}{x-1} < 1$

47. $10 \le \dfrac{6x}{x+2}$

44. $\dfrac{-2}{x+3} \ge 1$

48. $\dfrac{x}{2x+1} + 4 < 0$

45. $\dfrac{2}{x+1} < 4$

49. $x + \dfrac{5}{x} \le 7 + x$

46. $\dfrac{10}{x-3} \ge 2$

50. $x + \dfrac{5}{x} \le x - 7$

51. Find all values for x that satisfy *both* statements:

$$x + 2(x + 2) > 10 + x$$
$$5x - 2 < 2x + 13$$

52. Find all values for x that satisfy *both* statements:

$$5(2 - x) > 3x - 8$$
$$2(3x + 4) + 7 < 5 + x$$

53. Find all values of x that satisfy *both* statements:

$$3(x + 4) - 2x > 2(9 - x)$$
$$4x - 3 < 5 + 2x$$

54. Find all values of x that satisfy *both* statements:

$$5(x + 1) - 2 - x \le 0$$
$$2(x + 3) + 3x > x + 1$$

55. A student must have an average of 90% to 100% inclusive on five tests in a course to receive an A grade. If his grades on the first four tests were 93%, 86%, 82%, and 96%, what grade must he achieve on the fifth test in order to qualify for the A?

56. A student must have an average of 80% to 89% inclusive on five tests in a course to receive a B grade. Her grades on the first four tests were 98%, 76%, 86%, and 92%. Find the range of grades on the fifth test that would qualify her for a B in the course.

57. If a basketball team wins 30 of its first 35 games, what is the largest number of games it can lose in the remaining 45 games to have an average of at least 70% wins for the complete season?

58. A baseball team wins 40 of its first 50 games. What is the smallest number of games the team must win in the remaining 40 games in order to have an average of at least 60% wins for the complete season?

59. What temperatures in degrees Celsius correspond to 50 degrees $\le F \le$ 77 degrees, where F is the temperature in degrees Fahrenheit? $\left[\text{Use } F = \dfrac{9C + 160}{5}.\right]$

60. If a 12-year-old child has an I.Q. between 110 and 140, what are the possible mental ages of the child? (Use I.Q. $= \dfrac{100M}{C}$, where M is the mental age and C is the chronological age.)

6.3 QUADRATIC INEQUALITIES IN ONE VARIABLE

Statements having any one of the forms

$$ax^2 + bx + c > 0 \qquad ax^2 + bx + c < 0$$
$$ax^2 + bx + c \geq 0 \qquad ax^2 + bx + c \leq 0$$

where x is a variable and a, b, and c are constants with $a \neq 0$ are called **quadratic inequalities** in the variable x. The following discussion shows how quadratic inequalities are solved.

The multiplication of two real numbers can be separated into the following four cases:

1. Both factors are positive: $a > 0$ and $b > 0$, then $ab > 0$.
2. Both factors are negative: $a < 0$ and $b < 0$, then $ab > 0$.
3. The first factor is positive, the second is negative: $a > 0$ and $b < 0$, then $ab < 0$.
4. The first factor is negative, the second is positive: $a < 0$ and $b > 0$, then $ab < 0$.

Since these are the only cases possible, the following theorems may be stated.

THE POSITIVE PRODUCT THEOREM

If $ab > 0$, then either ($a > 0$ and $b > 0$) or ($a < 0$ and $b < 0$).

THE NEGATIVE PRODUCT THEOREM

If $ab < 0$, then either ($a > 0$ and $b < 0$) or ($a < 0$ and $b > 0$).

The positive product theorem states that if a product of two factors is positive, then both factors are positive or both factors are negative.

The negative product theorem states that if a product of two factors is negative, then one of the factors is positive and the other is negative.

EXAMPLE 1 Find the solution set of $(x + 2)(x + 3) > 0$.

Solution By the positive product theorem, there are two cases:

Case 1

$$x + 2 > 0 \quad \text{and} \quad x + 3 > 0$$

Case 2

$$x + 2 < 0 \quad \text{and} \quad x + 3 < 0$$

In set notation, these two cases are written

Case 1
$$(\{x|x + 2 > 0\} \cap \{x|x + 3 > 0\})$$

or

Case 2
$$(\{x|x + 2 < 0\} \cap \{x|x + 3 < 0\})$$

Case 1
If $x + 2 > 0$, then $x > -2$.
If $x + 3 > 0$, then $x > -3$. (See Figure 6.13.)

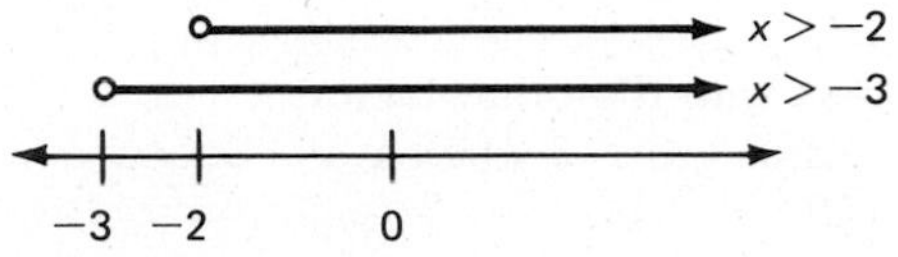

FIG. 6.13

$$\{x|x + 2 > 0\} \cap \{x|x + 3 > 0\} = \{x|x > -2\}$$

Case 2
If $x + 2 < 0$, then $x < -2$.
If $x + 3 < 0$, then $x < -3$. (See Figure 6.14.)

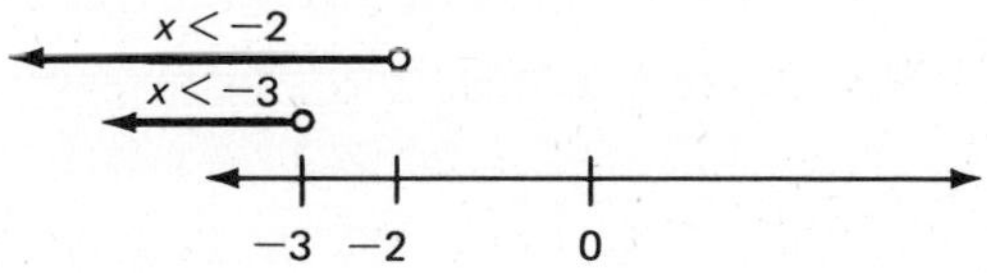

FIG. 6.14

$$\{x|x + 2 < 0\} \cap \{x|x + 3 < 0\} = \{x|x < -3\}$$
Therefore $\{x|x > -2\} \cup \{x|x < -3\}$ is the solution set—that is, $\{x|x < -3 \text{ or } x > -2\}$.

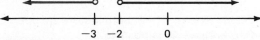

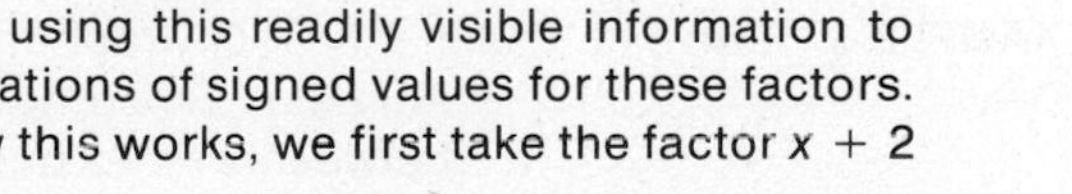

FIG. 6.15

A quicker way of handling these cases is by using a number line to mark the **regions** where each factor is positive or negative, and then using this readily visible information to determine the combinations of signed values for these factors.

To illustrate how this works, we first take the factor $x + 2$ from this example.

$x + 2 = 0$ when $x = -2$; $x + 2 > 0$ when $x > -2$; and $x + 2 < 0$ when $x < -2$.

On a number line (Figure 6.16), place a circle above -2, and then put plus signs to the right of -2 to indicate that the factor

$x + 2$ is positive to the right of -2; and put minus signs to the left of -2 to indicate that $x + 2$ is negative to the left of -2.

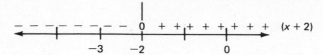

FIG. 6.16

Now repeat this process with the other factor, $x + 3$.
$x + 3 = 0$ when $x = -3$; $x + 3 > 0$ when $x > -3$; and $x + 3 < 0$ when $x < -3$.

Above the same number line put a circle over -3, put plus signs to the right of -3, and minus signs to the left of -3.

From Figure 6.17 we can readily see that to the left of -3 (that is, when $x < -3$) *both* factors are negative, and therefore the product $(x + 2)(x + 3)$ is positive. Between -3 and -2, one factor is positive and one factor is negative, and therefore the product $(x + 2)(x + 3)$ is negative.

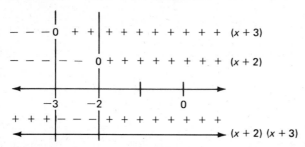

FIG. 6.17

To the right of -2, or when $x > -2$, both factors are positive and the product $(x + 2)(x + 3)$ is positive.

Using this information to find the solution set of $(x + 2)(x + 3) > 0$, we note that this is true when $x < -3$ or when $x > -2$. This result agrees with the solution we found by using cases. (See Figure 6.15.)

EXAMPLE 2   Solve $(x + 2)(x + 3) < 0$. Graph the solution set.

Solution   Referring to Figure 6.17, we see that between $x = -3$ and $x = -2$ the factor $x + 2$ is negative while the factor $x + 3$ is positive. The product of two factors with opposite signs is negative, and therefore the solution set of $(x + 2)(x + 3) < 0$ is $\{x \mid -3 < x < -2\}$, or as shown on the number line in Figure 6.18.

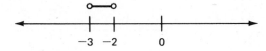

FIG. 6.18 $\{x|-3 < x < -2\}$

EXAMPLE 3 Find the solution set of the inequality $2x^2 - 5x < 3$ and represent the solution set on a line graph.

Solution

$$2x^2 - 5x < 3$$
$$2x^2 - 5x - 3 < 0 \qquad \text{(Addition theorem)}$$
$$(2x + 1)(x - 3) < 0 \qquad \text{(Factoring the polynomial)}$$

This product is negative when the factors $(2x + 1)$ and $(x - 3)$ are opposite in sign; that is, when one factor is positive and the other is negative. Again, the simplest method of solution is the number line method.

$2x + 1 = 0$ when $x = -\dfrac{1}{2}$; when x is to the right of $-\dfrac{1}{2}$, $2x + 1$ is positive, and when x is to the left of $-\dfrac{1}{2}$, $2x + 1$ is negative. (Try it. Use $x = 0$, a value to the right of $-\dfrac{1}{2}$. $2(0) + 1 = 1$, a positive number. Let $x = -2$, a value to the left of $-\dfrac{1}{2}$. Then $2(-2) + 1 = -3$, a negative number.)

$x - 3 = 0$ when $x = +3$. $x + 3$ is negative to the left of $+3$ and it is positive to the right of $+3$. Mark these regions above a number line as shown in Figure 6.19 with appropriate plus signs and minus signs. From the figure we can easily see that the factors are opposite in sign between $-\dfrac{1}{2}$ and $+3$.

Therefore the solution set of the inequality $(2x + 1)(x - 3) < 0$ is $\left\{x|-\dfrac{1}{2} < x < 3\right\}$ as illustrated in Figure 6.20.

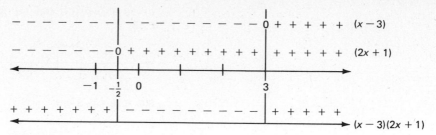

FIG. 6.19

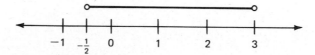

FIG. 6.20 $\left\{x \mid -\dfrac{1}{2} < x < 3\right\}$

Since division can be defined as multiplication

$$\frac{a}{b} = a \cdot \frac{1}{b}, \qquad b \neq 0$$

the product theorems for inequalities apply to the solution of expressions such as

$$\frac{f(x)}{g(x)} > 0 \quad \text{or} \quad \frac{f(x)}{g(x)} < 0$$

where $f(x)$ and $g(x)$ are polynomials in x, and $g(x) \neq 0$.

EXAMPLE 4   Find the solution set for

$$\frac{x + 2}{x - 3} \leq 5$$

and graph the solution set.

Solution

If $\dfrac{x + 2}{x - 3} \leq 5$,   then $\dfrac{x + 2}{x - 3} - 5 \leq 0$

Expressing the left side of the inequality as a single fraction,

$$\frac{x + 2 - 5(x - 3)}{x - 3} \leq 0$$

$$\frac{-4x + 17}{x - 3} \leq 0$$

The quotient is negative if and only if the numerator is positive and the denominator is negative, or if the numerator is negative and the denominator is positive. In other words, the quotient is negative if and only if the numerator and the denominator have opposite signs. (Remember that the denominator cannot be zero.)

Use the number line to mark the regions where the numerator is positive or negative. Be careful because $-4x + 17$ is *positive* to the *left* of $\frac{17}{4}$ and *negative* to the *right* of $\frac{17}{4}$, as shown below:

$$-4x + 17 = 0 \qquad\quad -4x + 17 > 0 \qquad\quad -4x + 17 < 0$$
$$-4x = -17 \qquad\qquad\quad -4x > -17 \qquad\qquad\quad -4x < -17$$
$$x = \frac{17}{4} \qquad\qquad\qquad\quad x < \frac{17}{4} \qquad\qquad\qquad\quad x > \frac{17}{4}$$

$$x - 3 - 0 \qquad x - 3 > 0 \qquad x - 3 < 0$$
$$x = 3 \qquad\quad x > 3 \qquad\quad x < 3$$

From Figure 6.21 we see that the numerator and denominator have opposite signs when x is less than 3 or when x is greater than $\frac{17}{4}$.

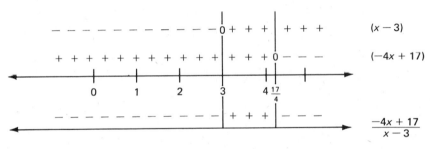

FIG. 6.21

The quotient *equals zero* only if the numerator is zero and the denominator is not zero. The numerator, $-4x + 17$, equals zero when $x = \frac{17}{4}$. Therefore the complete solution set is $\left\{x \mid x < 3 \text{ or } x \geq \frac{17}{4}\right\}$. Figure 6.22 is the graph of this solution set.

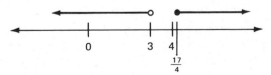

FIG. 6.22

EXAMPLE 5   Find the solution set for $x^2 + 4 \geq 4x$.

Solution   If $x^2 + 4 \geq 4x$, then
$$x^2 - 4x + 4 \geq 0$$
$$(x - 2)(x - 2) \geq 0$$
$$(x - 2)^2 \geq 0$$

We know that $x - 2$ is either a positive number, a negative number, or zero, depending on the value of x. We also know that the square of any positive number is positive; that the square of any negative number is positive; and that the square of zero is zero. Therefore no matter what value x is, the square of $x - 2$ will always be positive or zero. Thus the statement is true for *any* real value of x and the solution set is the set of real numbers R.

EXERCISES 6.3

Solve Exercises 1–40, represent the solution on a line graph, and write a symbolic statement describing the solution set.

1. $(x + 3)(x + 5) > 0$

2. $(x - 3)(x + 5) \geq 0$

3. $(x + 3)(x + 5) \leq 0$

4. $(x - 3)(x + 5) < 0$

5. $(2x - 1)(x + 4) \geq 0$

6. $(3x + 1)(x - 2) > 0$

7. $(2x - 1)(x + 4) < 0$

8. $(3x + 1)(x - 2) \leq 0$

9. $x(x + 4) > 0$

10. $3x(x + 5) < 0$

11. $x^2 \leq 2x$

12. $4x^2 \leq 3x$

13. $x^2 > 3x$

14. $5x^2 > x$

15. $x^2 - 3x - 4 < 0$

16. $x^2 - 5x + 6 \leq 0$

17. $x^2 < x + 12$

18. $x^2 < 4x + 12$

19. $2x^2 - 35 > 9x$

20. $4 - 5x^2 > 19x$

21. $x^2 + 9 \geq 6x$

22. $x^2 + 9 \leq 6x$

23. $(x + 1)^2 \leq 0$

24. $x^2 - 2x > 1$

25. $x^2 - 5 < 4x$

26. $(x - 2)^2 \geq 0$

27. $(x - 4)^2 > 25$

28. $x^2 + 3 > 4x$

29. $x + \dfrac{1}{x} > 2$

30. $x + \dfrac{9}{x} < 6$

31. $\dfrac{x - 3}{3 + x} \leq 0$

32. $\dfrac{3x^2 - 4}{3x} \geq x + 2$

33. $x^2 + 1 \leq 2x$

34. $\dfrac{x - 1}{x + 4} < \dfrac{x - 4}{x + 1}$

35. $\dfrac{1}{x + 3} \geq \dfrac{1}{x - 3}$

36. $\dfrac{1 - x}{x - 4} > 0$

37. $\dfrac{3x - 2}{3} \geq \dfrac{x^2 - 4}{x}$

39. $\dfrac{x}{x + 4} > \dfrac{x - 1}{x + 3}$

38. $\dfrac{3}{3x + 1} < \dfrac{4}{4x - 5}$

40. $\dfrac{1}{x} < \dfrac{1}{x - 5}$

In Exercises 41–44 consider the following: The product of three numbers is positive if all three numbers are positive, or if any two numbers are negative. The product of three numbers is negative if one of the numbers is negative and the other two are positive, or if all three numbers are negative. Use these facts plus the **number line approach** *shown in the examples to solve each of the following inequalities.*

41. $(x + 2)(x - 3)(x + 4) > 0$

43. $x(2x + 3)(3x - 1) < 0$

42. $(x + 2)(x - 3)(x + 4) < 0$

44. $x(2x + 3)(3x - 1) > 0$

For what values of x does each of the expressions in Exercises 45–50 designate a real number?

45. $\sqrt{5x + 10}$

48. $\sqrt{x^2 - 16}$

46. $\sqrt{3x - 2}$

49. $\sqrt{25 - x^2}$

47. $\sqrt{x^2 - 25}$

50. $\sqrt{16 - x^2}$

51. The braking distance s for a car decelerating at 22 ft/sec² from an initial velocity v is given by

$$s = \frac{v^2}{44}$$

Find the possible initial velocities v in (a) feet per second, and (b) miles per hour so that the car may stop in less than 99 ft. (*Hint:* 44 ft/sec = 30 mph.)

52. If a cubical box is lined with an insulating material t in. thick and if s is the length of a side of the interior of the box before insulation, then the loss of volume V due to the insulation may be approximated to within 1 unit by

$$V = 6ts^2 - 12t^2s \qquad \left(0 < t < \frac{1}{2}\right)$$

If $s = 6$ in., for what values of t will V be between 64 cu in. and 90 cu in.?

53. A traffic flow study produced the empirical relation

$$n = 5(60v - v^2)$$

where n = the number of vehicles per hour traveling on a certain bridge and v = the average speed in miles per hour maintained on the bridge. Find the values of v for which n is at most 4420 vehicles per hour.

54. If a missile is fired directly upward from the ground with an initial velocity of 800 ft/sec, then its distance s in feet above the ground t sec after the missile was fired may be approximated by

$$s = 800t - 16t^2$$

For what values of t is the missile more than 9600 ft above the ground?

6.4 ABSOLUTE VALUE: EQUALITIES

The number line has been used as a geometric model for the set of real numbers, and points on the number line correspond to real numbers. Any point x and its opposite, or additive inverse, $-x$, are the same distance from the origin but on opposite sides of the origin. The algebraic sign of a number indicates on which side of the origin the corresponding point is located. The distance of the point from the origin, regardless of the side on which it is located, is called the absolute value of the number and is denoted by the symbol $|x|$.

DEFINITION

The absolute value of a real number $|x|$:
$$|x| = x \quad \text{if} \quad x \geq 0$$
$$|x| = -x \quad \text{if} \quad x < 0$$

The definition indicates that the absolute value of a number is *never* negative, $|x| \geq 0$. In particular, $|0| = 0$. Figure 6.23 illustrates that the graphs of the number 3 and the number -3 are the same distance from the origin but on opposite sides.

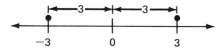

FIG. 6.23

If $x = 3$, then $x > 0$, and $|x| = x$ implies $|3| = 3$.
If $x = -3$, then $x < 0$, and $|x| = -x$ implies $|-3| = -(-3) = 3$.

THEOREM

If $P(x)$ is a polynomial in the variable x and a is a nonnegative real number, then the solution set of $|P(x)| = a$ is the **union** of the solution sets of $P(x) = a$ and $-P(x) = a$. Thus there are *two* cases to be considered.

If $|P(x)| = a$, then $P(x) = a \quad \text{or} \quad P(x) = -a$.

EXAMPLE 1   Solve for x: $|x| = 4$

Solution

Using the theorem, $|x| = 4$, then

$$x = 4 \quad \text{or} \quad x = -4$$

The solution set is $\{4, -4\}$.

EXAMPLE 2   Solve for x: $|x - 3| = 5$

Solution

$$|x - 3| = 5$$
$$x - 3 = 5 \quad \text{or} \quad x - 3 = -5$$
$$x = 8 \quad \text{or} \qquad x = -2$$

The solution set is $\{-2, 8\}$.

If a and b are any two real numbers, the *distance* between the points on the number line corresponding to a and b is defined in terms of absolute value.

DEFINITION

If A and B are two points on a number line with coordinates a and b, respectively, then the distance $|AB|$ between $A:(a)$ and $B:(b)$ is
$$|AB| = |b - a|$$

The notation $A:(a)$ means "the point on the number line corresponding to a."

If $a < b$, then by definition there exists a positive real number d so that $a + d = b$ and thus $d = b - a$. If $A:(a)$ and $B:(b)$ are two points on the number line such that A is to the left of B, then $a < b$ and $d = b - a$ is the length of the line segment AB (Figure 6.24).

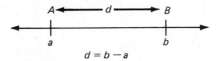

$$d = b - a$$

FIG. 6.24

In some problems it is not known whether A is to the left of B or not. Since $-(a - b) = b - a$, subtraction in the opposite order would produce a negative number. However, $|a - b| = |b - a|$, and thus a positive number can be obtained by using the concept of absolute value. In the special case that A is the same point as B, then $a = b$ and $a - b = b - a = 0$.

EXAMPLES

$$A:(3), \ B:(5), \ |AB| = |5 - 3| = 2$$
$$A:(3), \ B:(-5), \ |AB| = |-5 - 3| = |-8| = 8$$
$$A:(-3), \ B:(-5), \ |AB| = |-5 - (-3)| = |-5 + 3| = |-2| = 2$$

(See Figure 6.25.)

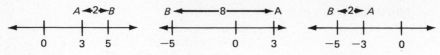

FIG. 6.25

EXAMPLE 3 Solve for x and graph the solution set on a number line: $|x + 2| + 1 = 4$

Solution

If $|x + 2| + 1 = 4$

then

$$|x + 2| = 3$$
$$x + 2 = 3 \quad \text{or} \quad x + 2 = -3$$
$$x = 1 \quad \text{or} \quad x = -5$$

The solution set is $\{-5, 1\}$, and its graph is shown in Figure 6.26.

FIG. 6.26

EXAMPLE 4 Solve for x: $|2x - 3| = -1$

Solution   By definition, the absolute value of a number is *never* negative. Therefore there is no value of x for which this statement is true, and the solution set is the empty set, $\emptyset$.

The following theorems are useful when working with absolute values and will be stated without proof.

THEOREMS

For every real number a and b,
$$|a||b| = |ab|$$
For every real number a and for every nonzero real number b,
$$\frac{|a|}{|b|} = \left|\frac{a}{b}\right|$$

EXAMPLES

$|3x| = 3|x|$, since $|3| = 3$; $|x + 2||x - 3| = |(x + 2)(x - 3)|$

$$\frac{|-5|}{|2|} = \left|\frac{-5}{2}\right|; \qquad \frac{|x + 2|}{|x|} = \left|\frac{x + 2}{x}\right| \text{ for } x \neq 0$$

EXERCISES 6.4

In Exercises 1–10 find the distance between A and B.

1. $A:(7), B:(3)$

2. $A:(3), B:(7)$

3. $A:(-7), B:(-2)$

4. $A:(4), B:(-3)$

5. $A:(-4), B:(3)$

6. $A:(-4), B:(-3)$

7. $A:(-5), B:(0)$

8. $A:(-3), B:(3)$

9. $A:(6), B:(-6)$

10. $A:(0), B:(-2)$

Solve Exercises 11–40 for all values of x.

11. $|x| = 3$

12. $|x| = 2$

13. $|x| = \dfrac{1}{2}$

14. $|x| = 0$

15. $|x| = -2$

16. $|x + 2| = 3$

17. $|-x| = 2$

18. $|x - 2| = 3$

19. $|x - 3| = 5$

20. $|2x + 1| = 5$

21. $|3x + 2| = 11$

22. $|4 - x| = 5$

23. $|2 - 3x| = 4$

24. $\left|\dfrac{2x - 4}{4}\right| = 2$

25. $\left|\dfrac{x + 3}{2}\right| = 1$

26. $3 - |x| = 4$

27. $4 - |x| = 2$

28. $\dfrac{3}{|x + 1|} = 4$

29. $\dfrac{2}{|x - 2|} = 1$

30. $\left|4 - \dfrac{x}{2}\right| = 3$

31. $\left|\dfrac{3x + 1}{2}\right| = 4$

32. $4 - |x + 1| = 3$

33. $2|x - 2| = 6$

34. $3|x + 1| = 1$

35. $\left|2x + \dfrac{1}{2}\right| - \dfrac{1}{4} = 0$

36. $\dfrac{1}{|x + 3|} = 4$

37. $\dfrac{|x|}{x} = 1$

38. $\dfrac{|x|}{x} = -1$

39. $|x - a| = b, b \geq 0$

40. $|a - x| = b, b \geq 0$

In Exercises 41–50 find the values of x for which each statement is true.

41. $|2x| = 2x$

42. $|x| = |-3|$

43. $|x| = |-4|$

44. $|2x - 1| = 2x - 1$

45. $|2x - 1| = |1 - 2x|$

46. $|2x - 1| = 1 - 2x$

47. $|2 - x| = x - 2$

48. $|x - 2| = 2 - x$

49. $|x - 2| = x - 2$

50. $|x - 2| = |2 - x|$

6.5 ABSOLUTE VALUE: INEQUALITIES

The solution of inequalities involving absolute value involves a closer look at what is meant by expressions such as $|x| > 3$ or $|x + 2| < 1$.

$|x| = 3$ means that $x = 3$ or $x = -3$—that is, x is 3 units to the right of the origin, or x is 3 units to the left of the origin on a horizontal number line.

$|x| > 3$ means that x is *more* than 3 units from the origin. This again raises two possibilities:

Case 1 x is *more* than 3 units to the *right* of the origin—that is, $x > 3$.

Case 2 x is *more* than 3 units to the *left* of the origin—that is, $x < -3$.

Thus $|x| > 3$ indicates that

$$x > 3 \quad \text{or} \quad x < -3$$

$|x + 2| < 1$ means that the distance between x and -2 is less than 1 unit. (Recall that $|a - b|$ is the distance between a and b, and that $|x + 2| = |x - (-2)|$.)

Thus, according to Figure 6.27,

$$-3 < x < -1$$

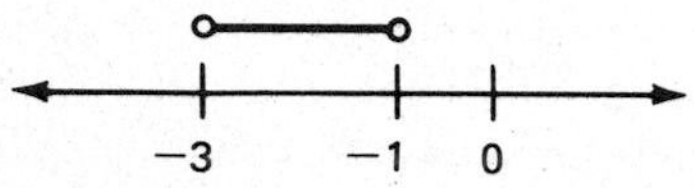

FIG. 6.27

The above intuitive arguments are formalized in the following theorems:

THEOREMS

Let $P(x)$ be an algebraic expression in the variable x and a be a real number such that $a \geq 0$.

1. $|P(x)| \leq a$ if and only if $-a \leq P(x) \leq a$.

2. $|P(x)| \geq a$ if and only if $P(x) \geq a$ or $P(x) \leq -a$.

EXAMPLE 1 Solve for x and graph the solution set: $|x| < 5$

Solution By Theorem 1, $|x| < 5$ means

$$-5 < x < 5$$

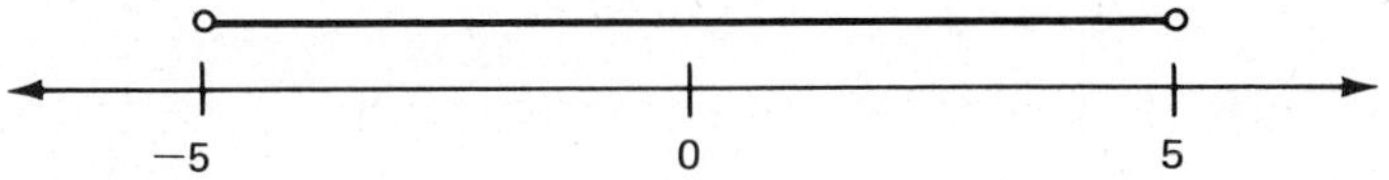

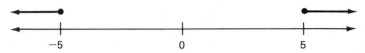

FIG. 6.28

EXAMPLE 2 Solve for x and graph the solution set: $|x| \geq 5$

Solution By Theorem 2, $|x| \geq 5$ means that
$$x \geq 5 \quad \text{or} \quad x \leq -5$$

FIG. 6.29

EXAMPLE 3 Solve for x and graph the solution set: $|x - 2| \leq 3$

Solution By Theorem 1, $|x - 2| \leq 3$ means
$$-3 \leq x - 2 \leq 3$$
Thus
$$-1 \leq x \leq 5 \quad \text{(Addition theorem, +2)}$$

FIG. 6.30

The solution set is $\{x \mid -1 \leq x \leq 5\}$. A verbal description of $|x - 2| \leq 3$ is that x is less than 3 units from 2, or x is exactly 3 units from 2.

EXAMPLE 4 Solve: $|x + 2| > 4$

Solution
$$x + 2 > 4 \quad \text{or} \quad x + 2 < -4 \quad \text{(Using Theorem 2)}$$
$$x > 2 \quad \text{or} \quad x < -6$$
The solution set is $\{x \mid x > 2 \text{ or } x < -6\}$.

FIG. 6.31

EXAMPLE 5 Solve for x: $\dfrac{3}{|x|} < 5$

Solution $|x|$ is never negative, so multiplication by $|x|$ does not change the sense of the inequality. Since $|x|$ is in the denominator, $x \neq 0$.

$$\frac{3}{|x|} < 5$$

$$3 < 5|x|$$

$$|x| > \frac{3}{5}$$

$$x > \frac{3}{5} \quad \text{or} \quad x < -\frac{3}{5} \qquad \text{(By Theorem 2)}$$

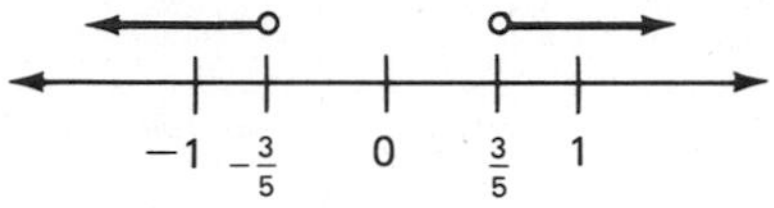

FIG. 6.32

EXAMPLE 6 Solve: $|2x + 1| > -2$

Solution By definition, the absolute value of a number is never negative. Since zero is greater than any negative number (0 is to the right of all negative numbers on a horizontal number line), and all positive numbers are greater than zero, it follows that the absolute value of a number is always greater than any negative number.

Therefore the solution set of $|2x + 1| > -2$ is the set of real numbers, R.

EXERCISES 6.5

In Exercises 1–40 solve and represent the solution set on a line graph.

1. $|x| < 2$

2. $|x| \leq 3$

3. $|x| \geq 2$

4. $|x| > 3$

5. $|2x| < 3$

6. $|3x| < 2$

7. $|x - 4| > 3$

8. $|x + 3| \geq 2$

9. $|x + 4| \leq 2$

10. $|x - 2| < 3$

11. $|3 - 2x| < 4$

12. $|2 - 3x| < 4$

13. $|3 - 2x| > 4$

14. $|2 - 3x| > 4$

15. $2 > |x - 1|$

16. $3 < |2 - x|$

17. $|4x - 1| \geq 0$

18. $|2x + 3| \geq 0$

19. $|x - 4| < 0$

20. $|2x - 1| < 0$

21. $|2x| + 3 > 4$

22. $\left|\dfrac{x - 2x}{3}\right| > 2$

23. $\dfrac{2}{|2 - x|} < 3$

24. $\dfrac{4}{|x|} > 2$

25. $|x| \leq -4$

26. $3 - |2 - x| < 6$

27. $|x - 3| > -2$
28. $|x - 3| < -2$
29. $\left|\dfrac{2x - 1}{4}\right| < 2$
30. $|x| + 2 \le 3$
31. $3 \le \left|4 - \dfrac{x}{2}\right|$
32. $\dfrac{3}{|x + 1|} > 4$
33. $|x| \le -2$

34. $2 + |x - 3| < 5$
35. $|x + 3| > -3$
36. $|x + 3| < -3$
37. $3|4x + 1| \le 6$
38. $\left|\dfrac{-2}{x}\right| > 3$
39. $\left|\dfrac{2 - 3x}{x}\right| > 1$
40. $|x - 2| < |x + 3|$

DIAGNOSTIC TEST

Complete each of the following by correctly filling in the blanks.

1. A symbolic statement for "x is greater than 2" is _____ .
2. A symbolic statement for "x is greater than 4 and x is less than 6, or x is equal to 6" is _____ .
3. $\{x \,|\, -2 < x \le 5\} \cap \{x \,|\, x \le -1\}$ is written as a single set as _____ .
4. The set operation "union" is symbolized _____ and corresponds to the English word _____ .
5. The set operation "intersection" is symbolized _____ and corresponds to the English word _____ .
6. If $a < b$ and $c > 0$, then ac _____ bc.
7. If $a < b$ and $c < 0$, then ac _____ bc.
8. If $x < 5$, then $-3x$ _____ -15.
9. If $3x - 5 < 4 + 7x$, then x _____ .
10. The solution set for $|x + 4| = 7$ is _____ .
11. The solution set for $|x + 4| < 7$ is _____ .
12. The solution set for $|x + 4| > 7$ is _____ .
13. If $(x + 3)(x - 2) > 0$, then the solution set is _____ .
14. If $\dfrac{x - 2}{3x + 1} \le 0$, then the solution set is _____ .
15. If $x^2 + 9 \le 6x$, then the solution set is _____ .

For each of the following problems, select the best answer. If the answer is "none of these" write the correct answer.

16. If $|x - 3| < -2$, then x is a real number such that
    (a) $x < -1$ or $x > 1$
    (b) $-1 < x < 5$
    (c) $x < 1$
    (d) $x > 0$
    (e) none of these

17. If $3x - 8 \geq 10$, then x is a real number such that

(a) $x \geq 6$ (d) $x \leq -6$

(b) $x \leq 6$ (e) none of these

(c) $x \geq -6$

18. The graph of the solution set of $|2x + 1| < 5$ includes all points on the number line that are

(a) between -5 and 5

(b) to the left of 2

(c) between -3 and 2

(d) to the left of -3 or to the right of 2

(e) none of these

19. If $|x - 3| = 3 - x$, then

(a) $x < 0$ (d) $x > 3$

(b) $x > 0$ (e) none of these

(c) $x < 3$

20. The solution set for $x^2 - 4x + 4 \geq 0$ is

(a) $x > 2$

(b) $x < 2$

(c) the empty set

(d) R, the set of real numbers

(e) none of these

REVIEW EXERCISES

Graph Exercises 1–10 on a number line and write a verbal statement describing the set. (Section 6.1)

1. $\{x \mid x < 4\}$ **6.** $\{x \mid x < -1\} \cup \{x \mid x < -3\}$

2. $\{x \mid x \geq -1\}$ **7.** $\{x \mid x > 2\} \cup \{x \mid x < -1\}$

3. $\{x \mid -2 < x \leq 5\}$ **8.** $\{x \mid x \geq -1\} \cap \{x \mid x > -3\}$

4. $\{x \mid -1 \leq x < 0\}$ **9.** $\{x \mid -1 < x < 3\} \cap \{x \mid 0 < x < 5\}$

5. $\{x \mid x < 2\} \cap \{x \mid x \geq -1\}$ **10.** $\{x \mid 0 < x < 5\} \cap \{x \mid 1 \leq x \leq 3\}$

Solve Exercises 11–30 and represent the solution set on a line graph.

11. $2x - 1 > x + 3$ **16.** $3 \leq \dfrac{x + 6}{x}$

12. $x^2 + 3x > 0$ **17.** $x^2 - x > 12$

13. $2x - 1 \leq 3x + 5$ **18.** $x^2 \leq 2x + 15$

14. $\dfrac{x}{x - 2} > 0$ **19.** $\dfrac{x - 1}{x + 3} \leq 5$

15. $\dfrac{2x + 3}{x - 2} < 0$ **20.** $|2x + 5| = 1$

21. $|5 - 3x| = 2$

22. $|x + 4| > 5$

23. $|3 - 2x| \leq 6$

24. $|2x - 6| \geq 4$

25. $\dfrac{3}{|x + 2|} \leq 5$

26. $4 + \dfrac{|x - 3|}{2} < 6$

27. $|4x - 7| \geq 0$

28. $|5x + 4| < 0$

29. $|3x + 2| > -3$

30. $|2x - 1| + 2 < 0$

31. For what values of x is $|2x - 5| = 2x - 5$?

32. For what values of x is $|2x - 5| = 5 - 2x$?

33. For what values of x is $|2x - 5| = |5 - 2x|$?

34. A student must have an average grade of 90% to 100% inclusive on six tests in a course to receive an A grade. If her grades on the first five tests are 86%, 84%, 97%, 92%, and 88%, what grade must she get on the sixth test to qualify for the A?

35. Suppose the student in Exercise 34 received 80% instead of 88% on the fifth test (the instructor made an error in grading the test), can the student still get an A? If so, what grade must she get on the sixth test to qualify for the A?

7
RELATIONS AND FUNCTIONS

One of the fundamental concepts of concern to the mathematician is how numbers, or their "real-world" applications, are related. For instance, profit from the sale of a certain commodity is related somehow to the number of items produced. The distance an object travels is related to the time it travels.

The mathematical term for these relationships is **relation.** A special type of relation, and one that is of great importance in mathematics, is called a **function.** Profit is a function of the number of items produced, and distance is a function of time. Not all relations are such that the term **function** can be applied. It is the purpose of this chapter to define the relation and function concepts and to introduce some special mathematical relations and functions.

7.1 ORDERED PAIRS, RELATIONS, AND FUNCTIONS

An **ordered pair** is an expression having the form (a, b), where a is called the **first component** (or first member) of the ordered pair and b is called the **second component** (or second member) of the ordered pair.

The order in which the components of an ordered pair are written is important. For example, the ordered pair $(3, 5)$ is not the same as the ordered pair $(5, 3)$.

DEFINITION—Relation

A set of ordered pairs is called a **relation.**

For example, the set $S = \{(3, 2), (4, 1), (2, 3)\}$ is a set whose elements are ordered pairs, and thus it is a relation. Also, the set $T = \{(x, y) | y = 3x + 1\}$ is a relation. To show *some* of the ordered pairs of the relation T described by this rule, let x take the values $0, \frac{1}{2}, -1$, and 4 and find the corresponding y value for each x value: when $x = 0$, $y = 3(0) + 1 = 1$, so the ordered pair is $(0, 1)$; when $x = \frac{1}{2}$, $y = 3\left(\frac{1}{2}\right) + 1 = \frac{5}{2}$, so the ordered pair is $\left(\frac{1}{2}, \frac{5}{2}\right)$; when $x = -1$, $y = 3(-1) + 1 = -2$, so the ordered pair is $(-1, -2)$; and when $x = 4$, $y = 3(4) + 1 = 13$; thus the ordered pair is $(4, 13)$.

DEFINITIONS—Domain and Range

The **domain** of a relation is the set of all first components of the relation (also called the **independent variable**).

The **range** of a relation is the set of all second components of the relation (also called the **dependent variable**).

EXAMPLE 1 Find the domain and the range of the relation
$$S = \{(1, 2), (3, -1), (4, 3), (-1, -2)\}$$
Solution From the definition, the domain is the set of all the first components, $\{1, 3, 4, -1\}$, and the range is the set of all the second components, $\{2, -1, 3, -2\}$.

EXAMPLE 2   What is the domain and the range of the relation
$T = \{(x, y)\,|\,y = x + 2\}$?

Solution   The largest domain of this relation in the set of real numbers is the set of all values of x for which y is defined in the set of real numbers. There are basically only two circumstances under which y will *not* be defined in the set of real numbers: *when a substitution for x results in division by zero, and when a substitution for x results in taking an even root of a negative number.* Since x does not appear in the denominator of a fraction, no division by zero will result. And since there are no radicals in this problem, we do not have to worry about even roots of negative numbers. Therefore any real number substituted in the expression for x will produce a real number for y.

The domain of T is the set of real numbers, and the range is also the set of real numbers. There are infinitely many ordered pairs that satisfy the rule stating this relation.

It is usually easier to find the domain of a relation stated by a rule than to find its range.

EXAMPLE 3   Find the domain and the range of the relation

$$P = \left\{(x, y)\,\Big|\,y = \frac{2}{x + 3}\right\}$$

Solution   Since we must find only the values of x for which y is defined, and the denominator of the fraction $\dfrac{2}{x + 3}$ is equal to zero when $x = -3$, the domain of the relation is the set of all real numbers *except* -3.

The range is more difficult, but we know, for example, that in order for a fraction to equal 0, the numerator must be 0 and the denominator cannot be 0. Since the numerator of the fraction is 2, and 2 can never equal 0, y cannot equal zero; therefore 0 is not in the range of the relation. Actually the range consists of all real numbers except $y = 0$.

As you see from the examples just given, there are several methods for specifying particular relations. One method is the **listing** of the ordered pairs, as in Example 1. Another method is the stating of the relation between the ordered pairs as an **equation** or **rule,** as in Examples 2 and 3. We found ordered pairs that satisfied the relation intuitively; thus we now state the formal definition of the solution of an equation in two variables.

DEFINITION

The solution set of an open equation in two variables x and y is the set of ordered pairs of the form (a, b) such that the equation becomes true when x is replaced by a and y is replaced by b.

A third way of specifying a relation is by a graph. The most common reference system for this purpose is a rectangular coordinate system as shown in Figure 7.1.

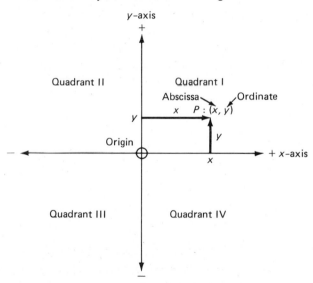

FIG. 7.1 Rectangular coordinate system

A **rectangular coordinate system** is obtained by taking two perpendicular number lines, called the **axes,** intersecting at their origins. The point of intersection of the axes is called the **origin** of the coordinate system.

It is customary to select one axis horizontal, called the **x-axis,** with its positive direction to the right, and the other axis vertical, called the **y-axis,** with its positive direction upward.

The axes separate the plane into four regions, called **quadrants,** that are numbered consecutively starting with the upper right quadrant and proceeding counterclockwise.

With each ordered pair (a, b) is associated a unique point P, the point of intersection of a vertical line through a point x = a on the x-axis and a horizontal line through a point y = b on the y-axis. The numbers of the ordered pair (a, b) are called the **coordinates** of P, with the first component a called the **abscissa** and the second component b called the **ordinate.**

If the axes on a graph are labeled by letters other than x and y, it is customary to take the domain elements (the first components of the ordered pairs) from the values on the horizontal axis and the range elements (the second components of the ordered pairs) from the vertical axis.

EXAMPLE 4 Given the relation: $\{(x, y)\,|\,y = 3x + 2\}$ plot the points (0, 2), (1, 5), (−1, −1), and (−2, −4). Note that these values are solutions of the equation $y = 3x + 2$ and that the points lie on a straight line. Join the points, and from the graph find what value of y corresponds to $x = \dfrac{1}{2}$ on this line. Does the point (2, 4) lie on the line?

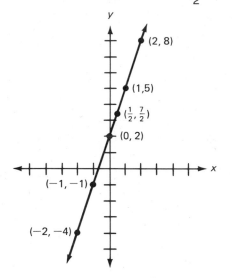

FIG. 7.2

Solution Figure 7.2 is the graph of the relation. From the figure we note that a vertical line through $x = \dfrac{1}{2}$ intersects the line joining the points at $y = 3\dfrac{1}{2}$. Therefore the point $\left(\dfrac{1}{2}, \dfrac{7}{2}\right)$ is on the graph, and the ordered pair $\left(\dfrac{1}{2}, \dfrac{7}{2}\right)$ is a solution to the equation $y = 3x + 2$.

To check, when $x = \dfrac{1}{2}$, $y = 3\left(\dfrac{1}{2}\right) + 2 = \dfrac{7}{2}$. The point (2, 4) does *not* lie on the line. When x = 2, we see from the graph that y = 8. Also, (2, 4) is *not* a solution to $y = 3x + 2$. By substituting these values, we get 4 = 3(2) + 2 = 8, and since 4 ≠ 8, this is a false statement.

It is important to recognize that solutions of equations, when read from a graph, are often only approximations and may be quite inaccurate. For example, a value of $x = \dfrac{3}{17}$ or y = 1.461 would be almost impossible to read with much accuracy from most graphs.

The graphs of many different relations are discussed in subsequent sections of this text.

A relation is defined as *any* set of ordered pairs. A special relation that associates with each first component in its domain *exactly* one second component in its range is now defined.

DEFINITION—Function

A **function** is a relation in which no two ordered pairs have the same first components and different second components.

If a relation consists of ordered pairs (x, y), then the relation is a function if and only if every x value in the domain of the relation corresponds to one and only one y value in the range of the relation.

Functions are often named by lowercase letters such as f, g, and h.

EXAMPLE 5 Which of the following relations are functions?

$$f = \{(2, 3), (4, 5), (1, 3), (6, 5)\}$$
$$g = \{(1, 1), (2, 2), (3, 3)\}$$
$$h = \{(3, 2), (5, 4), (3, 1), (5, 6)\}$$

Solution f is a function because each ordered pair has a different first component. g is a function because each ordered pair has a different first component. h is *not* a function because the pairs (3, 2) and (3, 1) have the first component, 3, associated with 2 and with 1. Also, 5 is paired with 4 and with 6.

EXAMPLE 6 State the domain and range of the functions f and g in Example 5.

Solution

Domain of $f = \{2, 4, 1, 6\} = \{1, 2, 4, 6\}$
Range of $f = \{3, 5\}$
Domain of $g = \{1, 2, 3\}$
Range of $g = \{1, 2, 3\}$

The notation $f(x)$, read "f of x," is used to designate the unique value of y that is paired with a given value of x in the domain of the function f.

For example, if

$$f = \{(x, y)\,|\,y = 2x + 1\}$$

then $f(x) = 2x + 1$, since $y = f(x)$.

The function f could also be expressed by

$$f = \{(x, f(x))\,|\,f(x) = 2x + 1\}$$

EXAMPLE 7 For $f(x) = 2x + 1$, find the ordered pairs in the function f for $-2 \le x \le 2$, where x is an integer and $f(x)$ is a real number.

    Solution

$$y = f(x) = 2x + 1$$
$$f(\ ) = 2(\ ) + 1$$
$$f(-2) = 2(-2) + 1 = -3, \quad \text{and} \quad (-2, -3)$$
$$f(-1) = 2(-1) + 1 = -1, \quad \text{and} \quad (-1, -1)$$
$$f(0) = 2(0) + 1 = 1, \quad \text{and} \quad (0, 1)$$
$$f(1) = 2(1) + 1 = 3, \quad \text{and} \quad (1, 3)$$
$$f(2) = 2(2) + 1 = 5, \quad \text{and} \quad (2, 5)$$

The ordered pairs $(-2, -3)$, $(-1, -1)$, $(0, 1)$, $(1, 3)$ and $(2, 5)$ are all elements of the function $f = \{(x, f(x)) \mid f(x) = 2x + 1\}$.

EXAMPLE 8 If $f = \{(x, y) \mid y = f(x) = x^2 + 2x + 1\}$, evaluate $f(1)$, $f(-1)$, $f(0)$, $f(2)$, $f(a)$, $f(a + h)$, $f(a + h) - f(a)$.

    Solution

$$f(x) = x^2 + 2x + 1$$
$$f(\ ) = (\ )^2 + 2(\ ) + 1$$
$$f(1) = (1)^2 + 2(1) + 1 = 4$$
$$f(-1) = (-1)^2 + 2(-1) + 1 = 0$$
$$f(0) = (0)^2 + 2(0) + 1 = 1$$
$$f(2) = (2)^2 + 2(2) + 1 = 9$$
$$f(a) = (a)^2 + 2(a) + 1 = a^2 + 2a + 1$$
$$f(a + h) = (a + h)^2 + 2(a + h) + 1$$
$$= a^2 + 2ah + h^2 + 2a + 2h + 1$$
$$f(a + h) - f(a) = (a^2 + 2ah + h^2 + 2a + 2h + 1) - (a^2 + 2a + 1)$$
$$= a^2 - a^2 + 2ah + h^2 + 2a - 2a + 2h + 1 - 1$$
$$= 2ah + h^2 + 2h$$

EXAMPLE 9 The equation $C = \dfrac{1}{2}x + 40$ represents the total cost C in dollars of manufacturing x items of a certain commodity. This equation defines a function f and $C = f(x)$.

(a) Use a table of values to specify this function for the cost of manufacturing x items if $0 \le x \le 10$. (Since x is the *number* of items, x must be a nonnegative integer.)

(b) Graph this function; draw a line through the points and continue the line, so that the cost of manufacturing up to 20 items can be read from the graph.

(c) From the graph, what is the cost of manufacturing 0 items? (This is the *fixed* cost.) What is the cost of manufacturing 15 items? 20 items?

    Solution

(a) $C = \dfrac{1}{2}x + 40$; $C = f(x)$

x	0	1	2	3	4	5	6	7	8	9	10
$f(x)$	40	40.5	41	41.5	42	42.5	43	43.5	44	44.5	45

(b)

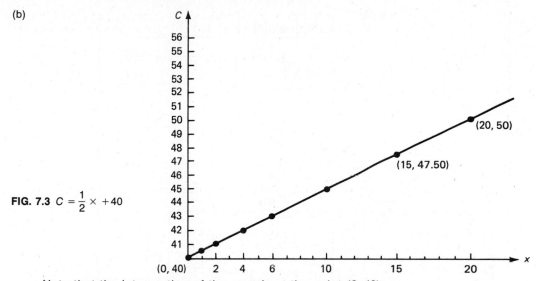

FIG. 7.3 $C = \frac{1}{2} \times + 40$

Note that the intersection of the axes is at the point (0, 40).
(c) The fixed cost is $40. The cost of manufacturing 15 items is $47.50. The cost of manufacturing 20 items is $55.00.

THE ORIGIN OF COORDINATES

The concept of a coordinate system probably originated with the ancient Egyptian surveyors. The hieroglyphic symbol used to designate the districts into which Egypt was divided was a grid symbol.

Records indicate that the Greeks used the ideas of longitude and latitude to locate points in the sky and on the earth. The Romans, who were noted for their surveying techniques, arranged the streets of their cities on a rectangular coordinate system.

The Arab and Persian mathematicians were the first to use geometric figures for algebraic problems. Examples are found in the works of the Arab al-Khowarizmi (ca. 825) and the Persian Omar Khayyam (ca. 1100). This usage is again found in the writings of Fibonacci (1220), Pacioli (1494), and Cardan (1545).

René Descartes (1596–1650) is credited with the invention of analytic geometry since he used a rectangular coordinate system to establish a relationship between equations and curves. Pierre de Fermat, another great French mathematician, formulated coordinate geometry at the same time and made a considerable contribution in this field. The modern terms **coordinates, abscissa,** and **ordinate** were contributed by the German mathematician Gottfried Wilhelm Leibniz in 1692.

CITY BLOCK MAP: A TYPE OF COORDINATE SYSTEM.

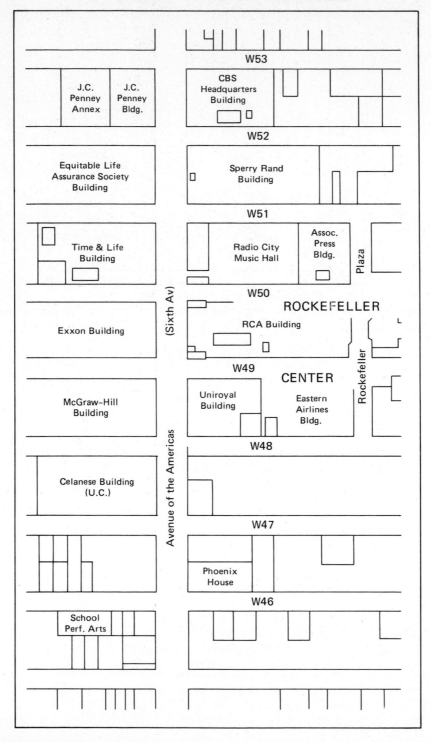

EXERCISES 7.1

For each of the relations in Exercises 1–10 state the domain and range. (Example 1)

1. {(1, 2), (3, 2), (5, 6), (2, 1)}
2. {(−1, −2), (0, 0), (3, −1)}
3. {(1, 2), (2, 2), (3, 3), (4, 4), (5, 5)}
4. {(10, 20), (12, 30), (14, 40), (16, 30)}
5. {(2, 4), (3, 1), (3, 2), (4, 2), (2, 4)}
6. {(2, 0), (−3, 1), (3, −2), (−4, 2)}
7. {(1, 1), (2, ½), (−2, −½)}
8. {(4, 0), (5, 0), (6, 0)}
9. {(1, 1), (2, 1), (3, 1), (4, 1), (5, 1)}
10. {(1, 1), (1, 2), (1, 3), (1, 4), (1, 5)}
11. Which of the relations in Exercises 1–10 are functions?

Each of the equations in Exercises 12–22 defines a relation. State the domain of each relation so that each corresponding range component is a real number. (Examples 2 and 3)

12. $y = \dfrac{x - 2}{x - 3}$

13. $y = \dfrac{2}{3x - x^2}$

14. $y = \sqrt{16 - x}$
15. $y = \sqrt{4 + x}$
16. $y = 5x - 3$
17. $y = x^2 - 4$

18. $y = \dfrac{|x|}{x}$

19. $y = \dfrac{x + 1}{x - 1}$

20. $y = \dfrac{x + 2}{x(x + 3)}$

21. $y = \sqrt{x - 16}$
22. $y = \sqrt{4 - x}$

23. From looking at the given equations, what, if anything, can you say about the range of each relation in Exercises 14, 15, 21, and 22?
24. Can you make any statement concerning the range of the relation defined in Exercise 13?

Which of the ordered pairs in Exercises 25–30 are solutions of $y = 2x + 3$? (Example 4)

25. (0, 3)
26. (2, 6)
27. (7, 2)

28. (3, 0)
29. (1, 5)
30. (5, 13)

If $f(x) = 3x^2 + 4x + 2$, find the designated values in Exercises 31–36. (Example 8)

31. $f(2)$
32. $f(-1)$
33. $f(0)$

34. $f(a)$
35. $f(a + 1)$
36. $f(a + 1) - f(a)$

If $g(x) = x^3 + x - 3$, find the designated values in Exercises 37–44. (Example 8)

37. $g(-3)$ **41.** $g(b)$
38. $g(0)$ **42.** $g(b + h)$
39. $2[g(1)]$ **43.** $g(b) + g(h)$
40. $g(2) + g(-2)$ **44.** $g(b + h) - g(b)$

If $f(x) = 2 + 3x - 4x^2$, find the designated values in Exercises 45–50. (Example 8)

45. $f(2)$ **48.** $f(a)$

46. $f(-1)$ **49.** $f\left(\dfrac{1}{a}\right)$

47. $f(0)$ **50.** $f(a + 2) - f(a)$

51. If $f(x) = 3x + 4$, find:
    (a) $f(x^2)$ (b) $[f(x)]^2$
52. If $f(x) = 2x + 1$ and $g(x) = x - 2$, find:
    (a) $f(x) + g(x)$ (b) $f(x) - g(x)$
53. The area of a triangle whose height is 5 in. is given by the equation $A = \dfrac{1}{2}(5)b$, where A represents the area of the triangle and b its base. This is a functional relation, so that $A = f(b)$. (The area is a function of the base.) Evaluate $f(b)$ for each of the following:
    (a) $f(3)$ (c) $f(b + h) - f(b)$

    (b) $f(b + h)$ (d) $\dfrac{f(b + h) - f(b)}{h}$

54. The intensity I of illumination on an object from a 120-candlepower source is a function of the distance x of the object from the source. Symbolically this particular function is written
$$I = I(x) = \frac{120}{x^2}$$
    (a) Find $I(1)$, $I(-1)$, $I(2)$, $I(-2)$, $I(10)$, $I(20)$.
    (b) Is $\left\{(x,\ I)\ \middle|\ I = \dfrac{120}{x^2},\ \text{when } x \text{ and } I \text{ are real numbers}\right\}$ a function? Why or why not?

55. Given the relation $\left\{(C, F)\ \middle|\ F = \dfrac{9}{5}C + 32\right\}$, which describes the conversion of degrees Celsius C to degrees Fahrenheit F. (Example 9)
    (a) Make a table of values for this relation for domain values $\{0, 5, 10, 15, 20, 25, 30, 35, 40\}$
    (b) Graph the relation for the specified values of C. (Let the horizontal axis be the C-axis.)
    (c) Draw a straight line through the points.
    (d) If normal body temperature is 98.6 degrees Fahrenheit, from the graph determine the corresponding temperature in degrees Celsius.
    (e) Is this relation a function? Why or why not?

56. The graph in Figure 7.4 relates the profit P in dollars for a certain commodity to the number of items x that are sold.

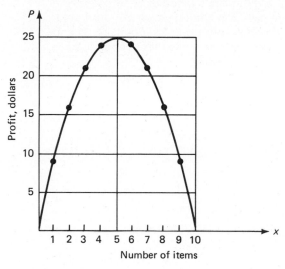

FIG. 7.4

(a) Using the graph, list the ordered pairs in the relation
$r = \{(x, P) | 0 \le x \le 10 \text{ and } x \text{ is an integer}\}$

(b) For what value of x is the profit the greatest? What is the greatest profit?

(c) Is the relation r a function? Why or why not?

7.2 LINEAR FUNCTIONS

Having defined a function as a set of ordered pairs for which no two ordered pairs have the same first component and a different second component, we are now ready to look at special functions in more detail. The first such category is the set of linear functions.

DEFINITION

A **linear function** is a function whose rule has the pattern
$$f(x) = ax + b$$
In set notation, a linear function f is the set
$$f = \{(x, y) | y = ax + b\} = \{(x, f(x)) | f(x) = ax + b\}$$

The graph of every linear function is a straight line. The equation of every nonvertical line is a linear function. The reason for excluding vertical lines should be obvious. All points on a vertical line have the same x value but different y values. For example, draw a line through the points (4, 2) and (4, 4) (see Figure 7.5). Since the only value that x can assume is 4, the equation of this line is $x = 4$.

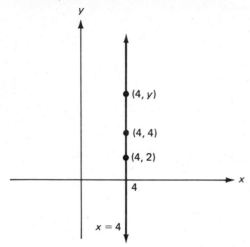

FIG. 7.5 Graph of $x = 4$

EXAMPLE 1 Graph the function $f = \{(x, y) | y = -2x + 6\}$.

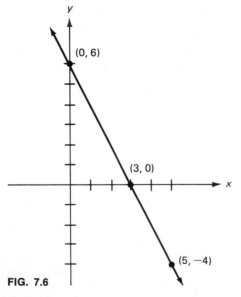

FIG. 7.6

Solution   Since any line is uniquely determined by two points, it is necessary only to find two solutions (ordered pairs) that satisfy a given linear function, plot the points corresponding to the coordinates, and draw a line through these points. This line represents the graph of the given equation. It is a good idea, however, to find a third solution as a check. Select *any x* value in the domain of f and compute the corresponding y value. It is convenient to select $x = 0$ for one point because the calculation is relatively easy. Selecting arbitrarily $x = 0$, $x = 3$, and $x = 5$, we compute

$$f(0) = -2(0) + 6 = 6; \quad (0, 6)$$
$$f(3) = -2(3) + 6 = 0; \quad (3, 0)$$
$$f(5) = -2(5) + 6 = -4; \quad (5, -4)$$

The points whose coordinates are given are plotted in Figure 7.6 and a straight line is drawn through them. The line continues beyond these points because they are only three points of infinitely many on the graph of this function.

A linear function of the form

$$y = ax + b$$

yields several helpful bits of information to aid in graphing the line. For example, if $x = 0$, $y = a(0) + b = b$, so the point $(0, b)$ is on the graph, and the line crosses the y-axis where $y = b$. This y value is called the **y-intercept** of the function.

EXAMPLE 2 If $y = 3x + 2$, find the y-intercept of the function.

Solution $b = 2$; therefore the y-intercept is 2, or the graph passes through the point $(0, 2)$.

EXAMPLE 3 If $f(x) = \dfrac{1}{2}x - \dfrac{1}{5}$, name the point where the graph of f crosses the vertical axis.

Solution The pattern still holds, and $b = \dfrac{1}{5}$; therefore the point where the graph of f crosses the vertical axis is $\left(0, \ -\dfrac{1}{5} \right)$.

An important characteristic of a line is its **slope.** Using any two points on a line—for example, the points whose coordinates are (a, b) and (c, d), respectively—we define the slope of the line passing through those points as follows.

DEFINITION

The slope of the line passing through the points whose coordinates are (a, b) and (c, d) is the ratio

$$\text{Slope} = \frac{b - d}{a - c}$$

if $a \neq c$.

It is important to notice that in the definition of slope, the *numerator* is the difference in the *y values* of the two points, and the *denominator* is the difference of the *x values* of the two points.

In using this formula to find the slope of a line, we may choose any two points on the graph. For example, in Figure 7.6 choose the points whose coordinates are $(3, 0)$ and $(5, -4)$. The y values are 0 and -4, and the x values are 3 and 5, corresponding to the y values.

$$\text{Slope} = \frac{0 - (-4)}{3 - 5} = \frac{4}{-2} = -2$$

The slope of a given line is the same for *all* points on the line, not just for the two specific points selected in the above illustration.

A definition for the slope of a general line can be conveniently expressed by using subscripts to indicate two points on the line. For example, P_1 (read "P sub one") and P_2 (read "P sub two") name two points. The numeral 1 at the lower right of P_1 and the numeral 2 at the lower right of P_2 are subscripts used to indicate that P_1 is the first point and P_2 is the second point.

Similarly, the coordinates of P_1 can be expressed as (x_1, y_1) and those of P_2 as (x_2, y_2).

DEFINITION

If $P_1:(x_1, y_1)$ and $P_2:(x_2, y_2)$ are any two points on a line and if $x_1 \neq x_2$, then the **slope** m of the line joining P_1 and P_2 is given by

$$m = \frac{y_2 - y_1}{x_2 - x_1}$$

See Figure 7.7.

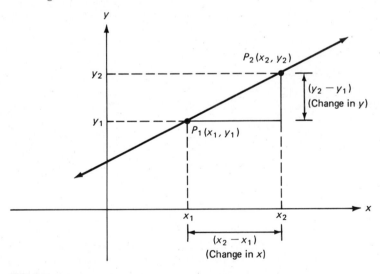

FIG. 7.7

If a line is vertical (parallel to the y-axis), then $x_1 = x_2$ and $x_2 - x_1 = 0$. Since the denominator of the slope ratio is zero, the *slope of a vertical line is undefined* (see Figure 7.8).

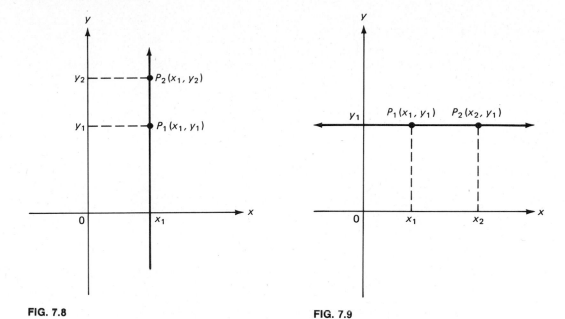

FIG. 7.8

FIG. 7.9

If a line is horizontal (parallel to the *x*-axis), then $y_1 = y_2$ and $y_2 - y_1 = 0$. In this case, the numerator of the slope ratio is zero, and *the slope of a horizontal line is 0* (see Figure 7.9).

EXAMPLE 4 Find the slope of the line passing through the points whose coordinates are (3, 2) and (5, 1), respectively.

Solution Using
$$m = \frac{y_2 - y_1}{x_2 - x_1} \quad \text{with} \quad (x_2, y_2) = (5, 1)$$
$$\text{and} \quad (x_1, y_1) = (3, 2)$$
$$m = \frac{1 - 2}{5 - 3} = -\frac{1}{2}$$

Alternate solution Using
$$m = \frac{y_2 - y_1}{x_2 - x_1} \quad \text{with} \quad (x_2, y_2) = (3, 2)$$
$$\text{and} \quad (x_1, y_1) = (5, 1)$$
$$m = \frac{2 - 1}{3 - 5} = -\frac{1}{2}$$

Note in Example 4 that the value of the slope does not depend on which point is called the first point and which the second.

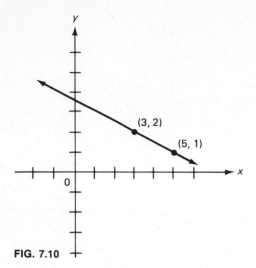

FIG. 7.10

Note also that the slope is negative, which means that the line "falls to the right" (see Figure 7.10).

Return now to the linear function $y = f(x) = ax + b$. It was established that b is the value of the y-intercept. It can now be stated that a is the slope of the line whose equation is $f(x) = ax + b$. Usually a is replaced by the letter m so that the linear function is stated

$$y = f(x) = mx + b$$

To verify that m is actually the slope, consider the following:

If $x = 0$, then $y = b$ and the ordered pair $(0, b)$ is a solution of $y = mx + b$. Selecting another value for x, let $x = 1$. Then $y = m + b$, and $(1, m + b)$ is a solution of $y = mx + b$. Since the slope of the line is determined by any two points on the line,

$$\text{Slope} = \frac{y_2 - y_1}{x_2 - x_1} = \frac{(m + b) - b}{1 - 0} = m$$

DEFINITION

The equation $y = mx + b$ is called the **slope-intercept form** of the line whose slope is m and whose y-intercept is b.

EXAMPLE 5 Find the slope and the y-intercept of the line described by the linear function $y = f(x) = \frac{1}{2}x - 6$.

Solution The slope $m = \frac{1}{2}$, and the y-intercept $b = -6$.

A linear function $f(x) = ax + b$ is called a constant function if $a = 0$. Thus the slope of a constant function is 0, and the general form of a constant function is $f(x) = b$.

DEFINITION

A function of the form $f(x) = b$, where b is a constant, is called a **constant function.**

The graph of a constant function is a horizontal line that crosses the y-axis at the point $(0, b)$.

EXAMPLE 6 Discuss and graph the function $\{(x, f(x)) \mid f(x) = 4\}$.

Solution $f(x) = 4$ is a constant function. The ordered pairs $(x, f(x))$ all have a second component, 4. All real numbers x are paired with 4. The graph is a line parallel to the x-axis and passing through the point $(0, 4)$ on the y-axis (see Figure 7.11).

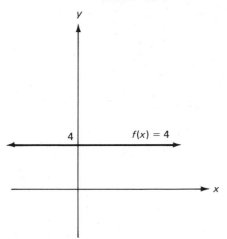

FIG. 7.11 Graph of $f(x) = 4$, a constant function

Another linear function that is important enough to have a special name is the function
$$f(x) = x$$

DEFINITION

The linear function f defined by
$$f(x) = x$$
is called the **identity function.**

Some ordered pairs of the identity function
$$f = \{(x, f(x)) \mid f(x) = x\}$$
are $\{(-3, -3), (0, 0), \left(\dfrac{1}{2}, \dfrac{1}{2}\right), (1, 1), (\sqrt{2}, \sqrt{2}), \ldots\}$. The first and second components of each ordered pair of the identity function are identical; hence the name of this function. The graph of the identity function is shown in Figure 7.12.

Although the slope of any given nonvertical line is uniquely determined, a given slope does not determine a unique line. Consider the equations

$$y = 2x + 1 \qquad\qquad (1)$$

and

$$y = 2x \qquad\qquad (2)$$

Select two points on the graph of Equation (1) and two points on the graph of Equation (2). The coordinates (0, 1) and (1, 3) satisfy Equation (1). (See Figure 7.13.) Slope of line (1) is $\frac{3 - 1}{1 - 0} = 2$. The coordinates (0, 0) and (1, 2) satisfy Equation (2). Slope of line (2) is $\frac{2 - 0}{1 - 0} = 2$. Both lines have the same slope, but from Figure 7.13 it is clear that the lines are distinct. They are in fact parallel, which leads to the following theorem.

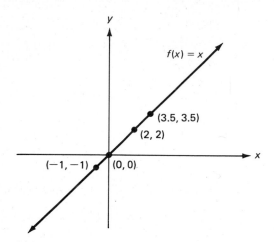

FIG. 7.12 The identity function, $f(x) = x$

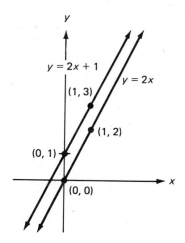

FIG. 7.13

THEOREM

Two distinct nonvertical lines in the same plane are *parallel* if and only if they have the same slope.

Often the linear function is not defined as y in terms of x, but is written in the form $ax + by = c$, or $ax + by + c = 0$, where a, b, and c are constants, not all equal to zero. We then solve for y in terms of x, or rewrite the expression in terms of $y = f(x)$.

EXAMPLE 7 If $3x + 2y + 6 = 0$, express y as a function of x and find the slope and y-intercept of the line described by this function.

Solution To express y as a function of x, solve the equation for y:

$$3x + 2y + 6 = 0$$
$$2y = -3x - 6$$
$$y = -\frac{3}{2}x - 3$$

Now the function is in the form $y = mx + b$, so that the slope $m = -\dfrac{3}{2}$ and the y-intercept $b = -3$.

EXAMPLE 8 Find the value of k so that the line through $(1, k)$ and $(5, 4k)$ has the property stated:
(a) Slope -6
(b) Slope 0
(c) Parallel to $3x - y - 2$

Solution

(a) $m = \dfrac{4k - k}{5 - 1} = \dfrac{3k}{4}$ and $m = -6$

$\dfrac{3k}{4} = -6,$ $3k = -24,$ $k = -8$

(b) $\dfrac{3k}{4} = 0,$ $k = 0$

(c) Solving $3x - y = 2$ for y,
$$3x = y + 2$$
$$y = 3x - 2$$
and the slope of this line is 3.

Since parallel lines have the same slope,

$\dfrac{3k}{4} = 3,$ $3k = 12,$ $k = 4$

EXERCISES 7.2

Graph each of the functions defined in Exercises 1–20 and state the slope and the y-intercept of each graph. (Examples 1 and 5)

1. $f(x) = 3x + 4$

2. $f(x) = 2x - 1$

3. $f(x) = \dfrac{1}{4}x + 2$

4. $f(x) = \dfrac{1}{2}x - 4$

5. $f(x) = 5$

6. $f(x) = -2$

7. $f(x) = -\dfrac{1}{4}x$

8. $f(x) = \dfrac{1}{2}x$

9. $f(x) = -x + 2$

10. $f(x) = \dfrac{3}{5}x + \dfrac{1}{2}$

11. $f(x) = 3x + 2$

12. $f(x) = -2x + 1$

13. $f(x) = -x - 1$

14. $f(x) = -x + 1$

15. $f(x) = -x$

16. $f(x) = 2x$

17. $f(x) = 0$

18. $f(x) = -2x + 2$

19. $f(x) = -\dfrac{1}{4}x - 1$

20. $f(x) = -x + 3$

For the points in Exercises 21–26 find the slope of the line passing through the points and graph each line. (Example 4)

21. $P_1:(2, 3), P_2:(1, 6)$

22. $P_1:(-2, 3), P_2:(-1, 6)$

23. $P_1:(-3, -4), P_2:(2, 5)$

24. $P_1:(1, 1), P_2:(4, 4)$

25. $P_1:(-2, -1), P_2:(-5, -2)$

26. $P_1:(0, -3), P_2:(-2, 0)$

In Exercises 27–40 write each equation in the slope-intercept form, $y = mx + b$, if possible. Determine the slope and y-intercept of the graph of the equation. Check the value of the slope by selecting any two points on the line and using the slope formula. (Example 7)

27. $5(x - y) + 3 = 0$

28. $2(x - y) - 3(x + y) = 5$

29. $3y - 15 = 0$

30. $3x - 9 = 0$

31. $y - 2x + 4 = 0$

32. $2y - 4x = 5$

33. $3x + 2y = 4$

34. $4x + 16 = 2y$

35. $3(x + 1) = 2y$

36. $x - (y - 1) + 5 = 0$

37. $2x + 8 = 0$

38. $4y + 8 = 0$

39. $y = x$

40. $x + y = 0$

For Exercises 41 and 42 state whether the two pairs of lines whose equations are given are parallel. If they are parallel, give the slope of the lines.

41. $x = 3y + 6$

$\quad\,\, 2x - 6y = 3$

42. $(x + y) - (x - y) = 6$

$\quad\,\, 2y - 8 = 0$

In Exercises 43–48 find the value of k so that the line through (2, k) and (5, 3k) has the property stated. (Example 8)

43. Slope 4

44. Slope $\dfrac{1}{2}$

45. Slope 0

46. Slope -2

47. Parallel to $y = x + 5$

48. Parallel to $2x + y = 4$

49. As an example of a slope, a highway grade expressed as a percentage means the number of feet the road changes in elevation for 100 ft measured horizontally.

(a) A certain highway has a $2\frac{1}{2}$% grade. How many feet does it rise in a 1-mi stretch (horizontal distance)? (*Note:* 1 mi = 5280 ft.)

(b) How many feet does a $-3\frac{1}{4}$% grade highway drop for a $\frac{1}{2}$-mi horizontal stretch?

50. A certain county specification requires that an inclined water pipe must have a slope greater than or equal to $\frac{1}{4}$. Which of the water pipes whose rises and runs are given below meets this specification?

(a) Rise = 20 ft, run = 64 ft
(b) Rise = 125 ft, run = 500 ft
(c) Rise = 60 ft, run = 250 ft

7.3 DETERMINING THE EQUATION OF A LINE

In the preceding section we discussed linear functions and their graphs. We were given an equation and from the equation we drew the graph of the corresponding line. Now consider the problem of finding an equation of a line whose geometric conditions are given.

In order to accomplish this task, one of the following conditions must be given:

1. One point on the line and the slope of the line, or
2. Two points on the line

If the coordinates of a point on a line are given and if the slope of the line is known, then the equation of the line can be found by using the definition of the slope. Since (x, y) is to be a general point on the line, then

$$m = \frac{y - y_1}{x - x_1}$$

where (x_1, y_1) and m are the given point and slope, respectively.

Multiplying both sides of this equation by $x - x_1$ yields the point-slope form of a line, $y - y_1 = m(x - x_1)$.

DEFINITION

The **point-slope form** of a line with slope m passing through the point (x_1, y_1) is
$$y - y_1 = m(x - x_1)$$

EXAMPLE 1 Find an equation of a line whose slope is -2, passing through the point $(3, 4)$.

Solution $m = -2$ and $(x_1, y_1) = (3, 4)$.
Using the point-slope form of a line,
$$y - y_1 = m(x - x_1)$$
$$y - 4 = -2(x - 3)$$
$$y - 4 = -2x + 6$$
$$y = -2x + 10 \quad \text{(Slope-intercept form)}$$
$$2x + y - 10 = 0 \quad \text{(Standard form)}$$
It is conventional to express the answer in **standard form**—that is, the linear equation is written in the form $Ax + By + C = 0$ where $A \geq 0$.

EXAMPLE 2 Find an equation of a line passing through the points $(3, 4)$ and $(2, 1)$.

Solution
1. Find the slope:
$$m = \frac{y_2 - y_1}{x_2 - x_1} = \frac{4 - 1}{3 - 2} = 3$$
2. Selecting (x_1, y_1) as $(2, 1)$ and using the point-slope form,
$$y - y_1 = m(x - x_1)$$
$$y - 1 = 3(x - 2)$$
$$y - 1 = 3x - 6$$
$$y = 3x - 5 \quad \text{(Slope-intercept form)}$$
$$3x - y - 5 = 0 \quad \text{(Standard form)}$$

Alternate solution and check Selecting (x_1, y_1) as $(3, 4)$ and (x_2, y_2) as $(2, 1)$, the slope,
$$m = \frac{y_2 - y_1}{x_2 - x_1} = \frac{1 - 4}{2 - 3} = \frac{-3}{-1} = 3$$
Using the point-slope form,
$$y - y_1 = m(x - x_1)$$
$$y - 4 = 3(x - 3)$$
$$y - 4 = 3x - 9$$
$$y = 3x - 5 \quad \text{(Slope-intercept form)}$$
$$3x - y - 5 = 0 \quad \text{(Standard form)}$$

Note in the preceding example that when two points on the line are given, it makes no difference which is called the first and which is called the second in finding the equation of the line. Since two possibilities are available, one choice can serve as a check for the other choice.

EXERCISES 7.3

In Exercises 1–10 write an equation of the line with the given slope and passing through the point whose coordinates are given. (Example 1)

1. $m = -2; (2, 4)$

2. $m = 3; (1, 5)$

3. $m = 6; (-1, 3)$

4. $m = -1; (2, -1)$

5. $m = \frac{1}{2}; (4, -2)$

6. $m = \frac{1}{3}; (-6, 1)$

7. $m = -\frac{3}{4}; (-3, -4)$

8. $m = -\frac{2}{5}; (-5, -12)$

9. $m = 0; (2, 7)$

10. $m = 0; (5, 0)$

In Exercises 11–20 find the slope, if it exists, and write an equation of the line that contains the points whose coordinates are given. (Example 2)

11. (2, 1) and (4, 2)

12. (2, 3) and (-2, -3)

13. (-2, -3) and (4, -1)

14. (-3, -4) and (-4, -3)

15. (5, 2) and (-3, 0)

16. (-2, 5) and (0, -3)

17. (2, 3) and (-5, 3)

18. (4, 6) and $\left(4, -\frac{1}{2}\right)$

19. (-2, 5) and (-2, -2)

20. (6, -7) and (-5, -7)

For Exercises 21–30 find an equation of the form $Ax + By + C = 0$ for the line satisfying the stated conditions.

21. Parallel to $3x + 2y = 6$ and passing through the point (1, 2)

22. Slope $-\frac{1}{2}$, y-intercept 3

23. Slope $-\frac{1}{2}$, passing through (3, 0)

24. Slope $\frac{3}{4}$, passing through the origin

25. Parallel to the x-axis, passing through (1, 2)

26. Parallel to the y-axis, passing through (1, 2)

27. Parallel to $2x - y + 3 = 0$ and passing through the point $(-2, -1)$

28. $m = \frac{4}{3}$, y-intercept -2

29. $m = -\dfrac{4}{3}$, passing through $(-2, 0)$

30. Passing through the origin and the point $(-3, 5)$

31. Do the points $A:(1, -2)$, $B:(3, 0)$, and $C:(0, 3)$ lie on a straight line? Why?

32. Do the points $P:(0, -1)$, $Q:(2, 0)$, and $R:(4, 1)$ lie on a straight line? Why?

33. For what value of k will the points $(8, 2)$, $(5, 3)$, $(6, k)$ lie on a straight line?

34. For what value of k will the points $(5, 0)$, $(-2, 4)$, $(k, -4)$ lie on a straight line?

35. The results of a study made by a physiologist for the purpose of trying to predict adult height from height as a child are shown in the following table, where C = the heights of two-year-old children in inches and A = adult heights in inches.

C	30	31	33	34	34	35	37	38
A	60	61	65	67	66	70	73	74

(a) Graph the ordered pairs, using the horizontal axis as the C-axis.

(b) Draw a straight line that comes closest to passing through all the points plotted. (This is called the *line of best fit.*)

(c) Write the equation for this line.

(d) Using the equation in (c), predict the adult height of a two-year-old child whose height is
 1. 32 in.   2. 36 in.   3. 39 in.

36. The "Bromine number" test is a test used in chemistry to determine the number of double bonds of carbon in a compound. Measurements made at the end of 5 min, 10 min, and 15 min for the chemical reaction yielded the results shown in the following table:

t Time in Minutes	B Bromine Number
5	35
10	50
15	65

(a) Graph the ordered pairs (t, B).

(b) Join the ordered pairs with a straight line and extend this line so it intersects the vertical B-axis.

(c) From the graph, find the value of B for $t = 0$. (This indicates the number of double bonds of carbon in the compound.)

(d) Check the result in (c) by writing the equation of the line in the graph and evaluating B for $t = 0$.

7.4 QUADRATIC FUNCTIONS

The first category or "family" of functions we discussed were linear functions. These functions are of the form $f(x) = ax + b$, and the variable x appears only to the first power. (a and b are constants as discussed previously.) The next category to be studied is functions of the form $f(x) = ax^2 + bx + c$, where a, b, and c are constants and a is not equal to zero. Note that the variable x appears to the second power, and the term ax^2 will not vanish since we placed the restriction on a that a cannot equal zero.

DEFINITION–Quadratic Function

A **quadratic function** is a function with the rule
$$f(x) = ax^2 + bx + c, \qquad a \neq 0$$

The graph of a quadratic function is *not* a straight line.

EXAMPLE 1 Draw the graph of the quadratic function
$$f(x) = 2x^2 - 4x + 1$$
by making a table of values for some selected integral values for x. Plot the points from this table of values and draw a smooth curve through the points. From the graph, determine the range of the function.

Solution

x	$f(x) = 2x^2 - 4x + 1$	(x, y)
-3	$2(-3)^2 - 4(-3) + 1 = 18 + 12 + 1 = 31$	$(-3, 31)$
-2	$2(-2)^2 - 4(-2) + 1 = 8 + 8 + 1 = 17$	$(-2, 17)$
-1	$2(-1)^2 - 4(-1) + 1 = 2 + 4 + 1 = 7$	$(-1, 7)$
0	$2(0)^2 - 4(0) + 1 = 1$	$(0, 1)$
1	$2(1)^2 - 4(1) + 1 = 2 - 4 + 1 + -1$	$(1, -1)$
2	$2(2)^2 - 4(2) + 1 = 8 - 8 + 1 = 1$	$(2, 1)$
3	$2(3)^2 - 4(3) + 1 = 18 - 12 + 1 = 7$	$(3, 7)$

Figure 7.14 is the graph of this function. It is evident from the graph that $f(x) \geq -1$. It will shortly be shown that the graph is correct in leading us to believe that the function value is always greater than or equal to -1, and we shall learn a method for finding the range of a quadratic function.

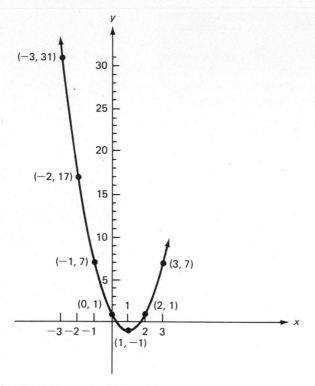

FIG. 7.14 Graph of $f(x) = 2x^2 - 4x + 1$

Clearly the graph of $f(x) = 2x^2 - 4x + 1$ is not a straight line. As a matter of fact, the graph of a quadratic function is a *smooth* curve called a **parabola.**

> The parabola that describes the graph of the function
> $$y = ax^2 + bx + c$$
> where a, b, c, x, and y are real numbers ($a \neq 0$), opens upward, $\cup$ if $a > 0$; or downward, $\cap$, if $a < 0$.

The turning point of the parabola is called its **vertex,** and the curve is symmetric with respect to a vertical line drawn through the vertex. This line is called the **axis of symmetry.** If the paper were to be folded along this axis, the part of the parabola on the right would coincide with the part on the left of the axis of symmetry.

The vertex always occurs on the axis of symmetry and is always either the low point or the high point of the graph of a quadratic function $y = ax^2 + bx + c$, depending on whether the parabola opens upward or downward.

Since the axis of symmetry is a vertical line, its equation has the form $x = k$, where k is constant.

7.4 QUADRATIC FUNCTIONS

To get a better understanding of the ideas of symmetry and vertex of a quadratic function, let us look at the simplest quadratic function, $f(x) = x^2$. We know that for any positive value of x, x^2 is positive, and for any negative value of x, x^2 is also positive. When $x = 0$, $x^2 = 0$. By selecting various ordered pairs that belong to the function—for example, $(-2, 4)$, $(-1, 1)$, $(0, 0)$, $(1, 1)$, $(2, 4)$—we can see that 0 is the smallest y value, and that the y-axis is the axis of symmetry. The equation of the y-axis is $x = 0$, and the point $(0, 0)$ is the vertex of the parabola. Figure 7.15 is the graph of this parabola. Note also that the coefficient of x^2 is 1, which is a positive number, so the parabola opens up.

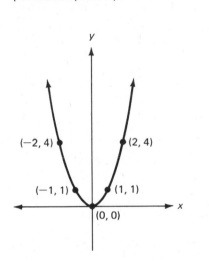

FIG. 7.15 Graph of $f(x) = x^2$

FIG. 7.16 Graph of $f(x) = x^2 + 2$

Figure 7.16 shows the graph of $f(x) = x^2 + 2$. This graph is exactly like the graph in Figure 7.15 except that it is shifted 2 units up on the y-axis, and its vertex is at $(0, 2)$.

In general the graph of $f(x) = x^2 + d$ has the same shape as the graph of $f(x) = x^2$, only the graph is shifted d units on the y-axis—upward if d is positive and downward if d is negative. Figure 7.17 shows several graphs with different values of d. The vertex of the graph of $f(x) = x^2 + d$ is at the point $(0, d)$. The equation of the axis of symmetry is $x = 0$.

Just as the value of d for $f(x) = x^2 + d$ has the effect of shifting the graph of $f(x) = x^2$ up or down on the y-axis, so the value of k in $f(x) = (x - k)^2$ has the effect of shifting the graph of $f(x) = x^2$, k units to the right or to the left of the y-axis, depending on whether k is positive or negative. Thus the x value of the vertex is moved to $x = k$, and the coordinates of the vertex of the graph of $f(x) = (x - k)^2 + d$ are (k, d).

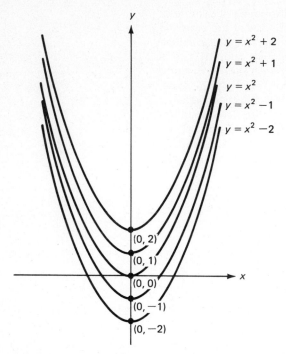

FIG. 7.17 Graph of $f(x) = x^2 + d$

By rewriting the quadratic function $f(x) = ax^2 + bx + c$ in the form $f(x) = a(x - k)^2 + d$, we can immediately determine the coordinates of the vertex (k, d), the equation of the axis of symmetry, $x = k$, and the range of the function, $y \geq d$ or $y \leq d$, depending on whether a is positive or negative.

EXAMPLE 2   Draw the graph of the function $f(x) = (x - 2)^2 - 1$.

From the above discussion, we know that the vertex of the graph is at $(2, -1)$ because $k = 2$ and $d = -1$. Note that
$$(x - 2)^2 - 1 = 1 \cdot (x - 2)^2 + (-1)$$
The graph opens up because $a = 1$. Figure 7.18 is the required graph.

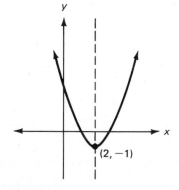

FIG. 7.18 Graph of $f(x) = (x - 2)^2 - 1$

EXAMPLE 3 Graph $f(x) = -2(x + 1)^2 + 3$.

Solution This parabola opens down because $a = -2$. Since $x + 1 = x - (-1)$, $k = -1$; $d = 3$. The coordinates of the vertex are $(-1, 3)$. Figure 7.19 is the graph. Notice that the shape of the parabola is a little different from those with $a = 1$. In order to determine the shape, select two or three points in addition to the vertex. For example, when $x = 0$, $y = 1$ yielding the point $(0, 1)$. Similarly, we obtain the points $(-2, 1)$, $(1, -5)$, and $(-3, -5)$.

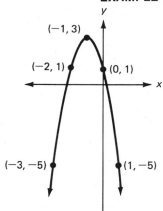

FIG. 7.19 Graph of $f(x) = -2(x + 1)^2 + 3$

EXAMPLE 4 Graph $y = x^2 + 4x - 5$.

Solution   Recognize that $y = f(x)$. We want to write the equation in the form $f(x) = a(x - k)^2 + d$. To do that, we "complete the square" for $x^2 + 4x$.

$$y = x^2 + 4x - 5$$
$$y = x^2 + 4x + 4 - 5 - 4 \qquad \text{(Add and subtract 4)}$$
$$y = (x + 2)^2 - 9 \qquad (x^2 + 4x + 4 = (x + 2)^2)$$

Now we see that the coordinates of the vertex are $(-2, -9)$. Since $a = 1$, the parabola opens up. The axis of symmetry is $x = -2$. The graph of this parabola is Figure 7.20.

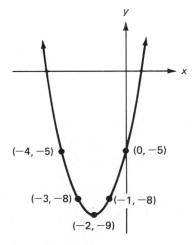

FIG. 7.20 Graph of $y = x^2 + 4x - 5$

The method of completing the square for

$$f(x) = ax^2 + bx + c$$

when $a \neq 1$ involves a couple of extra steps as shown in the following example.

EXAMPLE 5 Write the equation $y = 3x^2 - 18x + 4$ in the form $y = a(x - k)^2 + d$.

Solution   Since $a = 3$, divide both sides of the equation by 3.

$$\frac{y}{3} = x^2 - 6x + \frac{4}{3}$$

Now complete the square for $x^2 - 6x$ by adding 9 and then subtracting 9 so that the value of the right side of the equation remains unchanged.

$$\frac{y}{3} = x^2 - 6x + 9 + \frac{4}{3} - 9$$

$$\frac{y}{3} = (x - 3)^2 + \frac{4}{3} - 9 \qquad (x^2 - 6x + 9 = (x - 3)^2)$$

$$y = 3(x - 3)^2 + 4 - 27 \qquad \text{(Multiplying each side by 3)}$$
$$y = 3(x - 3)^2 - 23 \qquad \text{(In simplified form)}$$

The equation $y = 3x^2 - 18x + 4$ graphs as a parabola, opening up ($a = 3$, a positive number), and from the fact that $3x^2 - 18x + 4 = 3(x - 3)^2 - 23$ we also know that the vertex of the parabola has coordinates $(3, -23)$.

From the preceding examples it can be seen that the method of finding the coordinates of the vertex by "completing the square" is sometimes cumbersome. A quicker alternative approach is based on the following theorem.

THEOREM

If $y = f(x) = ax^2 + bx + c$, and $a \neq 0$, then the x-coordinate of the vertex of the graph of this function is

$$x = \frac{-b}{2a}$$

and the y-coordinate of the vertex is

$$y = f\left(\frac{-b}{2a}\right) = c - \frac{b^2}{4a}$$

The theorem may be justified by completing the square to obtain

$$y = a\left(x + \frac{b}{2a}\right)^2 + c - \frac{b^2}{4a}$$

and thus the x value of the vertex is $x = \frac{-b}{2a}$. It is usually easier to find the y value by evaluating $f(x)$ for $x = \frac{-b}{2a}$ than to remember that the y value is $c - \frac{b^2}{4a}$, as shown above.

EXAMPLE 6 Verify the coordinates of the vertex for Examples 4 and 5 by using the above theorem.

Solution In Example 4,
$$y = f(x) = x^2 + 4x - 5$$
Therefore
$$a = 1 \quad \text{and} \quad b = 4$$
From the theorem,
$$x = \frac{-b}{2a} = \frac{-4}{2(1)} = -2$$
$$y = f(-2) = (-2)^2 + 4(-2) - 5$$
$$= 4 - 8 - 5 = -9$$
The coordinates of the vertex are $(-2, -9)$, which checks with Example 4.

In Example 5,
$$y = f(x) = 3x^2 - 18x + 4$$
Therefore
$$a = 3 \quad \text{and} \quad b = -18$$
$$x = \frac{-b}{2a} = \frac{-(-18)}{2(3)} = 3$$
$$y = f(3) = 3(3)^2 - 18(3) + 4$$
$$= 27 - 54 + 4 = -23$$
The coordinates of the vertex are $(3, -23)$, and this checks with the result obtained in Example 5.

EXERCISES 7.4

Each equation in Exercises 1–20 states the rule for a quadratic function. Determine the coordinates of the vertex of the corresponding parabola, write the equation of the axis of symmetry, state the range of the function, and draw the graph. (Examples 2 and 3)

1. $y = 2x^2$

2. $y = 3x^2$

3. $y = -3x^2$

4. $y = -2x^2$

5. $y = x^2 + 4$

6. $y = x^2 - 1$

7. $y = 2 - x^2$

8. $y = 1 - 2x^2$

9. $y = \frac{1}{2}x^2$

10. $y = -\frac{2}{3}x^2$

11. $y = (x - 2)^2 + 1$

12. $y = (x + 3)^2 - 1$

13. $y = (x + 1)^2 - 2$

14. $y = (x - 3)^2 + 4$

15. $y = (x + 2)^2$

16. $y = (x + 1)^2$

17. $y = 3(x + 1)^2 - 5$

18. $y = 2(x + 1)^2 - 3$

19. $y = -2(x - 1)^2 + 2$

20. $y = -3(x + 2)^2 + 1$

For Exercises 21–40 find the coordinates of the vertex by using
$x = \dfrac{-b}{2a}, y = f\left(\dfrac{-b}{2a}\right)$ *as shown in Example 6. Graph the quadratic function, and label the vertex.*

21. $f(x) = x^2 + 2x + 4$

22. $f(x) = x^2 - 2x + 4$

23. $y = x^2 + 8x - 1$

24. $y = x^2 - 10x - 4$

25. $f(x) = x^2 + 3x + 1$

26. $f(x) = x^2 - x - 5$

27. $f(x) = x^2 - 5x - 3$

28. $f(x) = x^2 + 7x - 1$

29. $y = 2x^2 + 4x$

30. $y = 3x^2 - 9x$

31. $y = 2x^2 + 6x - 3$

32. $y = 4x^2 - 12x - 6$

33. $y = -x^2 - 2x + 3$

34. $y = -x^2 + 2x + 1$

35. $y = -3x^2 + 5x - 1$

36. $y = -2x^2 + 3x - 6$

37. $y = \dfrac{1}{2}x^2 + x - 4$

38. $y = \dfrac{1}{3}x^2 - 2x + 3$

39. $f(x) = 6x - x^2$

40. $f(x) = 10x - 2x^2$

41. After t sec the height of a ball tossed into the air at 48 ft/sec is given by the formula
$$h = 48t - 16t^2$$
   (a) Graph this function. (Let the t-axis be the horizontal axis and the h-axis be the vertical axis.)
   (b) From the graph determine the maximum height the ball reaches, the number of seconds it takes to reach that height, and the number of seconds it takes the ball to come to the ground again. (Assume the ball starts at $h = 0$.)

42. An arrow is shot vertically upward into the air. Its height h in feet after t sec is given by the formula
$$h = 112t - 16t^2$$
   (a) Graph the function. (Let the t-axis be the horizontal axis.)
   (b) From the graph determine the greatest height the arrow reaches and the number of seconds it takes to reach its maximum height.

7.5 QUADRATIC RELATIONS

In the preceding section we defined a quadratic function as a function with the rule $f(x) = ax^2 + bx + c$, $a \neq 0$. If we let $y = f(x)$, the functional rule becomes $y = ax^2 + bx + c$. By interchanging the role of x and y, we can state a rule, $x = ay^2 + by + c$, $a \neq 0$. The graph of this equation is also a parabola, but its axis of symmetry is parallel to the *x*-axis. This parabola opens to the right when *a* is positive, and it opens to the left when *a* is negative. By looking at the graph we can see that this new relation does *not* define a function in *x*.

Finding the vertex of the parabola is done in the same manner as finding the vertex of the graph of the quadratic function, but now $y = -\dfrac{b}{2a}$ and $x = c - \dfrac{b^2}{4a}$.

EXAMPLE 1   Graph $x = 2y^2 - 12y + 11$ by first finding the coordinates of the vertex of the parabola.

Solution

$$a = 2, \qquad b = -12, \qquad c = 11$$

$$y = \frac{-b}{2a} = \frac{-(-12)}{2(2)} = 3$$

$$x = c - \frac{b^2}{4a} = 11 - \frac{(-12)^2}{4(2)}$$

$$= 11 - \frac{144}{8}$$

$$= -7$$

The coordinates of the vertex are $(-7, 3)$. Since $a = 2$, a positive number, the parabola opens to the right. Note the reversal of the role of *x* and *y* in the coordinates.

We find some additional points on the parabola to aid in graphing.

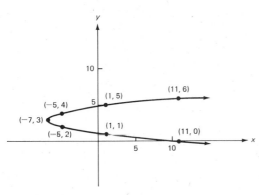

FIG. 7.21 Graph of $x = 2y^2 - 12y + 11$

y	$x = 2y^2 - 12y + 11$	(x, y)	
0	$x = 2(0) - 12(0) + 11 = 11$	$(11, 0)$	
1	$x = 2(1) - 12(1) + 11 = 1$	$(1, 1)$	
2	$x = 2(4) - 12(2) + 11 = -5$	$(-5, 2)$	
3	$x = 2(9) - 12(3) + 11 = -7$	$(-7, 3)$	← Vertex
4	$x = 2(16) - 12(4) + 11 = -5$	$(-5, 4)$	
5	$x = 2(25) - 12(5) + 11 = 1$	$(1, 5)$	
6	$x = 2(36) - 12(6) + 11 = 11$	$(11, 6)$	

Figure 7.21 is the graph of this equation. We see from the figure and from the ordered pairs in the table that this equation does *not* express a functional rule. For example, when $x = -5$, $y = 2$, or $y = 4$, there are two ordered pairs, $(-5, 2)$ and $(-5, 4)$ that have the same first element and different second elements. Since the solution set of the equation is a set of ordered pairs, it is a relation.

Thus the set $\{(x, y)\,|\,y = ax^2 + bx + c, a \neq 0\}$ defines a function, and the set $\{(x, y)\,|\,x = ay^2 + by + c, a \neq 0\}$ defines only a relation.

EXAMPLE 2 Graph the relation $x = 4y - y^2$.

Solution

$$a = -1, \qquad b = 4, \qquad c = 0$$

$$y = \frac{-4}{2(-1)} = 2$$

$$x = 0 - \frac{16}{4(-1)} = 4$$

The vertex is at $(4, 2)$.

Since $a = -1$, the parabola opens to the left. The table shows several ordered pairs that are in the solution set, and Figure 7.22 is the graph of the relation.

y	$x = 4y - y^2$	(x, y)	
-1	$x = 4(-1) - 1 = -5$	$(-5, -1)$	
0	$x = 4(0) - 0 = 0$	$(0, 0)$	
1	$x = 4 - 1 = 3$	$(3, 1)$	
2	$x = 4(2) - 4 = 4$	$(4, 2)$	← Vertex
3	$x = 4(3) - 9 = 3$	$(3, 3)$	
4	$x = 4(4) - 16 = 0$	$(0, 4)$	
5	$x = 4(5) - 25 = -5$	$(-5, 5)$	

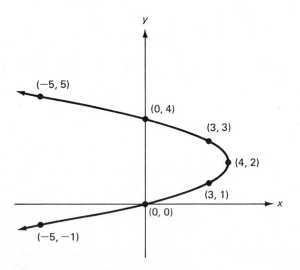

FIG. 7.22 Graph of $x = 4y - y^2$

The relation defined by $x = ay^2 + by + c$, $a \neq 0$ is not the only quadratic relation.

The **general quadratic equation in two variables** x and y has the form

$$Ax^2 + Bxy + Cy^2 + Dx + Ey + F = 0$$

where A, B, C, D, E, and F are constants and at least one of the constants A, B, and C is different from 0.

The second-degree terms are Ax^2, Bxy, and Cy^2.

The first-degree, or linear, terms are Dx and Ey.

The constant term F is said to have degree 0 because F can be considered as the coefficient of x^0—that is,

$$Fx^0 = F \cdot 1 = F. \ (x \neq 0)$$

Since the degree of a polynomial equation is defined as the greatest of the degrees of its terms, the general quadratic equation is an equation of second degree.

In this text we discuss only briefly some of the graphs that result from different values of these constants. In particular, we discuss the equation of a circle, an ellipse, and a hyperbola; we have already studied the quadratic function for which $B = 0$ and $C = 0$ and the quadratic relation for which $A = 0$ and $B = 0$.

To verify the above statement, when $B = 0$ and $C = 0$, $Ax^2 + Bxy + Cy^2 + Dx + Ey + F = 0$ becomes

$$Ax^2 + Dx + Ey + F = 0,$$

or $y = -\dfrac{1}{E} (Ax^2 + Dx + F)$. This equation, with appropriate changes for the constants, can be written $y = ax^2 + bx + c$. It is left for the student to examine the situation when A and B are zero.

EXERCISES 7.5

Graph each of the relations whose equation is given in Exercises 1–20 and state the coordinates of the vertex of the parabola. (Examples 1 and 2)

1. $x = y^2 - 4$

2. $x = y^2 + 4$

3. $x = -y^2$

4. $x = -y^2 + 2$

5. $x = 2 - 3y^2$

6. $x = 3 - 2y^2$

7. $x = y^2 + 5y$

8. $x = y^2 - 5y$

9. $x = 5y - y^2$

10. $x = y - 3y^2$

11. $x = y^2 + 8y + 21$

12. $x = y^2 - 8y + 12$

13. $x = y^2 - 10y + 16$

14. $x = y^2 + 4y - 3$

15. $x = 3y^2 + 12y - 4$

16. $x = 2y^2 - 3y + 2$

17. $x = 2y^2 + 2y - 4$

18. $x = 4y^2 - 8y + 2$

19. $x = 6 + 2y - 3y^2$

20. $x = 2 - 3y + 4y^2$

Each of Exercises 21–24 states a pair of equations. Graph each pair on one set of coordinate axes and state the coordinates where the graphs intersect (points that the graphs have in common).

21. $y = x^2$ and $x = y^2$

22. $y = x^2$ and $x = -y^2$

23. $x = y^2 - 4$ and $x - y = 2$

24. $x = y^2 + 3y$ and $y = x - 3$

7.6 THE DISTANCE FORMULA AND THE CIRCLE

The relation that defines a circle algebraically is very important, not only for its geometric and algebraic significance but also in the study of trigonometry. As a matter of fact, trigonometry is also called the study of **circular functions.**

DEFINITION

A **circle** is the set of points in a plane that are equidistant from a fixed point called the **center** of the circle.

The distance between any point on the circle and the center of the circle is called the **radius** of the circle.

In order to obtain an equation for the circle, it is first necessary to express the geometric concept of distance between two points as an algebraic statement relating the coordinates of the two points.

THE DISTANCE FORMULA

$$|P_1P_2| = \sqrt{(x_2 - x_1)^2 + (y_2 - y_1)^2}$$

The distance formula is based on the theorem of Pythagoras. (In a right triangle, the square of the length c of the hypotenuse is equal to the sum of the squares of the lengths a and b of the other two sides. In symbols, $c^2 = a^2 + b^2$.)

The distance between any two points in a coordinate plane may be expressed in terms of the coordinates of these points by constructing a right triangle whose hypotenuse is the line segment joining the two points, as in Figure 7.23.

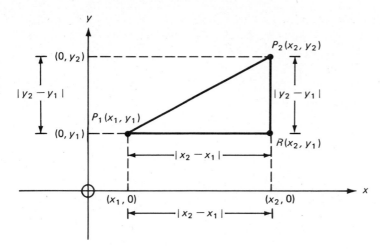

FIG. 7.23

If the points are $P_1:(x_1, y_1)$ and $P_2:(x_2, y_2)$ and R is the point where the horizontal line through P_1 intersects the vertical line through P_2, then the coordinates of R are (x_2, y_1).

Now the distance $|P_1R|$ is $|x_2 - x_1|$ and the distance $|P_2R|$ is $|y_2 - y_1|$. By the theorem of Pythagoras,

$$(|P_1P_2|)^2 = (|P_1R|)^2 + (|P_2R|)^2$$

or

$$(|P_1P_2|)^2 = |x_2 - x_1|^2 + |y_2 - y_1|^2$$

By taking the square root of each side, the distance formula is obtained.

EXAMPLE 1 Find the distance between $P_1:(2, -1)$ and $P_2:(5, 3)$.

Solution
$$|P_1P_2| = \sqrt{(5 - 2)^2 + (3 - (-1))^2}$$
$$= \sqrt{3^2 + 4^2} = \sqrt{25} = 5$$

In order to show how the distance formula leads to the general equation of a circle, look at Figure 7.24 and consider the following. By the definition of a circle, all points on the circle are the same distance from a fixed point, the center of the circle.

If $P:(x, y)$ designates any point on the circle and if $C:(a, b)$ is the center of the circle whose radius is r, then the equation of the circle is obtained by using the distance formula. Thus

$$\sqrt{(x - a)^2 + (y - b)^2} = r$$

By squaring both sides of this equation, the standard form of the equation of the circle is obtained.

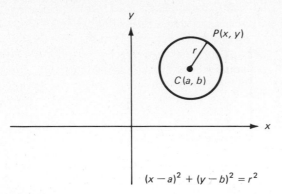

$$(x - a)^2 + (y - b)^2 = r^2$$

FIG. 7.24

EQUATION OF CIRCLE IN STANDARD FORM

$$(x - a)^2 + (y - b)^2 = r^2$$
Center is (a, b);   radius $= \sqrt{r^2} = r,$ $r > 0$

The standard form for the equation of the circle is useful because the radius and the coordinates of the center can be read directly from the equation. The circle is symmetric to every line through its center; in particular, the circle is symmetric to the lines $x = a$ and $y = b$.

EXAMPLE 2   Write the equation of the circle with center $(3, -4)$ and radius 5.

Solution
$$(x - 3)^2 + (y + 4)^2 = 25$$

EXAMPLE 3   Graph the circle whose equation is $x^2 + y^2 = 9$.

Solution   Since $x^2 + y^2 = 9$ can be written $(x - 0)^2 + (y - 0)^2 = 9$, the circle has its center at $(0, 0)$, the origin; the radius is 3. *Any equation of the form $x^2 + y^2 = r^2$ has as its graph a circle with center at the origin and radius r.* See Figure 7.25.

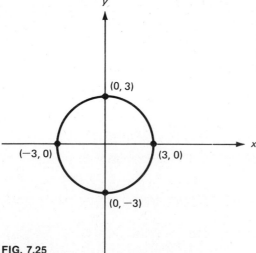

FIG. 7.25

EXAMPLE 4 Draw the graph of the relation
$\{(x, y) \mid (x + 2)^2 + (y - 3)^2 = 16\}$, a circle

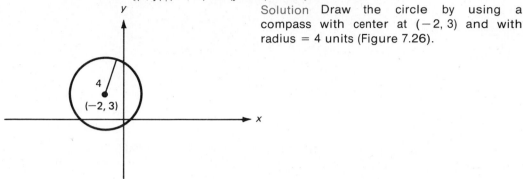

Solution Draw the circle by using a compass with center at $(-2, 3)$ and with radius = 4 units (Figure 7.26).

FIG. 7.26

EXAMPLE 5 Determine an equation of the circle with center $C:(2, 3)$ and passing through the point $P:(5, 6)$.

Solution Since $|CP|$ determines the radius of the circle,
$$r = \sqrt{(5 - 2)^2 + (6 - 3)^2}$$
$$r = \sqrt{18}$$
$$r^2 = 18$$
An equation of the circle is
$$(x - 2)^2 + (y - 3)^2 = 18.$$

EXAMPLE 6 Show that $x^2 + y^2 + 8x - 10y + 5 = 0$ is the equation of a circle and then determine its center and radius.

Solution Completing the squares for x and for y,
$$(x^2 + 8x \quad) + (y^2 - 10y \quad) = -5$$
$$(x^2 + 8x + 16) + (y^2 - 10y + 25) = -5 + 16 + 25$$
$$(x + 4)^2 + (y - 5)^2 = 36$$
Thus the equation is that for a circle with center $(-4, 5)$ and with radius = 6.

EXERCISES 7.6

For Exercises 1–10 find the distance between the points whose coordinates are given. (Example 1)

1. $(9, 5)$ and $(4, -7)$

2. $\left(\frac{1}{2}, -3\right)$ and $(2, -5)$

3. $(-6, 5)$ and $(-4, 1)$

4. $(9, 6)$ and $\left(\frac{11}{2}, -6\right)$

5. $(9, 6)$ and $\left(\frac{11}{2}, 6\right)$

6. $(2, 7)$ and $(10, 1)$

7. $(18, -2)$ and $(-6, 5)$

8. $(-4, 3)$ and $(-7, 0)$

9. $\left(3, \frac{9}{2}\right)$ and $(-3, 7)$

10. $(3, 1)$ and $(3, -4)$

For Exercises 11–20 write an equation of the circle with the given center and radius. (Example 2)

11. $(-1, 2); r = 6$

12. $(-2, 3); r = 4$

13. $(0, 0); r = 10$

14. $(-5, -6); r = 12$

15. $(2, 3); r = 3$

16. $(1, 4); r = 5$

17. $(3, -5); r = 5$

18. $(4, -3); r = 7$

19. $(-6, 2); r = \sqrt{3}$

20. $(0, 0); r = \sqrt{7}$

For Exercises 21–30 write an equation of the circle with center C and passing through point P. (Example 5)

21. $C:(1, 5), P:(2, 2)$

22. $C:(-2, -3), P:(1, 5)$

23. $C:(-4, 1), P:(-3, -6)$

24. $C:(0, 0), P:(3, 4)$

25. $C:(3, 4), P:(0, 0)$

26. $C:(2, 4), P:(3, 1)$

27. $C:(-1, -5), P:(-4, -2)$

28. $C:(-3, 4), P:(1, 1)$

29. $C:(1, 1), P:(-4, -5)$

30. $C:(-4, -5), P:(1, 1)$

Show that each of Exercises 31–40 is the equation of a circle and determine the center and radius of each. (Example 6)

31. $x^2 + y^2 - 4x + 6y - 3 = 0$

32. $x^2 + y^2 - 10x + 6y + 18 = 0$

33. $2x^2 + 2y^2 + 20x - 12y + 18 = 0$

34. $4x^2 + 4y^2 - 4x - 24y + 33 = 0$

35. $4x^2 + 4y^2 - 28x - 8y + 37 = 0$

36. $x^2 + y^2 - 4x + 2y - 6 = 0$

37. $x^2 + y^2 - 16x = 0$

38. $x^2 + y^2 + 20x = 0$

39. $x^2 + y^2 - 14y = 0$

40. $x^2 + y^2 - 12y = 0$

Graph each of the relations in Exercises 41–50. (Examples 3 and 4)

41. $\{(x, y)\,|\,(x - 2)^2 + (y - 3)^2 = 9\}$

42. $\{(x, y)\,|\,(x - 1)^2 + (y + 2)^2 = 16\}$

43. $\{(x, y)\,|\,(x + 1)^2 + (y + 2)^2 = 4\}$

44. $\{(x, y)\,|\,(x + 2)^2 + (y - 4)^2 = 1\}$

45. $\{(x, y)\,|\,x^2 + y^2 + 4x = 0\}$

46. $\{(x, y)\,|\,x^2 + y^2 + 6x = 0\}$

47. $\{(x, y)\,|\,3x^2 + 3y^2 - 27 = 0\}$

48. $\{(x, y)\,|\,4x^2 + 4y^2 - 64 = 0\}$

49. $\{(x, y)\,|\,x^2 + y^2 - 2y = 0\}$

50. $\{(x, y)\,|\,x^2 + y^2 - 10y = 0\}$

7.7 ELLIPSES AND HYPERBOLAS

There are two other types of quadratic graphs that are of interest because of their wide area of applications. These are the ellipse and the hyperbola.

DEFINITION

A standard **ellipse** is the graph of

$$\frac{x^2}{a^2} + \frac{y^2}{b^2} = 1 \qquad (a > 0 \text{ and } b > 0)$$

The graphs of the two types of standard ellipses are shown in Figure 7.27.

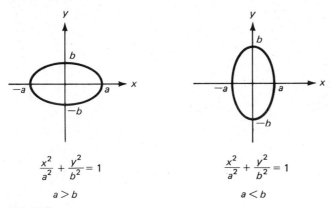

$$\frac{x^2}{a^2} + \frac{y^2}{b^2} = 1$$

$$a > b$$

$$\frac{x^2}{a^2} + \frac{y^2}{b^2} = 1$$

$$a < b$$

FIG. 7.27

The domain and range of an ellipse can be readily observed from the graphs shown in Figure 7.27.
The domain of a standard ellipse = $\{x \mid -a \leq x \leq a\}$.
The range of a standard ellipse = $\{y \mid -b \leq y \leq b\}$.

EXAMPLE 1 Graph $4x^2 + 9y^2 = 36$.

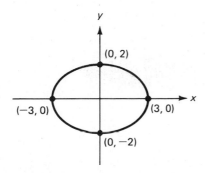

FIG. 7.28

Solution
1. Identify the curve as an ellipse. Dividing each side by 36,
$$\frac{x^2}{9} + \frac{y^2}{4} = 1$$
This is a standard ellipse with $a = 3$ and $b = 2$.
2. Since $a = 3$, the domain is $\{x \mid -3 \leq x \leq 3\}$. Since $b = 2$, the range is $\{y \mid -2 \leq y \leq 2\}$.
3. On the x-axis plot the points $(-3, 0)$ and $(3, 0)$, and on the y-axis plot the points $(0, -2)$ and $(0, 2)$. To obtain a **sketch** of the ellipse, join these points with smooth curves. In order to get more accurate values, solve for y,
$$y = \pm \frac{2}{3} \sqrt{9 - x^2}$$
and select values in the domain (that is, x values greater than -3 but less than 3) and obtain corresponding y values.

Figure 7.28 is a sketch of the ellipse.

EXAMPLE 2 Graph $4x^2 + y^2 - 100 = 0$.

Solution

1. Standard form:

$$\frac{x^2}{25} + \frac{y^2}{100} = 1$$

Thus the curve is an ellipse with $a = 5$ and $b = 10$.

Domain: $-5 \le x \le 5$

Range: $-10 \le y \le 10$

2. Plot the points $(-5, 0)$ and $(5, 0)$ on the x-axis and the points $(0, -10)$, and $(0, 10)$ on the y-axis, and sketch the curve, as in Figure 7.29.

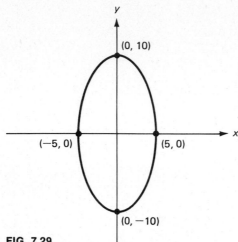

FIG. 7.29

DEFINITION

A standard **hyperbola** is the graph of

(1) $\dfrac{x^2}{a^2} - \dfrac{y^2}{b^2} = 1$ $(a > 0, b > 0)$

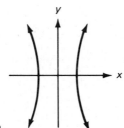

FIG. 730a

or

(2) $\dfrac{y^2}{b^2} - \dfrac{x^2}{a^2} = 1$ $(a > 0, b > 0)$

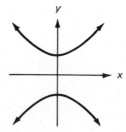

FIG. 7.30b

or

(3) $xy = c$

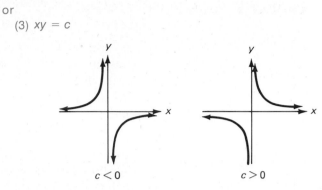

FIG. 7.30c

Once a hyperbola is recognized from the equation of a standard hyperbola, the graph of the hyperbola is obtained by finding the domain for the hyperbola and plotting points using values for x in the domain.

DOMAIN AND RANGE OF STANDARD HYPERBOLAS

Equation (1), solving for y, yields $y = \pm \dfrac{b}{a} \sqrt{x^2 - a^2}$; if y is to be a real number, $x^2 \geq a^2$, $|x| \geq a$ implies that $x \geq a$ or $x \leq -a$.

Equation (2), solving for y, yields $y = \pm \dfrac{b}{a} \sqrt{a^2 + x^2}$. In this equation y is a real number for all real values of x. Therefore the domain of (2) is all x.

Equation (3) is defined for all $x \neq 0$.

The range of standard hyperbolas is determined in a similar way.

Therefore

For $\dfrac{x^2}{a^2} - \dfrac{y^2}{b^2} = 1$,   the domain is $\{x \mid x \leq -a \text{ or } x \geq a\}$ and
the range is $\{y \mid y \text{ is a real number}\}$.

For $\dfrac{y^2}{b^2} - \dfrac{x^2}{a^2} = 1$,   the domain is $\{x \mid x \text{ is a real number}\}$ and
the range is $\{y \mid y \leq -b \text{ or } y \geq b\}$.

For $xy = c$,   the domain is $\{x \mid x \text{ is a real number and } x \neq 0\}$ and
the range is $\{y \mid y \text{ is a real number and } y \neq 0\}$.

EXAMPLE 3   Graph $25x^2 - 16y^2 = 400$.
Solution

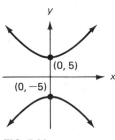

FIG. 7.31

1. Identify the curve by obtaining the equation of a standard hyperbola. Dividing both sides by 400,

$$\frac{x^2}{16} - \frac{y^2}{25} = 1 \qquad (a = 4, b = 5)$$

2. Domain: $y = \pm \dfrac{5}{4} \sqrt{x^2 - 16}$. Thus $|x| \geq 4$;

   that is, $x \leq -4$ or $x \geq 4$.

3. We now have the points $(-4, 0)$ and $(4, 0)$, and a general idea of the shape of the hyperbola. By selecting a few more points, we can get a reasonable sketch. Since the graph is symmetric to both the x-axis and the y-axis, every point on the graph in the first quadrant automatically yields three more points: one in the second quadrant, one in the third quadrant, and one in the fourth quadrant. Taking $x = 5$, we get the points $(5, 3.8)$ and $(-5, 3.8)$;   also $(-5, -3.8)$ and $(5, -3.8)$. Similarly, solving for $x = 6$, we get the points $(6, 5.6)$, $(-6, 5.6)$, $(-6, -5.6)$, and $(6, -5.6)$

Figure 7.31 is the graph of this relation.

EXAMPLE 4   Graph $25x^2 - 16y^2 + 400 = 0$.
Solution

1. $\dfrac{y^2}{25} - \dfrac{x^2}{16} = 1$. Thus the curve is a hyperbola with $b = 5$ and $a = 4$.

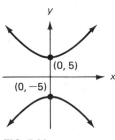

FIG. 7.32

2. Domain: all x since $y = \pm \dfrac{5}{4} \sqrt{x^2 + 16}$

   Range: $|y| \geq 5$ since $x = \pm \dfrac{4}{5} \sqrt{y^2 - 25}$

3. From the range values we get the points $(0, 5)$ and $(0, -5)$. Additional points on the graph are $(2, 5.6)$, $(-2, 5.6)$, $(-2, -5.6)$, and $(2, -5.6)$. Also $(3, 6.3)$, $(-3, 6.3)$, $(-3, -6.3)$, and $(3, -6.3)$.

Figure 7.32 is the hyperbola.

EXAMPLE 5 Graph $xy = 4$.

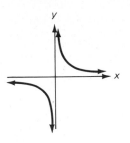

FIG. 7.33

Solution This equation matches the third example in Figure 7.30, with $c = 4$, a positive number. If the product of x and y is to be positive, then x and y must be either both positive or both negative. And since their product does not equal zero, neither x nor y can equal zero.

Solving for y, $y = \dfrac{4}{x}$. By selecting a few points, we can easily sketch this hyperbola, which has its graph in the first and third quadrants.

The following values satisfy the equation:

$$(1, 4),\ (-1, -4),\ (2, 2),\ (-2, -2),$$
$$(4, 1),\ (-4, -1)$$

Figure 7.33 is the graph of this relation. Is this relation a function? The answer is yes, since it can be seen from the graph that for every x value, there is one and only one y value, so no ordered pair in the relation has the same first component and a different second component.

EXERCISES 7.7

Discuss and graph each of the equations in Exercises 1–20.

1. $x^2 + 25y^2 = 100$
2. $16x^2 = 1600 - 25y^2$
3. $5x^2 + 2y^2 - 50 = 0$
4. $x^2 - y^2 = 9$
5. $2x^2 - 2y^2 + 18 = 0$
6. $9x^2 = 16y^2 - 576$
7. $xy = 12$
8. $xy = -6$
9. $y^2 - 4x^2 = 64$
10. $y^2 - 4x^2 = 0$

11. $9x^2 + y^2 = 225$
12. $y^2 = 52 - 4x^2$
13. $x^2 + 4y^2 - 169 = 0$
14. $4x^2 = y^2 + 16$
15. $4x^2 = y^2 - 16$
16. $xy = 8$
17. $xy = -4$
18. $4x^2 - 25y^2 = 0$
19. $x^2 + y^2 = 0$
20. $4x^2 - 9y^2 + 36 = 0$

Graph each pair of equations in Exercises 21–24 on the same set of axes, and determine the coordinates of the points of intersection from the graph. These values are common *solutions to the pair of equations since they belong to the solution set of each equation.*

21. $x^2 + 16y^2 = 169$ and $x^2 - y^2 = 16$
22. $x^2 - 2x - 5 = y$ and $x + y = 1$
23. $x^2 - y^2 = 4$ and $x^2 + y^2 = 4$
24. $x^2 - y^2 = 5$ and $3y^2 - x^2 = 3$

7.8 LINEAR AND QUADRATIC INEQUALITIES IN TWO VARIABLES

In the preceding sections it was shown that equations in two variables can be associated with points in a plane, using horizontal and vertical axes as references. Inequalities can also be graphed in a plane. We shall start our discussion with **linear inequalities.**

A linear inequality in two variables is a **relation** of the form $y < ax + b$ or $y > ax + b$. Since the solution set of this type of inequality is a set of ordered pairs, it is a relation.

Examples of linear inequalities are

$y < 3x - 2, y > 2x + 1$,

and $3x - 2y < 5$.

To graph a linear inequality, we go back to the graph of the linear equation, $y = ax + b$. This graph, a line, divides the plane into three mutually exclusive regions:

1. The *line* corresponding to the equation $y = ax + b$
2. The region, or *half-plane* **above** this line, which corresponds to the relation $y > ax + b$

and

3. The region, or *half-plane* **below** the line, which corresponds to the relation $y < ax + b$.

Figure 7.34 illustrates these regions.

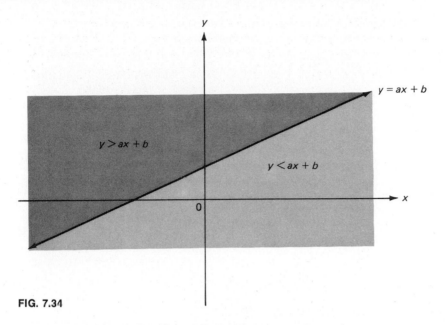

FIG. 7.34

To graph an inequality of the form $y < ax + b$ or
$y > ax + b$,
first graph the equation $y = ax + b$, using a broken line to indicate that the line is *not* part of the solution set, and then shade the desired region. Figure 7.35 is an example of the graph of the relation $y < 2x + 3$.

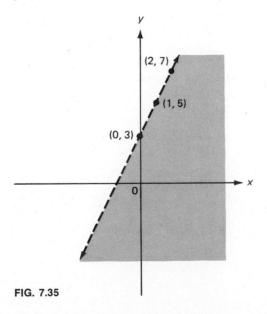

FIG. 7.35

EXAMPLE 1   Graph the relation $3x - 2y > x + y - 1$.
   Solution   First solve the inequality for y:

$$3x - 2y > x + y - 1$$
$$- 3y > -2x - 1$$
$$y < \frac{2}{3}x + \frac{1}{3}$$

Note that division by -3 changed the sense of the inequality.

Now graph the equality $y = \frac{2}{3}x + \frac{1}{3}$, using a broken line as shown in Figure 7.36. All points *below* the broken line are in the solution set

$$\left\{ (x, y) \middle| y < \frac{2}{3}x + \frac{1}{3} \right\}$$

Therefore the shaded region is the graphical solution.

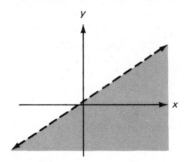

FIG. 7.36

EXAMPLE 2   Find the graphical solution for the relation defined by the inequality $y \geq -2x + 4$.

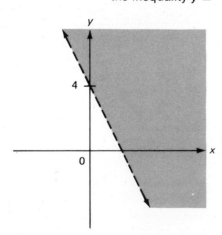

Solution   Graph the equation $y = -2x + 4$. Use a *solid* line since the line is part of the solution set. Shade the region *above* the line for the points

$$\{(x, y) \mid y > -2x + 4\}$$

The line and the shaded region in Figure 7.37 represent the solution set

$$\{(x, y) \mid y \geq -2x + 4\}$$

FIG. 7.37

EXAMPLE 3 Graph the relation $\{(x, y) \mid x > 3\}$.

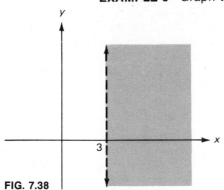

Solution The graph of the equation $x = 3$ is a vertical line through the point $(3, 0)$. All points (x, y) such that $x > 3$ lie to the *right* of this line (Figure 7.38).

FIG. 7.38

EXAMPLE 4 Graph the relation $\{(x, y) \mid y > 2\}$.

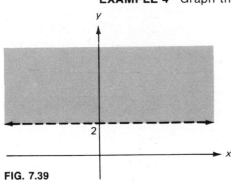

Solution The graph of the constant function $y = 2$ is a horizontal line through the point $(0, 2)$. Since $y > 2$, the shaded region in Figure 7.39 includes all points *above* this horizontal line.

FIG. 7.39

EXAMPLE 5 Find the graphical solution for the inequality

$$\{(x, y) \mid -2 \le x < 3\}$$

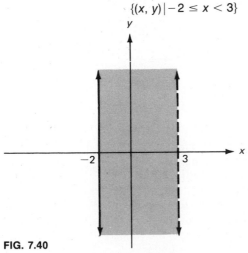

Solution First graph the vertical lines $x = -2$ and $x = 3$. Use a solid line to indicate the inclusion of $x = -2$ and a broken line to show that $x \ne 3$. Then shade the region *between* these lines since $-2 \le x < 3$ for *all* values of y. See Figure 7.40.

FIG. 7.40

EXAMPLE 6 Find the graphical solution for
$$\{(x, y)\,|\,y > 2x + 1\} \cap \{(x, y)\,|\,y < 3 - x\}$$

Solution   This calls for the *intersection* of the solution sets of each relation; in other words, we wish to find all points such that $y > 2x + 1$ *and* $y < 3 - x$.

Graph each relation separately, as shown in Figure 7.41. Since the solution set is the intersection of the two sets, it will be the region where the shading overlaps, as illustrated in Figure 7.42.

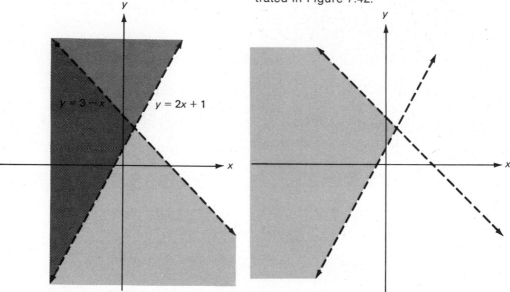

FIG. 7.41

FIG. 7.42 $\{(x, y)\,|\,y > 2x + 1\} \cap \{(x, y)\,|\,y < 3 - x\}$

We now look at some quadratic inequalities. Just as a line separates the plane into three mutually exclusive regions, so does a parabola, and so does a circle, and so does an ellipse.

Consider the case of a parabola first. Let
$$y = ax^2 + bx + c$$
and assume a is positive. Figure 7.43 illustrates the three regions. The region *inside* the parabola contains all points in the relation $y > ax^2 + bx + c$. The parabola corresponds to the equation $y = ax^2 + bx + c$, and the region *outside* the parabola contains all points in the relation $y < ax^2 + bx + c$.

If $a < 0$, then the situation is reversed, as illustrated in Figure 7.44. Now the region *inside* the parabola represents the solution $y < ax^2 + bx + c$, the parabola itself still corresponds

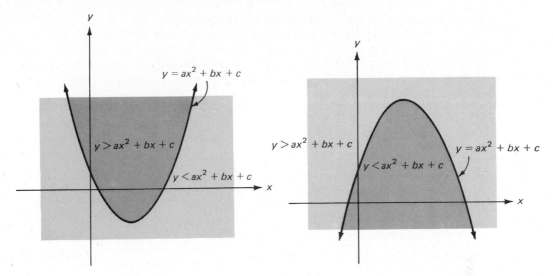

FIG. 7.43

FIG. 7.44

to $y = ax^2 + bx + c$, and the region *outside* the parabola corresponds to $y > ax^2 + bx + c$.

EXAMPLE 7   Graph the solution set of the relation
$$y \geq 2x^2 - 3x - 2$$

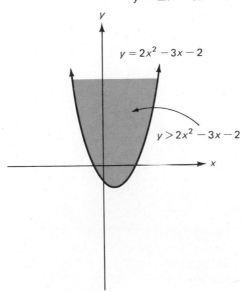

FIG. 7.45 $y \geq 2x^2 - 3x - 2$

Solution   Since the coefficient of x^2 is positive, the parabola opens up. Complete the square to find the vertex:
$$y = 2\left(x - \frac{3}{4}\right)^2 - \frac{25}{8}$$
Thus the coordinates of the vertex are $\left(\frac{3}{4}, -\frac{25}{8}\right)$. The region for which
$$y > 2x^2 - 3x - 2$$
is *inside* the parabola, as shown in Figure 7.45. The solution set is the parabola *and* the shaded region.

A circle very clearly divides the plane into three regions: the region outside the circle, the circle, and the region inside the circle.

Recall that the equation of a circle with radius r and center (a, b) is

$(x - a)^2 + (y - b)^2 = r^2$

If $(x - a)^2 + (y - b)^2 < r^2$, then the region *inside* the circle is described.

If $(x - a)^2 + (y - b)^2 > r^2$, then the region *outside* the circle is described.

Clearly $(x - a)^2 + (y - b)^2 = r^2$ describes the circle itself.

EXAMPLE 8   Graph and discuss each of the following regions:
(a) $x^2 + y^2 < 4$
(b) $x^2 + y^2 > 4$

Solution $x^2 + y^2 = 4$ is the equation of a circle with center at the origin and radius 2.
(a) If $x^2 + y^2 < 4$, then all the circles with center at the origin and radius *less* than 2 must be considered. This yields the entire region *inside* the circle.
(b) If $x^2 + y^2 > 4$, then all the circles with center at the origin and radius *greater* than 2 must be considered. This yields the entire region *outside* the circle (see Figure 7.46).

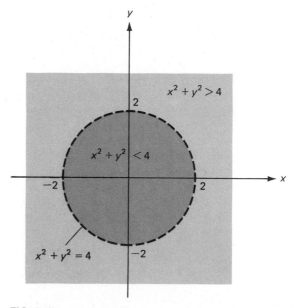

FIG. 7.46

EXERCISES 7.8

Show a graphical solution for each of the sets in Exercises 1–34.
(Examples 1–6)

1. $\{(x, y) \mid x \geq 2\}$

2. $\{(x, y) \mid x < -1\}$

3. $\{(x, y) \mid y > -1\}$

4. $\{(x, y) \mid y \leq 4\}$

5. $\{(x, y) \mid x > 4\}$

6. $\{(x, y) \mid x \leq -2\}$

7. $\{(x, y) \mid y \geq 4\}$

8. $\{(x, y) \mid y < 3\}$

9. $\{(x, y) \mid 1 \leq x \leq 3\}$

10. $\{(x, y) \mid 2 < y \leq 5\}$

11. $\{(x, y) \mid x < 0\}$

12. $\{(x, y) \mid y \geq 0\}$

13. $\{(x, y) \mid -2 < x \leq 1\}$

14. $\{(x, y) \mid 1 \leq x < 3\}$

15. $\{(x, y) \mid y < 3x + 2\}$

16. $\{(x, y) \mid y > x - 5\}$

17. $\{(x, y) \mid y \leq 2x - 3\}$

18. $\{(x, y) \mid y > x\}$

19. $\{(x, y) \mid x + y \leq 2\}$

20. $\{(x, y) \mid y \geq x + 2\}$

21. $\{(x, y) \mid y \leq x\}$

22. $\{(x, y) \mid x + 2y < 2\}$

23. $\{(x, y) \mid 4 \geq 2y - x\}$

24. $\{(x, y) \mid 3x \leq y + 2\}$

25. $\{(x, y) \mid x - y < 2\}$

26. $\{(x, y) \mid y \leq x + 2\}$

27. $\{(x, y) \mid y < x\} \cap \{(x, y) \mid y > -x\}$

28. $\{(x, y) \mid y > x\} \cap \{(x, y) \mid y < -x\}$

29. $\{(x, y) \mid x > 0\} \cap \{(x, y) \mid y > 0\} \cap \{(x, y) \mid x + y \leq 5\}$

30. $\{(x, y) \mid x < 0\} \cap \{(x, y) \mid y < 0\} \cap \{(x, y) \mid x + y \geq -3\}$

31. $\{(x, y) \mid y < x\} \cup \{(x, y) \mid x \geq 1\}$

32. $\{(x, y) \mid y > 2\} \cup \{(x, y) \mid x > 3\}$

33. $\{(x, y) \mid x + y \leq 2\} \cap \{(x, y) \mid y - x \leq 1\}$

34. $\{(x, y) \mid x + 3y \geq 3\} \cap \{(x, y) \mid 2x + 2y \geq 5\}$

Show a graphical solution for each of the relations in Exercises
35–46. (Example 7)

35. $y > x^2 - 5$

36. $y \leq x^2 + 2$

37. $y < 2 - x^2$

38. $y < -1 - x^2$

39. $y \geq 2x^2 - 4x$

40. $y \leq 4x^2 - 2x$

41. $y > 2x - 4x^2$

42. $y > x^2 + 2x + 1$

43. $y \leq x^2 - 4x + 4$

44. $y > -x^2 + 6x - 9$

45. $y < -x^2 + 4x - 4$

46. $y \leq 3x^2 - 4x + 1$

Graph each of the relations in Exercises 47–52. (Example 8)

47. $x^2 + y^2 < 9$

48. $x^2 + y^2 \geq 4$

49. $(x - 1)^2 + (y - 2)^2 > 1$

50. $(x + 2)^2 + (y - 1)^2 \geq 9$

51. $(x - 3)^2 + (y + 4)^2 \leq 16$

52. $(x - 2)^2 + (y + 1)^2 < 9$

7.9 ABSOLUTE VALUE FUNCTIONS AND RELATIONS

Several different functions have been discussed in this chapter: linear functions, quadratic functions, constant functions, the identity function. Another important function is the absolute value function $y = |x|$.

In order to graph the function
$$y = |x|$$
it is necessary to consider the definition of $|x|$:

If $x \geq 0$, $y = x$, and if $x < 0$, $-y = x$ and $y = -x$

The graph of Figure 7.47 is a composite of two graphs on the same set of axes. The graph in the first quadrant is the graph of $x \geq 0$ *and* $y = x$, including the origin, whereas the graph of $x < 0$ *and* $y = -x$ is the graph in the second quadrant. It is very important to break the absolute value function into two cases before graphing: The situation for which the quantity whose absolute value is being taken is positive or zero, and the situation for which this quantity is negative.

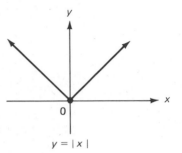

$y = |x|$

FIG. 7.47

EXAMPLE 1   Graph $y = |x| + 2$.

Solution
1. When $x \geq 0$, then $|x| = x$ and $y = x + 2$. Start graphing the line at $x = 0$ and draw the line only for positive values of x as shown in Figure 7.48.

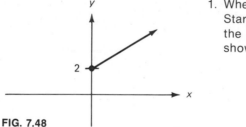

FIG. 7.48

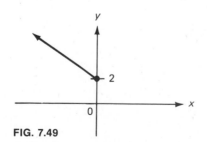

FIG. 7.49

2. When $x < 0$, then $|x| = -x$ and
$$y = -x + 2.$$
Graph the line only for negative values of x as shown in Figure 7.49.

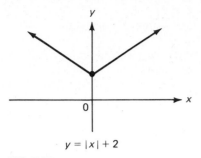

$y = |x| + 2$

FIG. 7.50

Figure 7.50 shows both of these graphs on one set of axes to give the complete graph of the solution set.

Absolute value inequalities are graphed in a manner similar to the inequality relations discussed in Section 7.8.

EXAMPLE 2   Graph the relation $y < |x + 2|$.

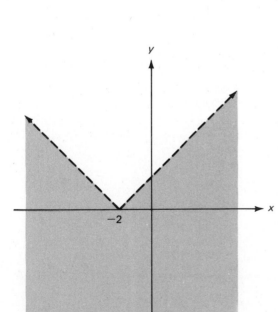

FIG. 7.51 $y < |x + 2|$

Solution   Use a broken line to graph the absolute value function,
$$y = |x + 2|$$
Case 1
$$x + 2 \geq 0 \quad \text{and} \quad y = x + 2$$
$$x \geq -2 \quad \text{and} \quad y = x + 2$$
This means that $y = x + 2$ *only* for values of x greater than -2 or equal to -2; in other words, going to the right from $x = -2$.
Case 2
$$x + 2 < 0 \quad \text{and} \quad y = -(x + 2)$$
$$x < -2 \quad \text{and} \quad y = -x - 2$$
Therefore for values of x less than -2, or to the left of $x = -2$, $y = -x - 2$. Since
$$y < |x + 2|,$$
the solution set is the shaded region *below* the broken line graph of $y = |x + 2|$, as illustrated in Figure 7.51.

EXAMPLE 3 Graph the relation $y \geq 2 + |x + 1|$.

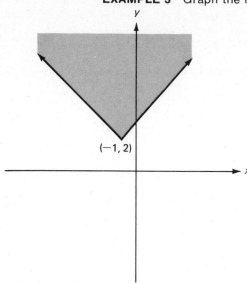

(−1, 2)

Solution This time the graph of
$$y = 2 + |x + 1|$$
is included, so its graph is a solid line. Again, take the two cases, $x + 1$ positive or zero, so that for x greater than or equal to -1, the equation to be graphed is $y = 2 + (x + 1)$, and the second case, $x + 1$ a negative quantity, so that for values of x less than -1, the equation to be graphed is $y = 2 - (x + 1)$.

Case 1
$$x + 1 \geq 0 \quad \text{and} \quad y = 2 + (x + 1)$$
$$x \geq -1 \quad \text{and} \quad y = x + 3$$

Case 2
$$x + 1 < 0 \quad \text{and} \quad y = 2 - (x + 1)$$
$$x < -1 \quad \text{and} \quad y = -x + 1$$

The relation $y > 2 + |x + 1|$ is the shaded region *above* the graph of $y = 2 + |x + 1|$, as illustrated in Figure 7.52.

FIG. 7.52 $y \geq 2 + |x + 1|$

EXERCISES 7.9

Graph each of the absolute value functions or relations in Exercises 1–24.

1. $y = |2x|$

2. $y = \left|\dfrac{x}{2}\right|$

3. $y = |x - 3|$

4. $y = |x + 5|$

5. $y = |x + 2|$

6. $y = |1 - x|$

7. $y = |2 - 3x|$

8. $y = 3|x + 1|$

9. $y > |x|$

10. $y \leq |2x|$

11. $y > |x - 3|$

12. $y > |x + 5|$

13. $y < |x - 3|$

14. $y < |x + 5|$

15. $y = 2 - |x + 4|$

16. $y = 3 + 2|3 - x|$

17. $y < 2 - |x + 4|$

18. $y < 3 + 2|3 - x|$

19. $y \geq 5 - |4x + 12|$

20. $y > -|x + 2|$

21. $y \geq 3 + |2x - 5|$

22. $y \leq 4 + 2|1 - x|$

23. $y > |x| + 2 - x$

24. $y < -2 - |3x + 6|$

Graph Exercises 25–30.

25. $\{(x, y) \mid |x| = x\}$

26. $\{(x, y) \mid |x| = -x\}$

27. $\{(x, y) \mid |x + 2| > y\}$

28. $\{(x, y) \mid |x + 2| < -y\}$

29. $\{(x, y) \mid x = |y|\}$

30. $\{(x, y) \mid |y| \geq 3\}$

DIAGNOSTIC TEST

Complete each of the following by correctly filling in the blank.

1. A set whose elements are ordered pairs is called _____.

2. If no two ordered pairs in a set of ordered pairs have the same first element and a different second element, the set is called _____.

3. If $f(x) = 3x^2 - 2x + 4$, then $f(2) =$ _____, and $f(-1) =$ _____.

4. For the function defined in Problem 3, $f(a + 2) =$ _____ and $f(a + 2) - f(a) =$ _____.

5. The slope of the line which passes through the points with coordinates $(3, -2)$ and $(-4, 5)$ is _____.

6. An equation of the line described in Problem 5 is _____.

7. The coordinates of the vertex or the parabola whose equation is $y = 2x^2 - 8x - 7$ are _____.

8. The graph of the equation $y = 12 - x^2$ is called a _____.

9. The graph of the equation $x^2 - 4y^2 = 12$ is called a _____.

10. The graph of the equation $(x - 2)^2 + (y + 3)^2 = 12$ is called a _____.

11. The distance between points A and B with coordinates $(2, 4)$ and $(-3, 1)$ is _____.

12. An equation of a circle with center at $C : (3, -1)$ and radius $r = 4$ is _____.

13. The domain of the relation $x^2 + 16y^2 = 144$ is _____.

14. The range of the relation defined in Problem 13 is _____.

15. The graph of the relation defined in Problem 13 is _____.

16. The graphical solution of $y > 3x - 5$ consists of all points _____ the line whose equation is $y = 3x - 5$.

17. The graphical solution of $x^2 + y^2 < 7$ consists of all points _____ the circle whose equation is $x^2 + y^2 = 7$.

18. The equation $xy - -3$ describes the graph of a _____ whose domain is _____ and whose range is _____.

19. The graph of the equation $y = 7$ is a _____ line, with slope $m =$ _____.

20. If a line passes through the points $(3, k)$ and $(5, -2)$, and is parallel to the line whose equation is $2x + 3y = 12$, then k equals _____.

REVIEW EXERCISES

For each of the relations in Exercises 1–5 state the domain and the range. If the relation is a function, say so. (Section 7.1)

1. $\{(3, 2), (-3, 2), (5, 4)\}$

2. $\{(-1, -1), (-4, -4), (4, 4), (-9, -9)\}$

3. $\{(2, 3), (2, -3), (4, 5)\}$

4. $\{(1, 2), (2, 2), (3, 2), (4, 2)\}$

5. $\{(2, 1), (2, 2), (2, 3), (2, 4)\}$

Each of the equations in Exercises 6–10 defines a relation. State the domain of each so that the corresponding range is a real number; if the relation is a function, say so. (Section 7.1)

6. $y = \dfrac{x + 3}{x + 2}$

7. $y = \dfrac{3}{2x + x^2}$

8. $y = |x + 3|$

9. $y = x^2 + 2x$

10. $y = \sqrt{x^2 - 16}$

11. State the range of each of the relations defined in Exercises 8, 9, and 10. (Section 7.1)

If $f(x) = x^2 - 3x + 2$, find Exercises 12–16. (Section 7.1)

12. $f(1)$

13. $f(0)$

14. $f(3 + h)$

15. $f(3 + h) - f(3)$

16. $f(x + 1)$

If $f(x) = \dfrac{1}{x}$ find Exercises 17–20. (Section 7.1)

17. $f(5)$

18. $f(x + 5)$

19. $f(x + 5) - f(5)$

20. $\dfrac{f(x + 5) - f(5)}{x}$

Graph each of the functions defined in Exercises 21–24 and state the slope and y-intercept of each graph. (Section 7.2)

21. $f(x) = 2x - 3$

22. $f(x) = 5$

23. $f(x) = 4 - x$

24. $f(x) = \dfrac{x}{2}$

For Exercises 25–28, see Section 7.3.

25. Find the slope of the line passing through the points $(-3, 2)$ and $(4, -1)$.

26. Write an equation of the line described in Exercise 25.

27. Determine k so that the line passing through the points $(3, k)$ and $(4, 5k)$ has a slope of $\dfrac{1}{2}$.

28. Determine k in Exercise 27 so that the line passing through these points is parallel to the line whose equation is $3x - 2y = 6$.

For Exercises 29–33 graph each parabola and state the coordinates of its vertex. (Sections 7.4 and 7.5)

29. $y = x^2 + 8x - 2$
30. $y = 2x - x^2$
31. $y = 3x^2 + 2x - 5$

32. $x = y^2 - 1$
33. $x = 3 + 2y - 4y^2$

For Exercises 34–37, see Section 7.6.

34. Find the distance between the points $A:(\ 3, 6)$ and $B:(-9, 4)$.
35. Write an equation of the circle with center $C:(-1, 3)$ and radius $r = 4$.
36. Write an equation of the circle with center $C:(2, -2)$ and passing through the point $P:(5, 1)$.
37. Find the center and the radius of the circle whose equation is $x^2 + y^2 - 6x + 10y + 9 = 0$.

For Exercises 38–47 identify the graph of each equation either as a line, a circle, an ellipse, a hyperbola, or a parabola. Sketch each graph. State whether the relation is a function, and give the domain and range of each.

38. $x^2 + 25y^2 = 100$
39. $4x^2 = y + 16$
40. $4y + 4 = 0$
41. $3x + 4y = 12$
42. $4x^2 = 9y^2 + 36$

43. $4x^2 = 36 - 9y^2$
44. $y^2 - 4x = 0$
45. $4x^2 + 4y^2 - 36 = 0$
46. $2x = 4y - 3$
47. $4x^2 = y^2 - 16$

Show a graphical solution for each of Exercises 48–59.

48. $y < 2x + 3$
49. $y \geq x - 1$
50. $\{(x, y)\,|-2 \leq x < 2\}$
51. $\{(x, y)\,|-2 \leq y < 2\}$
52. $y < x^2 + 3x + 2$
53. $y \geq x^2 - 6x - 7$

54. $y > 2x - 4x^2$
55. $y = |2x - 8|$
56. $y < 2 + |2x|$
57. $y \geq |3 - 2x|$
58. $x^2 + y^2 \leq 25$
59. $(x - 1)^2 + (y + 2)^2 > 4$

60. An object is thrown vertically upward according to the equation $h = 80t - 16t^2$, where h is the height at any given time and t is the time in seconds.
  (a) Graph the function.
  (b) From the graph, determine the maximum height the object reaches.
  (c) From the graph, determine how long it takes the object to return to earth if the initial height $h = 0$.

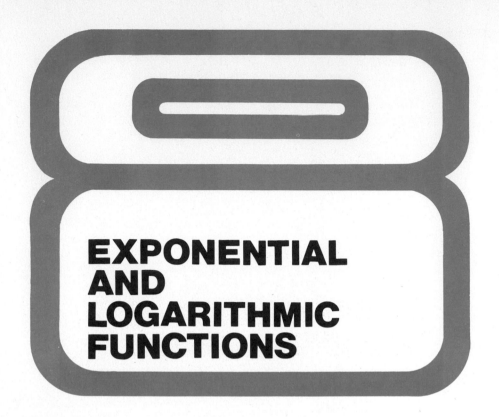

EXPONENTIAL AND LOGARITHMIC FUNCTIONS

A **relation** has been defined as a set of ordered pairs having a **domain,** the set of all first components of the ordered pairs in the relation, and a **range,** the set of all second components of the ordered pairs of the relation.

A **function** was defined as a special relation such that each first component in its domain is paired with exactly one second component in its range.

If the components of the ordered pairs of a relation r are interchanged, then another set of ordered pairs is obtained, and thus this set is also a relation. It is called the **inverse** of the relation r and is designated symbolically as r^{-1} (read "r inverse").

DEFINITION

The **inverse r^{-1} of a relation** r is the set of ordered pairs obtained by interchanging the components of r.

EXAMPLE 1   If $r = \{(2, 3), (4, 5), (6, 7)\}$, find r^{-1}.

Solution   By the definition of r^{-1}, the inverse of the relation, we interchange the components of each ordered pair in r.
$$r^{-1} = \{(3, 2), (5, 4), (7, 6)\}$$
The domain of $r = \{2, 4, 6\}$.
The range of $r^{-1} =$ the domain of r.
The range of $r = \{3, 5, 7\}$.
The domain of $r^{-1} =$ the range of r.

EXAMPLE 2   If $s = \{(x, y) \mid y = 4x^2 - 3\}$, find s^{-1}.

Solution   To find s^{-1}, we write the rule
$$y = 4x^2 - 3$$
and interchange x and y. Thus
$$s^{-1} = \{(x, y) \mid x = 4y^2 - 3\}$$
To verify that this interchanging of variables really gives the inverse relation, we list some of the elements of s and of s^{-1}.
$$s = \{(x, y) \mid y = 4x^2 - 3\}$$
when $x = 0$, $y = -3$
     $x = 1$, $y = 1$
     $x = 2$, $y = 13$
     $x = -1$, $y = 1$
     $x = 3$, $y = 33$
Therefore a partial listing of
$$s = \{(0, -3), (1, 1), (2, 13), (-1, 1), (33, 3), \ldots\}$$
Now look at $s^{-1} = \{(x, y) \mid x = 4y^2 - 3\}$
when $y = 0$, $x = -3$
     $y = 1$, $x = 1$
     $y = 2$, $x = 13$
     $y = -1$, $x = 1$
     $y = 3$, $x = 33$
Writing these values as ordered pairs, (x, y), we get a partial listing for
$$s^{-1} = \{(-3, 0), (1, 1), (13, 2), (1, -1), (3, 33), \ldots\}$$
and we can see that the definition is satisfied.

EXAMPLE 3   If $g = \{(x, y) \mid 5x - 3y = 4\}$, find g^{-1}.

Solution   By interchanging the roles of x and y in the rule for the relation,
$$g^{-1} = \{(x, y) \mid 5y - 3x = 4\}$$

If we graph a relation, then its inverse can be graphed very quickly by interchanging the components of the ordered pairs and plotting the new points.

HISTORICAL NOTE

Logarithms were devised as a method for rapid and accurate computations related to problems in astronomy, engineering, surveying, navigation, and other areas. One of the earliest contributions was the work of the Swabian Michael Stifel. In his *Arithmetica Integra,* published in Nuremburg in 1544, he stated the four laws of exponents for rational numbers and referred to the "upper numbers" as "exponents." He also presented the following table, which could be considered as a primitive table of logarithms.

x	-3	-2	-1	0	1	2	3	4	5	6
2^x	$\frac{1}{8}$	$\frac{1}{4}$	$\frac{1}{2}$	1	2	4	8	16	32	64

Since decimal fractions were not developed until after 1600, it would have been impossible for Stifel to make a table of logarithms suitable for practical calculations.

The Scotsman John Napier (1550–1617), who worked for about twenty years on the theory, is generally acknowledged as the founder of logarithms. Although he first used the term *artificial number,* he finally adopted the term *logarithm,* which in Greek literally means "ratio number."

Napier's work, *Mirifici logarithorum canonis descriptio* (*A Description of the Marvelous Law of Logarithms*), published in Edinburgh in 1614, contained the first real table of logarithms. The response was very enthusiastic. In a later work Napier also explained how to calculate a table of logarithms by a method which was essentially based on adding areas under the hyperbolic curve, $y = 1/x$.

About the same time but independently of Napier, the Swiss instrument maker Jobst Bürgi (1552–1632) calculated a logarithm table. His *Arithmetische und geometrische Progresstabuln* was published in Prague in 1620. This was actually a list of antilogarithms, with the logarithms written in red and the antilogarithms in black. Thus Bürgi referred to the logarithm as "Die Rothe Zahl" ("the red number"). Whereas Napier selected the base $b = 0.9999999 = 1 - 10^{-7}$ and used geometric methods, Bürgi selected $b = 1.0001$ and used algebraic methods. Both selected a base close to 1 so that the powers of b would be close together, and thus the antilogarithms could be listed in intervals of 0.0000001 or 0.0001, respectively.

Henry Briggs (1561–1631), a professor of geometry at Gresham College, London, as a result of a mutual agreement resulting from a conversation with Napier, developed a table of logarithms using the base 10. His *Arithmetica logarithmica,* published in 1624, contained fourteen-place tables for the integers from 1 to 20,000 and from 90,000 to 100,000. The interval from 20,000 to 90,000 was completed by the Dutch bookseller and publisher Adriaen Vlacq (1600–66), who published a complete fourteen-place table of logarithms in 1628.

It was Briggs who introduced the word *mantissa* (originally meaning an "addition" and later an "appendix") and who also suggested the term *characteristic.* In the early tables, the characteristic was printed and was not dropped until about the middle of the eighteenth century.

Tables more accurate than those of Briggs and Vlacq were not calculated until the years 1924–1949, when twenty-place tables were made.

Later developments were concerned with establishing the theory of logarithms on a logically sound foundation. The Swiss mathematician Leonhard Euler (1707–1783) made important contributions in his *Introductio.* It was Euler who introduced the letter *e* to represent the base of the Naperian or natural logarithms. Euler wrote, "Ponamus autem brevitatis gratia pro numero hoc 2.71828 . . . constanter litteram *e*" which is the Latin for "For the sake of brevity we shall let the literal constant *e* represent this number 2.71828. . . ."

The theory of logarithms was finally established through the works of the French mathematician Augustin Cauchy (1789–1857), particularly in his *Cours d'Analyse,* published in Paris in 1821.

EXAMPLE 4   Graph $g = \{(x, y) \mid 5x - y = 4\}$ and g^{-1} on the same set of axes.

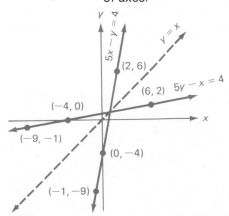

FIG. 8.1

Solution   The equation $5x - y = 4$ graphs as a line. Three points on this line have coordinates $(0, -4)$, $(2, 6)$, and $(-1, -9)$. Plot these points to obtain the graph of g.

To graph g^{-1}, we can see that interchanging x and y will not change the fact that the graph is a line, and furthermore, we know that it will be a line by the fact that the two graphs are symmetric to the line $y = x$.

Plot the points $(-4, 0)$, $(6, 2)$, and $(-9, -1)$, which were obtained by changing the order of the components in the ordered pairs from g. Figure 8.1 shows the graphs of g, g^{-1}, and the line $y = x$.

Since every set of ordered pairs defines a relation, the inverse of a relation is also a relation, but is the inverse of a function always a function? The answer is no. For example, consider the function $f = \{(2, 3), (3, 3), (4, 3)\}$. No two ordered pairs in f have the same first element, so f is a function. The inverse of f, $f^{-1} = \{(3, 2), (3, 3), (3, 4)\}$, which is clearly a relation, but *not* a function.

EXAMPLE 5   Given the function $f = \{(x, y) \,|\, y = x^2\}$. Graph f and f^{-1} on the same set of axes. Is f^{-1} a function?

Solution   The functional rule for f is $y = x^2$. We recognize the graph of this equation to be a parabola, opening up, with its vertex at the origin. Therefore the domain of f is the set of real numbers, and the range of f is all real values for $y \geq 0$.

Following the rule for finding f^{-1}, interchange x and y in the equation to get
$$f^{-1} = \{(x, y) \,|\, x = y^2\}$$
This is also a parabola, but it opens to the right, with its vertex at the origin. Interchanging the domain and range of f tells us that the domain of f^{-1} is all real values for $x \geq 0$, and the range of f^{-1} is all real numbers. But f^{-1} is *not* a function. We can readily see from the graph of $x = y^2$ that every x value except zero has two y values that satisfy the equation. Moreover, if we solve $x = y^2$ for y in terms of x, we get $y = \pm\sqrt{x}$, so that for any positive value of x, y is either the positive square root of that number or the negative square root of that number. Figure 8.2 is the graph of the function f, of the relation f^{-1}, and of the line $y = x$.

Example 5 shows that if a function has more than one value of x paired with the same y value, then the inverse of this function is *not* a function. Geometrically, this means that if a horizontal line intersects the graph of a function f at *more than one point*, then a vertical line will intersect the graph of the inverse of f at more than one point. Look at Figure 8.2 to verify this statement.

However, if a function has the property that there is a one-to-one correspondence between the elements in its domain and the elements in its range, then the function is said to be one-to-one, and its inverse is a function.

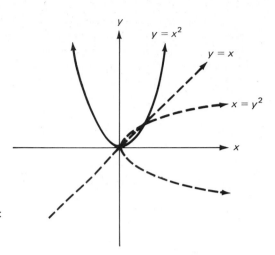

FIG. 8.2 Inverse of f is not a function:
$$f = \{(x, y)|y = x^2\}$$
Inverse of $f = \{(x, y)|y = \pm\sqrt{x}\}$

DEFINITION

The **function f is one-to-one** if and only if each element in its domain is paired with exactly one element in its range, and each element in its range is paired with exactly one element in its domain.

THE INVERSE FUNCTION THEOREM

If the function f is one-to-one, then its inverse is a function, designated as f^{-1} (read "f inverse").

EXAMPLE 6 Given the function f defined by the equation $y = x^3$. Graph this function by finding points on the graph corresponding to $x = -2, -1, 0, 1, 2$, and joining these points with a smooth curve. Graph f^{-1} on the same set of axes.

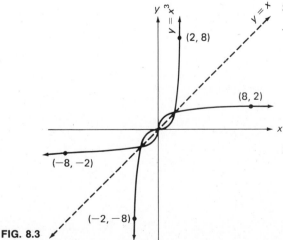

FIG. 8.3

Solution

1. The points on the graph of f to be plotted are $(-2, -8)$, $(-1, -1)$, $(0, 0)$, $(1, 1)$, and $(2, 8)$. From the x and y values of the coordinates we see that f is a one-to-one function. Also, from the graph in Figure 8.3 we see that a horizontal line anywhere on the graph of f will cross the graph in only one place.

2. From the above analysis we know that f^{-1} is also a function. Plot the points $(-8, -2)$, $(-1, -1)$, $(0, 0)$, $(1, 1)$, and $(8, 2)$ obtained by interchanging the first and second components in each ordered pair. Join the points with a smooth curve as in Figure 8.3.

It is often convenient to rename the equation defining a function f as $y = f(x)$. We have found this notation to be useful particularly when evaluating a function for different values of x. For example, if $y = f(x) = 3x + 2$, then the notation $f(1)$, $f(0)$, and $f(3)$ is useful and indicates that we are to evaluate the given function for $x = 1, 0,$ and 3, respectively. This same idea is used to express inverse functions but it involves an extra step, as shown in Example 7.

EXAMPLE 7   Given $f = \{(x, y)\,|\,y = 3x + 5\}$.
(a) Find $f(x)$.
(b) Find f^{-1}.
(c) Find $f^{-1}(x)$.

Solution
(a) $f(x) = 3x + 5$, since $y = f(x)$.
(b) $f^{-1} = \{(x, y)\,|\,x = 3y + 5\}$
The rule for f^{-1} was obtained by interchanging x and y in the defining equation.
(c) To find $f^{-1}(x)$, solve the equation
$$x = 3y + 5$$
for y in terms of x.
$$3y = x - 5$$
$$y = \frac{x - 5}{3}$$

Thus
$$f^{-1}(x) = \frac{x - 5}{3}$$

EXAMPLE 8   If f is a function such that $f(x) = x^2$, and we restrict the domain of f only to $x \geq 0$, then f is one-to-one, as shown in Figure 8.4. Find $f^{-1}(x)$ and graph f and f^{-1} on the same set of axes.

Solution   Let $y = f(x)$; then $y = x^2$. Interchange x and y to obtain $x = y^2$ and solve for y. $y = \sqrt{x}$ and $f^{-1}(x) = \sqrt{x}$. Note that the domain of f was restricted to *exclude* negative values of x; thus in the inverse function we obtain only nonnegative values in the range, and only the positive square root of x is to be considered.

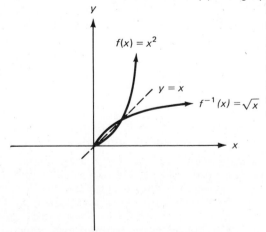

FIG. 8.4 Graph of a one-to-one function and its inverse function

The preceding example shows that it is possible to obtain a one-to-one function by restricting the domain of a function that is not one-to-one.

An important relation between a one-to-one function and its inverse is the following:

THE FUNCTION–INVERSE FUNCTION THEOREM

If f is a one-to-one function, then
$f^{-1}(f(x)) = x$ and $f(f^{-1}(x)) = x$

EXAMPLE 9   Given $f(x) = 5x - 2$; find $f^{-1}(x)$, $f(f^{-1}(x))$, and $f^{-1}(f(x))$.

Solution   Let $y = f(x)$; then $y = 5x - 2$.
Interchange x and y: $x = 5y - 2$
Obtain $f^{-1}(x)$ by solving this new equation for y:

$5y = x + 2$

$y = \dfrac{x + 2}{5}$ and thus $f^{-1}(x) = \dfrac{x + 2}{5}$

From the Function–Inverse Function Theorem, we know that $f(f^{-1}(x)) = x$ and that $f^{-1}(f(x)) = x$, but we can verify the result.

$f(x) = 5x - 2$

$f(f^{-1}(x)) = 5(f^{-1}(x)) - 2$ (Replacing x by $f^{-1}(x)$)

$= 5\left(\dfrac{x + 2}{5}\right) - 2$ $\left(\text{Since } f^{-1}(x) = \dfrac{x + 2}{5}\right)$

$= x + 2 - 2$

$= x$

Also,

$f^{-1}(x) = \dfrac{x + 2}{5}$

$f^{-1}(f(x)) = \dfrac{f(x) + 2}{5}$ (Replacing x by $f(x)$)

$= \dfrac{(5x - 2) + 2}{5}$ (Since $f(x) = 5x - 2$)

$= \dfrac{5x}{5}$

$= x$

EXERCISES 8.1

In Exercises 1–12 find the inverse of each relation and graph the relation and its inverse on the same set of axes. State the domain and range of each relation and inverse relation. (Examples 1–4)

1. $r = \{(x, y) \mid y = 2x - 1\}$
2. $r = \{(x, y) \mid y = x + 3\}$
3. $r = \{(x, y) \mid 3x + 2y = 4\}$

4. $r = \{(x, y)\,|\,2x - 3y = 6\}$
5. $r = \{(x, y)\,|\,y = x^2 + 1\}$
6. $r = \{(x, y)\,|\,y = 2 - x^2\}$
7. $r = \{(x, y)\,|\,y = x^2 + 3x - 1\}$
8. $r = \{(x, y)\,|\,y = x^2 + 4x\}$
9. $r = \{(x, y)\,|\,y = x^3 + 2\}$
10. $r = \{(x, y)\,|\,y = x^3 + 2x\}$
11. $r = \{(x, y)\,|\,y = |x|\}$
12. $r = \{(x, y)\,|\,y = |x + 2|\}$

In Exercises 13–20 $f(x)$ is given. Find $f^{-1}(x)$ and show that $f(f^{-1}(x)) = x$ and $f^{-1}(f(x)) = x$. (Examples 7, 8, and 9)

13. $f(x) = 2x - 1$

14. $f(x) = x + 3$

15. $f(x) = 5 - 2x$

16. $f(x) = 4 - 3x$

17. $f(x) = x^2 + 2,\ x \geq 0$

18. $f(x) = 3x^2,\ x \geq 0$

19. $f(x) = \sqrt{x}$

20. $f(x) = -\sqrt{x}$

21. Let $f(x) = 2^x$ and find $f(-3)$, $f(-2)$, $f(-1)$, $f(0)$, $f(1)$, $f(2)$, and $f(3)$. Plot the points $(x, f(x))$ on a graph and join the points with a smooth curve.

22. Graph f^{-1} for the function defined in Exercise 21 by interchanging the x and y values in each ordered pair.

23. For the function defined in Exercise 21 find $f^{-1}(2)$, $f^{-1}(1)$, $f^{-1}(8)$, $f^{-1}\left(\dfrac{1}{4}\right)$, and $f^{-1}\left(\dfrac{1}{8}\right)$.

24. Let $f(x) = 3^x$ and find $f(-3)$, $f(-2)$, $f(-1)$, $f(0)$, $f(1)$, $f(2)$, and $f(3)$. Plot the points $(x, f(x))$ on a graph and join the points with a smooth curve.

25. From the graph in Exercise 24, approximate the following:
$$f\left(-\frac{1}{2}\right) = 3^{-1/2} = \frac{1}{\sqrt{3}};\ f\left(\frac{1}{2}\right) = 3^{1/2} = \sqrt{3};\ f(1.5) = 3^{1.5} = 3\sqrt{3}.$$

26. Graph f^{-1} for the function defined in Exercise 24 by interchanging the x and y values in each ordered pair.

27. For the function defined in Exercise 24 find $f^{-1}\left(\dfrac{1}{9}\right)$, $f^{-1}(1)$, $f^{-1}(27)$, $f^{-1}\left(\dfrac{1}{27}\right)$, and $f^{-1}(9)$.

28. Let $f(x) = 4^{-x}$ and find $f(-2)$, $f\left(-\dfrac{3}{2}\right)$, $f(-1)$, $f\left(-\dfrac{1}{2}\right)$, $f(0)$, $f\left(\dfrac{1}{2}\right)$, $f(1)$, $f\left(\dfrac{3}{2}\right)$, and $f(2)$. Plot the points on a graph and join the points with a smooth curve.

29. Use the ordered pairs obtained in Exercise 28 to graph f^{-1} for $f(x) = 4^{-x}$ on the same set of axes.

30. Use the information from Exercises 28 and 29 to find $f^{-1}(2)$, $f^{-1}(8)$, $f^{-1}\left(\dfrac{1}{8}\right)$, $f^{-1}(16)$, $f^{-1}(1)$, and $f^{-1}\left(\dfrac{1}{2}\right)$ for $f(x) = 4^{-x}$.

8.2 EXPONENTIAL FUNCTIONS

In Chapter 4 it was shown that an expression such as b^x, where b is not zero and x is any rational number, is defined in the set of complex numbers. For b^x to be a real number, it was shown that b must be nonnegative whenever the rational exponent indicated that a square root, a fourth root, or any even root was to be taken. For the purpose of quick reference, the definitions and theorems for rational exponents are restated below. Note that we restrict the bases a and b to be positive numbers.

DEFINITIONS AND THEOREMS FOR RATIONAL EXPONENTS

Let a and b be positive real numbers.
Let n be a natural number and m an integer.
Let r and s be rational numbers.

Definitions

$$b^1 = b$$

$$b^n = b(b^{n-1})$$

$$b^0 = 1$$

$$b^{-n} = \frac{1}{b^n}$$

$$b^{1/n} = \sqrt[n]{b}$$

$$b^{m/n} = (\sqrt[n]{b})^m$$

Theorems

$$b^r b^s = b^{r+s}$$

$$\frac{b^r}{b^s} = b^{r-s}$$

$$(b^r)^s = b^{rs}$$

$$(ab)^r = a^r b^r$$

$$\left(\frac{a}{b}\right)^r = \frac{a^r}{b^r}$$

If b is any positive real number, then the expression b^x designates exactly one real number for every rational number x. Thus the equation $y = b^x$ defines a function whose domain is the set of rational numbers.

If $b = 1$, then $y = 1^x = 1$ for each rational number x, and the function defined by $y = 1^x$ is a constant function and obviously *not* one-to-one. For example, $1^5 = 1$, $1^{-5} = \frac{1}{1^5} = 1$, $1^{1/2} = \sqrt{1} = 1$, and so on. However, if $b \neq 1$, then exactly one value of y is paired with each value of x by the rule $y = b^x$, and the corresponding function *is* one-to-one.

We now ask the question, what happens if x is not rational? The answer is that b^x is a real number for all positive values of b and x *any* real number, even an irrational number.

The graphs of 2^x, 3^x, and 4^{-x}, which were developed in the exercises for Section 8.1, should be reexamined at this time to see what function values correspond to $x = \sqrt{2}$ for $f(x) = 2^x$ or $f(x) = 3^x$. The graph of 2^x is shown in Figure 8.5, taken from the ordered pairs indicated in the table below.

x	-2	$-\dfrac{3}{2}$	-1	$-\dfrac{1}{2}$	0	$\dfrac{1}{2}$	1	2	3
$y = 2^x$	$\dfrac{1}{4}$	$\dfrac{\sqrt{2}}{4}$	$\dfrac{1}{2}$	$\dfrac{\sqrt{2}}{2}$	1	$\sqrt{2}$	2	4	8

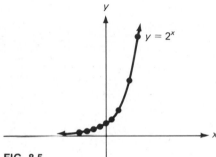

FIG. 8.5

Before we are ready to analyze the graph of $y = 2^x$ we need to state some properties of the function $f(x) = b^x$ when b is a *positive number other than 1* and the exponent x is a rational number.

THEOREMS

If $b > 0$ and $b \neq 1$, and if r and s are any rational numbers, then

1. $b^r = b^s$ if and only if $r = s$.
2. If $b > 1$, then $b^r > b^s$ if and only if $r > s$.
3. If $0 < b < 1$, then $b^r > b^s$ if and only if $r < s$.

By looking at the graph of $y = 2^x$ in Figure 8.5, we can see that the value of 2^x increases as x increases. This is in keeping with the second part of the theorem, since $b = 2$ and 2 is greater than 1.

We can now determine what value should be assigned to $2^{\sqrt{3}}$ by approximating $\sqrt{3}$ by rational numbers, and by assuming that for rational numbers r and s, if $r < \sqrt{3} < s$, then $2^r < 2^{\sqrt{3}} < 2^s$.

If $\quad 1 < \sqrt{3} < 2 \qquad\qquad$ then $\quad 2^1 < 2^{\sqrt{3}} < 2^2$
$\quad 1.7 < \sqrt{3} < 1.8 \qquad\qquad\qquad 2^{1.7} < 2^{\sqrt{3}} < 2^{1.8}$
$\quad 1.73 < \sqrt{3} < 1.74 \qquad\qquad 2^{1.73} < 2^{\sqrt{3}} < 2^{1.74}$
$\quad 1.732 < \sqrt{3} < 1.733 \qquad 2^{1.732} < 2^{\sqrt{3}} < 2^{1.733}$

$$\sqrt{3} \qquad\qquad\qquad\qquad 2^{\sqrt{3}}$$

As the approximations to $\sqrt{3}$ get better and better, the approximations to $2^{\sqrt{3}}$ get closer and closer to exactly one real number. This real number, called the limiting value, is assigned as the definition of $2^{\sqrt{3}}$.

In general, if x is an irrational real number and b is a positive real number, then b^x *is defined as the limiting value of the approximations to b^x,* as illustrated in the special case above.

By using the methods of higher mathematics, it can be shown that the theorems for rational exponents are still valid when the exponent is an irrational real number.

Accepting these results, the definition of an exponential function can now be stated.

DEFINITION—Exponential Function

If $b > 0$ and $b \neq 1$, and x is any real number, the exponential function to the base b is defined by the equation
$$y = b^x$$

Note that $y = b^x$ is defined not only as a function, but as a one-to-one function. By eliminating the possibility of $b = 1$, and thus eliminating the constant function, $y = b^x = 1$, the conditions for a one-to-one function are met. Therefore the inverse of the exponential function, which will be discussed in the next section, is also a function.

Figure 8.5 is the graph of $y = 2^x$. Since the base, 2, is greater than 1, y increases for successively larger values of x. This is an **increasing** function for all values of x, and was stated above by the rule: If $b > 1$ and $r > s$, then $b^r > b^s$. What about the function $y = 2^{-x}$? We can rewrite this equation as $y = \left(\dfrac{1}{2}\right)^x$, and we now have the situation where $0 < b < 1$, so that when $r > s$, $b^r < b^s$. The graph of $y = 2^{-x}$ is shown in Figure 8.6. Since the y values are successively smaller for increasing values of x, this is a **decreasing** function. Observe that both graphs, that of $y = 2^x$, and that of $y = 2^{-x}$, are one-to-one.

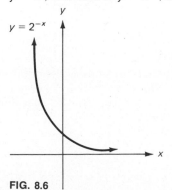

$y = 2^{-x}$

FIG. 8.6

The concepts stated above are illustrated in Figure 8.7, which is the graph of $y = b^x$ when $b > 1$, and on the same set of axes, the graph of $y = b^x$ when $0 < b < 1$. Notice that the graphs intersect at the point (0, 1) because $b^0 = 1$ for $b \neq 0$.

The fact that if $f(x) = b^x$ with the restrictions for an exponential function ($b > 0$, $b \neq 1$, x a real number), then $b^r = b^s$ if and only if $r = s$, can be useful in solving some exponential equations.

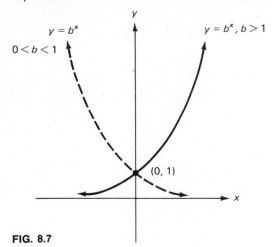

FIG. 8.7

EXAMPLE 1   Solve for x: $2^x = 8$

Solution $8 = 2^3$; therefore $2^x = 2^3$
and $x = 3$

EXAMPLE 2   Solve: $3^{1-x} = \dfrac{1}{81}$

Solution $3^{1-x} = \dfrac{1}{81} = \dfrac{1}{3^4} = 3^{-4}$

$$3^{1-x} = 3^{-4}$$
$$1 - x = -4$$
$$-x = -5$$
$$x = 5$$

EXAMPLE 3   Solve: $8^{2x+1} = 16$

Solution   In order to apply the theorem, both sides of the equation must be powers of the same base. We know that $8 = 2^3$ so that $8^{2x+1} = (2^3)^{2x+1} = 2^{6x+3}$ because $(a^m)^n = a^{mn}$. Also, $16 = 2^4$. Therefore $8^{2x+1} = 16$ implies that $2^{6x+3} = 2^4$

$$\text{and}\quad 6x + 3 = 4$$
$$6x = 1$$
$$x = \dfrac{1}{6}$$

EXERCISES 8.2

Graph Exercises 1–14 by plotting points corresponding to $x = -2$, -1, $-\dfrac{1}{2}$, 0, $\dfrac{1}{2}$, 1, and 2, and joining the points in a smooth curve.

1. $y = 3^x$

2. $y = 4^x$

3. $y = 3^{-x}$

4. $y = 4^{-x}$

5. $y = 5^x$

6. $y = 6^x$

7. $y = 5^{-x}$

8. $y = 6^{-x}$

9. $y = 10^x$

10. $y = \left(\dfrac{1}{2}\right)^{-x}$

11. $y = 10^{-x}$

12. $y = \left(\dfrac{3}{4}\right)^x$

13. $y = \left(\dfrac{1}{3}\right)^{-x}$

14. $y = \left(\dfrac{3}{4}\right)^{-x}$

In Exercises 15–20 use the coordinates of points from the applicable exercises above, and interchange the order of x and y to graph the following.

15. $y = f^{-1}(x)$, when $f(x) = 3^x$

16. $y = f^{-1}(x)$, when $f(x) = 4^x$

17. $y = f^{-1}(x)$, when $f(x) = 10^x$

18. $y = f^{-1}(x)$, when $f(x) = 4^{-x}$

19. $y = f^{-1}(x)$, when $f(x) = 3^{-x}$

20. $y = f^{-1}(x)$, when $f(x) = 10^{-x}$

Solve Exercises 21–40 by using the theorem $b^r = b^s$ if and only if $r = s$. (Examples 1, 2 and 3)

21. $5^x = 125$

22. $4^x = \dfrac{1}{64}$

23. $5^x = \dfrac{1}{25}$

24. $4^{2-x} = 16$

25. $5^x = 5\sqrt{5}$

26. $4^{-x} = \dfrac{1}{8}$

27. $2^{1-x} = \dfrac{1}{8}$

28. $8^{x+1} = 4$

29. $10^x = 1000$

30. $10^x = 10,000$

31. $10^x = 0.0001$

32. $10^x = 0.01$

33. $10^{-x} = 100$

34. $10^{-x} = 0.001$

35. $10^x = 1$

36. $10^{3-x} = 100$

37. $3^{x-2} = \sqrt{27}$

38. $3^{x-1} = \dfrac{1}{3}$

39. $8^{4-2x} = 4^{x+4}$

40. $9^{3x-12} = 1$

8.3   LOGARITHMIC FUNCTIONS

In the preceding section it was shown that the exponential function $y = b^x$, $b > 0$ and $b \neq 1$, and x any real number, is one-to-one and its inverse is also a function. This inverse function is called the logarithmic function.

DEFINITION—Logarithmic Function

The **logarithmic function** to the base b is a function whose rule is $x = b^y$, where $b > 0$, $b \neq 1$, and y any real number.

Since the range of the exponential function is the set of **positive** numbers, the **domain** of the logarithmic function is the set of positive numbers. This follows from the fact that b is a positive number, and a positive number raised to any power is always a positive number.

The domain of the exponential function is the range of the logarithmic function, that is, the set of real numbers.

In the previous problems, after interchanging x and y in the rule defining a function, we were able to solve the new equation for y and thus find $f^{-1}(x)$. There is no simple way to solve the equation $x = b^y$ for y to obtain $f^{-1}(x)$. So a definition is made instead: for $f(x) = b^x$, $f^{-1}(x) = \log_b x$ (read "the logarithm of x to the base b).

DEFINITION

For $b > 0$ and $b \neq 1$,
$$y = \log_b x \quad \text{if and only if} \quad x = b^y$$

It should be noted that the logarithm of a number to the base b is the exponent of b that must be used to produce the given number. It is useful to bear in mind that a **logarithm is an exponent.**

EXAMPLE 1   Write in logarithmic form: $10^3 = 1000$.

Solution   Since $b^y = x$ if and only if $\log_b x = y$,
$$10^3 = 1000 \text{ if and only if } \log_{10} 1000 = 3$$

EXAMPLE 2   Write in logarithmic form: $2^{-4} = \dfrac{1}{16}$

Solution

$$2^{-4} = \frac{1}{16} \quad \text{if and only if}$$

$$\log_2 \left(\frac{1}{16} \right) = -4$$

EXAMPLE 3   Write in logarithmic notation: $125^{-2/3} = \frac{1}{25}$

Solution

$$125^{-2/3} = \frac{1}{25} \quad \text{if and only if}$$

$$\log_{125} \left(\frac{1}{25} \right) = -\frac{2}{3}$$

Note the direct application of the definition in each of the examples just given: The *base* of the exponent b becomes the *base* of the logarithm, and the *exponent* is the logarithm.

EXAMPLE 4   Write in exponential form: $\log_{10} 100 = 2$

Solution

$$\log_{10} 100 = 2 \quad \text{if and only if}$$
$$100 = 10^2$$

This follows again directly from the definition:

$$\log_b x = y \quad \text{if and only if}$$
$$x = b^y$$

EXAMPLE 5   Write in exponential form: $\log_2 \left(\frac{1}{8} \right) = -3$

Solution   From the definition,

$$\log_2 \left(\frac{1}{8} \right) = -3 \quad \text{if and only if}$$

$$\frac{1}{8} = 2^{-3}$$

EXAMPLE 6   Write in exponential notation: $\log_{25} 5 = \frac{1}{2}$

Solution

$$\log_{25} 5 = \frac{1}{2} \quad \text{if and only if}$$
$$5 = 25^{1/2}$$

Before applying the definition in the solution of problems, it is useful to examine an important theorem, two immediate consequences of that theorem, and the graph of the exponential function and its inverse, the logarithmic function, drawn on one set of axes.

THE LOGARITHMIC FUNCTION THEOREM

$$b^{\log_b x} = x \quad \text{and} \quad \log_b b^x = x$$

Two important special cases of this theorem are the following:

$\log_b b = 1$ since for $x = 1$, $\log_b b^1 = 1$
$\log_b 1 = 0$ since for $x = 0$, $\log_b b^0 = 0$.

The graph of the exponential function with base $b > 1$ and its inverse, the logarithmic function with base $b > 1$, is illustrated in Figure 8.8.

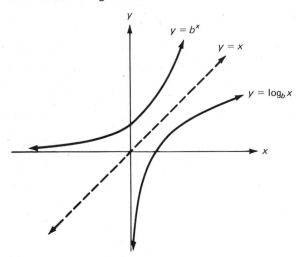

FIG. 8.8

EXAMPLE 7   Find each of the following:
$$3^{\log_3 7}, \quad 10^{\log_{10} 5}, \quad 10^{\log_{10} 3.2}$$
Solution
$$3^{\log_3 7} = 7, \qquad 10^{\log_{10} 5} = 5,$$
$$10^{\log_{10} 3.2} = 3.2$$

EXAMPLE 8   Find each of the following:
$$\log_4 4^3, \quad \log_{10} 10^2, \quad \log_{10} 10^{-1}$$
Solution
$$\log_4 4^3 = 3, \qquad \log_{10} 10^2 = 2,$$
$$\log_{10} 10^{-1} = -1$$

EXAMPLE 9   Find each of the following:
$$\log_3 1, \quad \log_5 1, \quad \log_{10} 1$$
Solution
$$\log_3 1 = 0, \qquad \log_5 1 = 0, \qquad \log_{10} 1 = 0$$

EXAMPLE 10 Find each of the following:
$\log_8 8, \qquad \log_5 5, \qquad \log_{10} 10$

Solution

$$\log_8 8 = 1, \qquad \log_5 5 = 1,$$
$$\log_{10} 10 = 1$$

Using the definition of the logarithmic function and the logarithmic function theorem and its special cases, we can now solve some equations involving logarithms.

EXAMPLE 11 Evaluate $\log_5 0.04$.

Solution Let $x = \log_5 0.04$. Then

$$5^x = 0.04 = \frac{4}{100} = \frac{1}{25} = 5^{-2}$$

$$5^x = 5^{-2}$$

Thus $x = -2$, since $b^x = b^y$ if and only if $x = y$. Therefore $\log_5 0.04 = -2$.

EXAMPLE 12 Solve for x: $\log_3 x = -4$

Solution $\text{Log}_3 x = -4$ if and only if $3^{-4} = x$.

Thus $x = \frac{1}{81}$.

EXAMPLE 13 Solve for x: $\log_x 8 = \frac{3}{2}$

Solution $\text{Log}_x 8 = \frac{3}{2}$ if and only if

$$x^{3/2} = 8$$
$$(x^{3/2})^{2/3} = 8^{2/3}$$
$$x = (\sqrt[3]{8})^2 = 4$$

EXERCISES 8.3

Write Exercises 1–16 in logarithmic form. (Examples 1, 2 and 3)

1. $10^4 = 10{,}000$

2. $10^0 = 1$

3. $9^{1/2} = 3$

4. $5^1 = 5$

5. $3^1 = 3$

6. $\left(\frac{1}{4}\right)^{-3/2} = 8$

7. $9^{-3/2} = \frac{1}{27}$

8. $5^{-4} = 0.0016$

9. $5^0 = 1$

10. $10^{-3} - 0.001$

11. $4^{-3} = \frac{1}{64}$

12. $16^{-1/2} = 0.25$

13. $64^{2/3} = 16$

14. $8^{1/3} = 2$

15. $\left(\frac{1}{6}\right)^{-2} = 36$

16. $125^{1/6} = \sqrt{5}$

Write Exercises 17–32 in exponential form. (Examples 4, 5 and 6)

17. $\log_3 9 = 2$

18. $\log_2 32 = 5$

19. $\log_{10} 10 = 1$

20. $\log_{16} 4 = \dfrac{1}{2}$

21. $\log_{36} 6 = \dfrac{1}{2}$

22. $\log_3 1 = 0$

23. $\log_{10} 0.1 = -1$

24. $\log_{10} 0.01 = -2$

25. $\log_{10} 1 = 0$

26. $\log_6 6 = 1$

27. $\log_2 0.125 = -3$

28. $\log_5 0.04 = -2$

29. $\log_{10} 100,000 = 5$

30. $\log_{10} 1,000,000 = 6$

31. $\log_8 0.25 = -\dfrac{2}{3}$

32. $\log_{25} 0.008 = -\dfrac{3}{2}$

Evaluate Exercises 33–48. (Examples 7–11)

33. $\log_{10} 1$

34. $\log_{10} 0.01$

35. $\log_2 0.125$

36. $\log_4 32$

37. $\log_4 \dfrac{1}{64}$

38. $\log_2 0.0625$

39. $5^{\log_5 2}$

40. $10^{\log_{10} 3}$

41. $\log_{10} 1000$

42. $\log_{10} 10$

43. $\log_5 0.0016$

44. $(\log_2 8)(\log_8 2)$

45. $\log_9 \dfrac{1}{27}$

46. $\log_3 \sqrt{243}$

47. $\log_{10} 10^{-2.5}$

48. $\log_5 5^{1.4}$

Solve Exercises 49–80 for x. (Examples 12 and 13)

49. $\log_2 64 = x$

50. $\log_3 x = -4$

51. $\log_{10} x = 0$

52. $\log_2 0.25 = x$

53. $\log_x 0.04 = -2$

54. $\log_x 16 = \dfrac{2}{3}$

55. $\log_5 x = 1$

56. $\log_{10} x = -3$

57. $\log_{10} x = 4$

58. $\log_{10} x = 1.5$

59. $\log_5 125 = x$

60. $\log_x 8 = -\dfrac{3}{2}$

61. $\log_x 6\sqrt{6} = \dfrac{3}{2}$

62. $\log_{27} x = -\dfrac{4}{3}$

63. $\log_x 9 = -2$

64. $\log_x 0.125 = 3$

65. $\log_5 0.008 = x$

66. $\log_4 x = \dfrac{3}{2}$

67. $\log_{10} x = 1$

68. $\log_{36} x = -1.5$

69. $\log_x 64 = -3$

70. $\log_x 9 = 0.5$

71. $\log_3 x = 0$

72. $\log_{10} x = -1.5$

73. $\log_{10} x = -1$

74. $\log_x 7 = \dfrac{1}{2}$

75. $\log_x 625 = 4$

76. $\log_x 5 = 1$

77. $\log_x 36 = -2$

78. $\log_x 10^{-3.7} = -3.7$

79. $\log_x 10^{1.24} = 1.24$

80. $\log_2 0.0625 = x$

8.4 PROPERTIES OF LOGARITHMS

Before the ready availability of computers, logarithms were used to speed mathematical computations. For example, problems such as

$$\frac{\sqrt{0.461 \times 0.00513 \times 0.0032}}{0.0035 \times (0.00571)^4}$$

were solved by finding the logarithms of the numbers, and then adding or subtracting the logarithms, or dividing the logarithm by 2 for finding a square root, or multiplying the logarithm by 4 to raise a number to the fourth power.

Now these problems are solved very quickly by using a calculator. The theories behind logarithms, however, are still useful in higher mathematics, and it is mainly for that reason that this section and the subsequent sections on logarithms are included.

THEOREMS—LOGARITHMIC OPERATIONS

Let x, y, and b be any *positive* real numbers, with $b \neq 1$. Let k be any positive integer. Then

1. $\log_b xy = \log_b x + \log_b y$

The logarithm of a *product* is the *sum* of the logarithms of the factors.

2. $\log_b \dfrac{x}{y} = \log_b x - \log_b y$

The logarithm of a *quotient* is the *difference* of the logarithms of the numerator and the denominator.

3. $\log_b x^y = y \log_b x$

The logarithm of a *power* is the *product of the exponent* and the logarithm of its base.

4. $\log_b \sqrt{x} = \dfrac{1}{k} \log_b x$

Since $\sqrt[k]{x} = x^{1/k}$, Theorem 4 is just a special case of Theorem 3.

To show how these theorems are derived, we must remember that a logarithm is an exponent, and that $b^{\log_b x} = x$. The proof of Theorem 1 is as follows (the other proofs are similar and will not be given at this time):

Proof of Theorem 1:

$$\log_b xy = \log_b (b^{\log_b x} b^{\log_b y})$$
$$\text{since } b^{\log_b x} = x \quad \text{for all real } x$$
$$= \log_b (b^{\log_b x + \log_b y})$$
$$\text{since } b^x b^y = b^{x+y}$$
$$= \log_b x + \log_b y$$
$$\text{since } \log_b b^x = x \quad \text{for all real } x$$

EXAMPLE 1

$$\log_3 20 = \log_3 (4)(5) = \log_3 4 + \log_3 5 \quad \text{(By Theorem 1)}$$
but $4 = 2^2$;
thus $\log_3 4 = \log_3 2^2 = 2 \log_3 2 \quad \text{(By Theorem 3)}$
Therefore
$$\log_3 20 = 2 \log_3 2 + \log_3 5$$

EXAMPLE 2

$$\log_6 \sqrt[3]{49} = \log_6 (49)^{1/3} = \log_6 (7^2)^{1/3}$$
$$= \log_6 7^{2/3} = \frac{2}{3} \log_6 7 \quad \text{(By Theorems 3 and 4)}$$

EXAMPLE 3

Express $\log_b \frac{16}{7}$ in terms of the logarithms of *prime* numbers.

Solution $\log_b \frac{16}{7} = \log_b 16 - \log_b 7 \quad \text{(By Theorem 2)}$
$$= \log_b 2^4 - \log_b 7$$
$$= 4 \log_b 2 - \log_b 7 \quad \text{(By Theorem 3)}$$

Since 2 and 7 are prime numbers, $\log_b \frac{16}{7}$ is now expressed in terms of the logarithms of prime numbers.

EXAMPLE 4

Express $\frac{1}{3} (\log_b 5 + \log_b 7 - \log_b 2)$ as the logarithm of a single number.

Solution $\frac{1}{3} (\log_b 5 + \log_b 7 - \log_b 2)$
$$= \frac{1}{3} (\log_b 35 - \log_b 2) \quad \text{(Theorem 1)}$$
$$= \frac{1}{3} \log_b \frac{35}{2} \quad \text{(Theorem 2)}$$
$$= \log_b \sqrt[3]{\frac{35}{2}} \quad \text{(Theorem 4)}$$

EXERCISES 8.4

By using the logarithmic operations theorems, express each of Exercises 1–20 in terms of the logarithms of prime numbers. (Examples 1, 2 and 3)

1. $\log_b 15$

2. $\log_b \dfrac{3}{5}$

3. $\log_b 3^5$

4. $\log_b \sqrt{3}$

5. $\log_b 6$

6. $\log_b \dfrac{3}{2}$

7. $\log_b 2^6$

8. $\log_b \sqrt[3]{2}$

9. $\log_b \dfrac{1}{6}$

10. $\log_b 12$

11. $\log_b \dfrac{27}{64}$

12. $\log_b \sqrt{1.5}$

13. $\log_b 4\sqrt{3}$

14. $\log_b 6\sqrt{6}$

15. $\log_b \sqrt{\dfrac{5}{3}}$

16. $\log_b \dfrac{1}{\sqrt[3]{5}}$

17. $\log_b \dfrac{81}{25}$

18. $\log_b 25\sqrt[4]{3}$

19. $\log_b \dfrac{5\sqrt{5}}{7}$

20. $\log_b \sqrt[3]{\dfrac{625}{9}}$

By using the logarithmic operations theorems, express each of Exercises 21–40 as the logarithm of a single number. (Example 4)

21. $\log_b 25 + \log_b 18$

22. $\log_b 25 - \log_b 18$

23. $\dfrac{1}{2} \log_b 5 + \dfrac{1}{3} \log_b 17$

24. $\log_b 7 + 5 \log_b 3$

25. $\log_b 8 + \log_b 9$

26. $\log_b 21 - \log_b 8$

27. $5 \log_b 3$

28. $\dfrac{1}{2} \log_b 17$

29. $2 \log_b 5 + 4 \log_b 3$

30. $\dfrac{1}{3} \log_b 7 - 2 \log_b 6$

31. $\dfrac{1}{3} (\log_b 56 - \log_b 45)$

32. $\log_b 1 - \log_b 7 - \log_b 10$

33. $3 \log_b b - \dfrac{1}{2} \log_b 25$

34. $\log_b \dfrac{5}{2} + \log_b \dfrac{1}{3} - \log_b 6$

35. $\dfrac{1}{5} \log_b 64 - \log_b \sqrt[5]{2}$

36. $\dfrac{1}{2} \log_b 5 + \dfrac{3}{4} \log_b 7 - \dfrac{3}{2} \log_b 6$

37. $\dfrac{1}{5} (\log_b 3 + 2 \log_b 5)$

38. $3(\log_b 4 - 3 \log_b 6)$

39. $2 \log_b 3 + \dfrac{1}{4} \log_b 5 - 3 \log_b 7$

40. $\dfrac{1}{3} (2 \log_b 9 + \log_b 4 - \log_b 85)$

8.5 COMMON LOGARITHMS

Since our numeral system is a base-10 positional system, the logarithmic function whose base is 10 is most useful for computations. The values of $\log_{10} x$ are called common logarithms or logarithms to the base 10. To reduce the amount of writing involved in a calculation using common logarithms, the numeral 10 designating the base is usually omitted.

CONVENTION

$$\log x = \log_{10} x$$

Finding the common logarithm of a number depends on the principle of scientific notation, restated below.

DEFINITION

Let r be a positive real number.
Let x be a real number between 1 and 10.
Let k be an integer.
Then the **scientific notation** for r is $x \times 10^k$.
$$r = x \times 10^k = x(10^k)$$

EXAMPLE 1 Express each of the following numbers in scientific notation:
(a) 285,000
(b) 0.032
(c) 7.2

Solution
(a) $285,000 = 2.85 \times 10^5$
(b) $0.032 = 3.2 \times 10^{-2}$
(c) $7.2 = 7.2 \times 10^0$

THE THEOREM FOR COMMON LOGARITHMS

$$\log x(10^k) = \log x + k$$

Proof

$\log x(10^k) = \log x + \log 10^k$ (By logarithmic operations, Theorem 1)

$= \log x + k \log 10$ (By logarithmic operations, Theorem 3)

$= \log x + k$ (Since $\log_{10} 10 = 1$)

8.5 COMMON LOGARITHMS

Since every positive real number can be expressed as the product of a number between 1 and 10 and an integral power of 10, the common logarithm of any positive real number can be determined from the logarithms of the numbers between 1 and 10. A table listing approximations to these values is calculated by using advanced mathematical techniques. An excerpt from the table inside the back cover of the book is shown below.

x	0	1	2	3	4	5	6	7	8	9
2.0	.3010	.3032	.3054	.3075	.3096	.3118	.3139	.3160	.3181	.3201
2.1	.3222	.3243	.3263	.3284	.3304	.3324	.3345	.3365	.3385	.3404
2.2	.3424	.3444	.3464	.3483	.3502	.3522	.3541	.3560	.3579	.3598
2.3	.3617	.3636	.3655	.3674	**.3692**	.3711	.3729	.3747	.3766	.3784
2.4	.3802	.3820	.3838	.3856	.3874	.3892	.3909	.3927	.3945	.3962

An approximation for log x for $1 \leq x < 10$ is found from the table by locating the entry in the row having the first two digits of x and in the column having the third digit of x. Thus log 2.34 is located in the row having 2.3 at the extreme left and in the column under 4. We find that

log 2.34 = 0.3692.

By using the theorem for common logarithms, the following logarithms can be obtained:

$$\log 0.00234 = \log 2.34 \times 10^{-3} = 0.3692 - 3$$
$$\log 0.0234 = \log 2.34 \times 10^{-2} = 0.3692 - 2$$
$$\log 0.234 = \log 2.34 \times 10^{-1} = 0.3692 - 1$$
$$\log 2.34 = \log 2.34 \times 10^{0} = 0.3692$$
$$\log 23.4 = \log 2.34 \times 10^{1} = 0.3692 + 1$$
$$\log 234 = \log 2.34 \times 10^{2} = 0.3692 + 2$$
$$\log 2340 = \log 2.34 \times 10^{3} = 0.3692 + 3$$

Observation of these special cases reveals that a common logarithm may be regarded as having two parts: an integral part called the **characteristic** and a positive (or zero) decimal part called the **mantissa.** The mantissa is found in a table of logarithms, and the characteristic is found by using the theorem for common logarithms.

EXAMPLE 2   Find log 6520.

Solution
$$\log 6520 = \log (6.52 \times 10^{3})$$
$$= \log 6.52 + 3$$
$$= 0.8142 + 3 \qquad \text{(Form for use in computations)}$$
$$= 3.8142 \qquad \text{(Form for use in formulas)}$$

From the preceding discussion it should be clear that the only difference between the logarithms of numbers which differ by a factor of 10 is the characteristic. Thus the following example shows that for log 6520, log 6.520, log 65.20, and log 0.00652 the mantissa is the same for each logarithm, but the characteristic is different.

EXAMPLE 3   Find the following logarithms:

log 6520,   log 6.520,   log 65.20,   log 0.00652

Solution   In Example 2 it was shown that
log 6.52 = 0.8142.

$$\log 6520 = \log (6.52 \times 10^3)$$
$$= 0.8142 + 3$$
$$\log 6.520 = 0.8142$$
$$\log 65.20 = \log (6.52 \times 10^1)$$
$$= 0.8142 + 1$$
$$\log 0.00652 = \log (6.52 \times 10^{-3})$$
$$= 0.8142 - 3$$

All the above logarithms are in computational form. When using a calculator to find the logarithm of a number, the result is shown in the form for use in formulas. Therefore

$$\log 6520 = 0.8142 + 3 = 3.8142$$
as shown in Example 2
$$\log 65.20 = 0.8142 + 1 = 1.8142$$
$$\log 0.00652 = 0.8142 - 3 = -2.1858$$

Example 4 illustrates another situation involving a negative characteristic.

EXAMPLE 4   Find log 0.347.

Solution

$$\log 0.347 = \log (3.47 \times 10^{-1})$$
$$= \log 3.47 - 1$$
$$= 0.5403 - 1 \quad \text{(Computational form)}$$
$$= -0.4597 \quad \text{(Formula form)}$$
$$(0.5403 - 1 = 0.5403 - 1.0000 = -0.4597)$$

The table of logarithms may also be used to find a number whose logarithm is given.

DEFINITION

If $\log x = n$, then x is the **antilogarithm** of n.

EXAMPLE 5 Find antilog (2.8089).

Solution

$$2.8089 = 0.8089 + 2$$
$$x = \text{antilog } 2.8089 = \text{antilog } (0.8089 + 2)$$
$$\log x = 0.8089 + 2$$
$$\log x = \log 6.44 + 2 \quad \text{(Using the table)}$$
$$\log x = \log (6.44 \times 10^2)$$
$$x = 6.44 \times 10^2 \quad \text{(Scientific notation)}$$
$$x = 644 \quad \text{(Ordinary notation)}$$

EXAMPLE 6 Find antilog (0.4518 − 1).

Solution

$$x = \text{antilog } (0.4518 - 1)$$
$$\log x = 0.4518 - 1$$
$$\log x = \log 2.83 - 1 \quad \text{(Using the table)}$$
$$\log x = \log (2.83 \times 10^{-1})$$
$$x = 2.83 \times 10^{-1} \quad \text{(Scientific notation)}$$
$$x = 0.283 \quad \text{(Ordinary notation)}$$

EXAMPLE 7 Find antilog (−3.0685).

Solution Since the mantissa is always positive, we must add and subtract 4. Thus

$$-3.0685 + 4 - 4 = 4 - 3.0685 - 4$$
$$= 0.9315 - 4$$

The problem now becomes one of finding antilog (0.9315 − 4).

$$x = \text{antilog } (0.9315 - 4)$$
$$\log x = 0.9315 - 4$$
$$\log x = \log 8.54 - 4 \quad \text{(Using the table)}$$
$$\log x = \log (8.54 \times 10^{-4})$$
$$x = 8.54 \times 10^{-4} \quad \text{(Scientific notation)}$$
$$x = 0.000854 \quad \text{(Ordinary notation)}$$

EXERCISES 8.5

In Exercises 1–32 express each logarithm in (a) *computational form and* (b) *formula form.* (*Examples 2–4*)

1. log 4.76
2. log 890
3. log 0.345
4. log 75,000
5. log 0.123

6. log 405,000
7. log 6.32
8. log 0.00419
9. log 0.0948
10. log 5

11. log 1.5
12. log 0.0842
13. log 0.007
14. log 6520
15. log 929
16. log 0.00053
17. log 0.564
18. log 2810
19. log 2.72
20. log 68.1
21. log 36,700

22. log 84.9
23. log 0.266
24. log 2,000
25. log 0.000151
26. log 0.35
27. log 0.0752
28. log 123,000,000
29. log 3.14
30. log 0.08
31. log 0.000 000 456
32. log 32.2

In Exercises 33–64 express each antilogarithm in (a) scientific notation and (b) ordinary notation. (Examples 5–7)

33. antilog 1.8407
34. antilog 4.6232
35. antilog (0.0682 − 1)
36. antilog 0.9750
37. antilog (0.8344 − 3)
38. antilog 6.9031
39. antilog (0.2900 − 2)
40. antilog 0.5740
41. antilog 2.5911
42. antilog 3.7832
43. antilog (0.2430 − 1)
44. antilog (0.5717 − 4)
45. antilog (0.4232 − 2)
46. antilog 1.8
47. antilog (0.0374)
48. antilog (− 1.2)

49. antilog (0.2330 − 2)
50. antilog 0.9991
51. antilog 5.7694
52. antilog 1.5065
53. antilog (0.3201 − 1)
54. antilog 3.6021
55. antilog (0.8500 − 3)
56. antilog 0.4346
57. antilog (0.1875 − 4)
58. antilog 2.85
59. antilog 0.48
60. antilog (− 0.52)
61. antilog (− 1.15)
62. antilog (− 2.71)
63. antilog (− 3.8962)
64. antilog (− 4.2)

8.6 LINEAR INTERPOLATION AND COMPUTATIONS

It was stated in the preceding section that the use of logarithms for speeding up calculations is no longer practical because of the ready availability of calculators. Similarly, the table inside the cover of this book that list at most three significant digits are rapidly going out of style because most calculators that calculate logarithms have capabilities that are far greater than those in the table. The material in this section is presented primarily for historical purposes, and all calculations may be checked by using a calculator.

A simple method to extend the three significant digits of the table in this text to four significant digits is a process called **linear interpolation.** Actually, this is only an approximation. The graph of the logarithmic function $\{(x, y) \mid y = \log x\}$ is approximated by the straight line joining two points on the curve whose ordinates are successive entries in the table. Figure 8.9 illustrates the procedure for determining log 2.346 by linear interpolation.

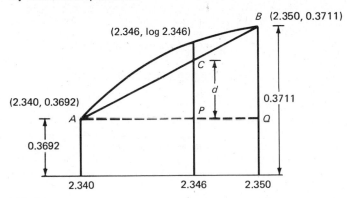

FIG. 8.9

The value of log 2.346 will be approximated by
log 2.340 + d = 0.3692 + d.
To determine d, an equation is found by using the fact from plane geometry that corresponding sides of similar triangles are proportional.

Triangle ACP is similar to triangle ABQ under the correspondence A to A, C to B, and P to Q. Thus

$$\frac{d}{BQ} = \frac{AP}{AQ}$$

or

$$\frac{d}{0.3711 - 0.3692} = \frac{2.346 - 2.340}{2.350 - 2.340}$$

$$\frac{d}{0.0019} = \frac{0.006}{0.010}$$

$$d = \frac{6}{10}(0.0019) = 0.00114$$

Since the decimal approximations in the table are valid for only four decimal places, d is rounded off to 0.0011.

Finally, log 2.346 = 0.3692 + 0.0011 = 0.3703.

EXPONENTIAL AND LOGARITHMIC FUNCTIONS

The following examples illustrate a more convenient arrangement of the work. In addition, the decimal points are omitted from the written work since this process can be performed mentally. After some practice, it is possible to perform all of the linear interpolation calculations mentally.

EXAMPLE 1   Find log 52.73.

Solution

$$
\begin{array}{ccc}
 & x & \log x \\
 & 52.70 & 1.7218 \\
3 & & \\
 & 52.73 & d \\
10 & & 8 \\
 & 52.80 & 1.7226 \\
\end{array}
$$

$$\frac{d}{8} = \frac{3}{10} \quad \text{or} \quad d = 0.3(8) = 2.4$$

Thus log 52.73 = 1.7218 + 0.0002 = 1.7220.

EXAMPLE 2   Find antilog 3.1888.

Solution

$$
\begin{array}{ccc}
 & x & \log x \\
 & 1540 & 3.1875 \\
d & & \\
 & & 3.1888 \quad 13 \\
10 & & 28 \\
 & 1550 & 3.1903 \\
\end{array}
$$

$$\frac{d}{10} = \frac{13}{28} \quad \text{or} \quad d = \frac{130}{28} = 5 \quad \text{approx.}$$

Thus antilog 3.1888 = 1545.

By using the theorems for logarithms, computations involving products, quotients, powers, and roots can be replaced by the simpler calculations involving sums, differences, products, and quotients.

EXAMPLE 3   Compute $(0.725)^4(34.7)$.

Solution

$$\text{Let } N = (0.725)^4(34.7).$$
$$\log N = 4 \log 0.725 + \log 34.7$$
$$\log 0.725 = 0.8603 - 1$$
$$4 \log 0.725 = 3.4412 - 4 = 0.4412 - 1$$
$$\log 34.7 = \underline{0.5403 + 1}$$
$$\text{Sum} = \quad \log N = 0.9815$$
$$N = 9.58 \quad \text{to three significant digits}$$

EXAMPLE 4   Compute $\sqrt[3]{\dfrac{1.380}{24.2}}$.

Solution   Let $N = \sqrt[3]{\dfrac{1.380}{24.2}}$.

$$\log N = \frac{1}{3}(\log 1.380 - \log 24.2)$$

$$\log 1.380 = 0.1399 = 2.1399 - 2$$
$$\underline{\log 24.2 = 1.3838 = 1.3838}$$
$$\text{Difference} = \qquad\qquad 0.7561 - 2$$
$$= \qquad\qquad 1.7561 - 3$$

$$\log N = \frac{1}{3}\text{difference} = \qquad 0.5854 - 1$$
$$N = 0.3849$$
$$N = 0.385 \quad \text{approximated to}$$
$$\text{three significant digits}$$

(2 is added and subtracted so that the subtraction will yield a positive decimal)

(1 is added and subtracted so that the division by 3 will yield an integer for the characteristic)

EXAMPLE 5   Compute $\dfrac{\sqrt[4]{0.566}}{68.5}$.

Solution   Let $N = \dfrac{\sqrt[4]{0.566}}{68.5}$.

$$\log N = \frac{1}{4}\log 0.566 - \log 68.5$$

$$\log 0.566 = 0.7528 - 1$$
$$+ 3 \qquad - 3$$
$$\log 0.566 = 3.7528 - 4$$
$$\frac{1}{4}\log 0.566 = 0.9382 - 1$$
$$\log 68.5 = 1.8357$$

(Adding and subtracting 3 to yield −4, exactly divisible by 4)

Now 1 must be added and subtracted from $\frac{1}{4}\log 0.566$ so the subsequent subtraction will yield a logarithm having a positive mantissa.

$$\frac{1}{4}\log 0.566 = 0.9382 - 1$$
$$+ 1 \qquad - 1$$
$$\frac{1}{4}\log 0.566 = 1.9382 - 2$$
$$\underline{\log 68.5 = 1.8357}$$
$$\log N = \text{difference} = 0.1025 - 2$$
$$N = 1.266(10^{-2}) = 0.01266$$
$$N = 0.0127 \quad \text{approximated to three significant digits}$$

EXERCISES 8.6

Find each logarithm in Exercises 1–20. (Example 1)

1. log 3.142

2. log 78.25

3. log 0.4356

4. log 0.09213

5. log 223.8

6. log 1234

7. log 5277

8. log 0.006 489

9. log 0.8461

10. log 6.125

11. log 29.56

12. log 0.01206

13. log 5989

14. log 179.6

15. log 0.004 563

16. log 70,420

17. log 900,900

18. log 0.1478

19. log 0.09346

20. log 0.004 904

Find each antilogarithm in Exercises 21–40. (Example 2)

21. antilog 0.5258

22. antilog 2.8714

23. antilog (0.6960 − 1)

24. antilog (0.4000 − 3)

25. antilog 1.7770

26. antilog 3.1780

27. antilog (0.2795 − 2)

28. antilog 0.3696

29. antilog (−0.9200)

30. antilog (−1.2847)

31. antilog (−2.8190)

32. antilog (−3.2956)

33. antilog 1.2180

34. antilog 1.9999

35. antilog (0.2590 − 1)

36. antilog (0.6205 − 3)

37. antilog 2.8480

38. antilog 3.9308

39. antilog (−1.6194)

40. antilog (−2.1495)

Compute Exercises 41–60 by using logarithms. (Examples 3, 4 and 5)

41. $(32.6)(0.854)$

42. $\dfrac{642}{79.1}$

43. $(2.43)^4$

44. $(0.589)^{-3}$

45. $\sqrt{608}$

46. $\sqrt{0.922}$

47. $\sqrt[3]{47.5}$

48. $\sqrt[3]{0.137}$

49. $\dfrac{(4.92)(0.0658)}{786}$

50. $(0.00729)(2.06)^8$

51. $\sqrt[4]{\dfrac{936}{288}}$

52. $\dfrac{\sqrt[3]{46.6}}{98.8}$

53. $\sqrt[5]{38.3}$

54. $\sqrt[5]{0.0383}$

55. $0.956\sqrt{0.645}$

56. $\dfrac{450\sqrt[3]{75.2}}{82.4}$

57. $3.142(24.65)^2$

58. $\sqrt{(6574)(21.25)}$

59. $\sqrt[3]{\dfrac{(72.8)^2}{-2610}}$

60. $\dfrac{(5.96)(7.82)(0.937)}{0.0568}$

8.7 LOGARITHMIC AND EXPONENTIAL EQUATIONS

An **exponential equation** is an equation in which the variable appears in an exponent. Exponential equations can be solved by using the property that the exponential function is one-to-one for $b > 0$ and $b \neq 1$—that is,

$$b^x = b^y \quad \text{if and only if} \quad x = y$$

and

$$\log_b x = \log_b y \quad \text{if and only if} \quad x = y$$

EXAMPLE 1 Solve $2^x = 0.125$.

Solution

$$0.125 = (0.5)^3 = \left(\frac{1}{2}\right)^3 = 2^{-3}$$

Thus $2^x = 2^{-3}$ and $x = \quad 3$.

EXAMPLE 2 Solve $2^x = 3$.

Solution Since 3 is not a rational power of 2, the logarithms of the two numbers are equated:

$$\log_{10} 2^x = \log_{10} 3$$
$$x \log 2 = \log 3$$
$$x = \frac{\log 3}{\log 2} \quad \text{exact answer}$$
$$x = \frac{0.4771}{0.3010} = 1.585 \quad \text{approx.}$$

A **logarithmic equation** is an equation that contains logarithms. It is also solved by using the one-to-one property of the logarithmic and exponential functions.

EXAMPLE 3 Solve $2 \log x - \log (x + 3) + \log 5 = \log 4$.

Solution

$$(\log x^2 + \log 5) - \log (x + 3) = \log 4$$
$$\log \frac{5x^2}{x + 3} = \log 4$$
$$\frac{5x^2}{x + 3} = 4$$
$$5x^2 - 4x - 12 = 0$$
$$(x - 2)(5x + 6) = 0$$
$$x = 2 \quad \text{or} \quad x = -\frac{6}{5}$$

Check $x = 2$: $2 \log 2 - \log 5 + \log 5 = 2 \log 2 = \log 2^2 = \log 4$. Thus 2 is a solution.

$$x = -\frac{6}{5}: 2 \log x = 2 \log \left(-\frac{6}{5}\right), \text{ which is undefined. Thus } -\frac{6}{5}$$

is *not* a solution.

Since the domain of the logarithmic function is the set of positive real numbers, the original equation requires the restriction that $x > 0$ and $x + 3 > 0$. *It is important, then, to check all proposed solutions of a logarithmic equation.*

EXAMPLE 4 Solve $x = \log_5 12$.

Solution Rewriting this equation in exponential form,
$$5^x = 12$$
$$x \log_{10} 5 = \log_{10} 12$$
$$x = \frac{\log 12}{\log 5} \quad \text{exact answer}$$
$$x = \frac{1.0792}{0.6990} = 1.54 \text{ correct to three significant digits}$$

EXERCISES 8.7

Solve for x in Exercises 1–36 using either tables or a calculator only when absolutely necessary.

1. $3^{x-2} = 243$

2. $4^x = \frac{1}{32}$

3. $5^x = 2$

4. $4^{1-x} = 64$

5. $6^{x-3} = 4.5$

6. $(2.5)^{3x} = 6.25$

7. $(3^x)^2 = \frac{1}{27}$

8. $7^{3x+5} = 1$

9. $9^{x+2} = 1$

10. $(125)^{2-x} = 0.04$

11. $2^{x-1} = \sqrt[3]{16}$

12. $8^{2x+1} = 15$

13. $2(5^{3x}) = 5$

14. $\left(\frac{1}{3}\right)^x = 81$

15. $2y = y(1.05)^x$

16. $\left(\frac{1}{8}\right)^{x-1} = 4^{1-2x}$

17. $3^{2 \log_3 5} = x$

18. $10^{-2 \log x} = 3$

19. $x - 2 = \log_3 10$

20. $x = \log_2 7$

21. $10^{-3 \log 2} = x$

22. $5^{2 \log_5 x} = 9$

23. $x = \log_2 26$

24. $x + 2 = \log 15$

25. $2 \log(x - 1) - \log (x - 4) = 4 \log 2$

26. $2 \log (x + 5) - 3 \log 2 = 0$

27. $2 \log (x + 3) - \log (x + 7) + 1 = \log 2$

28. $\log (x + 4) + \log (x - 4) - \log 9 = 0$

29. $0.5 \log 3 + \log x = 2 \log 5 - \log 2$

30. $3 \log x + \log 3 - 2 \log 5 = -3 + \log 15$

31. $\log x^2 - \log (x + 12) - \log 2 = 0$

32. $\log 64 - 3 \log x = 3 + \log 8$

33. $\log (x + 1) - \log 3 = \log (x + 2)$

34. $\log_3 (x + 1) - \log_3 x = \log_3 8$

35. $\log_7 (x + 1) - \log_7 x = \log_7 8$

36. $\log_b \left(\dfrac{x + 3}{x}\right) - \log_b (x + 3) - \log_b x = \log_b 1$

8.8 CHANGING THE BASE OF A LOGARITHM; NATURAL LOGARITHMS

Common logarithms, or logarithms to the base 10 (also called Briggsian after their inventor, Henry Briggs), are the most convenient for numerical computation. However, for more advanced mathematics, especially that involving calculus, **natural logarithms** (also called Naperian after their originator, John Napier), or logarithms to the base e, are more appropriate. The number e is irrational, with an approximate value of 2.71828.

Tables for common logarithms and tables for natural logarithms are readily available, and most calculators that are equipped to handle common logarithms are also equipped to handle natural logarithms. Tables for logarithms to bases other than 10 or e are not usually available, so it is necessary to change the base of a logarithm other than 10 or e to a base for which computation may be accomplished. Problems of this type were already done in the previous section, but we will now show how this base conversion is easily performed.

Let $\quad y = \log_b x; \quad$ then $\quad b^y = x$

We can now take the logarithm, to any suitable base, of both sides. Let a be the base desired ($a > 0$, $a \neq 1$). If $b^y = x$, then $\log_a b^y = \log_a x$.

Now solve for y:

$$y \log_a b = \log_a x \qquad \text{(By logarithm operations, Theorem 3)}$$

$$y = \frac{\log_a x}{\log_a b}$$

But $\quad y = \log_b x$; thus $\quad \log_b x = \dfrac{\log_a x}{\log_a b}$

We can now state the theorem for conversion of bases of a logarithm.

THE CONVERSION OF BASES THEOREM

$$\log_b x = \frac{\log_a x}{\log_a b}$$

For the special case that $b = e$ and $a = 10$,

$$\log_e x = \frac{\log_{10} x}{\log_{10} 2.71828} = \frac{\log_{10} x}{0.4343} = 2.303 \log_{10} x$$

Therefore *the natural logarithm of a number is obtained by multiplying its common logarithm by 2.303.*

EXAMPLE 1 Find $\log_e 54.6$.

Solution

$$\begin{aligned}
\log_e 54.6 &= 2.303 \log_{10} 54.6 \\
&= 2.303(1.7372) \\
&= 4.001 \quad \text{to four significant digits}
\end{aligned}$$

EXAMPLE 2 Find $\log_e 0.932$.

Solution

$$\begin{aligned}
\log_e 0.932 &= 2.303 \log_{10} 0.932 \\
&= 2.303(0.9694 - 1) \\
&= 2.303(-0.0306) \\
&= -0.0705 \quad \text{to three significant digits}
\end{aligned}$$

EXAMPLE 3 Find $\log_5 28$.

Solution

Using $\log_b x = \dfrac{\log_a x}{\log_a b}$,

$$\begin{aligned}
\log_5 28 &= \frac{\log_{10} 28}{\log_{10} 5} \\
&= \frac{1.4472}{0.6990} \\
&= 2.070 \quad \text{to four significant digits}
\end{aligned}$$

We have agreed that $\log_{10} x$ may be written simply as log x, and that if any other base is to be used, it will be indicated. This was done because the logarithm to the base 10 is so commonly used. The natural logarithm, the logarithm to the base e, is also written in a shortened form, usually as ln. Thus ln x is equivalent to writing $\log_e x$.

EXAMPLE 4 Find ln 8.4.

Solution By agreement, we know that ln $8.4 = \log_e 8.4$.
Therefore

$$\begin{aligned}
\ln 8.4 &= 2.303 \log 8.4 \\
&= 2.303(0.9243) \\
&= 2.129 \quad \text{to four significant digits}
\end{aligned}$$

EXERCISES 8.8

Find each logarithm in Exercises 1–30.

1. ln 10
2. ln 100
3. ln 1.56
4. ln 97
5. ln 0.2
6. ln 3.142
7. ln 0.012

8. ln e^3

9. ln e
10. ln $\sqrt{e}$
11. ln 25.2

12. ln $\dfrac{e}{2}$

13. ln 0.0825

14. ln 256
15. $\log_5 39$

16. $\log_{12} 7800$
17. $\log_{21} 36$
18. $\log_7 0.83$
19. $\log_8 4.3$
20. $\log_{21} 236$
21. $\log_3 48.5$
22. $\log_{12} 78$

23. $\log_{10} \dfrac{1}{e^2}$

24. $\log_{1/2} 750$
25. ln(2)(3)(4)(5)(6)
26. ln 3e

27. $\log_{1/3} 27$

28. ln $\dfrac{3}{5}$

29. ln $10^{\log 6}$
30. ln $e^{\ln e}$

8.9 APPLICATIONS

Some of the many applications of logarithms are presented in this section. The problems may be worked either by using the tables or by using a calculator. The student should be cautioned that use of a calculator may result in slightly different answers due to the inherent round-off errors. Generally speaking, the answers obtained by use of the calculator are more accurate than those obtained by use of the tables.

CHEMICAL pH

In chemistry the pH of a solution is a measure of the acidity or alkalinity of the solution. If [H^+] designates the hydrogen ion concentration measured in moles per liter, then the pH is defined as follows:

DEFINITION

$$pH = -\log_{10} [H^+]$$

EXAMPLE 1 Find the pH of a solution whose hydrogen ion concentration is 2.0×10^{-4}.

Solution
$$\begin{aligned} pH &= -\log_{10} (2.0 \times 10^{-4}) \\ &= -(0.3010 - 4) = -0.3010 + 4 \\ &= 3.6990 \\ &= 3.7 \quad \text{correct to the nearest tenth} \\ &\qquad \text{(pH values are usually stated} \\ &\qquad \text{correct to the nearest tenth)} \end{aligned}$$

EXAMPLE 2 Find the hydrogen ion concentration of a solution whose pH is 4.7.

Solution
$$\begin{aligned} 4.7 &= -\log_{10} [H^+] \\ \log_{10} [H^+] &= -4.7 = -5 + 0.3 \\ [H^+] &= 2.0 \times 10^{-5} \quad \text{approximated} \\ &\qquad \text{(usually calculated} \\ &\qquad \text{to two} \\ &\qquad \text{significant digits)} \end{aligned}$$

EXPONENTIAL GROWTH AND DECAY

Radioactive decay, population growth, and other phenomena that change at a rate directly proportional to the amount present at a given time are described by the exponential function $y = ae^{bt}$, where the natural logarithm base $e = 2.71828 \ldots$, the variable t measures the time, and a and b are constants. If $t = 0$, then $y = ae^{bt(0)} = ae^0 = a(1) = a$. Thus a measures y at the time when the measurements are begun — that is, when $t = 0$. For this reason, y_0 is often used instead of the letter a, or $y = y_0e^{bt}$.

EXAMPLE 3 The number of bacteria in a certain culture is determined from the relation $y = 500 \, e^{0.38t}$, where t is measured in hours. How many bacteria are present at the end of 15 hours?

Solution
$$\begin{aligned} y &= 500 \, e^{0.38(15)} = 500 \, e^{5.70} \\ \log y &= \log 500 + 5.70 \log e \\ &= 2.6990 + 5.7(0.4343) = 5.1745 \\ y &= 149{,}400 \quad \text{approx.} \end{aligned}$$

EXAMPLE 4 What is the half-life of a radioactive substance that decays according to the rule $y = y_0e^{-0.035t}$ if t is measured in years?

Solution   To find the half-life means to find t
when $y = \dfrac{1}{2} y_0$:

$$\frac{1}{2} y_0 = y_0 e^{-0.035t}$$

or
$$\frac{1}{2} = e^{-0.035t}$$

$$\log \frac{1}{2} = \log e^{-0.035t}$$

$$= -0.035t \log e$$

$$-0.035t \log e = -\log 2$$

$$t = \frac{\log 2}{0.035 \log e}$$

$$= \frac{0.3010}{0.035(0.4343)}$$

$$= 19.8 \text{ years} \quad \text{approx.}$$

COMPOUND INTEREST

If P designates the sum of money invested, i is the compound
interest rate per conversion period, and n is the number of conversion periods, then the amount A that the money is worth at
the end of n conversion periods is given by the formula
$$A = P(1 + i)^n$$
Four-place logarithm tables such as those in this book
are not accurate enough for most applications of this formula,
although they will yield useful approximations. For results
accurate to the nearest cent, there are tables of values of
$(1 + i)^n$ available, and of course, once again the use of calculators will provide a high degree of accuracy.

EXAMPLE 5   What will be the amount of $100 invested at 5% con-
verted quarterly at the end of 5 years?

Solution
$$A = P(1 + i)^n,$$

where $P = 100$, $i = \dfrac{0.05}{4} = 0.0125$,

$$n = 5 \times 4 = 20$$
$$A = 100(1.0125)^{20}$$
$$\log A = \log 100 + 20 \log 1.0125$$
$$= 2 + 0.1080$$
$$A \doteq 10^2 \times 1.28 = \$128 \quad \text{approx.}$$

Using the tables for $(1 + i)^n$ with $i = 1\dfrac{1}{4}\%$

and $n = 20$,
$$A = 100(1.2820) = \$128.20$$

correct to the nearest cent

EXAMPLE 6   How much money must be invested at 7% converted monthly to amount to $5000 at the end of 3 years?

Solution

$$A = P(1 + i)^n$$

where $A = 5000$, $i = \dfrac{0.07}{12}$, $n = 12(3) = 36$

$$P = A(1 + i)^{-n}$$

$$= 5000 \left(1 + \frac{0.07}{12}\right)^{-36}$$

$$\log P = \log 5000 - 36 \log 1.0058$$
$$= 3.6990 - 36(0.0025)$$
$$= 3.6990 - 0.0900$$
$$= 3.6090$$
$$P = \$4065 \quad \text{approx.}$$

EXAMPLE 7   How long will it take a sum of money to double itself if it is invested at 6% converted monthly?

Solution

$$A = P(1 + i)^n, \ A = 2P, \ i = \frac{0.06}{12} = 0.005$$

$$2P = P(1.005)^n \quad \text{or}$$
$$2 = (1.005)^n$$
$$n \log 1.005 = \log 2$$
$$n = \frac{\log 2}{\log 1.005}$$
$$= \frac{0.3010}{0.00215}$$
$$= 140 \text{ months} \quad \text{or}$$
$$11 \text{ years 8 months approx.}$$

ELECTRICAL CIRCUITS

In an RL electrical circuit consisting of a battery of E volts (V), a resistance of R ohms (Ω), an inductance of L henrys (H), and a switch S (Figure 8.10), then after the switch is closed, the current i in amperes is given by the formula

$$i = \frac{E}{R}(1 - e^{-Rt/L})$$

where the time t is measured in seconds. $\dfrac{E}{R}$ is called the steady-state current. If a steady current is flowing in the circuit and the battery is short-circuited, then the decay of the current is given by the equation

$$i = \frac{E}{R} e^{-Rt/L}$$

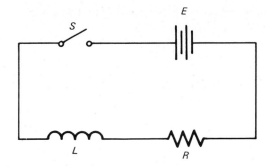

FIG. 8.10

EXAMPLE 8 If $E = 14$ V, $R = 5\ \Omega$, $L = 2.5$ H, find the current in an RL circuit 0.25 sec after the switch is closed.

Solution

$$i = \frac{14}{5}\,(1 - e^{-5(0.25)/2.5})$$
$$= 2.8(1 - e^{-0.5})$$
$$= 2.8 - 2.8e^{-0.5}$$
$$\log 2.8e^{-0.5} = \log 2.8 - 0.5 \log 2.718$$
$$= 0.4472 - 0.2172 = 0.2300$$
$$2.8e^{-0.5} = 1.70 \text{ amps and}$$
$$i = 2.8 - 1.7 = 1.1 \text{ amps}$$

EXAMPLE 9 If the battery of an RL circuit is short-circuited after a steady current is flowing, find $\dfrac{E}{R}$ for $i = 20$ amps, $R = 6\ \Omega$, $L = 2$ H, $t = 0.4$ sec.

Solution

$$20 = \frac{E}{R}\,e^{-6(0.4)/2} = \frac{E}{R}\,e^{-1.2}$$
$$\frac{E}{R} = 20e^{1.2} \quad \text{and}$$
$$\log \frac{E}{R} = \log 20 + 1.2 \log e = 1.8222$$
$$\frac{E}{R} = 66.4 \text{ amps}$$

EXERCISES 8.9

1. Find the pH of a solution whose hydrogen ion concentration is given below:
(a) 5.1×10^{-6}
(b) 3.4×10^{-2}
(c) 6.3×10^{-8}

2. Find the pH of a solution whose hydrogen ion concentration is given below:
   (a) 4.2×10^{-3}
   (b) 2.9×10^{-7}
   (c) 7.5×10^{-5}

3. Find the hydrogen ion concentration for each of the solutions whose pH is given below:
   (a) pH $= 7.0$
   (b) pH $= 2.6$
   (c) pH $= 8.7$

4. Find the hydrogen ion concentration for each of the solutions whose pH is given below:
   (a) pH $= 5.5$
   (b) pH $= 1.4$
   (c) pH $= 3.9$

5. Determine the half-life of the decay of thorium-234 into protactinium-234 if its decay equation is $y = y_0 10^{-0.0123t}$, where t is measured in years. If 100 g are present at $t = 0$, how much thorium is left at the end of 2 years?

6. Determine the half-life of the decay of polonium-218 into lead-214 if its decay relation is $y = y_0 10^{-0.0987t}$, where t is measured in minutes.
   If 1000 g are present at $t = 0$, how much polonium is left at the end of $\frac{1}{2}$ hour?

7. If the half-life of radium-226 is 1590 years, find the decay constant b so that $y = y_0 e^{-bt}$.

8. If the half-life of uranium-238 is 4.5×10^9 years, find the decay constant b so that $y = y_0 e^{-bt}$.

9. In a unimolecular chemical reaction, the number of molecules N present at time t is given by the equation $N = N_0 e^{-kt}$, where k is a positive constant called the *reaction rate*.
   (a) If the concentration of a dilute sugar solution is given by the equation $y = 0.01 e^{-kt}$, find the reaction rate k if the concentration is $\frac{1}{250}$ g/cm³ at the end of 5 hours.
   (b) Find the concentration at the end of 10 hours.

10. Bacteria in a certain culture grew according to the rule $y = 100 e^{bt}$. If y increased to 354 at the end of 1 hour, how many bacteria were in the culture at the end of $1\frac{1}{2}$ hours?

11. Find approximately to the nearest dollar the amount of money accumulated at the end of 5 years if $4000 is invested at 6% converted yearly.

12. What was the approximate value of the original investment that amounted to $3000 at the end of 6 years if the interest rate was 6% converted semiannually?

13. At what rate of compound interest will a sum of money double itself at the end of 10 years?

14. Approximately how long will it take a sum of money to double itself at 8% converted quarterly?

15. An RL circuit has an electromotive force E of 16.5 V, a resistance R of 0.1 Ω, and an inductance L of 0.05 H.

    (a) Find the current i $\frac{1}{4}$ sec after the switch is closed.

    (b) If the circuit is short-circuited after a steady current is flowing, find the current at the end of 1 sec.

16. The build-up of the current i in a circuit with a condenser of capacitance C farads (f), a resistance of R Ω, and a source of electromotive force of E V is expressed by the equation

    $$i = \frac{E}{R}e^{-t/CR}.$$

    Find E if $i = 0.03$ amp, $R = 250$ Ω, $C = 2 \times 10^{-6}$ f, and $t = 0.001$ sec.

17. The depreciation of a certain machine is calculated by the constant percentage method, $S = C(1 - r)^n$, where C is the original cost, S is the scrap value after a useful life of n years, and r is the constant percentage of depreciation. Find the scrap value of a machine costing $20,000 if its useful life is 10 years and it depreciates by a constant percentage of 18%.

18. Find the constant percentage of depreciation of a car initially costing $5000 if its scrap value is $250 at the end of a useful life of 15 years (see Exercise 17).

19. The gain G in decibels (a measure of loudness) due to an increase in power from w_1 to w_2 watts may be calculated from the equation $G = 10 \log_{10} w_2/w_1$.

    (a) What was the decibel gain of a television station that increased its power from 35 to 150 kilowatts?

    (b) If power (watts) $= \dfrac{(\text{volts})^2}{\text{ohms}}$, what is the decibel loss across a long radio transmission line if an input of 20 V across a resistance of 600 Ω is reduced to a voltage of 4 V measured across a resistance of 5000 Ω?

20. The population of a certain city increased from 100,000 in 1940 to 125,000 in 1960. Assume $P = P_0 10^{kt}$.

    (a) What will be the population in 1980?

    (b) In what year will the population be 200,000? (Let $t = 0$ for 1940.)

21. If a projectile fired vertically upward with initial velocity v_0 ft/sec is subject only to the force of gravity g and to the air resistance kv, then the maximum height H it reaches is given by

    $$H = \frac{1}{k}\left(v_0 - \frac{g}{k}\log_e \frac{g + v_0 k}{g}\right)$$

    Find H if $v_0 = 680$ ft/sec, $k = 2.40$, and $g = 32$ ($e = 2.71828.\ .\ .$).

22. By Newton's law of cooling, the temperature T of a body at time t is expressed by the equation

$$t = \frac{1}{k} \log_e \frac{T_0 - a}{T - a}$$

where a is the constant temperature of the surrounding air, T_0 is the temperature of the body at time $t = 0$, and k is the constant rate of cooling.

(a) If $a = 20$ degrees Celsius, find k if it takes 25 minutes for a substance to cool from 100 to 50 degrees Celsius.

(b) How long will it take the substance to cool to 25 degrees Celsius? to 20 degrees Celsius?

23. If

$$\frac{1}{\sqrt{5}} \left(\frac{1 + \sqrt{5}}{2} \right)^n$$

where n is a positive integer, is approximated to the nearest integer, then this nearest integer is the nth Fibonacci number. A Fibonacci number is the sum of the two preceding Fibonacci numbers, where 1, 1, 2, 3, 5, 8, 13, and 21 are the first eight Fibonacci numbers.

(a) Find the fifteenth Fibonacci number in two different ways.

(b) Find the twentieth Fibonacci number.

(One of the interesting features of Fibonacci numbers is the role they play in describing various kinds of biological growth.)

For quick reference the definitions and theorems of this chapter are listed below.

DEFINITIONS

Relation   A relation is a set of ordered pairs.

Function   A function is a relation for which each domain element is paired with exactly one element in the range.

One-to-One Function   A one-to-one function is a function with each element in its range paired with exactly one element in its domain.

Inverse of a Relation   The inverse r^{-1} of a relation r is the set of ordered pairs obtained by interchanging the components of the ordered pairs in r.

Inverse of a Function   The inverse of a function is a function if and only if the original function is one-to-one.

Exponential Function $f(x) = b^x$, $b > 0$, $b \neq 1$, and x any real number.

Logarithmic Function   For $b > 0$ and $b \neq 1$, $y = \log_b x$ if and only if $x = b^y$.

THEOREMS

(For all theorems, $b > 0$, $b \neq 1$, $x > 0$, and $y > 0$.)

1. $b^{\log_b x} = x$

2. $\log_b b = 1$

3. $\log_b (xy) = \log_b x + \log_b y$

4. $\log_b \dfrac{x}{y} = \log_b x - \log_b y$

5. $\log_b x^r = r \log_b x$, for r any rational number

DIAGNOSTIC TEST

Determine which of the following statements are true and which are false. If a statement is false, correct it.

1. The inverse of a relation is always a relation.

2. The inverse of a function is always a function.

3. If $f = \{(x, y) \,|\, 3x + 9y = 5\}$, then $f^{-1} = \{(x, y) \,|\, 3y + 9x = 5\}$.

4. If $f = \{(x, y) \,|\, y = b^x, b > 0$ and $b \neq 1\}$, then
$f^{-1} = \{(x, y) \,|\, y = \log_b x\}$.

5. The statements $y = b^x$ and $x = \log_b y$ are equivalent.

6. If $3^{x-2} = \dfrac{1}{81}$, then $x = -1$.

7. $4^{2 \log_4 x} = 16x$

8. $\log_b x^2 y = 2 \log_b x + \log_b y$

9. $\dfrac{\log 17}{\log 6} = \log 17 - \log 6$

10. If $\log_3 x = -4$, then $x = \dfrac{1}{81}$.

11. If $\log_b 8 = 1$, then $b = 8$.

12. If $\log_2 64 = x$, then $x = 6$.

13. If $\log_{10} 8.35 = 0.9217$, then $\log_{10} 0.000835 = -3.0783$.

14. If antilog $(2.8089) = 644$, then antilog $(1.8089) = 64.4$.

15. $\ln 4.7 = \dfrac{\log_{10} 4.7}{\log_{10} e}$

For Problems 16–20 select the best answer.

16. $\dfrac{1}{2} (\log_b x - \log_b y)$ equals

    (a) $\log_b \dfrac{1}{2} (x - y)$ (d) $(\log_b x - \log_b y)^2$

    (b) $\log_b \dfrac{x}{2y}$ (e) none of these

    (c) $\log_b \sqrt{\dfrac{x}{y}}$

17. If $\log_{10} 7.5 = 0.8751$, then $\log_{10} 0.075$ equals

(a) $0.8751 - 1$ (d) 2.8751

(b) $0.8751 - 2$ (e) $0.8751 + 2$

(c) 0.8751

18. $16^{1/4} = 2$ is written in logarithmic notation as

(a) $\log_2 \left(\dfrac{1}{4}\right) = 16$ (d) $\log_{1/4} 16 = 2$

(b) $\log_{1/4} 2 = 16$ (e) $\log_{16} 2 = \dfrac{1}{4}$

(c) $\log_2 16 = \dfrac{1}{4}$

19. If $2^x = 3$, then x equals

(a) $\log \dfrac{3}{2}$ (d) $\dfrac{\log 2}{\log 3}$

(b) $\log 3 - \log 2$ (e) none of these

(c) $\dfrac{\log 3}{\log 2}$

20. If $2 \log x - \log (x + 3) + \log 5 = \log 4$, then x equals

(a) 2

(b) $-\dfrac{6}{5}$

(c) both (a) and (b)

(d) cannot be found without a table

(e) none of these

REVIEW EXERCISES

In Exercises 1–5, (a) find the inverse of each relation, (b) graph the relation and its inverse on the same set of axes, and (c) state the domain and range of each relation and each inverse relation. (Section 8.1)

1. $r = \{(x, y) \,|\, y = 3x + 2\}$ **4.** $r = \{(x, y) \,|\, y = 5 - x^2\}$

2. $r = \{(x, y) \,|\, x + 5y = 2\}$ **5.** $r = \{(x, y) \,|\, y = |x - 3|\}$

3. $r = \{(x, y) \,|\, y = x^2 + 2x + 3\}$

In Exercises 6–10 $f(x)$ is given. Find $f^{-1}(x)$ and show that $f(f^{-1}(x)) = x$ and $f^{-1}(f(x)) = x$. (Section 8.1)

6. $f(x) = 3x + 2$ **9.** $f(x) = \sqrt{x} + 1, \, x \geq 0$

7. $f(x) = 4 - x$ **10.** $f(x) = x^3$

8. $f(x) = x^2 - 1, \, x \geq 0$

In Exercises 11–15 use the theorem $b^r = b^s$ if and only if $r = s$ to find x in each of the following. (Section 8.2)

11. $5^x = \dfrac{1}{125}$ **14.** $3^{2x} = \sqrt{27}$

12. $4^{x-2} = 32$ **15.** $8^{4x-3} = 1$

13. $10^{-x+1} = 100$

Write Exercises 16–22 in logarithmic form. (Section 8.3)

16. $10^3 = 1000$ **20.** $5^0 = 1$

17. $4^3 = 64$ **21.** $64^{1/3} = 4$

18. $64^{-1/2} = 0.125$ **22.** $\left(\dfrac{1}{2}\right)^{-3} = 8$

19. $10^1 = 10$

Evaluate Exercises 23–30 without using tables.

23. $\log_{10} 0.0001$ **27.** $\log_5 5^{-2}$

24. $\log_2 0.25$ **28.** $\log_3 \sqrt{3}$

25. $\log_4 128$ **29.** $10^{\log_{10} 7}$

26. $\log_{10} 0$ **30.** $e^{\ln 5.6}$

If $\log_b 2 = 0.4$, evaluate each of the following Exercises 31–36. (Section 8.4)

31. $\log_b 8$ **34.** $\log_b \sqrt[3]{0.25}$

32. $\log_b \dfrac{1}{2}$ **35.** $\log_b (\log_b \sqrt{b})$

33. $\log_b \sqrt{2}$ **36.** $\dfrac{\log_b 2}{\log_b b^{\sqrt{3}}}$

For Exercises 37–40 use the logarithmic operations theorems to write each in terms of the logarithms of prime numbers. (Section 8.4)

37. $\log_b 30$ **39.** $\log_b 16\sqrt{3}$

38. $\log_b 24$ **40.** $\log_b \dfrac{6\sqrt{6}}{15}$

For Exercises 41–46 express each logarithm in (a) computational form and (b) formula form. (Section 8.5)

41. $\log 57.4$ **44.** $\log 456.7$

42. $\log 0.000347$ **45.** $\log 0.003582$

43. $\log 234,000$ **46.** $\log 2001$

For Exercises 47–52 express each antilogarithm in (a) scientific notation and (b) ordinary notation. (Section 8.5)

47. antilog (2.4031)   **48.** antilog (3.9425)

49. antilog $(0.7259 - 2)$ **51.** antilog (-3.3215)

50. antilog $(0.9201 - 1)$ **52.** antilog (-1.7986)

53. (a) Write the logarithmic equation that would be used to compute

$$\sqrt[3]{\frac{(42.21)(23.50)}{5650}}$$

(b) Using tables, compute the number designated in (a).
(Section 8.6)

In Exercises 54–59 solve for x. (Sections 8.7 and 8.8)

54. $x = 5^{-4.6}$ **56.** $\log_3 25.4 = x$

55. $5^x = 3$ **57.** $\ln 5 = x$

58. $\log x = \log 15 + 3 \log 2 - \dfrac{1}{4} \log 81$

59. $2 \log (x + 3) - \log (-x - 1) = 0$

For Exercises 60–64 refer to Section 8.9.

60. Find the pH of a solution whose hydrogen ion concentration is 3.2×10^{-3}.

61. Find the hydrogen ion concentration for the solution whose pH is 6.4.

62. Find approximately to the nearest dollar the amount accumulated at the end of 6 years if $2000 is invested at 8.2% converted yearly.

63. For the investment in Exercise 62, find the amount accumulated at the end of 6 years if the interest is converted monthly.

64. The number of bacteria in a certain culture is determined from the relation $y = 400 \, e^{0.24t}$ where e is the base of the natural logarithm and t is measured in minutes. How many bacteria are present at the end of 15 min?

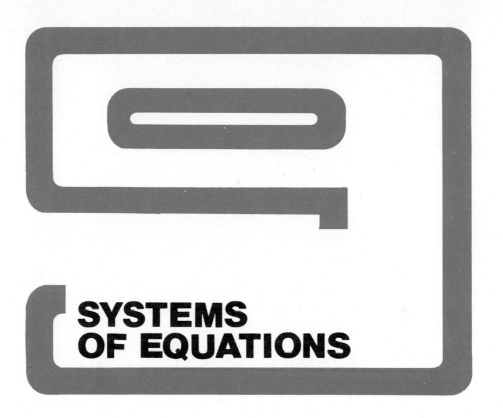

SYSTEMS
OF EQUATIONS

Many situations in real life involve relationships
between two or more variables. For example, in
economics a demand law relates the price of a
commodity to the number of units of that commodity demanded by the consumer, while a supply law relates the
price of a commodity to the number of units a manufacturer
can supply. These relationships are expressed as equations,
and the "market equilibrium" occurs at the **common solution**
of both equations. This is only one example of infinitely many
varieties of situations that demand the common solution of
several equations.

Algebraically, a set of two or more equations in two or
more variables is called a system of equations, and the solution
set of a system corresponds to the geometric points of intersection of the graphs of the equations.

DEFINITION

A **system of equations** is a set of two or more equations in two or more variables.

A system of equations is also called a set of **simultaneous equations.**

9.1 LINEAR SYSTEMS IN TWO VARIABLES

The simplest system of equations is a linear system in two variables.

DEFINITION

A **linear system** of equations in the variables x and y is a system each of whose equations has the form
$$ax + by + c = 0$$
where a, b, and c are constants with a and b not both 0.

DEFINITION

The **solution set of a system of equations in two variables** is the set of all ordered pairs that are common solutions to **all** the equations in the system.

Using the language of sets, the solution set of a system of equations in two variables is the **intersection** ($\cap$) of the solution set of one of the equations with the solution set of the other.

EXAMPLE 1 Show that $(9, -2)$ is a solution of the system
$$2x + 3y = 12 \quad \text{and} \quad x - 5y = 19$$
Solution For $x = 9$ and $y = -2$,
$$2x + 3y = 2(9) + 3(-2) = 18 - 6 = 12$$
$$x - 5y = 9 - 5(-2) = 9 + 10 = 19$$
Since $(9, -2)$ is a solution of each equation in the system, it is a solution of the system.

The *real solutions* of a system of equations in two variables can be determined graphically by graphing the equations

on the same set of axes. An ordered pair corresponding to a
point of intersection is a solution of the system.

EXAMPLE 2   Graphically solve the system $x + y = 5$ and $y = x + 2$.
   Solution

$x + y = 5$		$y = x + 2$	
x	y	x	y
0	5	0	2
5	0	1	3
1	4	2	4

From Figure 9.1, the solution is read as $\left(\dfrac{3}{2}, \dfrac{7}{2}\right)$.

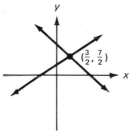

$A = \{(x, y) \mid x + y = 5\}$
$B = \{(x, y) \mid y = x + 2\}$
$A \cap B = \{(\tfrac{3}{2}, \tfrac{7}{2})\}$

FIG. 9.1

Check   For $x = \dfrac{3}{2}$ and $y = \dfrac{7}{2}$,

$x + y = 5$ becomes $\dfrac{3}{2} + \dfrac{7}{2} = 5, \dfrac{10}{2} = 5, 5 = 5,$ true.

$y = x + 2$ becomes $\dfrac{7}{2} = \dfrac{3}{2} + 2, \dfrac{7}{2} = \dfrac{3}{2} + \dfrac{4}{2}, \dfrac{7}{2} = \dfrac{7}{2},$ true.

EXAMPLE 3   Graphically solve the system $x + y = 5$ and $x + y = 2$.
   Solution

$x + y = 5$		$x + y = 2$	
x	y	x	y
0	5	0	2
5	0	2	0

In Figure 9.2 the lines are parallel and the solution set is empty.

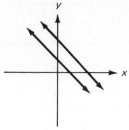

$A = \{(x, y) \mid x + y = 5\}$
$B = \{(x, y) \mid x + y = 2\}$
$A \cap B = \emptyset$

FIG. 9.2

Check Finding the slopes of each equation,
For $x + y = 5$; $y = -x + 5$, and the slope is -1.
For $x + y = 2$, $y = -x + 2$, and the slope is -1.
Nonvertical parallel lines always have the same slope, so the lines are parallel.

EXAMPLE 4 Graphically solve the system $x + y = 5$ and $3x + 3y = 15$.

Solution

$x + y = 5$		$3x + 3y = 15$	
x	y	x	y
0	5	0	5
5	0	5	0

In Figure 9.3 the lines coincide, and the solution set of the system is the solution set of either equation of the system.

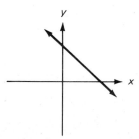

$A = \{(x, y) \mid x + y = 5\}$
$B = \{(x, y) \mid 3x + 3y = 15\}$
$A \cap B = \{(x, y) \mid x + y = 5\}$

FIG. 9.3

It can be seen from the graphic solutions that there are three possibilities for the solution of a system of two linear equations in two variables:

1. There is *exactly one* solution.
2. The solution set is the empty set (there are no solutions).
3. There are infinitely many solutions.

The geometric interpretations for these three possibilities are, respectively:

1. The lines intersect in *exactly one* point.
2. The lines are parallel.
3. The lines coincide.

To find the solution set of a system of two linear equations in two unknowns algebraically, two methods are used: substitution and addition.

In the **substitution method,** one of the equations is solved for one variable, and this solution is substituted in the second equation to create one equation in one variable, which can readily be solved.

EXAMPLE 5 Solve the system
$$3x + 2y = -5$$
$$x - 3y = 2$$
by the substitution method.

Solution   Look at the two equations and determine, if possible, which equation is easier to solve. In the given system, it is easy to take the second equation, $x - 3y = 2$, and solve for x in terms of y.
$$x - 3y = 2$$
$$x = 2 + 3y$$
Now replace x in the first equation by $(2 + 3y)$.
$$3(2 + 3y) + 2y = -5$$
$$6 + 9y + 2y = -5$$
$$11y = -11$$
$$y = -1$$
Now solve for x: $x = 2 + 3y$
$$x = 2 + 3(-1)$$
$$x = -1$$
The solution set is $\{(-1, -1)\}$. This solution should be checked in *both* equations.
Check
$$3x + 2y = 3(-1) + 2(-1) = -5$$
$$x - 3y = (-1) - 3(-1) = -1 + 3 = 2$$

Geometrically we can see that the lines whose equations are given in the system in Example 5 intersect at the point $(-1, -1)$.

The **addition method** is based on the following theorem.

THEOREM

If (p, q) is a solution of both
$$a_1x + b_1y = c_1 \quad \text{and} \quad a_2x + b_2y = c_2$$
then (p, q) is also a solution of
$$A(a_1x + b_1y - c_1) + B(a_2x + b_2y - c_2) = 0$$

In words, this theorem states that you can multiply each equation in the system by a constant, and add the resulting equations to each other. The plan is to choose the multiplier in such a way that, by adding the resulting equations, one of the variables will be eliminated.

EXAMPLE 6   Solve the system
$$3x + 2y = -5$$
$$x - 3y = 2$$
by the addition method.

Solution   We note that this is the same system we solved by substitution, so we know the correct solution is $(-1, -1)$.

By multiplying the second equation by -3, the x terms will be eliminated when adding the two equations. Be sure to remember to multiply *both* sides of the equation by the constant.

$$-3(x - 3y) = -3(2) \qquad \text{(Multiplying the second equation by } -3)$$

$$\begin{array}{rl} -3x + 9y = -6 & \\ 3x + 2y = -5 & \text{(Rewriting the first equation)} \\ \hline 11y = -11 & \text{(By adding the two equations)} \\ y = -1 & \end{array}$$

To find x, substitute the value of y in *either* equation, and then check the solution in *both* equations.

We could have eliminated the variable y by multiplying the first equation by 3 and the second equation by 2 in the following manner:

$$\begin{array}{rl} 3(3x + 2y) = & 3(-5) \\ 2(x - 3y) = & 2(2) \\ \hline 9x + 6y = -15 & \text{(Note that the coefficients of the } y \\ 2x - 6y = 4 & \text{terms, } +6 \text{ and } -6, \text{ are additive} \\ & \text{inverses)} \\ \hline 11x = -11 & \text{(Adding the two equations)} \\ x = -1 & \end{array}$$

EXAMPLE 7 Solve the system

$$x - 3y = 4$$
$$2x - 6y = -1$$

Solution   We can use either the substitution method to solve the first equation for x, or the addition method to eliminate the x terms by multiplying the first equation by -2. By substitution,

If $x - 3y = 4$,   then $x = 4 + 3y$

Substituting in the second equation,

$$2x - 6y = -1$$
$$2(4 + 3y) - 6y = -1$$
$$8 + 6y - 6y = -1$$
$$8 = -1$$

This result is not a true statement; thus the solution set for this system is the empty set, $\varnothing$. Geometrically, these lines whose equations are given must be parallel. If they are parallel, they have the same slope but *different y*-intercepts.

We can quickly check these facts by writing each equation in the slope-intercept form, $y = mx + b$.

For $x - 3y = 4$, we obtain $y = \dfrac{1}{3}x - \dfrac{4}{3}$.

For $2x - 6y = -1$, we obtain $y = \dfrac{1}{3}x + \dfrac{1}{6}$.

Therefore both lines have the same slope, $\dfrac{1}{3}$, but one line crosses the y-axis at $-\dfrac{4}{3}$, while the other line crosses the y-axis at $\dfrac{1}{6}$.

EXAMPLE 8 Solve the system

$$x - 3y = 4$$
$$2x - 6y = 8$$

Solution   Let us solve this system by the addition method, multiplying the first equation by -2 to eliminate the x terms.

$$-2(x - 3y) = -2(4) \quad \text{and}$$
$$\begin{array}{r} -2x + 6y = -8 \\ 2x - 6y = 8 \\ \hline 0 + 0 = 0 \\ 0 = 0 \end{array}$$

$0 = 0$ is always a true statement. Therefore the two equations have the same solution set,

which contains infinitely many solutions. *All
the solutions for one equation are also the solutions for the other.* Writing this in set notation, the solution set of the system is the set
$\{(x, y)\,|\,x - 3y = 4\}$. Note that we could also
have written the solution set
$$\{(x, y)\,|\,2x - 6y = 8\}$$
Geometrically, these two equations represent the same line, or the lines whose equations are given are *coincident*.

EXERCISES 9.1

*Show that each ordered pair in Exercises 1–8 is a solution of the
given system of equations; also verify the solution by graphing each
pair of equations. (Examples 1–4)*

1. $x + y = 8$
    $x - y = 2$ (5, 3)

2. $2x + y = 10$
       $y = x - 8$ (6, −2)

3. $3x + y = 1$
       $y = 2x + 6$ (−1, 4)

4. $5x - 2y\quad = 2$
    $3x + y + 12 = 0$ (−2, −6)

5. $x = y + 3$
    $y = 2x + 1$ (−4, −7)

6. $2x + y - 3 = 0$
    $3x - y - 2 = 0$ (1, 1)

7. $2x - 3y = 5$
    $3x + 2y = 9$ $\left(\dfrac{37}{13}, \dfrac{3}{13}\right)$

8. $4x - y + 6 = 0$
    $3x + 2y - 5 = 0$ $\left(\dfrac{-7}{11}, \dfrac{38}{11}\right)$

*Solve and check each system in Exercises 9–30 using either the sub-
stitution method or the addition method. (Examples 5–8)*

9. $x + y = 3$
    $x - y = 10$

10. $x + y = 5$
     $x - y = -8$

11. $2x - 3y = 15$
     $2x - 5y = 35$

12. $5x + y = 2$
     $4x - 3y = 13$

13. $x - y = 14$
     $4x \quad = y - 1$

14. $y - 2x = 6$
     $5y - 10x = 1$

15. $x = 5y - 2$
     $y = 3x - 1$

16. $x + 3 = 4y$
     $8y - 6 = 2x$

17. $2x + 3y - 4 = 0$
     $3x + 2y - 4 = 0$

18. $4x - 3 = y$
     $5x - 10 = 2y$

19. $3x - 2y = 6$
     $x - 2y = 10$

20. $3x + 2y = 7$
     $2x + y = 6$

21. $\dfrac{x}{2} + \dfrac{y}{3} = \dfrac{1}{6}$
     $\dfrac{x}{2} - \dfrac{y}{6} = \dfrac{-1}{3}$

22. $x + 3y = 2$
     $\dfrac{x}{3} - \dfrac{2y}{3} = 1$

23. $\dfrac{3x}{4} + \dfrac{2y}{3} = 5$
     $\dfrac{x}{2} - \dfrac{y}{4} = 2$

24. $14 - 3x = 8y$
    $22 - 7y = x$

25. $8x + 20y \quad = 6$
    $4x + 10y - 3 = 0$

26. $5x - 2y = 1$
    $5x \quad = y + 7$

27. $x = y + 6$
    $y = x + 6$

28. $3x - 4y = 2$
    $4y = 3x - 2$

29. $ax + by = 5$ (a and b are constants)
    $bx - ay = 8$

30. $ax - 2by = c$ (a, b, and c are constants)
    $2ax - by = c$

Use two variables to solve Exercises 31–40.

31. The sum of two numbers is 26 and their difference is 16. Find the two numbers.

32. The sum of two numbers is 105 and their difference is 10. Find the two numbers.

33. In a variety store 4 picture frames and 3 wall plaques cost $63, while 3 picture frames and 4 wall plaques cost $56. If the picture frames are all one price and the wall plaques are all another price, find the price of one picture frame and one wall plaque.

34. Three boxes of catfood and 2 jars of preserves cost $3.86, while 5 boxes of the same catfood and 3 jars of the same preserves cost $6.15. Find the price of one box of catfood and one jar of preserves.

35. Recalling the formula that distance traveled in uniform motion is equal to the rate at which the object travels times the amount of time spent in travel, $d = rt$, where d is the distance, r the rate, and t the time, solve the following problem. A crew rows 30 mi downstream in 4 hr. Returning, it covers the same distance in 6 hr and 40 min. If x represents the rate of the current and y represents the rate of the boat in still water, find these two rates.

36. Using the time-rate-distance formula from Exercise 35, solve the following problem. A plane with a headwind flies 300 mi from city A to city B in 2 hr. On the same day and under the same weather conditions, a plane with the same speed in still air flies from city B to city A in 1 hr and 30 min. Find the speed of the wind. (*Hint:* On the trip from B to A the plane has a tailwind.)

37. Twenty pounds of ore A combined with 30 lb of ore B produce 26.5 lb of silver. Ten pounds of ore A combined with 20 lb of ore B produce 15.5 lb of silver. Find the percentage of silver in each ore.

38. An alloy containing 74 g of gold is the combination of two other alloys containing gold: 40 g of alloy A and 70 g of alloy B. If 50 g of alloy A and 90 g of alloy B are combined, the resulting alloy contains 94 g of gold. Find the percentage of gold in alloy A and in alloy B.

39. A manufacturer must use two different machines for the production of two products. The number of hours required on each machine to produce the two products and the maximum number of hours each machine can be run per week are as follows:

Machine	Product A	Product B	Maximum hours for Machine
I	3	2	48
II	1	4	46

How many items of each product will be manufactured each week if the machines are used at their maximum capacity?

40. A mail order company charges a fixed fee for shipping and handling the first 10 lb of merchandise plus an additional fee for each pound over 10 lb. If the shipping and handling charge for 30 lb of merchandise is $2.49 and the charge for 42 lb is $3.45, find the fixed fee and the additional fee.

9.2 LINEAR SYSTEMS IN THREE OR MORE VARIABLES

A linear equation in three variables is defined as follows:

DEFINITIONS

A **linear equation in three variables** x, y, and z is an equation of the form $ax + by + cz = d$, where a, b, c, and d are constants.

A **solution of an equation in three variables** is an ordered triple of numbers (x, y, z) such that the open equation becomes true when x is replaced by the first member of the ordered triple, y by the second member, and z by the third member.

EXAMPLE 1 Show that $(2, -1, 3)$ is a solution of $5x + 2y - z = 5$.

Solution   If $x = 2$, $y = -1$, and $z = 3$, then
$$5x + 2y - z = 5(2) + 2(-1) - 3$$
$$= 10 - 2 - 3 = 5$$

DEFINITION

The **solution set of an equation in three variables** is the set of all solutions of the equation.

As is the case with an equation in two variables, there are infinitely many solutions in the solution set.

DEFINITION

The **solution set of a system of three linear equations in three variables** is the intersection of the solution sets of each of the three equations in the system.

In other words, the solution set of the system

$$a_1x + b_1y + c_1z = d_1$$
$$a_2x + b_2y + c_2z = d_2$$
$$a_3x + b_3y + c_3z = d_3$$

is

$$\{(x, y, z)\,|\,a_1x + b_1y + c_1z = d_1\}$$
$$\cap \{(x, y, z)\,|\,a_2x + b_2y + c_2z = d_2\}$$
$$\cap \{(x, y, z)\,|\,a_3x + b_3y + c_3z = d_3\}$$

The solution set of a system of three linear equations in three variables can be obtained by a method similar to the one used for systems of two linear equations in two variables. By adding constant multiples of two equations of the system, one of the variables can be eliminated. Repeating this operation with a different pair of equations, the same variable can be eliminated again.

The system has now been reduced to two equations in two variables and may be solved by the methods shown in Section 9.1. Again there are three possibilities for solution.

1. There is exactly one solution.
2. The solution set is the empty set.
3. There are infinitely many solutions.

Since there are three variables, the geometric interpretation is three-dimensional. It can be shown that a linear equation in three variables represents a plane. The solution set of a system of three equations in three variables is the geometrical intersection of the three planes. The possibilities are:

Algebraic Statement	Geometric Statement	Illustration
There is exactly one solution.	The three planes intersect at exactly one point P.	
The solution set is empty.	The three planes do not have a point in common.	
There are infinitely many solutions.	The three planes intersect on a line or the three planes are coincident.	

EXAMPLE 2   Solve the system

$$x + 2y - 3z = -11$$
$$x - y - z = 2$$
$$x + 3y + 2z = -4$$

Solution   It does not matter which variable is eliminated first, so look for the one that involves the least amount of work, if possible. In this system the variable x appears to be the simplest to eliminate.

1. Eliminate x by using the first and second equations:

$$x + 2y - 3z = -11 \qquad\qquad x + 2y - 3z = -11$$
$$-1(x - y - z) = -1(2) \qquad\qquad \underline{-x + y + z = -2}$$
$$3y - 2z = -13$$

2. Eliminate x by using the second and third equations:

$$x - y - z = 2 \qquad\qquad x - y - z = 2$$
$$-1(x + 3y + 2z) = -1(-4) \qquad\qquad \underline{-x - 3y - 2z = 4}$$
$$-4y - 3z = 6$$

3. Now solve the system

$$3y - 2z = -13$$
$$-4y - 3z = 6$$

$$-3(3y - 2z) = -3(-13) \qquad\qquad -9y + 6z = 39$$
$$2(-4y - 3z) = 2(6) \qquad\qquad \underline{-8y - 6z = 12}$$
$$-17y = 51$$
$$y = -3$$

Using the equation $3y - 2z = -13$ and $y = -3$ we get

$$3(-3) - 2z = -13$$
$$z = 2$$

4. Use any one of the three *original* equations to find x. Using the second equation, $x - y - z = 2$, with $y = -3$ and $z = 2$,

$$x - (-3) - (2) = 2$$
$$x = 1$$

5. The solution is $(1, -3, 2)$, and the solution set is $\{(1, -3, 2)\}$.

6. Check the solution in each equation:

$$x + 2y - 3z = 1 + 2(-3) - 3(2)$$
$$= 1 - 6 - 6 = 1 - 12 = -11$$
$$x - y - z = 1 - (-3) - 2 = 1 + 3 - 2 = 4 - 2 = 2$$

$$x + 3y + 2z = 1 + 3(-3) + 2(2)$$
$$= 1 - 9 + 4 = -8 + 4 = -4$$

Thus if

$$A = \{(x, y, z) | x + 2y - 3z = -11\}$$
$$B = \{(x, y, z) | x - y - z = 2\}$$
$$C = \{(x, y, z) | x + 3y + 2z = -4\}$$

then the solution set of the system is $A \cap B \cap C = \{(1, -3, 2)\}$.

EXAMPLE 3   Solve the system

$$2x + 2y - 2z = 1$$
$$5x - 2y + z = 2$$
$$3x + 3y - 3z = 5$$

Solution

1. $2x + 2y - 2z = 1$ $2x + 2y - 2z = 1$
   $2(5x - 2y + z) = 2(2)$ $\underline{10x - 4y + 2z = 4}$
                                 $12x - 2y = 5$

2. $3(5x - 2y + z) = 3(2)$ $15x - 6y + 3z = 6$
   $3x + 3y - 3z = 5$ $\underline{3x + 3y - 3z = 5}$
                                 $18x - 3y = 11$

3. $3(12x - 2y) = 3(5)$ $36x - 6y = 15$
   $-2(18x - 3y) = -2(11)$ $\underline{-36x + 6y = -22}$
                                 $0 = -7,$

a false statement

4. The solution set is the empty set, $\varnothing$.

EXAMPLE 4   Solve the system

$$x - 2y + 3z = 4$$
$$2x + y - z = 1$$
$$3x - y + 2z = 5$$

Solution

1. $x - 2y + 3z = 4$ $x - 2y + 3z = 4$
   $3(2x + y - z) = 3(1)$ $\underline{6x + 3y - 3z = 3}$
                                 $7x + y = 7$

2. $2(2x + y - z) - 2(1)$ $4x + 2y - 2z = 2$
   $3x - y + 2z = 5$ $\underline{3x - y + 2z = 5}$
                                 $7x + y = 7$

Since the same equation was obtained in both cases, the solution set is infinite. Solve this equation for y, $y = 7 - 7x$. Now replacing y by $7 - 7x$ in the second equation,

$$2x + (7 - 7x) - z = 1$$
$$-5x - z = -6$$
$$z = 6 - 5x$$

The solution set is $\{(x, 7 - 7x, 6 - 5x) | x \text{ is any real number}\}$.

Check

$$x - 2y + 3z = x - 2(7 - 7x) + 3(6 - 5x)$$
$$= x - 14 + 14x + 18 - 15x = 4$$
$$2x + y - z = 2x + (7 - 7x) - (6 - 5x)$$
$$= 2x + 7 - 7x - 6 + 5x = 1$$
$$3x - y + 2z = 3x - (7 - 7x) + 2(6 - 5x)$$
$$= 3x - 7 + 7x + 12 - 10x = 5$$

It is important to note that the solution for the system in Example 4 was written in terms of x. Had we eliminated a variable other than z as the first step, the solution might have been written in terms of y or z. For example, eliminate the variable x:

$$-2(x - 2y + 3z) = -2(4) \qquad -2x + 4y - 6z = -8$$
$$2x + y - z = 1 \qquad \underline{2x + y - z = 1}$$
$$5y - 7z = -7$$

$$-3(x - 2y + 3z) = -3(4) \qquad -3x + 6y - 9z = -12$$
$$3x - y + 2z = 5 \qquad \underline{3x - y + 2z = 5}$$
$$5y - 7z = -7$$

Again the same equation was obtained in both cases. We can now solve the equation either for y or for z. Solving for y we obtain $y = \dfrac{7z - 7}{5}$. Now replacing $y = \dfrac{7z - 7}{5}$ in the second equation,

$$2x + \frac{7z - 7}{5} - z = 1$$
$$10x + 7z - 7 - 5z = 5$$
$$x = \frac{-2z + 12}{10} = \frac{-z + 6}{5}$$

Thus the solution set may also be written as

$$\left\{ \left(\frac{-z + 6}{5}, \frac{7z - 7}{5}, z \right) \;\middle|\; z \text{ is any real number} \right\}$$

It is left for the student to see how the solution would be written if the first variable to be eliminated were y. Check the solution by replacing z by any real number, and substituting the resulting values for x, y, and z in the original equations.

A system of four linear equations in four variables can be solved by eliminating the same variable using three different pairs of equations. Each of the four equations should be represented in at least one pair. A system of three linear equations in three variables is thus obtained, and this may be solved by the preceding technique.

EXERCISES 9.2

Solve and check Exercises 1–24.

1. $x + y - z = 0$
$x - 2y + 3z = 0$
$2x - y - 2z = 0$

2. $x + 2y + z = 3$
$2x - y - 2z = 1$
$3x - 3y + 3z = 0$

3. $4x - y + 2z = 3$
$x + y - z = 4$
$x + 6y - 7z = 17$

4. $3x - y = z - 2$
$5x + 2y = 22 - z$
$2x + 4y - 5z = 0$

5. $x - 2y - 3z = 3$
$x + y - z = 2$
$2x - 3y - 5z = 5$

6. $x - y - z = 4$
$6x + y - 2z = 5$
$-3x + 3y + 3z = -7$

7. $x + y + 3z - 3 = 0$
$3x - y + 2z - 1 = 0$
$4x - 4y + z + 2 = 0$

8. $2r + 2s - t = 28$
$2r - 2s - t = 8$
$2r - 2s + t = 20$

9. $x + y + z = 5$
$2x - y - 2z = 3$
$3x - 3y - 5z = 1$

10. $2r + 2s - t = 28$
$2r - 2s - t = 8$
$6r - 2s - 3t = 44$

11. $x + 2y - 2z = 0$
$3x - 4y + 2z = 0$
$x + 12y - 10z = 0$

12. $x - 2y - z = 0$
$2x - y + z = 0$
$3x + y - z = 20$

13. $x + y + z = 9$
$x - y - z = 3$
$2x + 2y + 2z = 5$

14. $2x + y - z = 3$
$4x + 2y + 3z = 1$
$6x + 3y - 2z = 2$

15. $x + y + z = 9$
$x - y - z = 3$
$x + 3y + 3z = 15$

16. $x + y = z + 2$
$4x + 3z = y + 10$
$2x - 3y = 12 - z$

17. $x + y = 3$
$z - y = 3$
$2x + y = 8$

18. $x + y = 5$
$y + z = 10$
$x + z = 1$

19. $x + 2y - z = 5$
$3x - 4y + 2z = 15$
$2x - y + z = 13$

20. $x - y + 2z = 0$
$2x + y - z = 0$
$x + 5y - 8z = 0$

21. $x + y + z = 9$
$x - y - z = 3$
$x + y - z = 5$

22. $2r + 2s - t = 28$
$2r - 2s - t = 8$
$2r + s - t = 20$

23. $x + y - z = a$
$x - y + z = b$
$x - y - z = c$

24. $2x + 2y - z = 4a$
$2x - 2y - z = 4b$
$2x - 2y + z = 4c$

25. Find the solution set of the system
$x - 2y = 5$
$2x + 3y = 7$
$x + 5y = 2$

26. Find the solution set of the system
$x - 3y + z = 1$
$2x + y - 2z = -2$
(*Hint:* Express y and z in terms of x.)

27. Find the solution set of the system

$$
\begin{aligned}
x + y + z &= 4 \\
2x - y + 3z &= 14 \\
3x - y - 2z &= 1 \\
6x - y + 2z &= 19
\end{aligned}
$$

(*Hint:* Find the solution set for the system consisting of the first three equations. Show that the solution also satisfies the fourth equation.)

28. Find the solution set of the system

$$
\begin{aligned}
x + y + z &= 6 \\
2x - y + z &= 3 \\
5x + 2y - 2z &= 12 \\
3x - y - z &= 2
\end{aligned}
$$

29. Solve and check:

$$
\begin{aligned}
x + y + z + t &= 1 \\
x + 2y \quad\;\; - t &= 10 \\
2y - z + t &= 1 \\
x \quad\;\; + 2z - 2t &= 7
\end{aligned}
$$

30. Solve and check:

$$
\begin{aligned}
a - b + c - d &= 5 \\
2a \quad\;\; - 3c + d &= 2 \\
a \quad\quad\;\; - 3d &= 6 \\
2b - c + 2d &= -3
\end{aligned}
$$

31. Solve and check:

$$
\begin{aligned}
x + 2y - 3z + 2t &= 6 \\
2x + 3y + 4z + t &= -5 \\
3x - y + 2z - t &= 0 \\
2x + y + z - t &= -4
\end{aligned}
$$

32. Solve and check:

$$
\begin{aligned}
a \quad\quad + 2c \quad\quad\quad &= 11 \\
a + 3b \quad\;\; - 2d - 13 &= 0 \\
2a \quad\;\; - 3c - d + 4 &= 0 \\
3a - 3b - 2c + 2d - 1 &= 0
\end{aligned}
$$

9.3 DETERMINANTS

The solution of linear systems of equations may be found by other methods than those shown in Sections 9.1 and 9.2. Two such mechods are **matrix** solutions and solutions using **determinants.** In this section we define a determinant and state some theorems for evaluation of determinants, and in the next section we use determinants to solve linear systems. Matrix solutions will be reserved for a more advanced course in algebra.

In order to define a determinant, we first define a matrix.

A rectangular ordered array of numbers such as

$$\begin{pmatrix} 3 & 4 & 5 \\ 2 & -1 & 0 \\ -1 & 0 & 7 \\ 6 & 2 & 3 \end{pmatrix}$$

is called a **matrix.** The matrix in the illustration has four rows and three columns and is called a 4×3 matrix. The number of rows is always listed first. Thus an $m \times n$ matrix is a rectangular ordered array of numbers with m rows and n columns.

A square matrix is a matrix with the same number of rows as columns. Thus a square matrix can be said to be an $m \times m$ matrix, or an $n \times n$ matrix.

Associated with every square matrix is a real number, called the **determinant** of the matrix, and written in the same manner as a matrix but with two parallel bars enclosing it. Thus the 2×2 matrix,

$$\begin{pmatrix} a_1 & b_1 \\ a_2 & b_2 \end{pmatrix}$$

has a determinant

$$\begin{vmatrix} a_1 & b_1 \\ a_2 & b_2 \end{vmatrix}$$

Since we stated that the determinant is a real number associated with the matrix, we must have a method of finding that number, and for the 2×2 matrix, the determinant is found by the following rule:

$$\begin{vmatrix} a_1 & b_1 \\ a_2 & b_2 \end{vmatrix} = a_1 b_2 - a_2 b_1$$

The determinant of a 2×2 matrix is called a **second-order determinant,** and is formally defined below.

DEFINITION

A **determinant of the second order,** denoted by the symbol $\begin{vmatrix} a_1 & b_1 \\ a_2 & b_2 \end{vmatrix}$, where a_1, b_1, a_2, and b_2 are real numbers, is the real number, $a_1 b_2 - a_2 b_1$—that is,

$$\begin{vmatrix} a_1 & b_1 \\ a_2 & b_2 \end{vmatrix} = a_1 b_2 - a_2 b_1$$

EXAMPLE 1   Evaluate the determinant $\begin{vmatrix} 3 & -1 \\ 2 & 5 \end{vmatrix}$.

Solution   By the definition,

$$\begin{vmatrix} 3 & -1 \\ 2 & 5 \end{vmatrix} = 3(5) - (2)(-1)$$
$$= 15 + 2 = 17$$

The determinant for a 3×3 matrix is called a **third-order determinant** and is evaluated by first reducing the 3×3 matrix to one or more 2×2 matrices (the plural of matrix) which can then be evaluated in the manner defined for second-order determinants.

DEFINITION

A **determinant of the third order,** denoted by the symbol $\begin{vmatrix} a_1 & b_1 & c_1 \\ a_2 & b_2 & c_2 \\ a_3 & b_3 & c_3 \end{vmatrix}$, where the elements are real numbers, is the real

number $a_1 \begin{vmatrix} b_2 & c_2 \\ b_3 & c_3 \end{vmatrix} - a_2 \begin{vmatrix} b_1 & c_1 \\ b_3 & c_3 \end{vmatrix} + a_3 \begin{vmatrix} b_1 & c_1 \\ b_2 & c_2 \end{vmatrix}$ —that is,

$$\begin{vmatrix} a_1 & b_1 & c_1 \\ a_2 & b_2 & c_2 \\ a_3 & b_3 & c_3 \end{vmatrix} = a_1 \begin{vmatrix} b_2 & c_2 \\ b_3 & c_3 \end{vmatrix} - a_2 \begin{vmatrix} b_1 & c_1 \\ b_3 & c_3 \end{vmatrix} + a_3 \begin{vmatrix} b_1 & c_1 \\ b_2 & c_2 \end{vmatrix}$$

EXAMPLE 2 Evaluate the determinant $\begin{vmatrix} 1 & 2 & 3 \\ 3 & -1 & 2 \\ 4 & -3 & 1 \end{vmatrix}$.

Solution By the definition,

$$\begin{vmatrix} 1 & 2 & 3 \\ 3 & -1 & 2 \\ 4 & -3 & 1 \end{vmatrix} = 1 \begin{vmatrix} -1 & 2 \\ -3 & 1 \end{vmatrix} - 3 \begin{vmatrix} 2 & 3 \\ -3 & 1 \end{vmatrix} + 4 \begin{vmatrix} 2 & 3 \\ -1 & 2 \end{vmatrix}$$

$$= 1[(-1)(1) - (-3)(2)] - 3[(2)(1) - (-3)(3)]$$
$$\quad + 4[(2)(2) - (-1)(3)]$$
$$= 1(5) - 3(11) + 4(7)$$
$$= 5 - 33 + 28$$
$$= 0$$

The **order of a determinant** is the number of rows (or columns) in the square array associated with the determinant.

The **expansion of a determinant** is the indicated calculation in the definition of the determinant.

The **minor** of an element of a determinant is the determinant obtained by deleting the row and column in which the element lies.

For example, for a determinant of order 3,

$A_1 = \begin{vmatrix} b_2 & c_2 \\ b_3 & c_3 \end{vmatrix}$ is the minor of a_1. $\begin{vmatrix} a_1 & b_1 & c_1 \\ a_2 & b_2 & c_2 \\ a_3 & b_3 & c_3 \end{vmatrix}$

$A_2 = \begin{vmatrix} b_1 & c_1 \\ b_3 & c_3 \end{vmatrix}$ is the minor of a_2.

$$A_3 = \begin{vmatrix} b_1 & c_1 \\ b_2 & c_2 \end{vmatrix} \text{ is the minor of } a_3.$$

$$B_2 = \begin{vmatrix} a_1 & c_1 \\ a_3 & c_3 \end{vmatrix} \text{ is the minor of } b_2, \text{ and so on.}$$

Thus

$$D = \begin{vmatrix} a_1 & b_1 & c_1 \\ a_2 & b_2 & c_2 \\ a_3 & b_3 & c_3 \end{vmatrix} = a_1A_1 - a_2A_2 + a_3A_3$$

is called an expansion of D by the minors of the elements of the first column.

EXAMPLE 3 Evaluate the determinant $\begin{vmatrix} 1 & 2 & 3 \\ 3 & -1 & 2 \\ 4 & -3 & 1 \end{vmatrix}$ by expansion of the minors of the first column.

Solution

$$\begin{vmatrix} \mathbf{1} & 2 & 3 \\ \mathbf{3} & -1 & 2 \\ \mathbf{4} & -3 & 1 \end{vmatrix}$$

$$A_1 - \begin{vmatrix} -1 & 2 \\ -3 & 1 \end{vmatrix} = (-1)(1) - (-3)(2) = 5$$

$$A_2 = \begin{vmatrix} 2 & 3 \\ -3 & 1 \end{vmatrix} = (2)(1) - (-3)(3) = 11$$

$$A_3 = \begin{vmatrix} 2 & 3 \\ -1 & 2 \end{vmatrix} = (2)(2) - (-1)(3) = 7$$

$$\begin{aligned} D &= a_1A_1 - a_2A_2 + a_3A_3 \\ &= 1(5) - 3(11) + 4(7) \\ &= 5 - 33 + 28 = 0 \end{aligned}$$

(Compare this solution with that of Example 2.)

The following theorems provide us with tools for simplifying the work involved in evaluating third-order (and higher-order) determinants.

THEOREM 1

Any row or column may be used to expand a determinant of the third order—that is,

$$\begin{aligned} D &= a_1A_1 - a_2A_2 + a_3A_3 \\ &= -b_1B_1 + b_2B_2 - b_3B_3 \\ &= c_1C_1 - c_2C_2 + c_3C_3 \\ &= a_1A_1 - b_1B_1 + c_1C_1 \\ &= -a_2A_2 + b_2B_2 - c_2C_2 \\ &= a_3A_3 - b_3B_3 + c_3C_3 \end{aligned}$$

Whether a product in the expansion is to be multiplied by $+1$ or -1 is determined by the following array, called the checkerboard of signs:

$$\begin{vmatrix} + & - & + \\ - & + & - \\ + & - & + \end{vmatrix}$$

EXAMPLE 4   Expand $\begin{vmatrix} 2 & 3 & -1 \\ 1 & -2 & 2 \\ -4 & -1 & 5 \end{vmatrix}$ by row 1.

Solution

$$D = 2 \begin{vmatrix} -2 & 2 \\ -1 & 5 \end{vmatrix} - 3 \begin{vmatrix} 1 & 2 \\ -4 & 5 \end{vmatrix} + (-1) \begin{vmatrix} 1 & -2 \\ -4 & -1 \end{vmatrix}$$
$$= 2(-10 - (-2)) - 3(5 - (-8)) - (-1 - 8)$$
$$= 2(-8) - 3(13) - (-9)$$
$$= -16 - 39 + 9 = -46$$

EXAMPLE 5   Expand $\begin{vmatrix} 3 & 2 & -5 \\ 2 & 0 & 1 \\ 1 & 0 & 4 \end{vmatrix}$ by column 2.

Solution

$$D = -2 \begin{vmatrix} 2 & 1 \\ 1 & 4 \end{vmatrix} + 0 \begin{vmatrix} 3 & -5 \\ 1 & 4 \end{vmatrix} - 0 \begin{vmatrix} 3 & -5 \\ 2 & 1 \end{vmatrix}$$
$$= -2(8 - 1) + 0 - 0$$
$$= -14$$

The two preceding examples illustrate that the determinant is much easier to evaluate when there are zeros in a row or column. Thus it is desirable to establish some theorems indicating the transformations that can be performed on determinants in order to obtain an equal determinant with zeros as some of its elements.

THEOREM 2

Two determinants of the same order are equal if one is obtained from the other by *interchanging the rows and columns.*

In symbols,

$$\begin{vmatrix} a_1 & b_1 \\ a_2 & b_2 \end{vmatrix} = \begin{vmatrix} a_1 & a_2 \\ b_1 & b_2 \end{vmatrix} \quad \text{and} \quad \begin{vmatrix} a_1 & b_1 & c_1 \\ a_2 & b_2 & c_2 \\ a_3 & b_3 & c_3 \end{vmatrix} = \begin{vmatrix} a_1 & a_2 & a_3 \\ b_1 & b_2 & b_3 \\ c_1 & c_2 & c_3 \end{vmatrix}$$

THEOREM 3

If *two rows (or two columns) of a determinant are interchanged*, then the resulting determinant is the negative of the original one.

For example, let $D = \begin{vmatrix} a_1 & b_1 & c_1 \\ a_2 & b_2 & c_2 \\ a_3 & b_3 & c_3 \end{vmatrix}$ and let $E = \begin{vmatrix} c_1 & b_1 & a_1 \\ c_2 & b_2 & a_2 \\ c_3 & b_3 & a_3 \end{vmatrix}$.

Then $E = -D$.

THEOREM 4

If *each element of a row (or column) is multiplied by a constant,* then the determinant is multiplied by a constant.

For example,

$$\begin{vmatrix} ka_1 & kb_1 & kc_1 \\ a_2 & b_2 & c_2 \\ a_3 & b_3 & c_3 \end{vmatrix} = k \begin{vmatrix} a_1 & b_1 & c_1 \\ a_2 & b_2 & c_2 \\ a_3 & b_3 & c_3 \end{vmatrix}$$

and

$$\begin{vmatrix} 6 & 9 & -12 \\ 2 & 1 & 5 \\ 3 & -1 & 2 \end{vmatrix} = 3 \begin{vmatrix} 2 & 3 & -4 \\ 2 & 1 & 5 \\ 3 & -1 & 2 \end{vmatrix}$$

THEOREM 5

If the *corresponding elements of two rows (or columns) are equal,* then the value of the determinant is 0.

For example,

$$\begin{vmatrix} 1 & 4 & 1 \\ 3 & 1 & 3 \\ -2 & 5 & -2 \end{vmatrix} = 0 \quad \text{and} \quad \begin{vmatrix} 2 & 1 & 5 \\ 2 & 1 & 5 \\ 3 & 7 & 9 \end{vmatrix} = 0$$

THEOREM 6

If *each element of a row (or column) is multiplied by a constant and then added to the corresponding element of another row (or column),* then the resulting determinant is equal to the original determinant.

For example,

$$\begin{vmatrix} a_1 + ka_3 & b_1 + kb_3 & c_1 + kc_3 \\ a_2 & b_2 & c_2 \\ a_3 & b_3 & c_3 \end{vmatrix} = \begin{vmatrix} a_1 & b_1 & c_1 \\ a_2 & b_2 & c_2 \\ a_3 & b_3 & c_3 \end{vmatrix}$$

and

$$\begin{vmatrix} a_1 & b_1 + kc_1 & c_1 \\ a_2 & b_2 + kc_2 & c_2 \\ a_3 & b_3 + kc_3 & c_3 \end{vmatrix} = \begin{vmatrix} a_1 & b_1 & c_1 \\ a_2 & b_2 & c_2 \\ a_3 & b_3 & c_3 \end{vmatrix}$$

EXAMPLE 6 Show that the determinant $\begin{vmatrix} 3 & 2 & -1 \\ 2 & 1 & 0 \\ 1 & -1 & 1 \end{vmatrix}$ equals the determinant obtained by multiplying row 2 by -2 and adding it to row 1, and by writing row 3 as the sum of rows 2 and 3.

Solution $\begin{vmatrix} 3 & 2 & -1 \\ 2 & 1 & 0 \\ 1 & -1 & 1 \end{vmatrix}$ expanded about column 3 equals

$$-1 \begin{vmatrix} 2 & 1 \\ 1 & -1 \end{vmatrix} - 0 \begin{vmatrix} 3 & 2 \\ 1 & -1 \end{vmatrix} + 1 \begin{vmatrix} 3 & 2 \\ 2 & 1 \end{vmatrix}$$

$$= -1(-2 - 1) - 0(-3 - 2) + 1(3 - 4)$$
$$= 3 - 0 - 1$$
$$= 2$$

To compare with the second determinant, leave row 2 as is:

  2 1 0   **row 2**

Multiply row 2 by -2 and add to row 1:

$$-4 + 3 \quad = -1$$
$$-2 + 2 \quad = \quad 0$$
$$0 + (-1) = -1$$

Now row 1 reads:

  -1 0 -1 **row 1**

Row 3 becomes the sum of rows 2 and 3:

  3 0 1   **row 3**

The new determinant is written

$$\begin{vmatrix} -1 & 0 & -1 \\ 2 & 1 & 0 \\ 3 & 0 & 1 \end{vmatrix}$$

Expanding this determinant about column 2:

$$D = -0 \begin{vmatrix} 2 & 0 \\ 3 & 1 \end{vmatrix} + 1 \begin{vmatrix} -1 & -1 \\ 3 & 1 \end{vmatrix} - 0 \begin{vmatrix} -1 & -1 \\ 2 & 0 \end{vmatrix}$$

$$= 0(2 - 0) + 1(-1 - (-3)) - 0(0 - (-2))$$
$$= 0 + 2 - 0$$
$$= 2$$

The above examples show that any row or column that has zeros is quicker to compute when expanding. The use of the above theorems to create zeros is shown in the following example.

EXAMPLE 7 Find a determinant equal to $\begin{vmatrix} -2 & 3 & 1 \\ 1 & 2 & 3 \\ 2 & 3 & 3 \end{vmatrix}$ having zeros everywhere in column 3 except the first row. Expand the resulting determinant.

Solution $\begin{vmatrix} -2 & 3 & 1 \\ 1 & 2 & 3 \\ 2 & 3 & 3 \end{vmatrix} = \begin{vmatrix} -2 & 3 & 1 \\ 7 & -7 & 0 \\ 2 & 3 & 3 \end{vmatrix}$ (Multiply row 1 by -3 and add to row 2)

$$= \begin{vmatrix} -2 & 3 & 1 \\ 7 & -7 & 0 \\ 8 & -6 & 0 \end{vmatrix} \quad \text{(Multiply row 1 by } -3 \text{ and add to row 3)}$$

$$= 1 \begin{vmatrix} 7 & -7 \\ 8 & -6 \end{vmatrix} = 7(-6) - 8(-7) = -42 + 56 = 14$$

EXAMPLE 8 Expand the determinant having zeros everywhere in

row 2 except column 2 and equal to $\begin{vmatrix} 2 & -1 & 4 \\ 5 & 1 & -2 \\ 3 & -3 & -4 \end{vmatrix}$.

Solution

$$\begin{vmatrix} 2 & -1 & 4 \\ 5 & 1 & -2 \\ 3 & -3 & -4 \end{vmatrix} = \begin{vmatrix} 7 & -1 & 4 \\ 0 & 1 & -2 \\ 18 & -3 & -4 \end{vmatrix} \quad \text{(Multiply column 2 by } -5 \text{ and add to column 1)}$$

$$= \begin{vmatrix} 7 & -1 & 2 \\ 0 & 1 & 0 \\ 18 & -3 & 10 \end{vmatrix} \quad \text{(Multiply column 2 by 2 and add to column 3)}$$

$$= 1 \begin{vmatrix} 7 & 2 \\ 18 & -10 \end{vmatrix}$$

$$= 7(-10) - 18(2) = -70 - 36 = -106$$

EXERCISES 9.3

Evaluate each determinant in Exercises 1–10.

1. $\begin{vmatrix} 3 & 5 \\ 4 & 9 \end{vmatrix}$

2. $\begin{vmatrix} 2 & 3 \\ 4 & -5 \end{vmatrix}$

3. $\begin{vmatrix} 1 & 0 \\ 3 & -2 \end{vmatrix}$

4. $\begin{vmatrix} 2 & -5 \\ 5 & 2 \end{vmatrix}$

5. $\begin{vmatrix} -3 & -1 \\ -4 & -7 \end{vmatrix}$

6. $\begin{vmatrix} 0 & 2 \\ -9 & 10 \end{vmatrix}$

7. $\begin{vmatrix} 3 & 1 & 2 \\ 2 & 4 & 1 \\ 5 & 1 & 2 \end{vmatrix}$

8. $\begin{vmatrix} -1 & 0 & 3 \\ 2 & -2 & 1 \\ 3 & -1 & -2 \end{vmatrix}$

9. $\begin{vmatrix} 0 & 3 & 4 \\ 4 & 0 & -1 \\ 2 & 3 & 0 \end{vmatrix}$

10. $\begin{vmatrix} 2 & 3 & -1 \\ -2 & -3 & 1 \\ 3 & 1 & 4 \end{vmatrix}$

In Exercises 11–18 find a determinant equal to the given one and satisfying the stated conditions. Expand and evaluate each determinant.

11. $\begin{vmatrix} -2 & 3 & 1 \\ 1 & 2 & 3 \\ 2 & 3 & 3 \end{vmatrix}$ Zeros everywhere in row 2 except column 1

12. $\begin{vmatrix} 1 & 1 & 1 \\ 1 & 4 & 9 \\ 1 & 8 & 27 \end{vmatrix}$ Zeros everywhere in column 1 except row 1

13. $\begin{vmatrix} -1 & 4 & -4 \\ 2 & 3 & 2 \\ 1 & -1 & 2 \end{vmatrix}$ Zeros everywhere in column 3 except row 3

14. $\begin{vmatrix} 5 & -2 & 3 \\ -12 & 3 & 9 \\ 4 & 1 & -2 \end{vmatrix}$ Zeros everywhere in row 2 except column 2

15. $\begin{vmatrix} 3 & 1 & -1 \\ 2 & -1 & 2 \\ 8 & -2 & -1 \end{vmatrix}$ Zeros everywhere in row 1 except column 3

16. $\begin{vmatrix} 4 & 2 & -6 \\ 1 & -2 & 3 \\ 5 & -3 & -1 \end{vmatrix}$ Zeros everywhere in column 2 except row 1

17. $\begin{vmatrix} 2 & -4 & -3 \\ 1 & 5 & 2 \\ -5 & 2 & 1 \end{vmatrix}$ Zeros everywhere in row 3 except column 3

18. $\begin{vmatrix} 4 & 3 & -10 \\ 1 & -2 & 6 \\ 2 & -1 & 15 \end{vmatrix}$ Zeros everywhere in column 3 except row 2

For Exercises 19–20 the definition and theorems for third-order determinants can be generalized for fourth-order or higher-order determinants. Thus

$$\begin{vmatrix} a_1 & b_1 & c_1 & d_1 \\ a_2 & b_2 & c_2 & d_2 \\ a_3 & b_3 & c_3 & d_3 \\ a_4 & b_4 & c_4 & d_4 \end{vmatrix} = a_1A_1 - a_2A_2 + a_3A_3 - a_4A_4$$

Expand each determinant.

19. $\begin{vmatrix} 1 & 2 & -2 & 3 \\ 2 & 1 & 1 & -4 \\ 4 & 3 & -5 & 1 \\ 3 & -3 & 2 & -2 \end{vmatrix}$

20. $\begin{vmatrix} 2 & -4 & 7 & 1 \\ 1 & 3 & 1 & 2 \\ 5 & -1 & 2 & -1 \\ -3 & 2 & 6 & -6 \end{vmatrix}$

21. Without expanding, show that
$$\begin{vmatrix} 1 & 1 & 2x - 2y \\ 0 & x & x^2 - xy \\ x & y & x^2 - y^2 \end{vmatrix} = 0$$
State the reason for each step.

22. Without expanding, show that

$$\begin{vmatrix} 1 & 1 & 1 & 1 \\ 1 & a & a^2 & a^3 \\ 1 & b & b^2 & b^3 \\ 1 & c & c^2 & c^3 \end{vmatrix} = (a - 1)(b - 1)(c - 1)(b - a)(c - a)(c - b)$$

9.4 SOLUTION OF LINEAR SYSTEMS BY USE OF DETERMINANTS

As stated in Section 9.3 determinants may be used to solve linear systems of equations. The solution by using determinants is known as **Cramer's rule,** in honor of the Swiss mathematician Gabriel Cramer (1704–1752). We state Cramer's rule first for the solution of two equations in two variables.

To solve the system

$$a_1x + b_1y = k_1$$
$$a_2x + b_2y = k_2$$

by the addition method, multiply the first equation by b_2 and the second equation by $-b_1$ to get the equivalent system

$$a_1b_2x + b_1b_2y = b_2k_1$$
$$-a_2b_1x - b_1b_2y = -b_1k_2$$

By adding the two equations we obtain

$$(a_1b_2 - a_2b_1)x = b_2k_1 - b_1k_2$$

Now, to eliminate x, multiply the first equation by $-a_2$ and the second equation by a_1 to obtain the equivalent system

$$-a_1a_2x - a_2b_1y = -a_2k_1$$
$$a_1a_2x + a_1b_2y = a_1k_2$$

By adding the two equations we obtain

$$(a_1b_2 - a_2b_1)y = a_1k_2 - a_2k_1$$

Listing both of these derived equations, we have

$$(a_1b_2 - a_2b_1)x = b_2k_1 - b_1k_2$$
$$(a_1b_2 - a_2b_1)y = a_1k_2 - a_2k_1$$

Now letting

$$D = \begin{vmatrix} a_1 & b_1 \\ a_2 & b_2 \end{vmatrix}, \quad X = \begin{vmatrix} k_1 & b_1 \\ k_2 & b_2 \end{vmatrix}, \quad \text{and} \quad Y = \begin{vmatrix} a_1 & k_1 \\ a_2 & k_2 \end{vmatrix}$$

these equations become $Dx = X$, $Dy = Y$.

The determinant D is called the **determinant of the coefficients.**

If $D \neq 0$, then there is a unique solution, $\left(\dfrac{X}{D}, \dfrac{Y}{D}\right)$ (the case of two intersecting lines).

If $D = 0$ and $X = Y = 0$, then the solution set is infinite (the case of coincident lines).

If $D = 0$ and either $X \neq 0$ or $Y \neq 0$, then the solution set is empty (the case of parallel lines).

EXAMPLE 1   Use Cramer's rule to solve the system

$$3x + 2y = 5$$
$$x - 3y = 2$$

Solution   It is wise to find D, the determinant of coefficients, first because if $D = 0$, there is no unique solution to the system.

Note that D simply lists all the coefficients of the variables in the order in which they appear in the system.

$$D = \begin{vmatrix} 3 & 2 \\ 1 & -3 \end{vmatrix} = (3)(-3) - (1)(2)$$
$$= -9 - 2 = -11$$

Since $D = -11 \neq 0$, there is a unique solution. We find X by replacing the coefficients of the x terms by the constants. Similarly, to find Y we replace the coefficients of the y terms by the constants and evaluate both determinants.

$$X = \begin{vmatrix} 5 & 2 \\ 2 & -3 \end{vmatrix} = (5)(-3) - (2)(2)$$
$$= -15 - 4 = -19$$

$$Y = \begin{vmatrix} 3 & 5 \\ 1 & 2 \end{vmatrix} = (3)(2) - (1)(5)$$
$$= 6 - 5 = 1$$

$$\frac{X}{D} = \frac{-19}{-11} = \frac{19}{11}, \frac{Y}{D} = \frac{1}{-11} = -\frac{1}{11}$$

Thus $\left(\dfrac{19}{11}, -\dfrac{1}{11}\right)$ is the unique solution.

Check

$$3\left(\frac{19}{11}\right) + 2\left(-\frac{1}{11}\right) = \frac{57}{11} - \frac{2}{11} = \frac{55}{11} = 5$$

$$\frac{19}{11} - 3\left(-\frac{1}{11}\right) = \frac{19}{11} + \frac{3}{11} = \frac{22}{11} = 2$$

Cramer's rule can be extended to systems of three linear equations in three unknowns, four equations in four unknowns, and so on. For three equations in three unknowns, D, the determinant of coefficients, is obtained in the usual manner. If $D \neq 0$, there will be a unique solution to the system.

If the variables are x, y, and z, then $x = \dfrac{X}{D}$, $y = \dfrac{Y}{D}$, and $z = \dfrac{Z}{D}$, where the coefficients of the x terms are replaced by the constants in the determinant X, the coefficients of the y terms are replaced by the constants in the determinant Y, and the coefficients of the z terms are replaced by the constants in the determinant Z.

Thus for the system

$$a_1x + b_1y + c_1z = k_1$$
$$a_2x + b_2y + c_2z = k_2$$
$$a_3x + b_3y + c_3z = k_3$$

$$D = \begin{vmatrix} a_1 & b_1 & c_1 \\ a_2 & b_2 & c_2 \\ a_3 & b_3 & c_3 \end{vmatrix} \qquad X = \begin{vmatrix} k_1 & b_1 & c_1 \\ k_2 & b_2 & c_2 \\ k_3 & b_3 & c_3 \end{vmatrix}$$

$$Y = \begin{vmatrix} a_1 & k_1 & c_1 \\ a_2 & k_2 & c_2 \\ a_3 & k_3 & c_3 \end{vmatrix} \qquad Z = \begin{vmatrix} a_1 & b_1 & k_1 \\ a_2 & b_2 & k_2 \\ a_3 & b_3 & k_3 \end{vmatrix}$$

If $D \neq 0$, then there is exactly one solution: $\left(\dfrac{X}{D}, \dfrac{Y}{D}, \dfrac{Z}{D}\right)$.

If $D = 0$ and $X = Y = Z = 0$, then the solution set is infinite.

If $D = 0$ and either $X \neq 0$ or $Y \neq 0$ or $Z \neq 0$, then the solution set is empty.

EXAMPLE 2 Use Cramer's rule to solve the following system, and check the solution.

$$x + y + z = 4$$
$$2x - y - 2z = -1$$
$$x - 2y - z = 1$$

Solution

$$D = \begin{vmatrix} 1 & 1 & 1 \\ 2 & -1 & -2 \\ 1 & -2 & 1 \end{vmatrix} = \begin{vmatrix} 1 & 1 & 2 \\ 2 & -1 & 0 \\ 1 & -2 & 0 \end{vmatrix} = 2 \begin{vmatrix} 2 & -1 \\ 1 & -2 \end{vmatrix}$$

$$= 2(-4 + 1) = -6$$

$$X = \begin{vmatrix} 4 & 1 & 1 \\ -1 & -1 & -2 \\ 1 & -2 & -1 \end{vmatrix} = \begin{vmatrix} 4 & -3 & -7 \\ -1 & 0 & 0 \\ 1 & -3 & -3 \end{vmatrix} = \begin{vmatrix} -3 & -7 \\ -3 & -3 \end{vmatrix}$$

$$= 9 - 21 = -12$$

$$Y = \begin{vmatrix} 1 & 4 & 1 \\ 2 & -1 & -2 \\ 1 & 1 & -1 \end{vmatrix} = \begin{vmatrix} 1 & 4 & 1 \\ 0 & -3 & 0 \\ 1 & 1 & -1 \end{vmatrix} = -3 \begin{vmatrix} 1 & 1 \\ 1 & -1 \end{vmatrix}$$

$$= -3(-1 - 1) = 6$$

$$Z = \begin{vmatrix} 1 & 1 & 4 \\ 2 & -1 & -1 \\ 1 & -2 & 1 \end{vmatrix} = \begin{vmatrix} 1 & 1 & 4 \\ 2 & -1 & -1 \\ 3 & -3 & 0 \end{vmatrix} = \begin{vmatrix} 1 & 2 & 4 \\ 2 & 1 & -1 \\ 3 & 0 & 0 \end{vmatrix}$$

$$= 3 \begin{vmatrix} 2 & 4 \\ 1 & -1 \end{vmatrix} = -18$$

Thus

$$x = \frac{X}{D} = \frac{-12}{-6} = 2$$

$$y = \frac{Y}{D} = \frac{6}{-6} = -1$$

$$z = \frac{Z}{D} = \frac{-18}{-6} = 3$$

The solution set is $\{(2, -1, 3)\}$.

Check

$$x + y + z = 2 - 1 + 3 = 4$$
$$2x - y - 2z = 2(2) - (-1) - 2(3) = 4 + 1 - 6 = -1$$
$$x - 2y - z = 2 - 2(-1) - 3 = 2 + 2 - 3 = 1$$

EXAMPLE 3 Solve by using determinants:

$$2x - 4y + 2z = 3$$
$$x + y - z = 2$$
$$3x - 6y + 3z = 2$$

Solution

$$D = \begin{vmatrix} 2 & -4 & 2 \\ 1 & 1 & -1 \\ 3 & -6 & 3 \end{vmatrix} = \begin{vmatrix} 4 & -2 & 2 \\ 0 & 0 & -1 \\ 6 & -3 & 3 \end{vmatrix} = -(-1) \begin{vmatrix} 4 & -2 \\ 6 & -3 \end{vmatrix}$$

$$= -12 + 12 = 0$$

$$X = \begin{vmatrix} 3 & -4 & 2 \\ 2 & 1 & -1 \\ 2 & -6 & 3 \end{vmatrix} = \begin{vmatrix} 7 & -2 & 2 \\ 0 & 0 & -1 \\ 8 & -3 & 3 \end{vmatrix} = -(-1) \begin{vmatrix} 7 & -2 \\ 8 & -3 \end{vmatrix}$$

$$= -21 + 16 = -5$$

Since $D = 0$ and $X \neq 0$, the solution set is the empty set, $\emptyset$.

EXAMPLE 4 Solve by using determinants, and check:

$$x + 2y - 3z = 4$$
$$2x - y + z = 1$$
$$3x + y - 2z = 5$$

Solution

$$D = \begin{vmatrix} 1 & 2 & -3 \\ 2 & -1 & 1 \\ 3 & 1 & -2 \end{vmatrix} = \begin{vmatrix} 1 & 0 & 0 \\ 2 & -5 & 7 \\ 3 & -5 & 7 \end{vmatrix} = 0$$

$$X = \begin{vmatrix} 4 & 2 & -3 \\ 1 & -1 & 1 \\ 5 & 1 & -2 \end{vmatrix} = \begin{vmatrix} 6 & 2 & -1 \\ 0 & -1 & 0 \\ 6 & 1 & -1 \end{vmatrix} = -1 \begin{vmatrix} 6 & -1 \\ 6 & -1 \end{vmatrix} = 0$$

$$Y = \begin{vmatrix} 1 & 4 & -3 \\ 2 & 1 & 1 \\ 3 & 5 & -2 \end{vmatrix} = \begin{vmatrix} 1 & 1 & -3 \\ 2 & 2 & 1 \\ 3 & 3 & -2 \end{vmatrix} = 0$$

$$Z = \begin{vmatrix} 1 & 2 & 4 \\ 2 & -1 & 1 \\ 3 & 1 & 5 \end{vmatrix} = \begin{vmatrix} 1 & 2 & 4 \\ 3 & 1 & 5 \\ 3 & 1 & 5 \end{vmatrix} = 0$$

Thus the solution set is infinite. Now try to solve two equations for which the determinant of the coefficients of two of the variables, say x and y, is not zero. Using the first and second equations,

$$x + 2y = 4 + 3z$$
$$2x - y = 1 - z$$

$$D = \begin{vmatrix} 1 & 2 \\ 2 & -1 \end{vmatrix} = -1 - 4 = -5$$

$$X = \begin{vmatrix} 4 + 3z & 2 \\ 1 - z & -1 \end{vmatrix} = -6 - z$$

$$Y = \begin{vmatrix} 1 & 4 + 3z \\ 2 & 1 - z \end{vmatrix} = -7 - 7z$$

Thus

$$x = \frac{X}{D} = \frac{-6 - z}{-5} = \frac{6 + z}{5} \quad \text{and} \quad y = \frac{Y}{D} = \frac{-7 - 7z}{-5} = \frac{7 + 7z}{5}$$

The solution set is

$$\left\{ \left(\frac{6 + z}{5}, \frac{7 + 7z}{5}, z \right) \,\middle|\, z \text{ is any real number} \right\}$$

Check

$$x + 2y - 3z = \frac{6 + z}{5} + 2\left(\frac{7 + 7z}{5} \right) - 3z$$

$$= \frac{6 + 14 + z + 14z - 15z}{5} = 4$$

$$2x - y + z = 2\left(\frac{6 + z}{5} \right) - \frac{7 + 7z}{5} + z$$

$$= \frac{12 - 7 + 2z - 7z + 5z}{5} = 1$$

$$3x + y - 2z = 3\left(\frac{6 + z}{5} \right) + \frac{7 + 7z}{5} - 2z$$

$$= \frac{18 + 7 + 3z + 7z - 10z}{5} = 5$$

EXERCISES 9.4

Solve each of the systems in Exercises 1–24 by using Cramer's rule, and check.

1. $3x + 5y = 2$
  $x + 2y = 0$

3. $x - 4y + 6 = 0$
  $-3x - 7y + 1 = 0$

2. $2x - 3y = 7$
  $-5x + 4y = -14$

4. $10x + 3y - 4 = 0$
  $7x + 2y - 1 = 0$

5. $4x - 3y = 0$
$2x - 3y = 6$

6. $3x + 8y = 14$
$x + 7y = 22$

7. $2x + 3y - 5 = 0$
$x - 2y + 8 = 0$

8. $6x - 2y = 9$
$9x - 8y = 1$

9. $2x + y + 3z = 15$
$2x + 7z = 25$
$3x + 2y + 6z = 35$

10. $x + 2y = 2 + z$
$x - 4y = 5z - 7$
$x + 3y + 4z = 5$

11. $x + 2y = z + 7$
$4x + 3y = 1 - 2z$
$9x + 8y = 4 - 3z$

12. $5x + 2y - z + 1 = 0$
$2x - y + z = 0$
$3x + z - 1 = 0$

13. $2x - 4y + 7z = 5$
$3x + 2y - z = 2$
$x - 10y + 15z = 8$

14. $x - 2y + 3 = 0$
$4x + z - 2 = 0$
$3x + 2y - 2z = 5$

15. $x - 2y - 3z = -20$
$2x + 4y - 5z = 11$
$3x + 7y - 4z = 33$

16. $4x + y - 3z = 0$
$x - y + z = -7$
$3x + 2y - z = -5$

17. $8x + 2y = 5$
$4y - 3z = 0$
$4x + 6z = -1$

18. $x + y = z$
$2x + 2z = 3 - 2y$
$5x + 2y = z + 2$

19. $x + 2y - 4 = 0$
$4y - z + 2 = 0$
$2x + z = 10$

20. $6x + 3y - 3z = 2$
$2x + 2y - 2z = 5$
$3x - 3y + 3z = 7$

21. $2x + y - z + t = 1$
$x - y + 2z - t = 2$
$-x - y + z - t = -1$
$3x + y - z + 2t = 0$

22. $x - 2z = 4$
$y + z + t = 6$
$x + 3t = 6$
$x + y + z + t = 0$

23. $x + y + z + t = 2$
$2x + 2y - z = -3$
$y - 2z - 2t = 1$
$x + y - t = 5$

24. $x + y + z = 14$
$x + z + t = 13$
$y + z + t = 10$
$x - y + z - t = 9$

9.5 QUADRATIC SYSTEMS IN TWO VARIABLES: SOLUTION BY SUBSTITUTION

DEFINITION

A quadratic system of two equations in the variables x and y is a system one of whose equations is quadratic in x and y, and the other equation is linear or quadratic in x and y.

The solution set of a quadratic system, like the solution set of a linear system in two variables, is the set of all ordered pairs that are common solutions to both equations in the system.

In this section and the next, we shall discuss *all* the solutions of the systems; some of these solutions may be real numbers and some imaginary numbers, so that the ordered pairs will have components that are complex numbers.

The **real solutions** of a system of equations in two variables can be determined graphically by graphing the equations on the same set of axes, as demonstrated for the graphic solution of a system of two linear equations. An ordered pair corresponding to a point of intersection is a solution of the system.

EXAMPLE 1   Graphically solve the system
$$x^2 + y^2 = 25$$
$$y = 3x - 5$$

Solution   This is a quadratic system; the first equation is quadratic and the second equation is linear. The graph of the quadratic equation, $x^2 + y^2 = 25$, is a **circle** with center at the origin and radius 5. The graph of the second equation is a line passing through the points $(0, -5)$ and $(2, 1)$. From Figure 9.4 we read the solutions $(3, 4)$ and $(0, -5)$.

Check

For $(3, 4)$,

$x^2 + y^2 = 25$	$y = 3x - 5$
$3^2 + 4^2 = 25$	$4 = 3(3) - 5$
$9 + 16 = 25$	$4 = 9 - 5$
$25 = 25$	$4 = 4$

For $(0, -5)$,

$x^2 + y^2 = 25$	$y = 3x - 5$
$0 + (-5)^2 = 25$	$-5 = 0 - 5$
$25 = 25$	$-5 = -5$

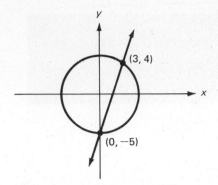

FIG. 9.4 $\{(x, y)|x^2 + y^2 = 25\} \cap \{(x, y)|y = 3x - 5\} = \{(3, 4), (0, -5)\}$

Another example of a quadratic system is shown in Example 2.

EXAMPLE 2 Graphically solve the system

$$x^2 + 16y^2 = 169$$
$$x^2 - y^2 = 16$$

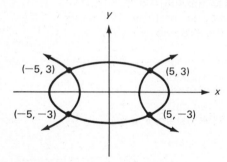

FIG. 9.5 $\{(x, y)|x^2 + 16y^2 = 169\} \cap$
$\{(x, y)|x^2 - y^2 = 16\} = \{(5, 3),$
$(-5, 3), (-5, -3), (5, -3)\}$

Solution   The equation $x^2 + 16y^2 = 169$ is equivalent to

$$\frac{x^2}{169} + \frac{y^2}{\frac{169}{16}} = 1$$

which is the equation of an ellipse.

$$x^2 - y^2 = 16$$

is the equation of a hyperbola,

$$\frac{x^2}{16} - \frac{y^2}{16} = 1.$$

Figure 9.5 is the graph of both equations and from the figure we approximate the solutions as $(-5, 3)$, $(-5, -3)$, $(5, -3)$, and $(5, 3)$. We must now check these proposed solutions. The check is illustrated for $(5, 3)$ but *all* solutions should be checked.

If $x = 5$ and $y = 3$, then

$$x^2 + 16y^2 = 5^2 + 16(3^2) = 25 + 144$$
$$= 169$$

and

$$x^2 - y^2 = 5^2 - 3^2 = 25 - 9 = 16$$

Thus $(5, 3)$ is a solution.

The graphic solution of a quadratic system yields the real solutions *only*. Moreover, due to the limitations imposed by eyesight and drawing techniques, these real solutions, as a general rule, can only be approximated. Thus it is essential to check a proposed solution obtained by the method of graphs.

The limitations of the graphic solution—only real solutions, and often extremely poor approximations—lead to the need for algebraic solutions for quadratic systems. One of these algebraic solution methods is **substitution.**

If one of the equations in a quadratic system can be solved for one variable, expressed explicitly as a function of the other variable, then the expression obtained may be *substituted* for the variable in the other equation. This replacement yields an equation in one variable. The solution is completed by solving this equation for the one variable, and then by using the function to obtain the corresponding values of the other variable.

EXAMPLE 3 (One linear and one quadratic equation) Solve and check:

$$x^2 + 4y^2 = 13, \qquad x + y = 2$$

Solution

1. Solve the linear equation for y: $y = 2 - x$
2. Replace y by $2 - x$ in the quadratic equation:
$$x^2 + 4(2 - x)^2 = 13$$
3. Solve for x:
$$x^2 + 4(4 - 4x + x^2) = 13$$
$$5x^2 - 16x + 3 = 0$$
$$(5x - 1)(x - 3) = 0$$
$$5x - 1 = 0 \quad \text{or} \quad x - 3 = 0$$
$$x = \frac{1}{5} \quad \text{or} \qquad x = 3$$
4. Solve for y, using the linear equation $y = 2 - x$:

If $x = \frac{1}{5}$, then $y = 2 - \frac{1}{5} = \frac{9}{5}$.

If $x = 3$, then $y = 2 - 3 = -1$.
5. Check each solution in both equations:

$\left(\frac{1}{5}, \frac{9}{5}\right)$: $x^2 + 4y^2 = \frac{1}{25} + 4\left(\frac{81}{25}\right)$

$$= \frac{325}{25} = 13$$

$$x + y = \frac{1}{5} + \frac{9}{5} = \frac{10}{5} = 2$$

$(3, -1)$: $x^2 + 4y^2 = 9 + 4(1) = 13$

$$x + y = 3 + (-1) = 2$$
6. State the solution set: $\left\{\left(\frac{1}{5}, \frac{9}{5}\right), (3, -1)\right\}$

EXAMPLE 4 (Two quadratic equations) Solve and check:
$$4x^2 + y^2 = 16, \qquad x^2 - y = 4$$

Solution

1. Solve $x^2 - y = 4$ for x^2: $x^2 = 4 + y$
2. Replace x^2 by $4 + y$ in the other equation:
$$4(4 + y) + y^2 = 16$$
3. Solve for y: $y^2 + 4y = 0$
$$y(y + 4) = 0$$
Thus $y = 0$ or $y = -4$.
4. Solve for x in the first equation used $(x^2 = 4 + y)$:
If $y = 0$, then $x^2 = 4 + 0 = 4$ and
$$x = 2 \quad \text{or} \quad x = -2.$$
If $y = -4$, then $x^2 = 4 - 4 = 0$.
Thus $x = 0$.
5. Check each solution in both equations:
$$(0, -4): \quad 4x^2 + y^2 = 4(0) + (-4)^2$$
$$= 0 + 16 = 16$$
$$x^2 - y = 0^2 - (-4) = 4$$
$$(2, 0): \quad 4x^2 + y^2 = 4(2^2) + 0^2$$
$$= 4(4) = 16$$
$$x^2 - y = 2^2 - 0 = 4$$
$$(-2, 0): \quad 4x^2 + y^2 = 4(-2)^2 + 0^2$$
$$= 4(4) = 16$$
$$x^2 - y = (-2)^2 - 0 = 4$$
6. State the solution set: $\{(0, -4), (2, 0), (-2, 0)\}$

EXAMPLE 5 (Two quadratic equations) Solve and check:
$$4x^2 - y^2 = 5, \qquad xy = 3$$

Solution

1. If $xy = 3$, then $y = \dfrac{3}{x}$ (for $x \neq 0$).

2. $\qquad 4x^2 - \left(\dfrac{3}{x}\right)^2 = 5$

$$4x^2 - \dfrac{9}{x^2} = 5$$
$$4x^4 - 9 = 5x^2$$
$$4x^4 - 5x^2 - 9 = 0$$
$$(4x^2 - 9)(x^2 + 1) = 0$$
$$4x^2 - 9 = 0 \quad \text{or} \quad x^2 + 1 = 0$$
$$x^2 = \dfrac{9}{4} \quad \text{or} \qquad x^2 = -1$$

Thus $x = \dfrac{3}{2}$, $x = -\dfrac{3}{2}$, $x = i$, or $x = -i$.

3. Using $y = \dfrac{3}{x}$ to find y,

  If $x = \dfrac{3}{2}$, then $y = \dfrac{3}{\frac{3}{2}} = 2$.

  If $x = -\dfrac{3}{2}$, then $y = -2$.

  If $x = i$, then $y = \dfrac{3}{i} = \dfrac{3i}{ii} = -3i$.

  If $x = -i$, then $y = 3i$.

4. Check each solution in both equations. (The check is left to the student.)

5. The solution set is

$$\left\{ \left(\dfrac{3}{2}, 2\right), \left(-\dfrac{3}{2}, -2\right), (i, -3i), (-i, 3i) \right\}.$$

EXERCISES 9.5

Graphically solve each system in Exercises 1–10. (Examples 1 and 2)

1. $x^2 + y^2 = 25$
   $x - y = 1$

2. $y = x^2 - 2x - 1$
   $x + y = 5$

3. $x^2 + y = 0$
   $x - y^2 = 0$

4. $y^2 + 4x + 2y = 8$
   $x + y - 2 = 0$

5. $4x^2 + 4y^2 = 169$
   $4y = x^2 - 26$

6. $y = x^2 + 4x + 4$
   $y^2 = 8x + 16$

7. $x^2 - 2x - y = 5$
   $x + y = 1$

8. $4x + y^2 = 36$
   $x^2 + y^2 = 81$

9. $x^2 - y^2 = 5$
   $3y^2 - x^2 = 3$

10. $x^2 + y^2 - 4x - 10y + 4 = 0$
    $3x + y = 6$

Solve each of the systems in Exercises 11–30 by the substitution method, and check. Leave irrational answers in simplified radical form and express imaginary answers in the a + bi form. (Examples 3, 4 and 5)

11. $x^2 + y^2 - 2x = 9$
    $x + 2y + 4 = 0$

12. $x^2 + 2x + 4 = y$
    $x = y - 16$

13. $x^2 - y^2 - 16 = 0$
    $3x + y = 8$

14. $4x^2 + y^2 = 25$
    $2x - y = 1$

15. $xy + 12 = 0$
    $2x + 3y = 6$

16. $y^2 = 4x$
    $x^2 + 3 = y^2$

17. $3x - y = 6$
    $y^2 + 3x - 8y = 0$

18. $xy = 4$
    $x + 3y = 10$

19. $2x^2 + y^2 = 12$
$\quad\quad x - 2y = 2$

20. $2x^2 - 2xy + y^2 = 10$
$\quad\quad\quad\quad 2x = y - 2$

21. $9x^2 - 4y^2 = 0$
$\quad\quad\quad y^2 = 4x + 1$

22. $xy + y = 1$
$\quad\quad xy - x = 4$

23. $\quad\quad\quad\quad y^2 - x = 9$
$\quad\quad y^2 + x - 2y + 5 = 0$

24. $\quad\quad\quad\quad y^2 - x^2 = 16$
$\quad\quad 2y + 2x^2 + 32 = 0$

25. $4a^2 + b^2 = 16$
$\quad\quad b + 4 = 2a^2$

26. $c^2 - 5d^2 = 1$
$\quad\quad\quad cd = 2$

27. $4x^2 + 4y^2 = 25$
$\quad\quad\quad 2y = 13 - 4x^2$

28. $4x = y^2$
$\quad\quad 3y = 7x - x^2$

29. $x^2 + y^2 = 100$
$\quad\quad x^2 = 4y + 40$

30. $x^2 - 4x + 9y = 18$
$\quad\quad\quad\quad x = y^2$

9.6 QUADRATIC SYSTEMS IN TWO VARIABLES: SOLUTION BY ADDITION

ELIMINATING ONE OF THE VARIABLES

If one equation of a quadratic system is linear, then the solution is easily obtained by the substitution method. However, when both equations are quadratic, the substitution method generally yields a complicated equation involving radicals, and upon simplification, a fourth-degree equation, which is not always easy to solve.

For certain special cases it is possible to eliminate one of the variables by the **addition method,** which is similar to that used for linear systems. Each equation of the system is multiplied by an appropriate constant, and the resulting equations are added. The procedure is justified by reasoning similar to that used for the linear systems.

EXAMPLE 1 Use the addition method to solve the system

$$3x^2 + 2y^2 = 23$$
$$x^2 - 3y^2 = -7$$

Solution

1. Multiply first equation by 3: $9x^2 + 6y^2 = 69$
   Multiply second equation by 2: $\underline{2x^2 - 6y^2 = -14}$
2. Add the new equations: $11x^2 = 55$

3. Solve for x: $x^2 = 5$. Thus $x = \pm\sqrt{5}$.
4. Multiply first equation by 1: $3x^2 + 2y^2 = 23$
   Multiply second equation by -3: $\underline{-3x^2 + 9y^2 = 21}$
5. Add the new equations: $11y^2 = 44$
6. Solve for y: $y^2 = 4$. Thus $y = \pm 2$.
7. List the solutions: $(\sqrt{5}, 2)$, $(\sqrt{5}, -2)$, $(-\sqrt{5}, 2)$, $(-\sqrt{5}, -2)$
8. Check each solution in both equations of the original system. (The check is left for the student.)

ELIMINATING THE SECOND-DEGREE TERMS

When it is not possible to eliminate one of the variables by adding multiples of the two equations, it may be possible to eliminate the second-degree terms and obtain a linear equation. Since each solution of the quadratic system must also be a solution of the linear equation, the linear equation may be solved simultaneously with either equation of the quadratic system.

EXAMPLE 2   Solve the system
$$x^2 + y^2 - 2x = 9$$
$$x^2 + y^2 - 4y = 1$$

Solution

1. Eliminate the second-degree terms by multiplying the first equation by -1.
$$-x^2 - y^2 + 2x \qquad = -9$$
$$\underline{x^2 + y^2 \qquad - 4y = 1}$$
2. Adding, $2x - 4y = -8$
   Divide by 2, $x - 2y = -4$
3. Now solve the equivalent system:
$$x^2 + y^2 - 4y = 1$$
$$x - 2y = -4$$
Using the substitution method,
$$x = 2y - 4$$
$$(2y - 4)^2 + y^2 - 4y = 1$$
$$5y^2 - 20y + 15 = 0$$
$$y^2 - 4y + 3 = 0$$
$$(y - 3)(y - 1) = 0$$
Thus $y = 3$ or $y = 1$.
If $y = 3$, then $x = 2y - 4 = 2(3) - 4 = 2$.
If $y = 1$, then $x = 2y - 4 = 2(1) - 4 = -2$.
Thus the solutions are $(2, 3)$, $(-2, 1)$.
The solution set is $\{(2, 3), (-2, 1)\}$.

ELIMINATING THE CONSTANTS

Sometimes the solutions of a quadratic system may be easily obtained by using the addition method to eliminate the constants.

EXAMPLE 3 Solve the system

$$x^2 + xy = 4$$
$$y^2 - xy = 6$$

Solution   To eliminate the constants, multiply the first equation by 3 and the second equation by -2.

$$3x^2 + 3xy \qquad = \quad 12$$
$$\underline{\qquad 2xy - 2y^2 = -12}$$

Adding, $\qquad 3x^2 + 5xy - 2y^2 = \quad 0$

Factoring, $\qquad (3x - y)(x + 2y) = \quad 0$

By the zero-product theorem,

$$3x - y = 0 \quad \text{or} \quad x + 2y = 0$$
$$y = 3x \quad \text{or} \qquad x = -2y$$

Now either equation of the *original system* can be replaced by this set of two linear equations, Thus the original system is equivalent to the *union* of the two systems

$$x^2 + xy = 4 \qquad \quad x^2 + xy = 4$$
$$\quad y = 3x \qquad \qquad \quad x = -2y$$

Each of these systems can now be solved by the substitution method:

$$x^2 + x(3x) = 4 \qquad (-2y)^2 + (-2y)y = 4$$
$$4x^2 = 4 \qquad \qquad \qquad 2y^2 = 4$$
$$x^2 = 1 \qquad \qquad \qquad \quad y^2 = 2$$
$$x = \pm 1 \qquad \qquad \qquad \quad y = \pm\sqrt{2}$$

If $x = 1$, then $y = 3x = 3$. $\quad$ If $y = \sqrt{2}$, $x = -2y = -2\sqrt{2}$.

If $x = -1$, then $y = -3$. $\qquad$ If $y = -\sqrt{2}$, then $x = 2\sqrt{2}$.

Thus the solution set is $\{(1, 3), (-1, -3), (-2\sqrt{2}, \sqrt{2}), (2\sqrt{2}, -\sqrt{2})\}$. Each of these solutions should be checked in both equations of the original system.

EXERCISES 9.6

Solve Exercises 1–20 by the addition method using either elimination of one of the variables, elimination of the second-degree terms, or elimination of the constant. Check the solutions.

1. $x^2 + y^2 = 20$
$\quad y^2 - x^2 = 12$

2. $x^2 + 3y^2 = 31$
$\quad x^2 \qquad = 8y^2 + 9$

3. $4x^2 + y^2 = 24$
$\quad 2x^2 + y \; = 12$

4. $6x^2 - 5y^2 + 21 = 0$
$\quad 7x^2 - 4y^2 + \; 8 = 0$

5. $2x^2 + y^2 = 13$
   $3x^2 + y^2 = 17$

6. $x^2 - y^2 + 7 = 0$
   $3x^2 + 2y^2 = 24$

7. $x^2 - 5y^2 + 3 = 0$
   $2x^2 + y^2 - 5 = 0$

8. $3x^2 - 2y^2 - 6 = 0$
   $5x^2 - 3y^2 - 7 = 0$

9. $3xy + y^2 = 28$
   $4x^2 + xy = 8$

10. $xy - 4x - 4y = 0$
    $xy = -8$

11. $x^2 + y^2 - 6x - 2y = 6$
    $x^2 + y^2 - 2x - 2y = 10$

12. $x^2 + 5xy + 2x = 8$
    $2x^2 + 3xy + 4x = 9$

13. $xy = 48$
    $xy - 10x + 5y = 98$

14. $x^2 + xy = 24$
    $xy - y^2 = 4$

15. $y^2 + xy = 12$
    $2x^2 - xy = 24$

16. $3xy + y^2 = 10$
    $xy = 3$

17. $x^2 - y^2 + 2x - y = -9$
    $x^2 - y^2 + 4x - 2y = -9$

18. $4x^2 - xy = 4$
    $3y^2 + xy = 3$

19. $xy = 120$
    $xy + x - 4y = 124$

20. $xy - x = 3$
    $xy + y = 8$

9.7 APPLICATIONS

There are many applications of linear systems involving three or more variables. Some problems require the solution of 100 linear equations in 100 variables! Fortunately, a computer may be programmed to perform the calculations for the solutions to these systems. Linear systems are used to solve problems in areas such as physics and chemistry; for example, in quantum mechanics and atomic structure, in biology related to genetics and heredity, in sociology and psychology where data analysis is required, in economics and industry pertaining to cost analysis, in management, in military tactics using the strategy of game theory, and so on.

Quadratic systems also have many practical applications. Arches and dams are often parabolic, elliptical, or hyperbolic. Orbits of satellites and planets are elliptical, while the orbits of comets and meteors may be parabolic, elliptical, or hyperbolic.

A parabolic surface is obtained by rotating a parabola about its axis. Radar screens, searchlights, and headlights have parabolic reflectors. Other types of parabolic reflectors are used to concentrate rays of heat or light from a distant source; for example, rays of heat from the sun have been focused in this manner and then have been used to drive machinery.

An elliptical surface is obtained by rotating an ellipse about its major axis. If a lamp is placed at one focus, rays of light striking the surface will be reflected toward the other focus. "Whispering galleries" are rooms with domed ceilings in the form of a semiellipse. A person at one focus of the ellipse can be heard distinctly by a person at the other focus, although he cannot be heard by people in between.

A hyperbolic surface is obtained by rotating one branch of a hyperbola about its transverse axis. If rays of light are directed toward one focus, the rays will be reflected toward the other focus. A Cassegrainian telescope is made by using both a hyperbolic mirror and a parabolic mirror.

Hyperbolas are used to determine positions of airplanes and to determine enemy positions by using properties of sound and the property that a hyperbola is the set of points such that the difference of the distances from any point on the hyperbola to the foci is a constant. The system of air navigation using this property is called *loran*.

The equation of the path of a projectile shot from the origin with initial velocity v ft/sec directed along the line $y = mx$ is the parabola (Figure 9.6)

$$y = \frac{-16(m^2 + 1)}{v^2} x^2 + mx$$

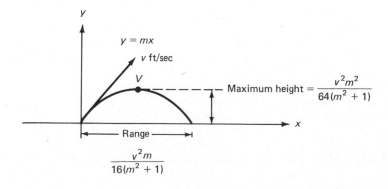

FIG. 9.6

EXERCISES 9.7

Exercises 1–12 may be solved by using linear systems.

1. There are 47 coins in a collection of nickels, dimes, and quarters. The total value of the collection is $5. The total number of nickels and quarters is 3 less than the number of dimes. How many coins of each kind are there?

2. Three machines, working together, require 20 min to complete a certain job. If the first two machines require 30 min to complete the job when working together, and if the first and the third require 36 min, find the time it would take for each machine alone.

3. Tickets to a certain campus show are 50 cents for students, 75 cents for faculty, and $1 for the general public. The total sales from 550 people amounted to $337.50. If there are 4 times as many student tickets sold as general public tickets, how many tickets of each type were sold?

4. Two men A and B when working together can do a certain job in 4 days. If B and a young boy C work together 6 days, then B can complete the job alone in 2 more days. If A and C work together, they can complete the job in 6 days. How long would it take for each person working alone to complete the job?

5. Three machines A, B, and C each are able to produce three different products P, Q, and R. The following table indicates the number of hours required by each machine to make each product. If machine A can operate 14 hr a day, B, 16 hr, and C, 11 hr,

	Product		
Machine	P	Q	R
A	2	1	3
B	4	2	2
C	1	3	2

find the number of items of each product the factory is able to produce when all three machines work to full capacity.

6. Find the amounts of the following three foods that will provide precisely the daily minimum vitamin and mineral requirements

	Food			
Vitamin	I	II	III	Daily Minimum Requirement
Thiamine	0.2	0.2	0.5	2.00
Niacin	90	30	30	450.00
Iron	1.5	2	1.5	11.0

indicated in the last column of the array. The content of each food is given in milligrams per ounce.

7. Applying Kirchhoff's laws to the electrical circuit in Figure 9.7, the following equations are obtained.

$$I_1 - I_2 - I_3 = 0$$
$$5I_1 \qquad + 20I_3 = 100$$
$$15I_2 - 20I_3 = 15$$
$$5I_1 + 15I_2 \qquad = 115$$

Solve this system by showing that the common solution of the first three equations is also a solution of the fourth equation.

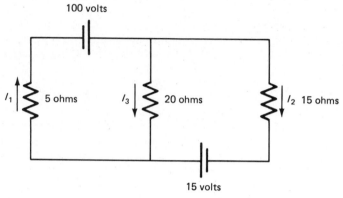

FIG. 9.7

8. Applying Kirchhoff's laws to the electrical circuit in Figure 9.8, the following equations are obtained. Solve for I_1, I_2, and I_3.

$$I_1 - I_2 - I_3 = 0$$
$$2I_1 + 6I_2 \qquad = 22$$
$$6I_2 - 4I_3 = 0$$
$$2I_1 \qquad + 4I_3 = 22$$

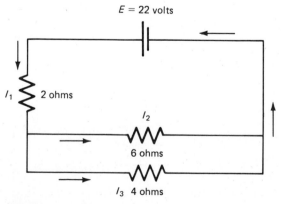

FIG. 9.8

9. A person went to a bus station in a taxicab averaging 25 mph and took a bus averaging 40 mph to the airport. At the airport she boarded a plane which flew her to her destination. The plane averaged 600 mph. The *entire* trip of 1469 mi required 3 hr and 12 min. The plane trip took 3 times as long as the other two trips combined. How much time was spent in each type of travel?

10. The copper content of coins in a certain country is 95% for the 1-unit coin (A), 75% for the 5-unit coin (B), and 10% for the 10-unit coin (C). For a total of 82 lb of coins, all of the B coins and one-third of the C coins were made from 50 lb of ore containing 50% copper. The rest of the C coins and all of the A coins were made from 50 lb of ore containing 42% copper. How many pounds of each type of coin were there?

11. Find all possible amounts of the following four foods that will provide precisely the amounts of nutrients indicated in the last column of the array if each contains the amounts per unit as indicated.

| Nutrient | Food | | | | |
	I	*II*	*III*	*IV*	Total
A	1	3	3	0	12
B	2	2	0	3	26
C	3	1	5	4	44
D	3	9	1	2	32

If food *I* costs 40 cents per unit, food *II* 40 cents per unit, food *III* 10 cents per unit, and food *IV* 20 cents per unit, is there a solution costing exactly $1? exactly $3?

12. A factory has raw materials of types A, B, and C to produce products P, Q, and R. There are on hand 2400 lb of A, 310 lb of B, and 28 lb of C. Product P requires 25 lb of A and 5 lb of B. Product Q requires 20 lb of A, 2 lb of B, and 1 lb of C. Product R requires 150 lb of A, 10 lb of B, and $\frac{1}{2}$ lb of C. How many items of each product should be produced in order to use all of the raw material?

Exercises 13–20 may be solved by using quadratic systems.

13. A rectangular piece of metal has an area of 200 sq in. A square whose side is 3 in. is cut from each of the four corners. The flaps are then turned up to form an open box whose volume is 168 cu in. Find the original dimensions of the sheet of metal.

14. A rectangular picture has a frame of uniform width and its area is equal to the area of the picture. If the width of the frame is 3 in.

and the area of the picture is 240 sq in., find the dimensions of the picture.

15. A roadway 400 ft long is supported by a parabolic cable. The cable is 100 ft above the roadway at the ends and 4 ft above at the center. Find the lengths of the vertical supporting cables at 50-ft intervals along the roadway (Figure 9.9). (*Hint:* Let $y = kx^2 + 4$. Find k and then find y for $x = 50, 100, 150$.)

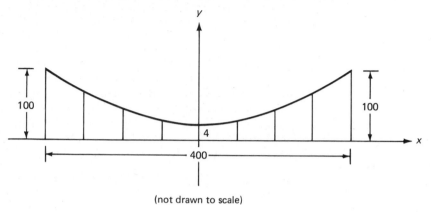

(not drawn to scale)

FIG. 9.9

16. When two resistances x and y are connected in series, they have a joint resistance of 25 ohms—that is,

$x + y = 25$

When the same two resistances are connected in parallel, they have a joint resistance of 6 ohms—that is,

$$\frac{1}{x} + \frac{1}{y} = \frac{1}{6}$$

Find the number of ohms in each resistance.

17. Determine the values of k so that the line $y = 2x + k$ will be tangent to the circle $x^2 + y^2 = 20$. (A tangent line intersects the circle in exactly one point.) (*Hint:* Use the substitution method to eliminate y. The roots of the resulting quadratic equation must be equal, and thus its discriminant must equal 0.)

18. The total cost y in dollars of producing x units of a commodity is given by

$y = x^2 + 2x$

The total revenue in dollars received from the sale of x units of the commodity is given by

$y = 20x - 2x^2$

(a) Graph each of the equations on the same set of axes. Note that $x \geq 0$ and $y \geq 0$.

(b) Determine the values of x for which the revenue is greater than the cost.

(c) The profit equation, where profit = revenue − cost, is given by

$$y = (20x - 2x^2) - (x^2 + 2x)$$

Graph this equation and determine the value of x for which the profit is maximum.

19. A *demand law* is an equation relating the price y and the number of units x of a commodity demanded by consumers. A *supply law* is an equation relating the price y and the number of units x of the commodity that the manufacturer can supply. A point of intersection of a demand curve and a supply curve corresponds to "market equilibrium." The value of x and the value of y are called the equilibrium quantity and equilibrium price, respectively. Find the equilibrium quantity and the equilibrium price for the following demand and supply laws:

Demand law: $xy = 45$	$(x \geq 0, y \geq 0)$
Supply law: $4x = y^2 + 3y - 4$	$(x \geq 0, y \geq 0)$

20. A square park is surrounded by a sidewalk of uniform width. The sidewalk is surrounded by a paved road whose width is 8 times that of the sidewalk. If the area of the sidewalk is 4500 sq ft and the area of the road is 43,200 sq ft, find the length of a side of the park.

DIAGNOSTIC TEST

Select the best answer for each of the following problems. If the answer is "none of these," write a correct answer.

1. If the system of equations

   $$a_1 x + b_1 y = c_1$$
   $$a_2 x + b_2 y = c_2$$

   has no solutions, then the graph of this system consists of
   (a) two coincident lines
   (b) two parallel lines
   (c) two lines intersecting at one point
   (d) all of these
   (e) none of these

2. If the solution set for a system of equations is
   $\left\{ \left(x, \dfrac{x-3}{2}, \dfrac{x+3}{4} \right) \;\middle|\; x \text{ any real number} \right\}$, then
   (a) the system has exactly one real solution
   (b) the system has no real solutions
   (c) (5, 1, 2) is a solution
   (d) graphically, the system may be described as three parallel planes
   (e) none of these

3. The solution set of the system
$$2x + 3y - z = 12$$
$$3x + 5y - 3z = 6$$
$$x - 2y + 2z = 18$$
   is
   (a) $\varnothing$ (d) $\{(6, 3, 9)\}$
   (b) infinite (e) none of these
   (c) $\{(4, 2, 2)\}$

4. The solution set of the system
$$x + 2y - 2z = 0$$
$$x + 12y - 10z = 0$$
$$3x - 4y + 2z = 0$$
   is

   (a) $\left\{ \left(x, 2x, \dfrac{5x}{2} \right) \,\middle|\, x \text{ any real number} \right\}$

   (b) $\left\{ \left(\dfrac{y}{2}, y, \dfrac{5y}{4} \right) \,\middle|\, y \text{ any real number} \right\}$

   (c) $\left\{ \left(\dfrac{2z}{5}, \dfrac{4z}{5}, z \right) \,\middle|\, z \text{ any real number} \right\}$

   (d) all of these
   (e) none of these

5. A determinant is
   (a) a matrix
   (b) a square matrix
   (c) a function defined in terms of a square matrix
   (d) all of these
   (e) none of these

6. The determinant $\begin{vmatrix} 1 & 2 & 3 \\ 4 & -3 & 1 \\ 3 & -1 & 2 \end{vmatrix}$ equals

   (a) 0 (d) -30
   (b) 25 (e) none of these
   (c) 56

7. If
$$A = \begin{vmatrix} 5 & 7 & 1 \\ -2 & 4 & -3 \\ 1 & -1 & 5 \end{vmatrix} \quad \text{and} \quad B = \begin{vmatrix} 2 & -2 & 10 \\ -4 & 8 & -6 \\ 10 & 14 & 2 \end{vmatrix}$$
   then
   (a) $A = 2B$ (d) $B = -2A$
   (b) $B = 2A$ (e) none of these
   (c) $A = -2B$

8. If a system consists of two equations, one of which graphs as an ellipse and the other as a straight line, the number of possible real solutions is
   (a) 0 (d) all of these
   (b) 1 (e) more than 2
   (c) 2

9. The solution set for the system $x^2 - 4y^2 = 5$ and $xy = 3$ is

(a) $\left\{(3, 1), \left(2i, -\dfrac{3}{2}i\right)\right\}$

(b) $\{(3, 1), (-3, -1)\}$

(c) $\left\{(3, 1), (-3, -1), \left(2i, \dfrac{3}{2}i\right), \left(-2i, -\dfrac{3}{2}i\right)\right\}$

(d) $\left\{(3, 1), (-3, -1), \left(2i, -\dfrac{3}{2}i\right), \left(-2i, \dfrac{3}{2}i\right)\right\}$

(e) none of these

10. The system

$$x^2 + y^2 + 4x = 1$$
$$x^2 + y^2 + 2y = 9$$

has

(a) no solutions   (d) three solutions

(b) one solution   (e) four solutions

(c) two solutions

Problems 11 and 12 refer to the system in Problem 10. Correctly complete the problems by filling in the blanks.

11. The solution set for the system in Problem 10 is _____.

12. The graph of the system in Problem 10 consists of _____.

Problems 13, 14, and 15 refer to the following problem: In using Cramer's rule to solve the system of equations

$$x + y - z = 2$$
$$2x - 4y + 2z = 3$$
$$3x - 2y + z = 1$$

13. The determinant of coefficients, D, is written _____ and equals _____.

14. The determinant X used to solve for x is written _____ and equals _____.

15. Using the above information, we find that $x =$ _____, and that the solution set of the system is _____.

REVIEW EXERCISES

Solve each of the linear systems in Exercises 1–10 without using determinants, and check. (Sections 9.1 and 9.2)

1. $3x - 2y = 4$
   $2x - y = 0$

2. $x = 4y - 3$
   $y = 3x + 9$

3. $5x + 2y - 1 = 0$
$\quad\quad 4y + 10x = 2$

4. $2x - 3 = 8y$
$\quad\quad 5y - 7 = x$

5. $x - y + 2z = 3$
$\quad 2x + 3y + 2z = 10$
$\quad x + 2y - 6z = -15$

6. $3x - y + 2z = 10$
$\quad 6x + 3y - 8z = 21$
$\quad 2x + y - 3z = 8$

7. $x + y + z = 0$
$\quad 2x + 5y + 3z = 3$
$\quad 3x - y + 2z = -5$

8. $5x + y + 2z - 0$
$\quad 3x - 2y + 6z = 0$
$\quad 2x + 3y - 4z = 0$

9. $x + y + z = 1$
$\quad x + 2y - 3z = -9$
$\quad -3x - 4y + 2z = 10$

10. $-x + y + 3z + 2t = -2$
$\quad x + 3y - 2z - t = 9$
$\quad 4x + y + z + 2t = 2$
$\quad 3x + y - z + t = 5$

11. Find an equal determinant having zeros everywhere in column 1 except in row 1. Then expand the determinant. (Section 9.3)

$$\begin{vmatrix} 1 & 3 & 2 & -2 \\ 2 & 3 & -1 & 1 \\ -3 & 2 & 1 & 6 \\ -1 & 4 & 5 & 2 \end{vmatrix}$$

Solve Exercises 12–15 by using Cramer's rule. (Section 9.4)

12. $2x - 4y - 2z = 5$
$\quad 3x + 2y - z = 3$
$\quad 4x - 3y - 2z = 1$

13. $x + y = 8$
$\quad y + z = 1$
$\quad x + z = 0$

14. $3x + 2y + 3z = 4$
$\quad 6x - 2y + 9z = 6$
$\quad 3x - 4y + 3z = 1$

15. $5x = 8 - 3z$
$\quad 3x = 10 + 4y$
$\quad x = 14 - 2y + 6z$

Solve Exercises 16–20 graphically and algebraically. (Sections 9.5 and 9.6)

16. $x^2 + y^2 = 25$
$\quad\quad y^2 = x + 5$

17. $4y^2 - 4x^2 = 25$
$\quad\quad x^2 + 4y = 2$

18. $x^2 + 4y^2 = 36$
$\quad\quad x^2 - y^2 = 1$

19. $4x^2 + y^2 = 100$
$\quad\quad xy = 24$

20. $x^2 - 4y^2 = 9$
$\quad\quad 2y^2 = 3 - x$

Solve Exercises 21–27 algebraically. (Sections 9.5 and 9.6)

21. $4x^2 + y^2 = 25$
$\quad\quad x^2 - y = 1$

22. $y^2 - x^2 = xy + 1$
$\quad\quad x^2 - y^2 = xy - 5$

23. $x^2 - 3xy - y^2 = 9$
$\quad 2x^2 - 4xy - 7y^2 = 9$

24. $12x^2 - y^2 = 12$
$\quad\quad 2y = x^2 + 8$

25. $x^2 + y^2 + 4y = 6$
$\quad x^2 + y^2 + 2x = 4$

26. $4x^2 - 3y^2 = 1$
$\quad\quad 2x^2 + y = 3$

27. $xy = 480$
$\quad xy + 16x - 5y = 560$

28. The amounts per unit and the total amounts on hand of three different chemicals needed to manufacture three different products are given in the following table. How many units of each product should be made if all of the chemicals are used? (Section 9.7)

	Product			
Chemical	P	Q	R	Total
A	3	4	1	330
B	5	0	2	330
C	1	2	3	240

10

THE BINOMIAL THEOREM

Before leaving the study of intermediate algebra, there is another theorem that is a useful tool in mathematics and the applications of mathematics. This theorem is the **binomial theorem.** We have learned how to expand $(a + b)^2$ and $(a + b)^3$, and the binomial theorem provides a relatively simple method for expanding the binomial $a + b$ to any power n, where n is a natural number.

In order to understand the theorem, which will be stated presently, let us look at some binomial expansions.

The expansions of the following powers of the binomial $(a + b)^n$ may be obtained by performing the indicated multiplications:

$$(a + b)^1 = a + b$$
$$(a + b)^2 = a^2 + 2ab + b^2$$
$$(a + b)^3 = a^3 + 3a^2b + 3ab^2 + b^3$$
$$(a + b)^4 = a^4 + 4a^3b + 6a^2b^2 + 4ab^3 + b^4$$
$$(a + b)^5 = a^5 + 5a^4b + 10a^3b^2 + 10a^2b^3 + 5ab^4 + b^5$$

By examining these special cases, the following properties may be observed for $n = 1, 2, 3, 4,$ or 5.

THE BINOMIAL THEOREM

1. Each expansion has $(n + 1)$ terms.
2. The first term is a^n and the last term is b^n.
3. The coefficient of the second term and the next to the last term is n.
4. The exponent of a is one less than that of the preceding term.
5. The exponent of b is one more than that of the preceding term.
6. The sum of the exponents of a and b in each term is n.
7. The coefficient of any term after the first can be obtained by multiplying the coefficient of the preceding term by its exponent of a and dividing by the number of this preceding term.

Assuming that these properties are valid for any natural number n, an expansion can be written for $(a + b)^n$. This statement is called the *binomial theorem*.

THE BINOMIAL THEOREM

Let n be any natural number. Then

$$(a + b)^n = a^n + na^{n-1}b + \frac{n(n - 1)}{1(2)} a^{n-2}b^2$$
$$+ \frac{n(n - 1)(n - 2)}{1(2)(3)} a^{n-3}b^3 + \cdots$$
$$+ \frac{n(n - 1)(n - 2) \cdots (n - r + 1)}{1(2)(3) \cdots (r)} a^{n-r}b^r + \cdots + b^n$$

EXAMPLE 1 Expand and simplify $(x - 2)^6$.

Solution $a = x$, $b = -2$, and $n = 6$. Substituting in $(a + b)^6$,

$$(x - 2)^6 = x^6 + 6x^5(-2) + \frac{6(5)}{1(2)} x^4(-2)^2 + \frac{6(5)(4)}{1(2)(3)} x^3(-2)^3$$
$$+ \frac{6(5)(4)(3)}{1(2)(3)(4)} x^2(-2)^4 + \frac{6(5)(4)(3)(2)}{1(2)(3)(4)(5)} x(-2)^5 + (-2)^6$$

Simplifying,

$$(x - 2)^6 = x^6 - 12x^5 + 60x^4 - 160x^3 + 240x^2 - 192x + 64$$

EXAMPLE 2 Find the 9th term of $(5y + 0.2)^{15}$.

Solution

$$(r + 1)\text{st term} = \frac{n(n - 1) \cdots (n - r + 1)}{1(2)(3) \cdots (r)} a^{n-r}b^r$$

Since $r + 1 = 9$, then $r = 8$. Also $a = 5y$, $b = 0.2$, $n = 15$, and $n - r + 1 = 15 - 8 + 1 = 8$.

Substituting,

$$9\text{th term} = \frac{15(14)(13)(12)(11)(10)(9)(8)}{1(2)(3)(4)(5)(6)(7)(8)} (5y)^{15-8}(0.2)^8$$
$$= 1287 \, y^7$$

PASCAL'S TRIANGLE

By including $(a + b)^0 = 1$, the coefficients of the terms of the bionomial expansion form an interesting pattern, known as **Pascal's triangle:**

$(a + b)^0$							1						
$(a + b)^1$						1		1					
$(a + b)^2$					1		2		1				
$(a + b)^3$				1		3		3		1			
$(a + b)^4$			1		4		6		4		1		
$(a + b)^5$		1		5		10		10		5		1	
$(a + b)^6$	1		6		15		20		15		6		1

Each of the numbers different from 1 may be obtained by adding the two numbers to its left and right in the row immediately above. Thus $10 = 4 + 6$ and $15 = 5 + 10$, and so on.

HISTORICAL NOTE

The French mathematician Blaise Pascal (1623–1662) discovered many properties of this triangular array which appear in his *Traité de triangle arithmétique* written in 1653. In this work is to be found the binomial theorem for positive integral exponents and one of the first acceptable uses of the method of mathematical induction.

Although the "arithmetical triangle" is named after Pascal because of his work on it, Pascal did not discover this array of numbers. It appears in one of the works of the great Chinese algebraist Chu Shï-kié written around 1303. The array appeared in European publications more than 100 years before Pascal's time.

The great English mathematician Sir Isaac Newton (1642–1727) generalized the binomial theorem for rational values of the exponent. His work appears in letters he wrote in 1676.

EXAMPLE 3 Find the third term of $(2x + 3y)^6$.

Solution By using a combination of the binomial theorem and Pascal's triangle, we obtain the following:

The third term of $(a + b)^6 = 15a^4b^2$.

For $a = 2x$ and $b = 3y$,

$$a^4 = (2x)^4 = 16x^4$$
$$b^2 = (3y)^2 = 9y^2$$

Therefore

$$15a^4b^2 = 15(16x^4)(9y^2) = 2160x^4y^2$$

The student should verify this result by the

$$(r + 1)\text{st term} = \frac{n(n - 1) \cdots (n - r + 1)}{1(2)(3) \cdots (r)} a^{n-r}b^r$$

with $r + 1 = 3$, $n = 6$, $a = 2x$, and $b = 3y$.

EXERCISES

Expand Exercises 1–16 by using the binomial theorem and simplify.

1. $(x + 1)^3$
2. $(x + 1)^4$
3. $(x - 1)^3$
4. $(x - 1)^4$
5. $(x + 2)^5$
6. $(y - 2)^5$
7. $(x + 2)^7$
8. $(y - 1)^8$

9. $(b - 1)^6$
10. $(x + y)^5$
11. $(x^2 + 3)^3$
12. $(1 - y^2)^5$
13. $\left(x + \dfrac{1}{2}\right)^4$
14. $(a^2 - b^2)^3$
15. $(2x + 5)^3$
16. $(3x - 2)^4$

Write the first four terms of the expansions of each of Exercises 17–20.

17. $(x - y)^{34}$
18. $(t + 1)^{100}$

19. $(y^2 - \sqrt{2})^{15}$
20. $\left(\dfrac{2x + 5}{10}\right)^5$

Express in simplified form the specified term in the expansion of each of Exercises 21–28.

21. 4th term of $(y + 4)^{12}$
22. 6th term of $\left(x^2 - \dfrac{1}{2}\right)^{20}$
23. 4th term of $\left(r^3 + \dfrac{1}{s^3}\right)^9$
24. 8th term of $\left(\dfrac{2}{x} - \dfrac{y}{2}\right)^{15}$

25. 5th term of $(x - 3)^{10}$
26. 4th term of $\left(\dfrac{x}{2} + 1\right)^{16}$
27. 6th term of $(1 - 0.02y)^8$
28. 7th term of $\left(x + \dfrac{1}{x}\right)^{12}$

29. Approximate $(1.02)^{12}$ by finding the sum of the first four terms of the expansion of $(1 + 0.02)^{12}$.
30. Approximate $(0.98)^8$ correct to the nearest hundredth. Use $(1 - 0.02)^8$.
31. Approximate $(0.97)^{10}$ by using the first four terms of $(1 - 0.03)^{10}$.
32. Approximate $(1.01)^{10}$ correct to the nearest thousandth.
33. Evaluate $(1 - i)^8$ where $i^2 = -1$.
34. Evaluate
$$\left(\dfrac{1}{2} + \dfrac{\sqrt{3}}{2}\, i\right)^9$$
where $i^2 = -1$.

ANSWERS

EXERCISES 1.1, pp. 8

1. {9}

3. ∅

5. {5, 6, 7, . . .}

7. {0, 1}

9. {10, 12, 14}

11. {3, 5, 7}

13. ∅

15. {23, 29}

17. {2}

19. ∅

21. $\{p \mid p$ is a prime number$\}$

23. $\left\{ \dfrac{-1}{n} \mid n \text{ is a natural number} \right\}$

25. $\{x \mid x$ is an element of $S\}$

27. $\left\{ \dfrac{1}{m} \mid m \text{ is an even natural number} \right\}$

29. $\{x \mid x$ is a letter of the English alphabet$\}$

31. {3}

33. {−7, 0, 3}

35. $\left\{ -\sqrt{11}, \sqrt[3]{5}, \dfrac{\sqrt{10}}{4} \right\}$

37. 4

39. $\dfrac{2}{3}$

41. 2

43. 17

45. 0

47. -4

49. $\dfrac{3}{5}$

51. $-\sqrt{3}$

53. 0

55. 5

57. True

59. True

61. False; the sets are mutually exclusive

EXERCISES 1.2, pp. 14

1. $23 > 18$

3. $-3 < 2$

5. $-4 > -6$

7. $4 > 0$

9. $|3| = 3, |-4| = 4; 3 < 4$

11. $|-6| = |6|$

13. $0 < |-3|$

15. $-|-3| < 0$

17.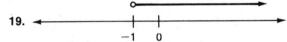

19.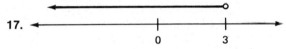

21.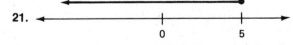

23.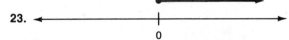

25.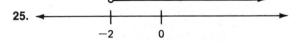

27.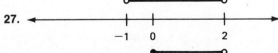

29.

31. Transitive, (c)

33. Reflexive, (a)

35. Substitution, (d)

37. Symmetric, (b)

39. Transitive, (c)

EXERCISES 1.3, pp. 20

1. Commutative axiom, addition, (b)
3. Additive inverse, (f)
5. Distributive axiom, (d)
7. Commutative axiom, addition (b)
9. Negative of a negative, $-(-5) = 5$; and additive inverse, (h) and (f)
11. Additive inverse, $-2 + 2 = 0$; and multiplication by 0, (f) and (g)
13. Multiplicative inverse, (*Note:* $x \neq 0$), (f)
15. Commutative axiom, addition; and closure, (b) and (a)
17. Commutative axiom, addition, (b)
19. Closure, multiplication, (a)
21. Negative of a negative; multiplication by 0, (h) and (g)
23. Associative axiom, multiplication, (c)
25. Distributive axiom, (d)
27. Commutative axiom, addition and multiplication, (b)
29. Associative and commutative axioms, addition, (c) and (b)
31. $\quad x + (3 + (-3))$ Associative axiom, addition
$= x + 0$ Additive inverse
$= x$ Zero is the additive identity
33. $(-7)\left(-\dfrac{1}{7}\right) = 1$ Multiplicative inverse (reciprocal)
35. $\quad 57(3 + 2)$ Distributive axiom
$= 57(5) = 285$
37. $\quad (64 + 75) + (25 + 36)$
$= (64 + 36) + (75 + 25)$ Associative and commutative axioms, addition
$= 100 + 100 = 200$
39. $14n$; associative and distributive axioms
41. $11y$ 47. Unlike terms; cannot be simplified
43. $x^2 + 5x + 5$ 49. $7x^2y + 2$
45. $3x^2y + xy^2$

EXERCISES 1.4, pp. 27

1. -30	21. 4	41. -9
3. 5	23. 10	43. -10
5. 0	25. -10	45. 59
7. -7	27. 0	47. 46
9. -20	29. -14	49. 50
11. -9	31. -28	51. 15
13. -5	33. 10	53. -9
15. -10	35. -2	55. 35
17. -40	37. 7	57. -5
19. -2	39. -23	59. -9

61. -20
63. 21
65. 24
67. -24
69. 15
71. -16
73. 0
75. 21
77. -27

79. -24
81. -8
83. -3
85. 3
87. 0
89. -7
91. 5
93. -1
95. -60

97. -10
99. -1
101. 6
103. -2
105. -1
107. Undefined
109. -2

EXERCISES 1.5, pp. 33

1. 74
3. 180
5. 0
7. 2
9. 168
11. 45
13. 8
15. 64
17. 34

19. 5
21. 39
23. 2
25. 213
27. -7
29. 9
31. 7
33. 35
35. -4

37. 4
39. -2
41. $\dfrac{45}{2} = 22\dfrac{1}{2}$
43. -22
45. 28
47. $-5°C$
49. 25 cm
51. 60 cm

53. (a) 6 ft-lb   (b) -50 joules   (c) 575 joules   (d) -20 Btu   (e) 340 Btu
55. (a) -843.75 (b) 300
57. -2.5 mm; depressed

EXERCISES 1.6, pp. 45

1. $\{2\}$
3. $\{3\}$
5. $\{1\}$
7. $\{3\}$
9. $\varnothing$
11. $\{1\}$
13. All real numbers, R
15. $\left\{\dfrac{1}{2}\right\}$
17. $\{5\}$
19. $\{2\}$
21. $\{-3\}$

23. $\{-5\}$
25. $\{2\}$
27. All real numbers, R
29. $\{2\}$
31. $\{0\}$
33. $\{4\}$
35. $\{10\}$
37. $\{2\}$
39. $\{24\}$
41. $\varnothing$
43. $\{7\}$

45. $\{-2\}$
47. $\{2\}$
49. All real numbers, R
51. $\{3\}$
53. $\varnothing$
55. $\{-6\}$
57. $\{7\}$
59. $\{2\}$
61. $b = \dfrac{3V}{h}$
63. $r = \dfrac{P}{l^2}$
65. $h = \dfrac{2A}{a + b}$

67. $b = P - a - c$

69. $A = \dfrac{D}{n - 1}$

71. $y = 10 + 5x$

73. $x = 12 - 4y$

75. $x = -y - 1$

77. $y = \dfrac{6 - 3x}{2}$

79. $y = \dfrac{-ax - c}{b}$

CHAPTER 1 DIAGNOSTIC TEST, pp. 46

1. False; the natural numbers start with 1; $N = \{1, 2, 3, \ldots\}$

2. True; $I = \{\ldots, -4, -3, -2, -1, 0, 1, 2, 3, \ldots\}$

3. False; 1 and 15 are not the only factors of 15. 3 and 5 are also factors.

4. True; $|-3| = 3$; $|3| = 3$

5. True; $|-3| = 3$ and $3 > 0$

6. True; $-3 - (-3) = -3 + 3 = 0$

7. False; $3 + (-3) = 0$, and division by zero is not defined.

8. True; $3 + (-3) = 0$, $3 - (-3) = 3 + 3 = 6$; $\dfrac{0}{6} = 0$

9. True; $(-5)^2 = (-5)(-5) = 25$

10. True; $-5^2 = -(5)(5) = -25$

11. True

12. True; $3(2)^2 - 2(2) + 4 = 3(4) - 4 + 4 = 12$

13. True; $3(-2)^2 - 2(-2) + 4 = 3(4) + 4 + 4 = 20$

14. False; $3(3 - 4)^2 = 3(-1)^2 = 3(1) = 3$

15. True; $(2)^2 + 2(2)(-3) + (-3)^2 = 4 - 12 + 9 = 1$

16. False; $3(2)^2 + 2(2)(-1) - (-1)(3)^2 = 3(4) - 4 - (-9) = 12 - 4 + 9 = 17$

17. True; $F = \dfrac{9}{5}(25) + 32 = 45 + 32 = 77$

18. True; $F = \dfrac{9}{5}(-5) + 32 = -9 + 32 = 23$

19. True; $C = \dfrac{5}{9}(41 - 32) = \dfrac{5}{9}(9) = 5$

20. False; $\dfrac{5 - 9}{9 - 5} = \dfrac{-4}{4} = -1$

21. $\{4\}$

22. $\{-2\}$

23. $\{9\}$

24. $\{11\}$

25. $y = \dfrac{x - 3}{2}$

26. $h = \dfrac{3V}{b}$

27. $r = \dfrac{A - P}{Pt}$

28.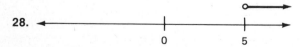

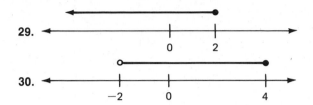

29.

30.

CHAPTER 1 REVIEW EXERCISES, pp. 47

1. {13, 14, 15, 16, 17}
2. {0, 1}
3. {12, 14}

4. {2, 3, 5, 7, 11, 13}

5. {2}
6. {1, 2, 3, 4, . . .}
7. {0, 1, 2, 3, 4}
8. {3, 5, 7, 11, 13}
9. $\varnothing$
10. {23, 29}
11. {0, 4, 20}
12. {4, 20}
13. {−2, 0, 4, 20}

14. $\left\{-\sqrt{2},\ \sqrt{8},\ \dfrac{12}{\sqrt{2}}\right\}$

15. $\left\{-2,\ -\dfrac{1}{2},\ 0,\ \dfrac{3}{5},\ 4,\ 20\right\}$

16. S
17. 3
18. −2
19. 17
20. 0
21. 4
22. −6

23. $\dfrac{1}{2}$

24. −2
25. $-\sqrt{5}$

26. >
27. =
28. =
29. <
30. >

31.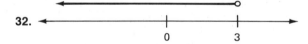

32.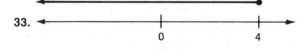

33.

34.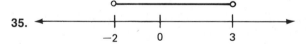

35.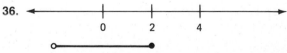

36.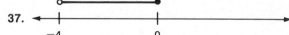

37.

38. Distributive axiom
39. Multiplicative inverse (reciprocal)
40. Associative axiom, multiplication

41. Commutative axiom, multiplication
42. Closure, multiplication
43. Reciprocal of a reciprocal
44. Associative axiom, addition
45. Associative and commutative axioms, multiplication
46. Distributive axiom; commutative axiom, multiplication
47. Associative axiom, addition; additive inverse; addition identity
48. Negative of a negative; additive inverse
49. 9
50. 14
51. -17
52. 3
53. 9
54. -1
55. 0
56. Undefined
57. 0
58. -18
59. -9
60. 64
61. 162
62. 21
63. -9
64. -9
65. $\dfrac{3}{2}$
66. -8
67. (a) 14   (b) -6 (c) 24
68. (a) -16 (b) 0   (c) -1
69. (a) 6   (b) 4   (c) -1
70. (a) 9   (b) -7 (c) -15
71. (a) 5   (b) -2 (c) -3
72. (a) 12   (b) -4 (c) 36
73. (a) -56 (b) 0   (c) 20
74. (a) 13   (b) -29
75. (a) 9   (b) 1
76. (a) -25 (b) -10
77. (a) 700   (b) -24
78. $-\dfrac{1}{2}$

79. -550
80. 80 mg
81. $\dfrac{1}{2}$
82. $x = 5$
83. $x = -6$
84. $y = -6$
85. $y = 15$
86. $x = 4$
87. $x = -3$
88. $z = -2$
89. All real numbers, R
90. $x = 2$
91. $x = -6$
92. $x = 7$
93. $x = -5$
94. $\varnothing$
95. $x = -9$
96. $m = 3$
97. $x = 4$
98. $x = 2$
99. $x = 0$
100. $x = 4$
101. $y = 3x - 5$
102. $x = \dfrac{y + 5}{3}$
103. $y = -2x - 5$
104. $x = \dfrac{-y - 5}{2}$
105. $x = \dfrac{2y + 10}{5}$
106. $y = \dfrac{5x - 10}{2}$
107. $x = \dfrac{y - 2}{3}$
108. $y = \dfrac{x - 2}{3}$

109. $y = \dfrac{-ax + c}{b}$

110. $x = \dfrac{-by + c}{a}$

111. $d = \dfrac{C}{\pi}$

112. $t = \dfrac{A - P}{Pr}$

113. $y = P - x - z$

114. $b = \dfrac{S - 2a}{2}$

115. $t = T - \dfrac{g}{R}$ or $t = \dfrac{TR - g}{R}$

116. $d = \dfrac{1}{2}\left(\dfrac{M}{n} - n\right)$ or $d = \dfrac{M - n^2}{2n}$

117. $A = 2(15 - B)$ or $A = 30 - 2B$

118. $H = \dfrac{W + 190}{5}$

119. $B = \dfrac{3C - A + 20}{2}$

120. $C = \dfrac{A + 2B - 20}{3}$

EXERCISES 2.1, pp. 58

1. x^8

3. $3y^4$

5. $6x^{12}$

7. x^4

9. t^{12}

11. $625t^4$

13. $\dfrac{-n^5}{32}$

15. $\dfrac{81}{y^4}$

17. $10^4 = 10,000$

19. $24a^2b^2$

21. $-z^6$

23. $15x^3yz$

25. x^8

27. x^6

29. $5z^6$

31. y^{12}

33. xy^{11}

35. $-x^6$

37. ab^{21}

39. $-y^4$

41. $2^5 = 32$

43. $2^6 = 64$

45. $3^6 = 729$

47. $(-2)^6 = 64$

49. $4^3 = 64$

51. $3^3 = 27$

53. $4^2 = (2^2)^2 = 2^4$; $2^7 = 128$

55. $2^6 = 64$

57. $\dfrac{1}{2^7} = \dfrac{1}{128}$

59. $-125x^6$

61. $-2y^{12}$

63. a^3b^8

65. $3y^4$

67. $3y^8$

69. a^8b^8

71. $-x^4$

73. $10^2 = 100$

75. $125x^3y^6$

77. $-3a^4b^6c^8$

79. $\dfrac{36a^2c^2}{25b^4}$

81. $\dfrac{3x^2y^2}{4}$

83. -1

85. $\dfrac{1}{10^4} = \dfrac{1}{10,000}$

87. $10^3 = 1000$

89. $\dfrac{9}{2}$

91. $40,000$

93. $6^6 = 46,656$

95. x^{n+1}

97. 3^n

99. y^{4n}

101. $x^{2n}y^n$

103. 1

105. x^{2n+1}

107. x^{2n+1}

EXERCISES 2.2, pp. 64

1. $8x^2 - 4x - 4$
3. $-6y - 8$
5. $y^2 - y + 1$
7. $-2r - 2s + 2t$
9. $-1 + x - x^4$
11. $7x + 4y - 5z$
13. $10x - 20$
15. $2a^2 - 2a^2b^2 + 4b^2$
17. $2y - 2$
19. 0
21. $4x^2 + 1$; $107 + (-2) + 40 = 145$
23. $2t - 9$; $(-45) - (-46) = 10 - 9 = 1$
25. $x^3 - 2x^2 - 2x - 1$; $(-7) + 2 = -5$
27. $-a^2 - 3ab + 4b^2$; $-(2)^2 - 3(2)(3) + 4(3)^2 = -4 - 18 + 36 = 14$
29. $a + 5b - c$; $(-6) - (-18) = 12$
31. $-5x^2 - 2x + 10$; $(-1) - (-4) = 3$
33. $6x^4 + x^3 - x^2 + 21x - 6$; $(5 + 1) - (-15) = 21$
35. $-3x^4 + 11x^3 - 11x^2 + x - 1$
37. $-2a^3 + 3b^3 + a^2b - 4ab^2$
39. $3x^3 - 10x^2y - 2xy^2 + y^3 + 4$

41. $P(0) = -4$
$P(-1) = -6$
$P(2) = 18$
43. $P(0) = 7$
$P(-1) = 12$
$P(2) = 21$
45. $P(0) = -2$
$P(-1) = -7$
$P(2) = -4$
47. $P(0) = 0$
$P(-1) = -11$
$P(2) = 28$
49. $P(0) = -2$
$P(-1) = 2$
$P(2) = 26$

EXERCISES 2.3, pp. 70

1. $x^3 + 3x$
3. $2y^4 - 3y^3 + y^2$
5. $3x^5y - 2x^4y^2 + x^3y^3 - x^2y^4$
7. $x^2 - 2x - 15$
9. $6x^2 - x - 2$
11. $x^3 + 5x^2 + 2x - 8$
13. $2x^3 + 9x^2 - 18x$
15. $x^3 + 216y^3$
17. $4t^3 - 20t^2 - 81t + 405$
19. $x^4 + 2x^3 - 8x^2 - 13x + 6$
21. $x^2 - 3x - 10$
23. $x^2 + 7x + 12$
25. $x^2 - x - 12$
27. $6x^2 - x - 2$
29. $6x^2 + 7x + 2$
31. $9x^2 - 25$
33. $x^2 + 6x + 9$
35. $6x^2 - 13x + 6$
37. $x^4 - 9$
39. $2x^2 - 13x + 15$
41. $16x^2 + 40x + 25$
43. $9 - 6x + x^2$
45. $4 + 28x + 49x^2$
47. $16 + 24a + 9a^2$
49. $x^2 + 2xy + y^2$
51. $4x^2 + 20xy + 25y^2$
53. $4x^2 - 20xy + 25y^2$
55. $x^3 + 9x^2 + 27x + 27$
57. $x^3 - 9x^2 + 27x - 27$
59. $8a^3 + 60a^2 + 150a + 125$
61. $729 + 486y + 108y^2 + 8y^3$
63. $y^2 - 1$
65. $9x^2 - 4$
67. $16a^2 - 25b^2$
69. $4x^4 - 9$
71. $25x^3 - 9x$
73. $48x^3 + 24x^2 + 3x$
75. $4x^3 + 8x^2 - 35x - 75$
77. $x^2 - y^2 - 4y - 4$
79. $x^{2n} + 4x^n + 4$

EXERCISES 2.4, pp. 74

1. $4(x + 3)$
3. $3(2x - 1)$
5. $3(x^2 + 2x + 3)$
7. $4(2ab - 3a + 1)$
9. $4y^2(y - 25)$
11. $a(y - 1)$
13. $x^3(x - 1)$
15. $x^4(1 - x)$
17. $x^2(24x - 30 + y)$
19. $cn^2(c + n)$

21. $a(x + y - z)$
23. $ax(a - x)$
25. $6(4x^2 + 2x + 1)$
27. $c(c^3 - c^2 + c - 2)$
29. $-5x(5x^2 + 3)$
31. $-xy(y^2 + y + 1)$
33. $rs(r - s - 4)$
35. $9(3a^2b^2 + 2ab - 7)$
37. $-x^2(x^2 - 2x + 6)$
39. $3p^2q(1 + 3q - 4q^2)$

41. $(x - 3)(x + 3y)$
43. $(y + 2)(x^2 - 1)$
45. $(b + c)(b^2 - t)$
47. $(x + y)(a + 3)$
49. $(x + 5)(x^2 - 2y^2)$
51. $(t^2 + 4)(5a^2 - 1)$
53. $(t^3 + 1)(t - 1)$
55. $(a + b)(4x^2 + 2x + 1)$
57. $x^n(x^n - 1)$
59. $y^n(y^{3n} - z)$

EXERCISES 2.5, pp. 80

1. $(x - 2)(x - 4)$
3. $(x + 2)(x - 3)$
5. $(y + 1)(y + 8)$
7. Not factorable
9. $(x - 4)(x - 5)$
11. $(r - 4)(r + 8)$
13. $(x + 2)(x + 3)$
15. $(y - 3)(y - 4)$
17. Not factorable
19. $(x + y)^2$
21. $3(x - 1)(x - 2)$
23. $10(x - 7)(x + 5)$
25. $150x^2(x - 2)(x - 5)$
27. $4ay(a + 4)(a - 3)$
29. $20(y^2 + 7y - 6)$
31. $(2a - 11)(4a - 3)$
33. $(2x + 3)^2$
35. $(2p - 3)(4p + 7)$
37. $(x - 1)(2x - 1)$
39. $(4x - 1)(3x - 5)$
41. $(2x - 1)(6x - 5)$

43. $(3a + 7)(a - 1)$
45. $(3a + 1)(a - 7)$
47. Not factorable
49. $(3y + 4)^2$
51. $(2x + 7)(5x - 1)$
53. $(2x + 1)(x + 1)$
55. $(2x + 5)(2x + 3)$
57. $(4x + 1)(x + 1)$
59. $(4x - 3)(x - 2)$
61. $2(3a + 1)(a + 1)$
63. $(8y - 5)(2y + 1)$
65. $a(3x + 1)(3x - 8)$
67. $(3a - b)(a - 2b)$
69. $(2m - n)(m + 3n)$
71. $(3x - y)(5x + y)$
73. $3x(2a^2 + 4a - 7)$
75. $x^3(3x - 1)(x + 7)$
77. $2x^2y^2(x + 5)(x - 2)$
79. $6(5x + 2)(x - 3)$
81. $4, -4, 5, -5$
83. $21, -21, 36, -36, 69, -69$

EXERCISES 2.6, pp. 86

1. 9
3. 49
5. 25

7. $6x$
9. $36y^2$
11. $(x + 9)(x - 9)$

13. $(2x + 7)(2x - 7)$
15. $(6x + 5y)(6x - 5y)$
17. $(4 + 3a)(4 - 3a)$
19. Not factorable
21. $(x + 1)(x^2 - x + 1)$
23. $(x - a)(x^2 + ax + a^2)$
25. $(2y + 1)(4y^2 - 2y + 1)$
27. $(p + 3)(p^2 - 3p + 9)$
29. $x(x + 4)(x^2 - 4x + 16)$
31. $(x + 4)^2$
33. Not factorable
35. $(x - 7y)^2$
37. Not factorable

39. Not factorable
41. $(x + 2)^2$
43. $(x + 9)^2$
45. $(y + 8)^2$
47. $(4p - 1)^2$
49. $(8y - 3)^2$
51. $(4x - 5y^2)(16x^2 + 20xy^2 + 25y^4)$
53. $x(8x - 3)(8x + 3)$
55. Not factorable
57. $8y^2(y^2 - 2)$
59. $x(x + 6)(x^2 - 6x + 36)$
61. $2yz(y - 3z)(y^2 + 3yz + 9z^2)$

EXERCISES 2.7, pp. 90

1. $(x + 5)(x + 4)$
3. $(x + 6)(x + 2)$
5. $(t - 1)(t + 4)$
7. $(x^2 + y - 9)(x^2 - y + 9)$
9. $(x + 2)(x^2 + 3)$
11. $r(r - 9)$
13. $(s - 20)^2$
15. $(y - 4)(y + 1)(y - 1)$
17. $(x + 2)(x - 2)(3x - 2)$
19. $(x + 2)(x + 5)(x - 5)$
21. $(x - 1)(x + 3)(x^2 - 3x + 9)$
23. $(x + y + 8)(x - y - 8)$
25. $(y^2 + 5x + 2)(y^2 - 5x + 2)$
27. $(x + 5)(x - 5)(x^2 + 10)$

29. $(c + 3 - 4x)(c + 3 + 4x)$
31. $(a - 3)(2a - 1)(2a + 1)$
33. $(x - 3)^2(a + b)(a^2 - ab + b^2)$
35. $-(x - 5)^2$
37. $(x - y)(x + y)(x^2 + y^2)(x^4 + y^4)$
39. $p^3(r - s + 1)[(r - s)^2 - (r - s) + 1]$
41. $(x + 3)(2x - 1)(2x + 5)$
43. $(x + 1)(x - 1)(x^4 + x^2 + 1)$
45. $(1 - x - y)(1 + x + y)$
47. $(z + x - y)(z - x + y)$
49. $(a^2 + 2a - 1)(a^2 + 2a - 5)$
51. $(x + 3y)(x - 3y + 5)$
53. $(x + y)(2x + y + 3)$

EXERCISES 2.8, pp. 94

1. $\{-2, 3\}$
3. $\left\{\dfrac{3}{2}, -2\right\}$
5. $\{-5, -2\}$
7. $\left\{\dfrac{1}{4}, -\dfrac{3}{2}\right\}$
9. $\left\{-2, -\dfrac{3}{2}\right\}$

11. $\{1, 2\}$
13. $\{0, 5\}$
15. $\{0, -2\}$
17. $\{3, -9\}$
19. $\{-3, 11\}$

21. $\{3\}$
23. $\{3, -1\}$
25. $\{3, -2\}$
27. $\{-2, -5\}$
29. $\{15, -3\}$

ANSWERS

31. $\{-3, -4\}$ **37.** $\{-3, 3\}$ **43.** $\{-6, 6\}$

33. $\{9, -1\}$ **39.** $\{-10, 10\}$ **45.** $\{0\}$

35. $\{0, 3, -2\}$ **41.** $\{-4, 4\}$ **47.** $\{0, -1\}$

49. $\{-2, 2\}$

51. $x^2 = 6x + 91$; 13 or -7

53. $x(x + 2) = 255$; 15, 17 or $-17, -15$

55. $10x - 18 = \dfrac{x^2}{2}$; 2 or 18

CHAPTER 2 DIAGNOSTIC TEST, pp. 95

1. (d); $(-2y^2)^3 = (-2)^3(y^2)^3 = -8y^6$

2. (b); $4x(3x^2y)^2 = 4x(3^2)(x^2)^2y^2 = 36x^5y^2$

3. (a); $P(-2) = 4(-2)^3 - 3(-2)^2 - 2(-2) + 5$
$$= 4(-8) - 3(4) + 4 + 5$$
$$= -32 - 12 + 4 + 5 = -35$$

4. (b); $(2x + 5)(x - 2) = 2x^2 - 4x + 5x - 10 = 2x^2 + x - 10$

5. (d); $(x - 3)^2 = x^2 - 2(3x) + 3^2 = x^2 - 6x + 9$

6. (c); $(3 + y)(5 - y) = 15 - 3y + 5y - y^2 = 15 + 2y - y^2$

7. (e); this expression cannot be factored over the integers.

8. (e); $x^2 - 4x - 12 = (x - 6)(x + 2)$

9. (b); $x^2 + 2x + 4$ because $(x - 2)(x^2 + 2x + 4) = x^3 - 8$

10. (c); see Answer 9 above.

11. (c); $(-x)^3 = (-x)(-x)(-x) = -x^3$

12. (c); by the zero-product theorem,
$$x + 3 = 0 \quad \text{or} \quad x - 1 = 0$$
$$x = -3 \quad \text{or} \quad x = 1$$

13. (d);
$$2x^2 + x - 3 = 0$$
$$(2x + 3)(x - 1) = 0$$
$$2x + 3 = 0 \quad \text{or} \quad x - 1 = 0$$
$$x = -\frac{3}{2} \quad \text{or} \quad x = 1$$

14. (e); the zero-product theorem applies only when the product is 0.
$$(x - 4)(x + 3) = -6$$
$$x^2 - x - 12 = -6$$
$$x^2 - x - 6 = 0$$
$$(x - 3)(x + 2) = 0$$
$$x - 3 = 0 \quad \text{or} \quad x + 2 = 0$$
$$x = 3 \quad \text{or} \quad x = -2$$

15. (b); if x is the first odd integer, then $x + 2$ is the next consecutive odd integer; thus $x(x + 2) = 195$.

CHAPTER REVIEW EXERCISES, pp. 96

1. x^8
2. $24x^6y$
3. $a^{11}b^9$
4. 243
5. $-x^4$
6. $\dfrac{4}{x}$
7. $27x^6y^3$
8. $\dfrac{16x^{12}}{y^8}$
9. 5
10. $\dfrac{1}{243}$
11. 36
12. $32x^5$
13. $-6x^4y^6$
14. -243
15. 243
16. -243
17. 243
18. x^{2n}
19. $\dfrac{1}{x^{2n-1}}$
20. $\dfrac{y^n}{x^{n^2-2n}}$
21. $x^3 + x^2 - 3x + 17$
22. $-3x^2 + 3x + 6$
23. $-a^2b + 7ab - ab^2 + 3b^2$
24. $4x^2y - 4xy - 4$
25. $-4x^3 + 7x^2 - 3x$
$\quad\quad 8 - 14 + 1 = -5$
$\quad\quad 16 + 10 - 3 = 23$
$\quad\quad 40 - 12 + 2 - 2 = 28$
$\quad\quad (-5 + 23) - 28 = -10$
26. $P(-2) = -37$
$\quad P(1) = 2$
$\quad P(2) = 11$
27. $R(-1) = -14$
$\quad R(0) = -6$
$\quad R(3) = 78$
28. $x^3 + 3x^2 + 2x$
29. $x^4 + 2x^2 - x$
30. $6x^3 - 8x^2 - 2x$
31. $4ax^3 - 4bx^2 + 4cx$
32. $x^2 - 5x + 6$

33. $y^2 - 3y - 4$
34. $6a^2 + 7a - 3$
35. $a^2 + 10a + 25$
36. $25x^2 - 1$
37. $4y^2 - 12y + 9$
38. $2x^3 - 3x^2 - 14x$
39. $2x^5 + x^4 - 2x^3$
40. $-3x^7 - 2x^5 + 4x^4$
41. $6x^2 - 29xy - 16y^2$
42. $2x^3 + 3x^2 - 20x$
43. $18x^3 + 24x^2 + 8x$
44. $x^3 + 8x^2 + 21x + 18$
45. $x^3 - x^2 - 10x - 8$
46. $6x^3 - x^2 + 8x + 3$
47. $x^4 + 2x^3 - 8x^2 - 13x + 6$
48. $4x^2 + 16x + 16$
49. $9x^2 - 12x + 4$
50. $a^3 + 6a^2 + 12a + 8$
51. $a^3 - 6a^2 + 12a - 8$
52. $27y^3 + 27y^2 + 9y + 1$
53. $125x^3 + 150x^2 + 60x + 8$
54. $4x^2 - 25$
55. $9x^2 - 4$
56. $12x^3 + 60x^2 + 75x$
57. $a^{2n} - 1$
58. $x^{2n} - 6x^n + 9$
59. $b^{3n} + b^{2n} - b^n + b^{n+2} + b^2 - 1$
60. $x^{3n} + 3x^{2n} + 3x^n + 1$
61. $3x(x - 2 + 8y)$
62. $(t + 3)(t + 8)$
63. $(x + 5)^2$
64. $(7y^2 - 1)(7y^2 + 1)$
65. $(x - 1)(x^2 + x + 1)$
66. $(a + 1)(a^2 - a + 1)$
67. $-3x(x^2 - 9) = -3x(x - 3)(x + 3)$
68. $(3a + 1)(2a - 3)$
69. $3x(12 + x^2 - x^4)$
$\quad = -3x(x^4 - x^2 - 12)$
$\quad = -3x(x^2 - 4)(x^2 + 3)$
$\quad = -3x(x - 2)(x + 2)(x^2 + 3)$

70. $25(4p^2 - q^2) = 25(2p - q)(2p + q)$

71. $4y(16y^2 - 9) = 4y(4y - 3)(4y + 3)$

72. $(x^3 - y^3)(x^3 + y^3) = (x - y)(x + y)(x^2 + xy + y^2)(x^2 - xy + y^2)$

73. $2y(6x^2 - 11x - 10) = 2y(2x - 5)(3x + 2)$

74. $(a + b)(8a - 3b)$

75. $2y(y - 3)(y + 3)(y^2 + 9)$

76. $3ax^4(5x^2 + 14x - 3) = 3ax^4(x + 3)(5x - 1)$

77. $(5r + 1)^2$

78. $(6x - 1)^2$

79. $(2p - 5q)^2$

80. $(x + 3)(y + 2)$

81. $(p + 5)(x - 3)$

82. $-x^3 - 4x^2 + 3x = -x(x^2 + 4x - 3)$

83. $yz(y[a - 1] + 1) = yz(ay - y + 1)$

84. $(a + 2)[x(a + 2) + y] = (a + 2)(ax + 2x + y)$

85. $(x + y)(3 - 2[x + y]) = (x + y)(3 - 2x - 2y)$

86. $(x + 30)^2$

87. $y^2(x + 1)(x^2 - x + 1)$

88. $(7x + 7a + 3y)(x + a - 2y)$

89. $(r - 5s)(r + 5s + 1)$

90. $(a + c)(x + ab)$

91. $(x - 3)(x^2 + 3x + 9)$

92. $(x - 1)(x^3 + x^2 - 1)$

93. $(x - 1)(x - 3)(x + 2)$

94. $(b + 1)(b^2 - 4b + 7)$

95. $(x + 1)(x + 5)(x^2 - 5x + 25)$

96. $(m - 1)(m + 3)(m^2 + 2m + 5)$

97. $5(x^n + 1)(x^n - 1)$

98. $x^n(x^n + 1)(x^n - 1)$

99. $y^n(y + 5)(y - 3)$

100. $x(x^n + 1)(x^n - 1)(x^{4n} + x^{2n} + 1)$

101. $\left\{-2, \dfrac{7}{2}\right\}$

102. $\{0, -7\}$

103. $\{5, -12\}$

104. $\{0, 3\}$

105. $\{2, 9\}$

106. $\{-1\}$

107. $\{-1\}$

108. $\{1, 6\}$

109. $\{0, 3\}$

110. $\{8, -8\}$

111. $\{3, -2, 1\}$

112. $\left\{\dfrac{3}{2}, \dfrac{3}{4}\right\}$

113. $\{-8, -3\}$

114. $\{6\}$

115. $x(x + 2) = 15(x + 2) + 18$; 16, 18

116. $5x^2 + 10x = 315$; 7 or -9

117. (a) $b^2 = (c - a)(c + a)$

    (b) 1: $b^2 = (85 - 84)(85 + 84) = 169$; $b = 13$

        2: $b^2 = (37 - 35)(37 + 35) = 144$; $b = 12$

118. (a) $A = 2\pi r(h + r)$ (b) $A = 2\left(\dfrac{22}{7}\right)(15)(35) = 3300$

119. (a) $D = ckw(L^2 - 4cL + 4c^2) = ckw(L - 2c)^2$

    (b) $D = \dfrac{1}{4}\left(\dfrac{1}{500}\right)(200)\left(12 - \dfrac{1}{2}\right)^2 = \dfrac{529}{40} = 13.225$

120. (a) $1200x - x^2 = 0$

        $x = 0$ or $x = 1200$

    (b) $320,000 = 1200x - x^2$

        $x = 400$ or $x = 800$

121. $x = 2$ or $x = \dfrac{2}{3}$; since the number of moles that react must be less than 1, the only

    answer is $x = \dfrac{2}{3}$.

EXERCISES 3.1, pp. 105

1. 1

3. $\dfrac{1}{3x^2}$

5. $\dfrac{-1}{m^2n}$

7. $\dfrac{x + 3}{x}$

9. $\dfrac{x - 4}{x - 2}$

11. $\dfrac{y - 8}{y}$

13. $\dfrac{x - y}{x + y}$

15. $\dfrac{y}{5}$

17. $\dfrac{c}{2d}$

19. $\dfrac{r}{4}$

21. $\dfrac{a + b}{a - b}$

23. $\dfrac{m^2 + 1}{m + 1}$

25. $2y - 3$

27. $\dfrac{y}{3}$

29. $\dfrac{c - d}{6}$

31. $\dfrac{y}{3}$

33. $\dfrac{x - y}{x + y}$

35. $\dfrac{a^2 + 1}{a + 1}$

37. $x + 2$

39. -1

41. $\dfrac{x + 1}{x + 4}$

43. $\dfrac{2(t + 2)}{3(t - 3)}$

45. $\dfrac{a + b}{a - b}$

47. $\dfrac{4a}{a + b}$

49. $\dfrac{2(n + 1)}{n - 2}$

51. $\dfrac{-1}{x + 7}$

53. $\dfrac{n - 8}{n - 1}$

55. $\dfrac{y(2y - 1)}{3}$

57. $\dfrac{1}{x - 3}$

59. $\dfrac{-15x^3}{40x^2}$

61. $\dfrac{-18x^2y^2}{48xyz}$

63. $\dfrac{10}{2x + 2}$

65. $\dfrac{5x - 15}{5x + 20}$

ANSWERS

67. $\dfrac{2x^2}{2x^2 - 10x}$

69. $\dfrac{3y^2 - 9y}{y^2 - 9}$

71. $\dfrac{x^2 + 3x + 2}{x + 1}$

73. $\dfrac{x}{x - 7}$

75. $\dfrac{-y^2 - 4y}{y^2 - 16}$

77. $\dfrac{a^2 + 2a + 1}{a^2 - 2a - 3}$

79. $\dfrac{t^2 + 5t - 14}{14 + 9t + t^2}$

81. $\dfrac{4x^2 + 2x}{4x^2 - 1}$

83. $\dfrac{r^2 - 3r}{r^2 - r - 12}$

85. $\dfrac{6x^2 - 17x + 10}{x - 2}$

87. $\dfrac{t}{t - 1}$

89. $\dfrac{2r + 7}{4r^2 - 49}$

91. $\dfrac{y^2 - 7y + 6}{12 - 8y + y^2}$

93. $\dfrac{5x^2 + 10x}{5x + 10}$

95. $\dfrac{x^3 - 25x^2}{3x^2 - 75x}$

97. $\dfrac{12a^3 + 18a^2}{10a^2 + 13a - 3}$

99. $\dfrac{-2a^3 + 5a^2}{4a^2 - 10a}$

EXERCISES 3.2, pp. 110

1. $\dfrac{6xy}{(x - 3)(y - 2)}$

3. $\dfrac{3}{11b}$

5. $\dfrac{4x}{9acy}$

7. $\dfrac{x^2 - y^2}{16}$

9. $\dfrac{5y}{x^3}$

11. $\dfrac{-b}{az}$

13. $\dfrac{2a(a - 5)}{a - 1}$

15. $\dfrac{x + 3}{3x}$

17. $\dfrac{(a - b)^2}{a}$

19. $\dfrac{2(n - 2)}{n(n + 3)}$

21. 1

23. $\dfrac{8}{5}$

25. $\dfrac{7x(x + 7)}{(x - 7)^3}$

27. $\dfrac{12y}{y + 4}$

29. $\dfrac{3x(5x - 1)}{x + 1}$

31. $\dfrac{2x}{x - 3}$

33. $\dfrac{x(x - 1)}{x - 9}$

35. 1

37. $\dfrac{y(x - 3)}{1 - 3xy}$

39. $\dfrac{y}{(y + 1)^2}$

41. $\dfrac{(x - 1)(x - 3)}{(x + 1)(x + 3)}$

43. $\dfrac{(x - 2)(x^2 - 3x + 9)}{x^2}$

45. -1

47. $(y - 1)(y^3 + 1)$

49. $x - y$

EXERCISES 3.3, pp. 116

1. $2x^2 + 3 + \dfrac{1}{x - 5}$

3. $x^3 + x^2 + x + 1$

5. $2x^3 + 4 + \dfrac{7}{x^2 - 3}$

7. $5a^2 - 2ab + b^2 - \dfrac{b^3}{5a + 2b}$

9. $x + y + 3$

17. $5x^2 + \dfrac{2x - 1}{x^2 - 6}$

11. $2x^2 - 3x + 9 - \dfrac{29}{x + 4}$

19. $Q(x) = 2x^2 + x + 6, R = 7$

13. $2x^2 - 3x - 2$

21. $Q(x) = x^2 - 6x + 8, R = -5$

15. $2x^2 - 2x - 1 - \dfrac{7}{2x + 7}$

23. $Q(x) = x^3 - x^2 + x - 3, R = 4$

25. $Q(x) = x^3 - 3x^2 + 10x - 30, R = 84$
27. $Q(x) = 2x^3 + 10x^2 + 50x + 50, R = 256$
29. $Q(x) = 5x^3 - 8x^2 + 2x - 8, R = 0$
31. $Q(x) = x^4 - x^3 + x^2 - x + 1, R = 0$
33. $Q(x) = 5x^4 + 17x^3 + x^2 + 3x + 8, R = 25$

35. $Q(x) = 4x^2 - 8x - 3, R = -\dfrac{5}{2}$

37. $Q(x) = 2x^2 - 4x + 3, R = -2$
39. $Q(x) = x^2 - 2x - 1, R = 0$
41. $p = 56$
43. $R = 8, P(2) = 8$

EXERCISES 3.4, pp. 121

1. $R = 80$
3. $R = 80$
5. $R = 0$
7. $R = 5$

9. $R = 0$

11. $P(2) = 0$, yes
13. $P(-2) = 14$, no
15. $P(-3) = 0$, yes
17. $P(1) = 0$, yes
19. $P(-1) = 0$, yes
21. $P(1) = -18$
23. $P(2) = -28$
25. $P(3) = 0$
27. $P(x) = (x - 3)(x + 2)(x^2 + x + 1)$
29. $P(-1) = 30$

31. $P(-5) = 10$
33. $P(-2) = 49$
35. (a) $P(1) = -2, \quad P(-1) = 0$
$P(2) = 0, \quad P(-2) = -20$
$P\left(\dfrac{1}{2}\right) = 0, \quad P\left(-\dfrac{1}{2}\right) = \dfrac{5}{2}$
(b) $P(x) = (x + 1)(x - 2)(2x - 1)$
37. (a) $k = -10$
(b) $k = -20$
(c) $k = -16$
39. $k = 2$
41. $k = 1$
43. $p = -3$
45. $p = 0$
47. $R = 339$

EXERCISES 3.5, pp. 127

1. $24x^2$
3. $(x - 1)^2(x + 1)$
5. $2x(x + 2)$
7. $(x - 6)(x + 2)(2x^2 - 9x - 12)$
9. $(x - 1)(x + 1)(x^2 + x + 1)$ or $(x^3 - 1)(x + 1)$

ANSWERS

11. $\dfrac{x}{8y^2}$

13. $\dfrac{131}{180}$

15. $\dfrac{57}{60} = \dfrac{19}{20}$

17. $\dfrac{1}{3x}$

19. $\dfrac{3x^2 + 8x - 1}{6x^3}$

21. $\dfrac{1}{b(b - 1)}$

23. $\dfrac{y + 5}{(y + 1)(y + 2)}$

25. $\dfrac{18y^2 + 4y - 3}{30y^3}$

27. $\dfrac{-x}{y(y + 1)}$

29. $\dfrac{-x^2 + 3x - 4}{(x - 1)(x - 2)}$

31. $\dfrac{a^2 + b^2}{ab}$

33. $\dfrac{10y^2 + 39y - 72}{(2y - 3)^2(2y + 3)}$

35. $\dfrac{x^2 + 2x}{x + 1}$

37. $\dfrac{x^2 + 6x + 12}{(x + 1)(x + 3)(x + 4)}$

39. $\dfrac{4y^2 - y + 9}{(y - 2)(y - 3)(2y + 1)}$

41. $\dfrac{5a - 13}{6(a + 1)}$

43. $\dfrac{13x}{2(x - 4)(x + 6)}$

45. $\dfrac{1}{y(y - 5)}$

47. $\dfrac{x + 1}{x}$

49. $\dfrac{x - y}{x + y}$

51. -1

53. $\dfrac{y + 15}{(4y + 3)(5y - 1)}$

55. $\dfrac{2}{(a - 1)(a + 1)^2}$

57. $\dfrac{3}{t + 1}$

59. -2

61. $\dfrac{-8}{(x - 3)(x + 5)(x^2 + 4)}$

EXERCISES 3.6, pp. 131

1. $\dfrac{22}{31}$

3. $\dfrac{32}{73}$

5. $\dfrac{20}{13}$

7. 20

9. 1

11. $-\dfrac{1}{13}$

13. $\dfrac{10x + 2y^2}{xy^2}$

15. $y - 3$

17. $\dfrac{y + 3}{y - 4}$

19. $\dfrac{x - 2}{x - 5}$

21. $\dfrac{xy}{x + y}$

23. $\dfrac{ab}{3a + 2b}$

25. x

27. $\dfrac{x}{x + 3}$

29. $-\dfrac{3}{5}$

31. $\dfrac{13}{8}$

33. $1 - x$

35. $\dfrac{7}{10}$

37. $\dfrac{4}{x^2 - 3}$

39. $\dfrac{x^2 - 1}{x^3}$

EXERCISES 3.7, pp. 136

1. $\{-4\}$ $x \neq 0$

3. $\{-11\}$ $a \neq 0, -5$

5. $\{3\}$ $x \neq -2, \dfrac{7}{3}$

7. $\left\{-\dfrac{3}{2}\right\}$ $y \neq 0, 1, -1$

9. $\left\{\dfrac{1}{4}\right\}$ $x \neq 0$

11. $\{6\}$ $y \neq -\dfrac{1}{2}$

13. $\varnothing$, $x \neq 0$

15. $\varnothing$, $x \neq -1$, -6

17. $\{12\}$ $y \neq 4$, -4

19. $\{-5\}$ $x \neq 0$

21. $\{-2\}$ $x \neq 3$, $-\dfrac{1}{2}$

23. $\left\{\dfrac{1}{6}\right\}$ $t \neq 0$

25. $\{1\}$ $t \neq 5$, -4

27. $\{13\}$ $x \neq -1$

29. $\left\{\dfrac{5}{2}\right\}$ $x \neq -\dfrac{3}{5}$, $\dfrac{1}{2}$

31. $\{-2\}$ $x \neq 0$, 1

33. $\{3, -2\}$ $x \neq -3$, -1

35. $x = 2a - 3b$, $x \neq b$

37. $x = \dfrac{a}{5}$; $x \neq 0$, a, $-a$

39. $S = \dfrac{RT}{T - R}$

41. $R = \dfrac{ST}{S + T}$

43. $O = I - E I$

45. $V = \dfrac{Fv}{F - f}$

47. $p = \dfrac{100q}{Pq - 100}$

EXERCISES 3.8, pp. 142

1. $40,000, $16,000

3. (a) H: $11\dfrac{1}{9}$ g, O: $88\dfrac{8}{9}$ g   (b) H: 27.8 g, O: 222.2 g

5. 18×20 ft

7. 15×10.8 cm

9. $350

11. $F = kd$

13. $u = \dfrac{k}{v}$

15. $P = \dfrac{k}{t^3}$

17. $y = kxz^2$

19. $A = kab$

21. $Y = 245$

23. $D = 1457.14$ km

25. $F = 75$ lb

27. 6.4 in.

29. 1.28 lumens

31. $W = 80$

33. $p = 14.6$ lb

35. $e = 280$

37. 40

39. 3 and 12, or -3 and -12

CHAPTER 3 DIAGNOSTIC TEST, pp. 145

1. False; $\dfrac{3x + 6}{3x} = \dfrac{3(x + 2)}{3x} = \dfrac{x + 2}{x}$. This fraction cannot be reduced any further.

2. False; $\dfrac{x^2 - y^2}{x - y} = \dfrac{(x - y)(x + y)}{x - y} = x + y$

3. True; $\dfrac{x^2 - x - 6}{2x^2 - 5x - 3} \div \dfrac{x^2 + x - 2}{2x^2 - x - 1} = \dfrac{(x - 3)(x + 2)}{(2x + 1)(x - 3)} \cdot \dfrac{(2x + 1)(x - 1)}{(x + 2)(x - 1)} = 1$

4. False; the remainder is 3. By synthetic division,

$$\begin{array}{r|rrrr} 1 & 1 & 4 & 0 & -2 \\ & & 1 & 5 & 5 \\ \hline & 1 & 5 & 5 & \boxed{3} \ \text{Remainder} \end{array}$$

5. False; x and 3 are not factors of $x + 3$. Therefore the common denominator is $3x(x + 3)$.

6. False; even though the solution $x = 3$ appears after eliminating the fractions and solving the resulting equation, $x = 3$ is *not* a solution of the original equation since that value would result in division by zero, and division by zero cannot be done.

7. True; by direct substitution, $\dfrac{0 + 5}{0 + 2} = \dfrac{5}{2}$.

8. False; $\dfrac{2}{3} = \dfrac{4}{x}$, $2x = 12$, and $x = 6$. Therefore the fourth term of this proportion is 6.

9. True; $\dfrac{x + a}{x - b} = \dfrac{x - b}{x - a}$

$$(x + a)(x - a) = (x - b)(x - b)$$
$$x^2 - a^2 = x^2 - 2bx + b^2$$
$$2bx = a^2 + b^2$$
$$x = \dfrac{a^2 + b^2}{2b}$$

10. True; $9x^2$ is the least common denominator of all the denominators, so that

$$\dfrac{9x^2 \left(\dfrac{1}{9} + \dfrac{3}{x^2} \right)}{9x^2 \left(\dfrac{x}{3} - \dfrac{2}{x} \right)} = \dfrac{x^2 + 27}{3x^3 - 18x}$$

which is not a complex fraction.

11. $-(x + 4)$; $\dfrac{x^2 - 16}{4 - x} = -\dfrac{(x + 4)(x - 4)}{x - 4} = -(x + 4)$

12. $4x^3 - 4x^2$; $\dfrac{4x^3 - 4x^2}{4x^2 - 4x} = \dfrac{x(4x^2 - 4x)}{(4x^2 - 4x)} = x$

13. 14; by synthetic division,

$$\begin{array}{r|rrr} 2 & 3 & 4 & -p \\ & & 6 & 20 \\ \hline & 3 & 10 & \boxed{20 - p} \end{array} \quad \text{Remainder}$$

but $20 - p = 6$; therefore $p = 14$.

14. $(x - 3)^2(x + 3)$; $-3x^2 + 11x + 6$;

$$\dfrac{2}{x^2 - 6x + 9} + \dfrac{3x}{9 - x^2} = \dfrac{2}{x^2 - 6x + 9} - \dfrac{3x}{x^2 - 9}$$
$$= \dfrac{2}{(x - 3)(x - 3)} - \dfrac{3x}{(x - 3)(x + 3)}$$
$$= \dfrac{2(x + 3) - 3x(x - 3)}{(x - 3)^2(x + 3)}$$
$$= \dfrac{-3x^2 + 11x + 6}{(x - 3)^2(x + 3)}$$

15. $-\dfrac{3}{5}$, 0, and $\dfrac{1}{2}$. If $5x + 3 = 0$, then $x = -\dfrac{3}{5}$. If $1 - 2x = 0$, then $x = \dfrac{1}{2}$.

$10x^2 + x - 3 = (5x + 3)(2x - 1) = 0$ if $x = -\dfrac{5}{3}$ or $x = \dfrac{1}{2}$,

and since x appears as a factor in a denominator, $x = 0$ is also an excluded value.

16. Means; extremes

17. Inversely proportional

18. 5; $y = kx^2$; $20 = k(2)^2 = 4k$ and $k = 5$

19. $\dfrac{1}{7}$; $\dfrac{\dfrac{2}{3} - \dfrac{1}{2}}{\dfrac{2}{3} + \dfrac{1}{2}} = \dfrac{6\left(\dfrac{2}{3} - \dfrac{1}{2}\right)}{6\left(\dfrac{2}{3} + \dfrac{1}{2}\right)} = \dfrac{4 - 3}{4 + 3} = \dfrac{1}{7}$

20. The set of real numbers *excluding* $x = 0$. By writing the left side of the equation as a single fraction, its numerator is $3 + 2x$ and its denominator is $3x$ so that the equation looks like an identity. But since x is a factor in the denominator, $x \neq 0$.

CHAPTER 3 REVIEW EXERCISES, pp. 146

1. $\dfrac{5x^2}{9yz}$

2. $\dfrac{x + 3}{x}$

3. -1

4. $\dfrac{9(x + 2y)}{x - 2y}$

5. $\dfrac{2x - 15}{2x - 3}$

6. $-\dfrac{4x^3(x - 1)}{3(x + 1)}$

7. 1

8. $\dfrac{x - 1}{3}$

9. $-\dfrac{a + 2}{a - 4}$

10. a

11. $\dfrac{5 - x}{x + 3}$

12. $\dfrac{6x^2 - 3x}{x^2 - 9}$

13. $\dfrac{4x - 15}{x^2 - 6x + 5}$

14. $\dfrac{4}{x(x + 1)(x - 1)}$

15. $\dfrac{x^2 + 14x + 4}{2x(x + 2)}$

16. $\dfrac{1}{(x + 3)(x + 4)}$

17. $\dfrac{16}{(x - 2)^2(x + 2)^2}$

18. $\dfrac{xy + 3}{xy - 3}$

19. $\dfrac{-1}{3(x + 3)}$

20. $\dfrac{5(5x + 2)}{12(3x + 1)}$

21. $\dfrac{9x - 2}{3}$

22. Undefined; division by zero

23. $x^2 - x - \dfrac{1}{x - 3}$

24. $x^2 - 2 + \dfrac{4}{x^2 + 3}$

25. $3x^2 - 2x + 4 - \dfrac{1}{5x + 10}$

26. $a^2 + a + 1$

27. $5x^2 - 2x - 3 - \dfrac{7}{2x + 1}$

28. $Q(x) = 2x^4 - 8x^3 + 2x^2 + x + 1$, $R = -7$

29. $Q(x) = 3x^3 + 12x^2 + 50x + 199, R = 800$
30. $Q(x) = x^3 - 3x^2 + 9x - 27, R = 162$
31. $Q(x) = x^2 - 3x + 2, R = 3$
32. $Q(x) = 3x^4 + 4x^3 + 8x^2 + 17x + 33, R = 75$

33. (28): $P(4) = -7 = R$ (31): $P\left(\dfrac{-3}{4}\right) = 3 = R$

     (29): $P(4) = 800 = R$ (32): $P(2) = 75 = R$
     (30): $P(-3) = 162 = R$

34. $P(3) = -18 \neq 0$; therefore $x - 3$ is not a factor.
35. $P(1) = 0, P(-1) = 0, P(2) = -12, P(-2) = 0, P(3) = 0, P(-3) = 48$
36. $P(x) = (x - 1)(x + 1)(x + 2)(x - 3)$
37. (a) $k = -20$ (b) $k = -75$ (c) $k = 9$

38. $m = \dfrac{1}{2}$

39. $\left\{-\dfrac{1}{2}\right\} x \neq 6, -6$

40. $\{0\}$ $x \neq -2$
41. $\{9\}$ $x \neq 3, -3$

42. $\left\{-\dfrac{1}{4}\right\} x \neq 1, -3$

43. $\left\{\dfrac{1}{7}\right\} x \neq -\dfrac{1}{3}, 0$

44. $\varnothing$; $x \neq 4, -4$
45. All real numbers except $x = 1$

46. $\{-5\}$ $x \neq 3$

47. $x = \dfrac{a - b}{2}$; $a \neq -b, x \neq a, -b$

48. $x = \dfrac{p - 1}{c + 1}$; $c \neq -1, x \neq 0$

49. $x = \dfrac{-(b + c)}{2}$; $x \neq -b, -c$

50. $b = \dfrac{2ac}{a - c}$; $a \neq c, a \neq 0, b \neq 0, c \neq 0$

51. 25, 5

52. 40, 140
53. $4.50

54. 15

55. 3

56. $k = 7.2$
57. 43.2 lb

58. $F = \dfrac{2}{3}$

59. $A = 42$ square units

60. Shorten 10 in. to a length of 8 in.

61. (a) $x = -3$ (b) $k = 3$ (c) $x = 2$

EXERCISES 4.1, pp. 155

1. $\dfrac{1}{2}$

3. $\dfrac{1}{10} = 0.1$

5. 16

7. 125

9. 9

11. 8

13. $\dfrac{1}{1000} = 0.001$

15. $\dfrac{1}{10,000} = 0.0001$

17. $\dfrac{81}{10,000} = 0.0081$

19. $\dfrac{-16}{3}$

21. $\dfrac{1}{16}$

23. $\dfrac{1}{100} = 0.01$

25. $\dfrac{1}{100,000} = 0.00001$

27. 1

29. 1,000,000

31. $10^9 = 1,000,000,000$

33. 1

35. 100

37. $\dfrac{1}{2^6} = \dfrac{1}{64}$

39. $\dfrac{1}{2^2 \cdot 3^4} = \dfrac{1}{324}$

41. -4

43. -5

45. $\dfrac{3^5}{2} = \dfrac{243}{2}$

47. $\dfrac{5}{81}$

49. 1

51. 1

53. -15

55. 29

57. 10

59. 10,000

61. $\dfrac{1}{1000} = 0.001$

63. 1

65. $\dfrac{1}{8x^3}$

67. $\dfrac{-3}{x^8}$

69. $\dfrac{16}{x^2}$

71. $\dfrac{10^6}{x^6}$

73. $4x^2$

75. $\dfrac{25x^2}{x^2 + 25}$

77. y^2

79. $\dfrac{y^2}{x^3}$

81. 1

83. $\dfrac{y^2}{x^3}$

85. $\dfrac{-1}{xy}$

87. $\dfrac{a^5b^6 - b}{a^6}$

89. $\dfrac{1}{x^4 + 4x^2 + 4}$

91. $\dfrac{-2}{x^2y^5}$

93. 1

95. $2^{2n} - 2^{-2n} = \dfrac{2^{4n} - 1}{2^{2n}}$

97. $\dfrac{x^n}{y^n}$

99. $\dfrac{1}{2^{5n} \cdot 3^n}$

EXERCISES 4.2, pp. 164

1. 4

3. 3

5. 20

7. 1

9. 17

11. 13

13. 6.40

15. 1.41

17. 2.83

19. 1.41

21. $t^2 - 1$

23. 5

25. -6

27. 6 or -6

29. No solution

31. 4

33. -5

35. 2

37. -8

39. 7

41. 15

43. 2

45. $-3x^3y^6$

47. -41

49. 1

51. 4

53. 4

55. 2 or -2

57. No solution

59. 10

ANSWERS

EXERCISES 4.3, pp. 169

1. 2

3. 5

5. 2

7. 3

9. -2

11. $\dfrac{3}{5}$

13. 16

15. $\dfrac{1}{4}$

17. $\dfrac{1}{1000} = 0.001$

19. 25

21. $\dfrac{1}{8}$

23. 4

25. $\dfrac{1}{4}$

27. 15,625

29. $\sqrt{10}$

31. $\sqrt[4]{14}$

33. $\sqrt[3]{5}$

35. $\sqrt{10}$

37. 36

39. $2^{3/4} = \sqrt[4]{8}$

41. $\dfrac{8}{\sqrt[3]{x^2}}$

43. $\dfrac{1}{64x^2}$

45. $\dfrac{3y}{x^2}$

47. $x + y + 2\sqrt{xy}$

49. $2x + 5$

51. $25 - x$

53. $\dfrac{1}{x} + 1$

55. 3

57. $2^{7/4} = \sqrt[4]{128}$

59. 4

61. 3.86

63. 1.97

65. 126.01

67. 27.54

69. 0.02

EXERCISES 4.4, pp. 172

1. 10

3. $6x$

5. $2\sqrt{3}$

7. $9\sqrt{5}$

9. $2x\sqrt{6}$

11. x^2

13. $6x\sqrt{x}$

15. $4\sqrt{2x}$

17. $7x\sqrt{7x}$

19. x^4

21. $xy^2\sqrt{xy}$

23. $6\sqrt{3y}$

25. $4x\sqrt{7x}$

27. $-4\sqrt[3]{2}$

29. $5\sqrt[3]{5}$

31. 6

33. $2\sqrt[3]{3x^2}$

35. $2y\sqrt[3]{7y^2}$

37. $5y^3\sqrt[3]{5}$

39. $7x$

41. $2\sqrt[4]{6}$

43. $2xy\sqrt[5]{4x^2y}$

45. $6x^3\sqrt{6}$

47. 3

49. 2

51. 6

53. $\dfrac{1}{2}$

55. $4x - y^2$

EXERCISES 4.5, pp. 175

1. $\dfrac{\sqrt{2}}{2}$

3. $\dfrac{\sqrt{15}}{6}$

5. $\dfrac{\sqrt{5}}{10}$

7. $\dfrac{\sqrt{6}}{10}$

9. $\dfrac{\sqrt{6}}{4}$

11. $\dfrac{\sqrt{2}}{6}$

13. $\dfrac{\sqrt{6}}{8}$

15. $\dfrac{\sqrt{2}}{10}$

17. $\dfrac{\sqrt{3x}}{3x}$

19. $\dfrac{\sqrt{2x}}{2x}$

21. $2\sqrt{7}$

23. $\dfrac{\sqrt{2x}}{x}$

25. $\dfrac{\sqrt{2x}}{2x^2}$

27. $\dfrac{\sqrt{2x}}{2x}$

29. $\dfrac{\sqrt{5}}{5}$

31. $\dfrac{\sqrt{21} + 3}{4}$

33. $\sqrt{7} - \sqrt{3}$

35. $\dfrac{\sqrt{11} - 1}{2}$

37. $\sqrt{x} + \sqrt{y}$

39. $\dfrac{\sqrt{x}}{x}$

41. $\dfrac{3}{2\sqrt{3}}$

43. $\dfrac{1}{\sqrt{3}}$

45. $\dfrac{x - 1}{x\sqrt{x - 1}}$

47. $\dfrac{-1}{3(\sqrt{2} + \sqrt{3})}$

49. $\dfrac{1}{\sqrt{x} + \sqrt{y}}$

EXERCISES 4.6, pp. 179

1. $8\sqrt{5}$

3. $6\sqrt{6}$

5. $3\sqrt{5} + 5\sqrt{3}$

7. $2\sqrt{14x}$

9. $4\sqrt{3}$

11. $12\sqrt{6}$

13. $\sqrt{13x}$

15. $6x$

17. $\dfrac{3\sqrt{10}}{10}$

19. $5\sqrt{5} + 2\sqrt{3}$

21. $\dfrac{9\sqrt{42}}{20}$

23. $\sqrt{6} + 4\sqrt{2}$

25. $\dfrac{\sqrt{30}}{6} - \dfrac{\sqrt{14}}{4}$

27. $\dfrac{\sqrt{3}}{3} + 21$

29. $21 + 4\sqrt{5}$

31. $-2 - 2\sqrt{6}$

33. $2x - 2\sqrt{x} - 12$

35. $4\sqrt{14} - \dfrac{\sqrt{2}}{2}$

37. $30 - 12\sqrt{6}$

39. $9\sqrt{2} + 6$

41. 0

43. $7 + 2\sqrt{6} - \sqrt{2} - \sqrt{3}$

EXERCISES 4.7, pp. 181

1. $\sqrt{5}$

3. $\sqrt{7}$

5. $\sqrt{3t}$

7. $2\sqrt{5}$

9. $\sqrt{3x}$

11. $\sqrt[3]{6x}$

13. $4\sqrt{t}$

15. $\sqrt[4]{8r^3}$

17. $\sqrt[3]{3}$

19. $7\sqrt[6]{7}$

21. $\sqrt[3]{2c}$

23. $\sqrt[4]{216}$

25. $\sqrt[3]{25}$

27. 3

29. $\dfrac{\sqrt[4]{2}}{2}$

31. $\sqrt[6]{40}$

33. $\sqrt[6]{5}$

35. $\dfrac{\sqrt[3]{6}}{6}$

37. $\sqrt{5}$
39. $\sqrt[3]{4y^2}$
41. $\sqrt[3]{4}$
43. 3

45. 27
47. $\sqrt[3]{x}$
49. $5\sqrt[9]{5}$
51. $5\sqrt{3}$

53. 0
55. $6\sqrt{2x} + 2\sqrt{5x}$
57. $\sqrt[3]{100y^2} + \sqrt{10y}$

EXERCISES 4.8, pp. 184

1. 9.29×10^7
3. 1.14×10^7
5. 2.4×10^{-9}
7. 6.1×10^{-4}
9. 2.205×10^{-3}
11. 2×10^9
13. 3×10^{11}
15. 1.1×10^{-5}
17. 3×10^{-2}

19. 6.03×10^{23}
21. 2300
23. 0.000 018
25. 0.000 000 0303
27. 0.16667
29. 1,870,000,000
31. 49,000,000,000
33. 83,100,000
35. 0.000 000 01

37. 0.025
39. 0.000 000 5
41. $3.96 \times 10^{-1} = 0.396$
43. $5 \times 10 = 50$
45. $7 \times 10^{-3} = 0.007$
47. $1 \times 10^{-2} = 0.01$
49. $3 \times 10^5 = 300,000$
51. 1.047 radians
53. Approx. 8.9 hours

CHAPTER 4 DIAGNOSTIC TEST, pp. 186

1. False; $3x^{-2} = 3 \cdot \dfrac{1}{x^2} = \dfrac{3}{x^2}$

2. False; $(x^{-1} + 3)^2 = (x^{-1})^2 + 2(x^{-1})(3) + 3^2 = x^{-2} + 6x^{-1} + 9 = \dfrac{1}{x^2} + \dfrac{6}{x} + 9$

3. True; $\dfrac{x - y^{-1}}{x^{-1} - y} = \dfrac{xy(x - y^{-1})}{xy(x^{-1} - y)} = \dfrac{x^2y - x}{y - xy^2} = -\dfrac{x(xy - 1)}{y(xy - 1)} = -\dfrac{x}{y}$

4. False; $\sqrt{64 + 36} = \sqrt{100} = 10$

5. True; $\left(\dfrac{1}{8}\right)^{-2/3} = 8^{2/3} = (8^{1/3})^2 = 2^2 = 4$

6. False; $(3^{0.5})(3^{-1.5}) = 3^{-1} = \dfrac{1}{3}$

7. True; $(x^{1/2} + 3^{1/2})(x^{1/2} - 3^{1/2}) = (x^{1/2})^2 - (3^{1/2})^2 = x - 3$

8. False; $(2x + \sqrt{5})^2 = (2x + \sqrt{5})(2x + \sqrt{5}) = 4x^2 + 4\sqrt{5}x + 5$

9. True; $\sqrt{128x^3}\sqrt{2xy^3} = \sqrt{(64)(2)(2)x^4y^3} = 16x^2y\sqrt{y}$

10. False; $\dfrac{2}{(x - \sqrt{3})} \dfrac{(x + \sqrt{3})}{(x + \sqrt{3})} = \dfrac{2(x + \sqrt{3})}{x^2 - 3}$

11. False; $\dfrac{(5 + \sqrt{3})}{2} \dfrac{(5 - \sqrt{3})}{(5 - \sqrt{3})} = \dfrac{25 - 3}{2(5 - \sqrt{3})} = \dfrac{22}{2(5 - \sqrt{3})} = \dfrac{11}{5 - \sqrt{3}}$

12. True; $x = -2 + \sqrt{6}$; $x^2 = (-2 + \sqrt{6})^2 = 4 - 4\sqrt{6} + 6 = 10 - 4\sqrt{6}$
$x^2 + 4x - 2 = 10 - 4\sqrt{6} + 4(-2 + \sqrt{6}) - 2$
$= 10 - 4\sqrt{6} - 8 + 4\sqrt{6} - 2 = 0$

13. True; $\sqrt[3]{8\sqrt{8}} = (8 \cdot 8^{1/2})^{1/3} = (8^{3/2})^{1/3} = (8)^{1/2} = (4 \cdot 2)^{1/2} = 4^{1/2}2^{1/2} = 2\sqrt{2}$

14. True

15. True

16. False; $1.42 \times 10^{-3} = 0.00142$

17. True; $3.415 \times 10^2 = 341.5$

18. False; let c = length of the hypotenuse. Then

$a^2 + b^2 = c^2$; $2^2 + 5^2 = c^2$ and $c^2 = 4 + 25 = 29$

Therefore $c = \sqrt{29}$.

19. True; $4^2 + b^2 = 10^2$

$$b^2 = 10^2 - 4^2 = 100 - 16 = 84$$
$$b = \sqrt{84} = 2\sqrt{21} \text{ units}$$

20. False; every *nonnegative* real number has at least one real square root.

CHAPTER 4 REVIEW EXERCISES, pp. 187

1. (a) $2\sqrt{3}$ (b) 10 (c) $7\sqrt{2}$

2. (a) 8.54 (b) 0.02 (c) 2.92 (d) 0.03

3. (a) $-3\sqrt[3]{3}$ (b) 7 (c) $\dfrac{2}{3}$ (d) 0.0013 (e) $2xy$

4. (a) $8\sqrt{3}$ (b) $6\sqrt{5}$ (c) $2\sqrt[3]{18}$ (d) $12\sqrt{15}$ (e) $2x\sqrt[4]{2xy^3}$

5. (a) $\dfrac{3\sqrt{5}}{5}$ (b) $\dfrac{\sqrt{15}}{5}$ (c) $6\sqrt{3}$ (d) $2\sqrt{2} - 5$ (e) $-(\sqrt{3} + \sqrt{6})$

(f) $\dfrac{3\sqrt{x} - 6}{x - 4}$

6. (a) $5\sqrt{6}$ (b) $5\sqrt{2} - 20$ (c) $8\sqrt{7} - 3\sqrt{2}$ (d) 288 (e) $-\dfrac{\sqrt{6}}{2}$

(f) $\dfrac{31\sqrt{30}}{30} + 3\sqrt{3}$

7. (a) a^8b^{13} (b) $-\dfrac{18a^8b^7}{25m^8n^3}$ (c) $\dfrac{a^8b^4}{192}$ (d) $-y^4$

8. (a) $\dfrac{1}{m^4}$ (b) p^2 (c) $\dfrac{x^2z}{y}$ (d) $\dfrac{(a + b)}{ab}$ (e) $\dfrac{x^2y^2}{(x + y)^2}$

9. (a) $\sqrt[4]{x}$ (b) $2\sqrt{a}$ (c) $\sqrt[4]{x^3}$ (d) $\dfrac{\sqrt[3]{xy}}{xy}$ (e) $3\sqrt{x}$

10. (a) $y^{1/3}$ (b) $x^{1/2}y^{5/6}$ (c) $xy^{1/2}$ (d) $(xy)^{2/3}$

11. (a) 1 (b) 5 (c) $\dfrac{1}{8}$ (d) $\dfrac{29}{6}$

12. (a) 0 (b) 0 (c) 18

13. (a) 7   (b) $3\sqrt{5}$ (c) $4\sqrt{6}$

14. (a) 64   (b) $\dfrac{399}{4}$

15. (a) 0.09   (b) 80,000

16. (a) $5x\sqrt{2x}$ (b) $2\sqrt{5}$ (c) $\sqrt{3x}$

EXERCISES 5.1, pp.195

1. $2i$

3. xi

5. $i\sqrt{2}$

7. $4i$

9. $\dfrac{6}{7}i$

11. $-20i$

13. $12i\sqrt{3}$

15. $40i$

17. $6i\sqrt{2}$

19. $21xi$

21. $1 - 6i$

23. $4 - 12i\sqrt{2}$

25. $0 - 10i$

27. $7 + 6i$

29. $5 + 0i$

31. $0 + 2i$

33. $0 + 6i$

35. $x = 4, y = 2$

37. $x = 0, y = 2$

39. $x = 3, y = -1$

41. $7 + 2i$

43. $-2 - 3i$

45. $4 - 2i$

47. $-2 + 6i$

49. $7 - 2i$

51. $-1 + i\sqrt{2}$

53. $-4 + 0i = -4$

55. $0 + 4i = 4i$

57. a is any real number, $b = 0$

EXERCISES 5.2, pp. 200

1. -10

3. $6i$

5. $-6i$

7. $6 + 15i$

9. $12 - 5i$

11. 20

13. $5 + 5i$

15. $5 + 12i$

17. $17 + i$

19. $3 + i$

21. $5 - 15i$

23. 0

25. 61

27. $-20 + 20i$

29. -53

31. $25 + 25i$

33. i

35. $-i$

37. $-i$

39. 0

41. $5 - 7i$

43. $2 - i$

45. $-i$

47. 3

49. 0

51. $\dfrac{3}{13} - \dfrac{2}{13}i$

53. $\dfrac{6}{13} + \dfrac{4}{13}i$

55. $\dfrac{4}{13} + \dfrac{6}{13}i$

57. $0 + i = i$

59. $1 + 0i = 1$

61. $\dfrac{\sqrt{6} - 2}{7} + \dfrac{\sqrt{3} + 2\sqrt{2}}{7}i$

63. $\dfrac{3}{5} + \dfrac{3}{10}i$

65. $(-2 - \sqrt{3}) + 0i = -2 - \sqrt{3}$

67. $\dfrac{-\sqrt{6} - 2\sqrt{3} - 3\sqrt{2} - 6}{6}$

69. $\dfrac{8}{5} + 0i = \dfrac{8}{5}$

71. $(3 + i)^2 - 6(3 + i) + 10 = 9 + 6i + i^2 - 18 - 6i + 10$
$$= 0$$
$(3 - i)^2 - 6(3 - i) + 10 = 9 - 6i + i^2 - 18 + 6i + 10$
$$= 0$$

73. $(2 + i\sqrt{5})^2 - 4(2 + i\sqrt{5}) + 9 = 4 + 4i\sqrt{5} + 5i^2 - 8 - 4i\sqrt{5} + 9$
$$= 13 - 13 = 0$$
$(2 - i\sqrt{5})^2 - 4(2 - i\sqrt{5}) + 9 = 4 - 4i\sqrt{5} + 5i^2 - 8 + 4i\sqrt{5} + 9$
$$= 13 - 13 = 0$$

EXERCISES 5.3, pp. 207

1. $\{1, -5\}$

3. $\{0, 2\}$

5. $\{2, -3\}$

7. $\{0, 3\}$

9. $\{-3i, 3i\}$

11. $\{0, -9\}$

13. $\{-2, -1\}$

15. $\{-3, 2\}$

17. $\{-5, 4\}$

19. $\{10, -3\}$

21. $\left\{0, \dfrac{5}{2}\right\}$

23. $\left\{-\dfrac{1}{2}, 1\right\}$

25. $\{5\}$

27. $\{3, 1\}$

29. $\left\{\dfrac{-b + c}{a}, \dfrac{-b - c}{a}\right\}$

31. $\left\{\dfrac{-3}{2}, 5\right\}$

33. $\{0, 1, 5\}$

35. $\{-4, -3, -1, 1\}$

37. $\{-3, -2, 2, 3\}$

39. $\{-1, -3\}$

41. $\{4\}\ x \neq -2$

43. $\{4\}\ x \neq 0$

45. $\{-4, 4\}$

47. $\left\{\dfrac{-9}{2}, 2\right\}$

49. $\{3, 4\}$

EXERCISES 5.4, pp. 211

1. $\{5 + \sqrt{2}, 5 - \sqrt{2}\}$

3. $\{5 + i\sqrt{2}, 5 - i\sqrt{2}\}$

5. $\left\{\dfrac{3 + \sqrt{7}}{2}, \dfrac{3 - \sqrt{7}}{2}\right\}$

7. $\{2 + \sqrt{3}, 2 - \sqrt{3}\}$

9. $\{-1, 3\}$

11. $\{5 + \sqrt{5}, 5 - \sqrt{5}\}$

13. $\{3 + \sqrt{6}, 3 - \sqrt{6}\}$

15. $\{-3 + \sqrt{6}, -3 - \sqrt{6}\}$

17. $\{4 + i, 4 - i\}$

19. $\left\{\dfrac{1 + \sqrt{5}}{2}, \dfrac{1 - \sqrt{5}}{2}\right\}$

21. $\{1 + i, 1 - i\}$

23. $\{0, -4\}$

25. $\left\{-\dfrac{1}{2}, 1\right\}$

27. $\{2 + i\sqrt{5}, 2 - i\sqrt{5}\}$

29. $\left\{-\dfrac{5}{2}, \dfrac{3}{2}\right\}$

31. $\left\{-4, \dfrac{3}{2}\right\}$

33. $\left\{\dfrac{3 + i\sqrt{7}}{2}, \dfrac{3 - i\sqrt{7}}{2}\right\}$

35. $\{1 + 2i, 1 - 2i\}$

37. $\{1 + \sqrt{2}, 1 - \sqrt{2}\}$

39. $\{5 + i\sqrt{70}, 5 - i\sqrt{70}\}$

41. $\{5 + i, 5 - i\}$

43. $\{5y, y\}$

45. $\left\{\dfrac{y + y\sqrt{5}}{2}, \dfrac{y - y\sqrt{5}}{2}\right\}$

47. $\left\{\dfrac{2}{5}\sqrt{25 - x^2}, -\dfrac{2}{5}\sqrt{25 - x^2}\right\}$

49. $\left\{\dfrac{-b + \sqrt{b^2 - 4ac}}{2a}, \dfrac{-b - \sqrt{b^2 - 4ac}}{2a}\right\}$

EXERCISES 5.5, pp. 217

1. (a) $2x^2 - 3x - 1 = 0$; $a = 2, b = -3, c = -1$ (b) Discriminant = 17   (c) Real and unequal

3. (a) $x^2 - 5 = 0$; $a = 1, b = 0, c = -5$ (b) Discriminant = 20   (c) Real and unequal

5. (a) $x^2 - 4x + 8 = 0$; $a = 1, b = -4, c = 8$ (b) Discriminant = -16 (c) Imaginary

7. (a) $x^2 + x - 2 = 0$; $a = 1, b = 1, c = -2$ (b) Discriminant = 9   (c) Real and unequal

9. (a) $x^2 - x + 20 = 0$; $a = 1, b = -1, c = 20$ (b) Discriminant = -79 (c) Imaginary

11. $\left\{-1, -\dfrac{2}{3}\right\}$

13. $\left\{\dfrac{3 + \sqrt{5}}{2}, \dfrac{3 - \sqrt{5}}{2}\right\}$

15. $\left\{\dfrac{3 + i\sqrt{3}}{2}, \dfrac{3 - i\sqrt{3}}{2}\right\}$

17. $\left\{-1, -\dfrac{3}{4}\right\}$

19. $\left\{-2, \dfrac{1}{4}\right\}$

21. $\{2 + 2\sqrt{2}, 2 - 2\sqrt{2}\}$

23. $\left\{\dfrac{5 + \sqrt{5}}{10}, \dfrac{5 - \sqrt{5}}{10}\right\}$

25. $\left\{\dfrac{2 + \sqrt{7}}{2}, \dfrac{2 - \sqrt{7}}{2}\right\}$

27. $\left\{\dfrac{-1 + \sqrt{7}}{3}, \dfrac{-1 - \sqrt{7}}{3}\right\}$

29. $\left\{0, \dfrac{5}{3}\right\}$

31. $\left\{\dfrac{5 + \sqrt{5}}{2}, \dfrac{5 - \sqrt{5}}{2}\right\}$

33. $\left\{-2, \dfrac{1}{2}\right\}$

35. $\{3 + \sqrt{3}, 3 - \sqrt{3}\}$

37. $\left\{\dfrac{4}{5}\right\}$

39. $\{-p + \sqrt{p^2 - q}, \, -p - \sqrt{p^2 - q}\}$

41. $\left\{\dfrac{y}{3}, \, -3y\right\}$

43. $\left\{\dfrac{-v + \sqrt{v^2 + 2gs}}{g}, \, \dfrac{-v - \sqrt{v^2 + 2gs}}{g}\right\}$

45. $\left\{\dfrac{-b + \sqrt{b^2 - ac}}{c}, \, \dfrac{-b - \sqrt{b^2 - ac}}{c}\right\}$

47. $\left\{\dfrac{3y + y\sqrt{5}}{4}, \, \dfrac{3y - y\sqrt{5}}{4}\right\}$

EXERCISES 5.6, pp. 221

1. $\{3, -3, \sqrt{5}, -\sqrt{5}\}$
3. $\{\sqrt{2}, -\sqrt{2}, i\sqrt{3}, -i\sqrt{3}\}$

5. $\{2, -1 + i\sqrt{3}, -1 - i\sqrt{3}\}$

7. $\{-2, 5\}$

9. $\left\{i, -i, \dfrac{3}{2}, -\dfrac{3}{2}\right\}$

11. $\{81, 256\}$
13. $\{-64, 8\}$

15. $\{0, 1, 5, 6\}$
17. $\{-9, 1, -4 + 2\sqrt{3}, -4 - 2\sqrt{3}\}$

19. $\left\{\dfrac{1}{3}, \dfrac{5}{4}\right\}$

21. $\left\{-1, 2, \dfrac{1}{2}\right\}$

23. $\{3, 3 + \sqrt{35}, 3 - \sqrt{35}\}$
25. $\{2\sqrt{2}, -2\sqrt{2}, 2, -2\}$

EXERCISES 5.7, pp. 225

1. $\{11\}$
3. $\{-4\}$
5. $\varnothing$
7. $\{0\}$
9. $\{2\}$
11. $\varnothing$

13. $\{2\}$
15. $\{2, -1\}$
17. $\{10\}$
19. $\{-5, -4\}$
21. $\{18\}$
23. $\left\{1, \dfrac{9}{8}\right\}$

25. $\{-1\}$
27. $\{1, 2\}$
29. $\{20\}$
31. $\{13\}$
33. $\{-4\}$

EXERCISES 5.8, pp. 228

1. $\{-3, -1, 1\}$
3. $\left\{-\dfrac{2}{3}, \dfrac{2}{3}, -1, 1\right\}$

5. $\{-3, 3, -2, 1 + i\sqrt{3}, 1 - i\sqrt{3}\}$

7. $\left\{-\dfrac{2}{3}, 1, -2\right\}$

9. $\left\{1, \dfrac{-1 + i\sqrt{3}}{2}, \dfrac{-1 - i\sqrt{3}}{2}\right\}$

11. $\left\{4, \dfrac{-1 + i\sqrt{15}}{2}, \dfrac{-1 - i\sqrt{15}}{2}\right\}$

13. $\{3, -2, i\sqrt{6}, -i\sqrt{6}\}$

15. $\{2, 3, -5\}$

17. $\{-5, \sqrt{2}, -\sqrt{2}\}$

19. $\left\{-2, 2, -\dfrac{1}{2}i, \dfrac{1}{2}i\right\}$

21. $\left\{1, -1, 3, \dfrac{-3 + 3i\sqrt{3}}{2}, \dfrac{-3 - 3i\sqrt{3}}{2}\right\}$

23. $\left\{-5, \dfrac{-1 + \sqrt{5}}{2}, \dfrac{-1 - \sqrt{5}}{2}\right\}$

25. $\{-2, 2, 3\}$

27. $\{-1, 3\}$

29. $\{1\}$

EXERCISES 5.9, pp. 232

1. $x^2 + x^2 = 3^2$; $\dfrac{3\sqrt{2}}{2}$ = approx. 2.12 units

3. $x^2 + (x + 7)^2 = 13^2$; 5 in., 12 in.

5. $\dfrac{1}{x} - \dfrac{1}{7 - x} - \dfrac{1}{12}$; (3 and 4) or (28 and -21)

7. $(x - 4)(2x - 4)(2) = 320$; 12 in. $\times$ 24 in.

9. $\dfrac{3}{x} + \dfrac{3}{x - 8} = 1$; Faster man, 4 days for $200
      Slower man, 12 days for $240
      Together, 3 days for $210

11. $15\left(\dfrac{1}{x}\right) + 15\left(\dfrac{1}{x + 12}\right) - 15\left(\dfrac{1}{45}\right) = 1$; 18 min

13. $\dfrac{2610}{x - 20} - \dfrac{1}{3} = \dfrac{2610}{x + 20}$; 560 mph

15. $x^2 = 90^2 + 90^2$; $x = 90\sqrt{2}$, approx. 127.3 ft
       $127.3 - 60.5 = 66.8$ ft

17. $x^2 = 16^2 + 30^2$; 34 ft

19. $80^2 = 400h - 4h^2$; 20 ft

21. $25{,}000 = 110\,I - 0.04\,I^2$; $I = 2500$ amps

23. $x^2 + 32^2 = (x + 8)^2$; $x = 60$ in., or 5 ft

CHAPTER 5 DIAGNOSTIC TEST, pp. 235

1. Conjugate pairs

2. Real

3. 4, -2; $3x = 12$ and $x = 4$, $-5y = 10$ and $y = -2$

4. $-1 - i$; since the coefficients of $x^2 + 2x + 2$ are real numbers, if one root of the equation is imaginary, the other root is its conjugate.

5. 64; $a = 3$, $b = 2$, $c = -5$; $b^2 - 4ac = 2^2 - 4(3)(-5) = 64$

6. Real, rational, and unequal; $\sqrt{64} = 8$, a rational number.

7. $u = x^{1/3}$; the equation then becomes $u^2 - 3u - 4 = 0$.

8. $k = 2$; $x^2 + 4x + 6 - k = 0$; $a = 1$, $b = 4$, $c = (6 - k)$
$b^2 - 4ac = 16 - 4(1)(6 - k) = 16 - 24 + 4k = -8 + 4k$
The roots are equal if $b^2 - 4ac = 0$.
Therefore $-8 + 4k = 0$ and $k = 2$.

9. $7 - (6\sqrt{2})i$; $(3 - \sqrt{-2})^2 = (3 - i\sqrt{2})(3 - i\sqrt{2}) = 9 - 6i\sqrt{2} + 2i^2$
$\qquad = 9 - 6i\sqrt{2} + 2(-1)\ = 7 - 6i\sqrt{2} = 7 - (6\sqrt{2})i$

10. 29; $x^2 - 6x + 9 = 20 + 9$, $(x - 3)^2 = 29$

11. (d); $x^2 + 1 = 0$, $x^2 = -1$, and $x = \pm\sqrt{-1} = \pm i$

12. (a); $x + \sqrt{x - 1} = 3$
$\qquad\qquad \sqrt{x - 1} = 3 - x$
$\qquad\qquad\quad x - 1 = 9 - 6x + x^2$
$\quad x^2 - 7x + 10 = 0$
$(x - 2)(x - 5) = 0$; $x = 2$ or $x = 5$
For $x = 2$, $2 + \sqrt{2 - 1} = 2 + 1 = 3$.
For $x = 5$, $5 + \sqrt{5 - 1} = 5 + \sqrt{4} = 5 + 2 = 7 \neq 3$.
Therefore the only solution is $x = 2$.

13. (a); $(3 + 2\sqrt{-3})(2 - 3\sqrt{-3}) = (3 + 2i\sqrt{3})(2 - 3i\sqrt{3}) = 6 - 9i\sqrt{3} + 4i\sqrt{3} - 6i^2(3)$
$\qquad = 6 - 5i\sqrt{3} + 18 = 24 - 5i\sqrt{3}$

14. (d); $(1 + i)^3 = 1^3 + 3(1^2)(i) + 3(1)(i^2) + i^3 = 1 + 3i + 3i^2 + i^3$
$\qquad = 1 + 3i - 3 - i = -2 + 2i$

15. (c); $(\sqrt{5x - 4})^2 = (\sqrt{x} + 3)^2$
$\qquad\quad 5x - 4 = x + 6\sqrt{x} + 9$

16. (a); let $u = x^2$ and $u^2 = x^4$; then $u^2 - 11u + 24 = 0$
$\qquad\qquad\qquad\qquad (u - 3)(u - 8) = 0$
$\qquad\qquad\qquad\qquad\qquad u = 3 \quad \text{or} \quad u = 8$
Therefore $x^2 = 3$ and $x = \pm\sqrt{3}$, or $x^2 = 8$ and $x = \pm\sqrt{8} = \pm 2\sqrt{2}$

17. (a); $a = 3$, $b = 5$, $c = -8$, $b^2 - 4ac = 25 - 4(3)(-8) = 25 + 96 = 121$
Since $121 = n^2$ for $n = 11$, the roots are real, rational, and unequal.

18. (d); $3 = 3 + 0i$. The conjugate of $3 + 0i = 3 - 0i = 3$.

CHAPTER 5 REVIEW EXERCISES, pp. 236

1. $2 + 4i$

2. $1 + 8i$

3. $2 + i\sqrt{3}$

4. $18 + i$

5. $-48 - 14i$

6. $\dfrac{-1}{13} - \dfrac{5}{13}i$

7. $x = \dfrac{5}{3}, y = 0$

8. $\{-4, 10\}$

9. $\left\{-\dfrac{1}{6}, -\dfrac{5}{2}\right\}$

10. $\left\{0, \dfrac{4}{3}\right\}$

11. $\{-3i, 3i\}$

12. $\{-3, -1, 1\}$

13. $\{-1, 1, -i, i\}$

14. $\{-5, -3\}$

15. $\{1 + i\sqrt{2}, 1 - i\sqrt{2}\}$

16. $\{2 + \sqrt{5}, 2 - \sqrt{5}\}$

17. $\left\{\dfrac{2 + 3\sqrt{3}}{2}, \dfrac{2 - 3\sqrt{3}}{2}\right\}$

18. $\left\{\dfrac{-5 + \sqrt{33}}{4}, \dfrac{-5 - \sqrt{33}}{4}\right\}$

19. $\left\{-\dfrac{4}{3}, \dfrac{5}{2}\right\}$

20. $\left\{\dfrac{1 + \sqrt{17}}{4}, \dfrac{1 - \sqrt{17}}{4}\right\}$

21. $\left\{\dfrac{8 + 2\sqrt{10}}{3}, \dfrac{8 - 2\sqrt{10}}{3}\right\}$

22. $\left\{\dfrac{1 + \sqrt{29}}{2}, \dfrac{1 - \sqrt{29}}{2}\right\}$

23. $\left\{\dfrac{-3 + \sqrt{89}}{10}, \dfrac{-3 - \sqrt{89}}{10}\right\}$

24. $\left\{0, -\dfrac{3}{8}\right\}$

25. $\left\{\dfrac{7\sqrt{3}}{3}, \dfrac{-7\sqrt{3}}{3}\right\}$

26. $\left\{\dfrac{\sqrt{42}}{6}, \dfrac{-\sqrt{42}}{6}\right\}$

27. $\{-2 + 2i, -2 - 2i\}$

28. $\left\{\dfrac{5}{2}, -2\right\}$

29. $\left\{\dfrac{-5 + \sqrt{13}}{6}, \dfrac{-5 - \sqrt{13}}{6}\right\}$

30. $\{-4, -1\}$

31. $k = 9$

32. $k = 2$ or -2

33. $k = \dfrac{5}{4}$

34. $k = 4$ or -4

35. $k = 1$

36. $y = 5x, y = -2x$

37. $y = (-1 + \sqrt{2})x^2$ or $y = (-1 - \sqrt{2})x^2$

38. $\left\{\dfrac{\sqrt{39}}{3}, \dfrac{-\sqrt{39}}{3}, i\sqrt{6}, -i\sqrt{6}\right\}$

39. $\{79\}$

40. $\{1, 125\}$

41. $\left\{-\dfrac{1}{2}, 5\right\}$

42. $\{4\}$

43. $\{10\}$

44. $\dfrac{15}{x} + \dfrac{25}{x + 20} = 1$; 30 min

45. $\dfrac{200}{x + 20} + \dfrac{200}{x - 30} = \dfrac{7}{3}$; 180 mph

46. $x(45 - x) = 450$; 15×30 ft

47. $x^2 + (x + 9)^2 = 45^2$; 27 in., 36 in.

48. $t = \dfrac{cv + \sqrt{c^2v^2 + 2gH}}{g}$

49. $r = \dfrac{-\pi h + \sqrt{\pi^2 h^2 + 2\pi T}}{2\pi}$

50. $\dfrac{L}{W} = \dfrac{1 + \sqrt{5}}{2}; \dfrac{W}{L} = \dfrac{-1 + \sqrt{5}}{2}$

EXERCISES 6.1, pp. 243

1. 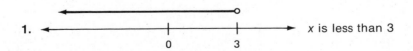 x is less than 3

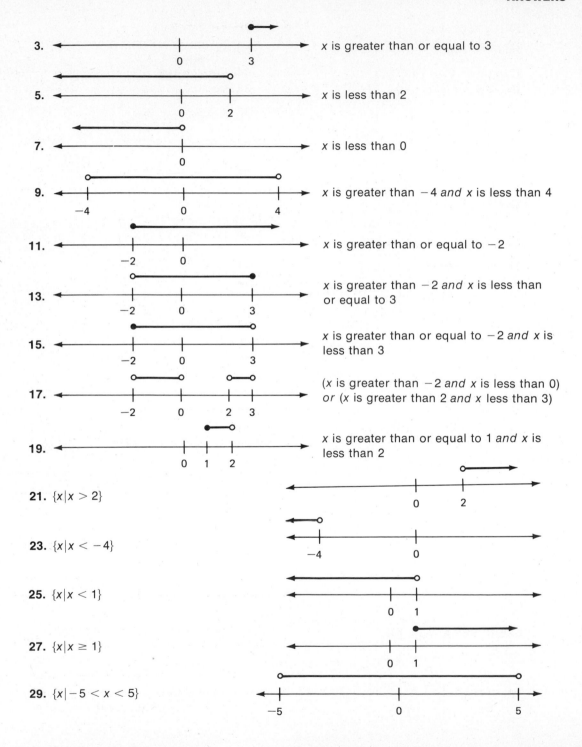

3. *x* is greater than or equal to 3

5. *x* is less than 2

7. *x* is less than 0

9. *x* is greater than −4 *and* *x* is less than 4

11. *x* is greater than or equal to −2

13. *x* is greater than −2 *and* *x* is less than or equal to 3

15. *x* is greater than or equal to −2 *and* *x* is less than 3

17. (*x* is greater than −2 *and* *x* is less than 0) *or* (*x* is greater than 2 *and* *x* less than 3)

19. *x* is greater than or equal to 1 *and* *x* is less than 2

21. $\{x \mid x > 2\}$

23. $\{x \mid x < -4\}$

25. $\{x \mid x < 1\}$

27. $\{x \mid x \geq 1\}$

29. $\{x \mid -5 < x < 5\}$

31. $\{x|x \le -1\} \cap \{x|x < 2\}$

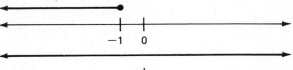

33. $\{x|x > -5\} \cup \{x|x < 5\}$

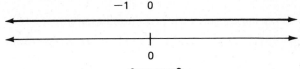

35. $\{x|2 \le x\} \cap \{x|x < 4\}$

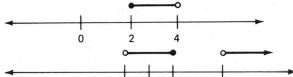

37. $\{x|-1 < x \le 1\} \cup \{x|x > 3\}$

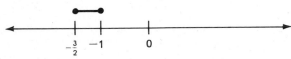

39. $\{x|-2 \le x \le -1\} \cap \left\{x \ \middle| \ -\dfrac{3}{2} \le x < 5\right\}$

EXERCISES 6.2, pp. 248

1. $\{x|x > 2\}$

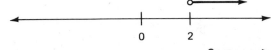

3. $\{x|x > 3\}$

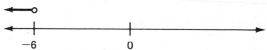

5. $\{x|x < -6\}$

7. $\{x|x \ge 2\}$

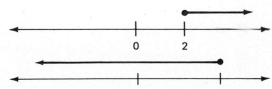

9. $\{x|x \le 5\}$

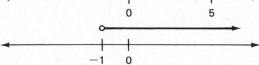

11. $\{x|x > -1\}$

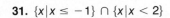

13. $\{x \mid x < -3\}$

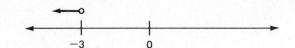

15. R, the set of all real numbers

17. $\{x \mid x \geq 0\}$

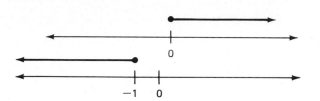

19. $\{x \mid x \leq -1\}$

21. $\{x \mid x \geq 3\}$

23. $\{x \mid x \leq 12\}$

25. $\{x \mid x \geq -3\}$

27. $\{x \mid x > 8\}$

29. $\{x \mid x > 46\}$

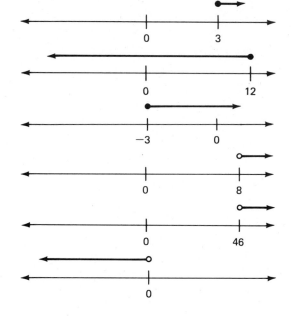

31. $\{x \mid x < 0\}$

33. $\varnothing$

35. $\{x \mid x < 2\}$

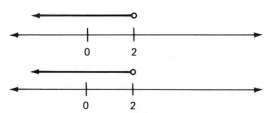

37. $\{x \mid x < 2\}$

39. $\varnothing$

41. $\left\{x \mid x < 0 \text{ or } x \geq \dfrac{3}{2}\right\}$

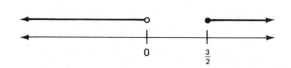

43. $\{x \mid x < 1 \text{ or } x > 2\}$

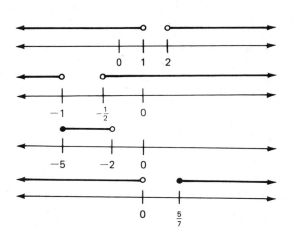

45. $\left\{x \mid x < -1 \text{ or } x > -\dfrac{1}{2}\right\}$

47. $\{x \mid -5 \le x < -2\}$

49. $\left\{x \mid x < 0 \text{ or } x \ge \dfrac{5}{7}\right\}$

51. $\{x \mid 3 < x < 5\}$

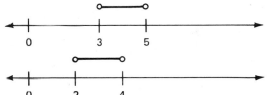

53. $\{x \mid 2 < x < 4\}$

55. $90 \le \dfrac{357 + x}{5} \le 100$; $\{x \mid 93 \le x \le 100\}$

57. $\dfrac{5 + y}{80} \le \dfrac{30}{100}$; 19 games

59. $50 \le \dfrac{9C + 160}{5} \le 77$; $\{C \mid 10 \le C \le 25\}$

EXERCISES 6.3, pp. 256

1. $\{x \mid x < -5 \text{ or } x > -3\}$

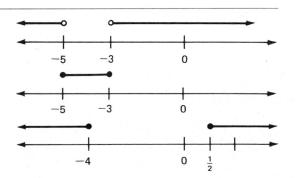

3. $\{x \mid -5 \le x \le -3\}$

5. $\left\{x \mid x \le -4 \text{ or } x \ge \dfrac{1}{2}\right\}$

7. $\left\{x \mid -4 < x < \dfrac{1}{2}\right\}$

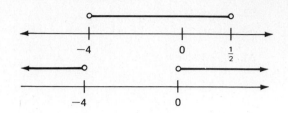

9. $\{x \mid x < -4 \text{ or } x > 0\}$

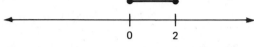

11. $\{x \mid 0 \le x \le 2\}$

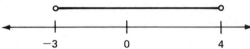

13. $\{x \mid x < 0 \text{ or } x > 3\}$

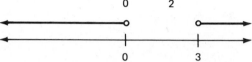

15. $\{x \mid -1 < x < 4\}$

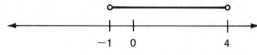

17. $\{x \mid -3 < x < 4\}$

19. $\left\{x \mid x < -\dfrac{5}{2} \text{ or } x > 7\right\}$

21. $\{x \mid x \text{ is any real number}\}$

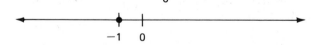

23. $\{x \mid x = -1\}$

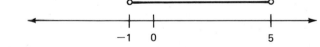

25. $\{x \mid -1 < x < 5\}$

27. $\{x \mid x < -1 \text{ or } x > 9\}$

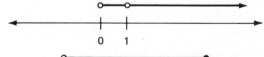

29. $\{x \mid x > 0 \text{ and } x \ne 1\}$

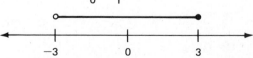

31. $\{x \mid -3 < x \le 3\}$

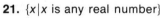

ANSWERS

33. $\{x \mid x = 1\}$

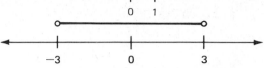

35. $\{x \mid -3 < x < 3\}$

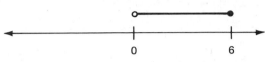

37. $\{x \mid 0 < x \le 6\}$

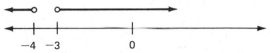

39. $\{x \mid x < -4 \text{ or } x > -3\}$

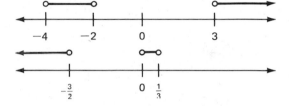

41. $\{x \mid -4 < x < -2 \text{ or } x > 3\}$

43. $\left\{x \mid x < -\dfrac{3}{2} \text{ or } 0 < x < \dfrac{1}{3}\right\}$

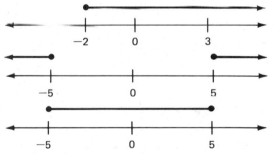

45. $\{x \mid x \ge -2\}$

47. $\{x \mid x \le -5 \text{ or } x \ge 5\}$

49. $\{x \mid -5 \le x \le 5\}$

51. (a) $0 < v < 66$ ft/sec   (b) $0 < v < 45$ mph
53. $0 \le v \le 26 \text{ or } v \ge 34$

EXERCISES 6.4, pp. 261

1. 4

3. 5

5. 7
7. 5
9. 12

11. $\{-3, 3\}$
13. $\left\{-\dfrac{1}{2}, \dfrac{1}{2}\right\}$
15. $\varnothing$
17. $\{-2, 2\}$
19. $\{-2, 8\}$

21. $\left\{-\dfrac{13}{3}, 3\right\}$

23. $\left\{-\dfrac{2}{3}, 2\right\}$

25. $\{-5, -1\}$

27. $\{-2, 2\}$

29. $\{0, 4\}$

31. $\left\{-3, \dfrac{7}{3}\right\}$

33. $\{-1, 5\}$

35. $\left\{-\dfrac{3}{8}, -\dfrac{1}{8}\right\}$

37. $\{x \mid x > 0\}$

39. $\{a - b, a + b\}$

41. $\{x \mid x \geq 0\}$

43. $\{-4, 4\}$

45. $\{x \mid x \text{ is any real number}\}$

47. $\{x \mid x \geq 2\}$

49. $\{x \mid x \geq 2\}$

EXERCISES 6.5, pp. 264

1. $\{x \mid -2 < x < 2\}$

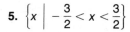

3. $\{x \mid x \leq -2 \text{ or } x \geq 2\}$

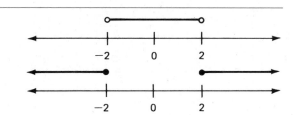

5. $\left\{x \mid -\dfrac{3}{2} < x < \dfrac{3}{2}\right\}$

7. $\{x \mid x < 1 \text{ or } x > 7\}$

9. $\{x \mid -6 \leq x \leq -2\}$

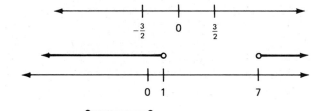

11. $\left\{x \mid -\dfrac{1}{2} < x < \dfrac{7}{2}\right\}$

13. $\left\{x \mid x < -\dfrac{1}{2} \text{ or } x > \dfrac{7}{2}\right\}$

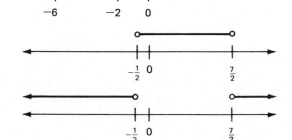

15. $\{x \mid -1 < x < 3\}$

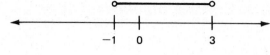

17. $\{x \mid x \text{ is any real number}\}$

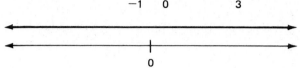

19. $\varnothing$

21. $\left\{x \mid x > \dfrac{1}{2} \text{ or } x < -\dfrac{1}{2}\right\}$

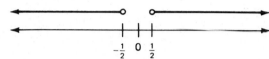

23. $\left\{x \mid x < \dfrac{4}{3} \text{ or } x > \dfrac{8}{3}\right\}$

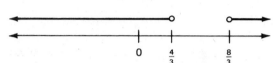

25. $\varnothing$

27. $\{x \mid x \text{ is any real number}\}$

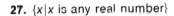

29. $\left\{x \mid -\dfrac{7}{2} < x < \dfrac{9}{2}\right\}$

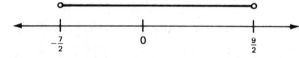

31. $\{x \mid x \leq 2 \text{ or } x \geq 14\}$

33. $\varnothing$

35. $\{x \mid x \text{ is any real number}\}$

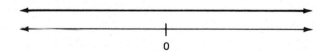

37. $\left\{x \mid -\dfrac{3}{4} \leq x \leq \dfrac{1}{4}\right\}$

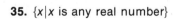

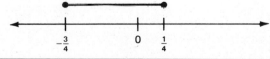

39. $\left\{x \mid x < 0 \text{ or } 0 < x < \dfrac{1}{2} \text{ or } x > 1\right\}$

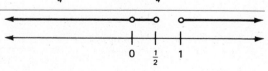

CHAPTER 6 DIAGNOSTIC TEST, pp. 265

1. $x > 2$
2. $4 < x \le 6$
3. $\{x \mid -2 < x \le -1\}$; this is the intersection of the two sets and contains all numbers that the two sets have in common.
4. $\cup$, or
5. $\cap$, and
6. $<$; multiplication by a positive number does not change the sense of the inequality.
7. $>$; multiplication by a negative number reverses the sense of the inequality.
8. $>$; see answer 7.
9. $x > -\dfrac{9}{4}$; $3x - 5 < 4 + 7x$

$$-4x < 9$$
$$x > -\frac{9}{4}$$

10. $\{-11, 3\}$; if $|x + 4| = 7$, then $x + 4 = 7$ or $x + 4 = -7$
$$x = 3 \quad \text{or} \quad x = -11$$

11. $\{x \mid -11 < x < 3\}$; if $|x + 4| < 7$, then $-7 < x + 4 < 7$
$$-11 < x \quad < 3$$

12. $\{x \mid x < -11 \quad \text{or} \quad x > 3\}$; if $|x + 4| > 7$, then $x + 4 > 7$ or $x + 4 < -7$
$$x > 3 \quad \text{or} \quad x < -11$$

13. $\{x \mid x < -3 \quad \text{or} \quad x > 2\}$;

14. $\left\{x \mid -\dfrac{1}{3} < x \le 2\right\}$;

The fraction can equal zero only when the numerator is zero; that is, $x - 2 = 0$ and $x = 2$. We therefore include 2 in the solution set.

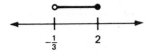

15. $\{3\}$; $\quad x^2 + 9 \le 6x$
$$x^2 - 6x + 9 \le 0$$
$$(x - 3)^2 \le 0$$

Since the square of a real number is never negative, the only solution occurs when $x - 3 = 0$, and $x = 3$.

16. (e); The solution set is empty, $\varnothing$. By definition the absolute value of an expression is always greater than or equal to zero.

17. (a); $3x - 8 \ge 10$
$$3x \ge 18$$
$$x \ge 6$$

18. (c); $\quad |2x + 1| < 5$
$$-5 < 2x + 1 < 5$$
$$-6 < 2x < 4$$
$$-3 < x < 2$$

19. (c); $3 - x = -(x - 3)$; $|x - 3| = -(x - 3)$ if $x - 3 < 0$. Therefore $x < 3$.

20. (d); $x^2 - 4x + 4 \ge 0$
$$(x - 2)^2 \ge 0$$

This is a true statement for all real numbers because the square of a real number is always positive or zero. (Compare with Problem 15.)

CHAPTER 6 REVIEW EXERCISES, pp. 266

1. x is less than 4

2. x is greater than or equal to -1

3. x is greater than -2 *and* x is less than or equal to 5

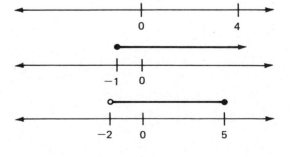

4. *x* is greater than or equal to −1 *and x* is less than 0

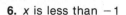

5. *x* is greater than or equal to −1 *and x* is less than 2

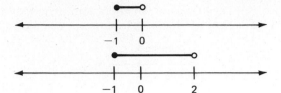

6. *x* is less than −1

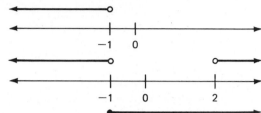

7. *x* is less than −1 *or x* is greater than 2

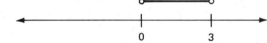

8. *x* is greater than or equal to −1

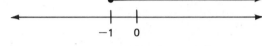

9. *x* is greater than 0 *and x* is less than 3

10. *x* is greater than or equal to 1 *and x* is less than or equal to 3

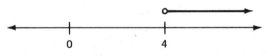

11. $\{x \mid x > 4\}$

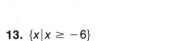

12. $\{x \mid x < -3 \text{ or } x > 0\}$

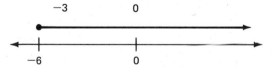

13. $\{x \mid x \geq -6\}$

14. $\{x \mid x < 0 \text{ or } x > 2\}$

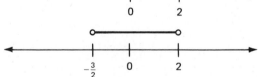

15. $\left\{x \mid -\dfrac{3}{2} < x < 2\right\}$

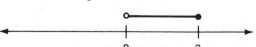

16. $\{x \mid 0 < x \leq 3\}$

ANSWERS

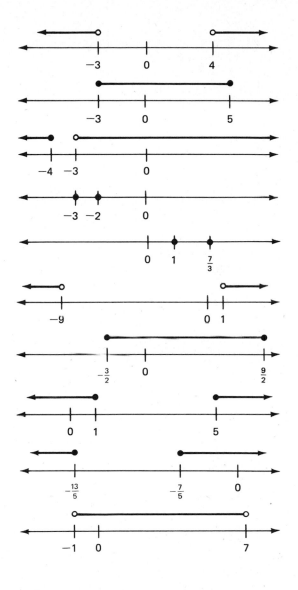

17. $\{x \mid x < -3 \text{ or } x > 4\}$

18. $\{x \mid -3 \le x \le 5\}$

19. $\{x \mid x \le -4 \text{ or } x > -3\}$

20. $\{x \mid x = -2 \text{ or } x = -3\}$

21. $\left\{x \mid x = 1 \text{ or } x = \dfrac{7}{3}\right\}$

22. $\{x \mid x < -9 \text{ or } x > 1\}$

23. $\left\{x \mid -\dfrac{3}{2} \le x \le \dfrac{9}{2}\right\}$

24. $\{x \mid x \le 1 \text{ or } x \ge 5\}$

25. $\left\{x \mid x \le -\dfrac{13}{5} \text{ or } x \ge -\dfrac{7}{5}\right\}$

26. $\{x \mid -1 < x < 7\}$

27. $\{x \mid x \text{ is any real number}\}$
28. $\varnothing$
29. $\{x \mid x \text{ is any real number}\}$
30. $\varnothing$
31. $x \ge \dfrac{5}{2}$
32. $x \le \dfrac{5}{2}$
33. x is any real number

34. $90 \leq \dfrac{447 + x}{6} \leq 100$; at least 93

35. $90 \leq \dfrac{439 + x}{6} \leq 100$; not possible; would need a grade of 101 (or a generous instructor!)

EXERCISES 7.1, pp. 277

1. Domain: $\{1, 3, 5, 2\}$, range: $\{2, 6, 1\}$

3. Domain: $\{1, 2, 3, 4, 5\}$, range: $\{2, 3, 4, 5\}$

5. Domain: $\{2, 3, 4\}$, range: $\{4, 1, 2\}$

7. Domain: $\{1, 2, -2\}$, range $\left\{1, \dfrac{1}{2}, -\dfrac{1}{2}\right\}$

9. Domain: $\{1, 2, 3, 4, 5\}$, range: $\{1\}$

11. All except problems 5 and 10 are functions

13. Domain: $\{x \,|\, x \neq 0, 3\}$

15. Domain: $\{x \,|\, x \geq -4\}$

17. Domain: all real numbers

19. Domain: $\{x \,|\, x \neq 1\}$

21. Domain: $\{x \,|\, x \geq 16\}$

23. The range for 14, 15, 21, and 22 is always nonnegative; that is, $\{y \,|\, y \geq 0\}$

25. Yes

27. No

29. Yes

31. $f(2) = 22$

33. $f(0) = 2$

35. $f(a + 1) = 3(a + 1)^2 + 4(a + 1) + 2 = 3a^2 + 10a + 9$

37. $g(-3) = -33$

39. $2[g(1)] = -2$

41. $g(b) = b^3 + b - 3$

43. $g(b) + g(h) = b^3 + b + h^3 + h - 6$

45. $f(2) = -8$

47. $f(0) = 2$

49. $f\left(\dfrac{1}{a}\right) = 2 + \dfrac{3}{a} - \dfrac{4}{a^2}$

51. (a) $f(x^2) = 3x^2 + 4$ \qquad (b) $[f(x)]^2 = (3x + 4)^2 = 9x^2 + 24x + 16$

53. (a) $f(3) = \dfrac{15}{2}$ \qquad (b) $f(b + h) = \dfrac{5}{2}(b + h)$ \qquad (c) $f(b + h) - f(b) = \dfrac{5h}{2}$

(d) $\dfrac{f(b + h) - f(b)}{h} = \dfrac{5}{2}$

55. (a)

C	0	5	10	15	20	25	30	35	40
F	32	41	50	59	68	77	86	95	104

ANSWERS

(b)
(c)

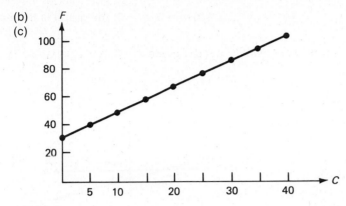

(d) 37°C
(e) Yes. For every value in the domain there is exactly one value in the range.

EXERCISES 7.2, pp. 287

1. $m = 3$, $b = 4$

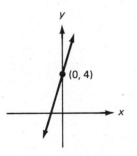

3. $m = \dfrac{1}{4}$, $b = 2$

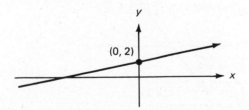

5. $m = 0$, $b = 5$

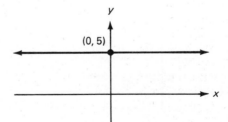

7. $m = -\dfrac{1}{4}$, $b = 0$

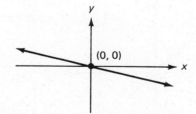

475

9. $m = -1, b = 2$

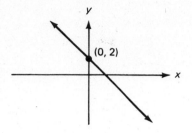

(0, 2)

11. $m = 3, b = 2$

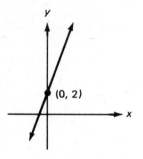

(0, 2)

13. $m = -1, b = -1$

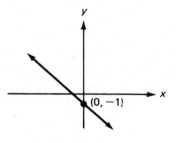

(0, −1)

15. $m = -1, b = 0$

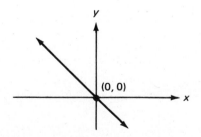

(0, 0)

17. $m = 0, b = 0$; the graph is the x-axis

19. $m = -\dfrac{1}{4}, b = -1$

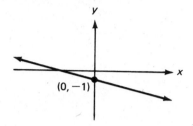

(0, −1)

21. $m = \dfrac{6-3}{1-2} = -3$

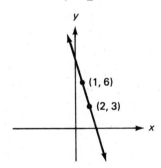

(1, 6)

(2, 3)

23. $m = \dfrac{5+4}{2+3} = \dfrac{9}{5}$

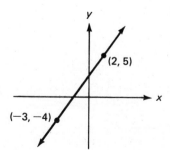

(2, 5)

(−3, −4)

ANSWERS

25. $m = \dfrac{-2 + 1}{-5 + 2} = \dfrac{1}{3}$

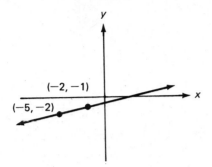

27. $m = 1, b = \dfrac{3}{5}$

29. $m = 0, b = 5$

31. $m = 2, b = -4$

33. $m = -\dfrac{3}{2}, b = 2$

35. $m = \dfrac{3}{2}, b = \dfrac{3}{2}$

37. Slope undefined; no y-intercept

39. $m = 1, b = 0$

41. Parallel; $m = \dfrac{1}{3}$

43. $k = 6$

45. $k = 0$

47. $k = \dfrac{3}{2}$

49. (a) 132 ft   (b) 85.8 ft

EXERCISES 7.3, pp. 291

1. $2x + y - 8 = 0$

3. $6x - y + 9 = 0$

5. $x - 2y - 8 = 0$

7. $3x + 4y + 25 = 0$

9. $y = 7$

11. $m = \dfrac{1}{2}; x - 2y = 0$

13. $m = \dfrac{1}{3}; x - 3y - 7 = 0$

15. $m = \dfrac{1}{4}; x - 4y + 3 = 0$

17. $m = 0; y = 3$

19. Slope undefined; $x = -2$

21. $3x + 2y - 7 = 0$

23. $x + 2y - 3 = 0$

25. $y - 2 = 0$

27. $2x - y + 3 = 0$

29. $4x + 3y + 8 = 0$

31. No, slope of line $AB \neq$ slope of line BC

33. $K = \dfrac{8}{3}$

35. (a) and (b)

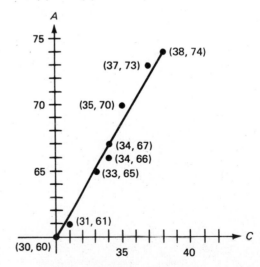

(c) $7C - 4A + 30 = 0$

(d) $63\dfrac{1}{2}$ in., $70\dfrac{1}{2}$ in., $75\dfrac{3}{4}$ in.

EXERCISES 7.4, pp. 299

1. Vertex (0, 0); $x = 0$; range $\{y | y \geq 0\}$

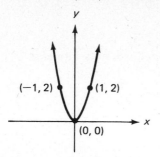

7. Vertex (0, 2); $x = 0$; range $\{y | y \leq 2\}$

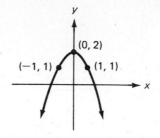

3. Vertex (0, 0); $x = 0$; range $\{y | y \leq 0\}$

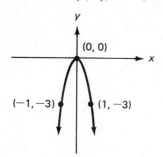

9. Vertex (0, 0); $x = 0$; range $\{y | y \geq 0\}$

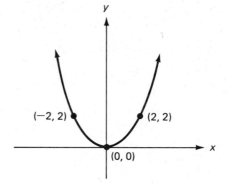

5. Vertex (0, 4); $x = 0$; range $\{y | y \geq 4\}$

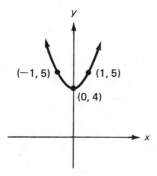

11. Vertex (2, 1); $x = 2$; range $\{y | y \geq 1\}$

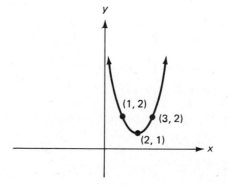

13. Vertex $(-1, -2)$; $x = -1$;
 range $\{y\,|\,y \geq -2\}$

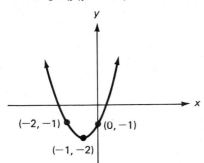

15. Vertex $(-2, 0)$; $x = -2$; range $\{y\,|\,y \geq 0\}$

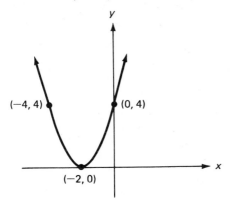

17. Vertex $(-1, -5)$; $x = -1$;
 range $\{y\,|\,y \geq -5\}$

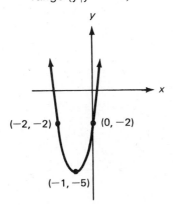

19. Vertex $(1, 2)$; $x = 1$; range $\{y\,|\,y \leq 2\}$

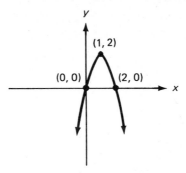

21. $(-1, 3)$

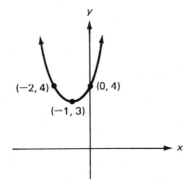

23. $(-4, -17)$

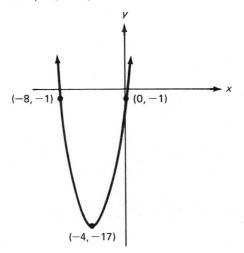

25. $\left(-\dfrac{3}{2}, -\dfrac{5}{4}\right)$

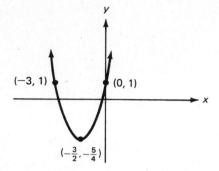

27. $\left(\dfrac{5}{2}, -\dfrac{37}{4}\right)$

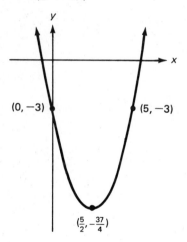

29. $(-1, -2)$

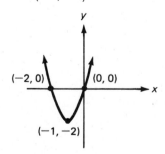

31. $\left(-\dfrac{3}{2}, -\dfrac{15}{2}\right)$

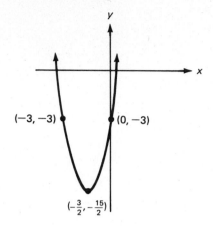

33. $(-1, 4)$

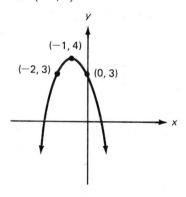

35. $\left(\dfrac{5}{6}, \dfrac{13}{12}\right)$

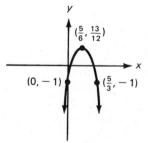

37. $\left(-1, -\dfrac{9}{2}\right)$

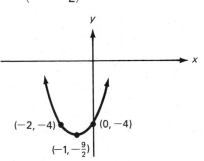

39. (3, 9)

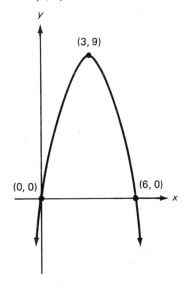

41. (a)

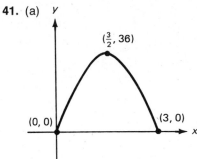

(b) 36 ft, $1\dfrac{1}{2}$ sec, 3 sec

EXERCISES 7.5, pp. 303

1.

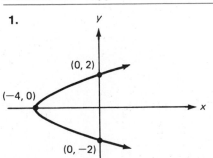

3.

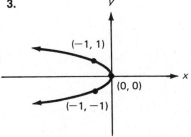

5.

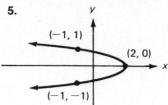

(−1, 1)
(2, 0)
(−1, −1)

7.

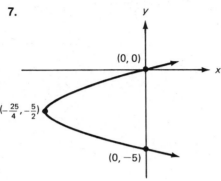

(0, 0)
$\left(-\frac{25}{4}, -\frac{5}{2}\right)$
(0, −5)

9.

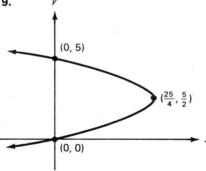

(0, 5)
$\left(\frac{25}{4}, \frac{5}{2}\right)$
(0, 0)

11.

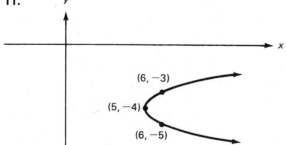

(6, −3)
(5, −4)
(6, −5)

13.

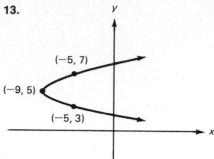

(−5, 7)
(−9, 5)
(−5, 3)

15.

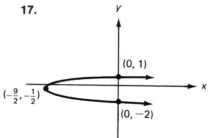

(−4, 0)
(−16, −2)
(−4, −4)

17.

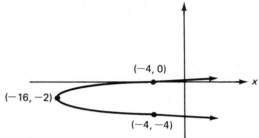

(0, 1)
$\left(-\frac{9}{2}, -\frac{1}{2}\right)$
(0, −2)

19.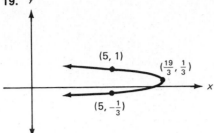

21.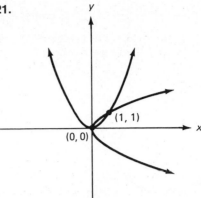

23.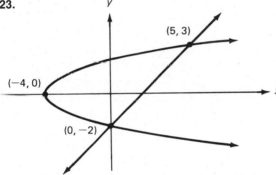

EXERCISES 7.6, pp. 307

1. 13

3. $\sqrt{20} = 2\sqrt{5}$

5. $\dfrac{7}{2}$

7. 25

9. $\dfrac{13}{2}$

11. $(x + 1)^2 + (y - 2)^2 = 36$

13. $x^2 + y^2 = 100$

29. $(x - 1)^2 + (y - 1)^2 = 61$

31. $(x - 2)^2 + (y + 3)^2 = 16$; center $(2, -3)$, radius = 4

33. $(x + 5)^2 + (y - 3)^2 = 25$; center $(-5, 3)$, radius = 5

15. $(x - 2)^2 + (y - 3)^2 = 9$

17. $(x - 3)^2 + (y + 5)^2 = 25$

19. $(x + 6)^2 + (y - 2)^2 = 3$

21. $(x - 1)^2 + (y - 5)^2 = 10$

23. $(x + 4)^2 + (y - 1)^2 = 50$

25. $(x - 3)^2 + (y - 4)^2 = 25$

27. $(x + 1)^2 + (y + 5)^2 = 18$

35. $\left(x - \dfrac{7}{2}\right)^2 + (y - 1)^2 = 4$; center $\left(\dfrac{7}{2}, 1\right)$, radius = 2

37. $(x - 8)^2 + y^2 = 64$; center (8, 0), radius = 8

39. $x^2 + (y - 7)^2 = 49$; center (0, 7), radius = 7

41.
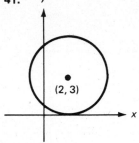

47.
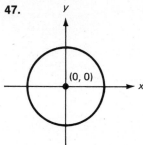

43.
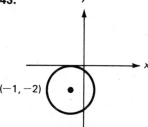

49.
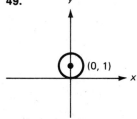

45.
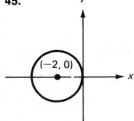

EXERCISES 7.7, pp. 313

1. Ellipse; domain: $|x| \leq 10$; range: $|y| \leq 2$

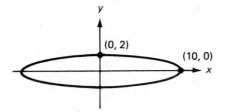

3. Ellipse; domain: $|x| \leq \sqrt{10}$; range: $|y| \leq 5$

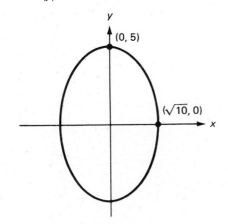

5. Hyperbola; domain: all R; range: $|y| \geq 3$

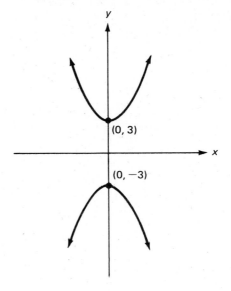

7. Hyperbola; domain: all $x \neq 0$; range: all $y \neq 0$

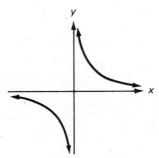

9. Hyperbola; domain: all R; range:
$|y| \geq 8$

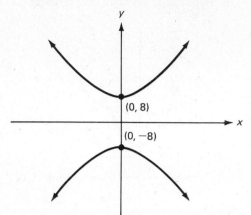

(0, 8)

(0, −8)

13. Ellipse; domain: $|x| \leq 13$; range:
$|y| \leq \dfrac{13}{2}$

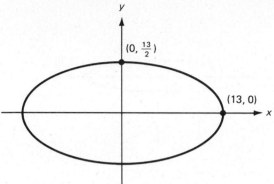

$(0, \frac{13}{2})$

(13, 0)

11. Ellipse; domain: $|x| \leq 5$; range:
$|y| \leq 15$

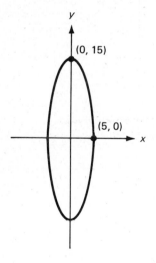

(0, 15)

(5, 0)

15. Hyperbola; domain: all real numbers,
range: $|y| \geq 4$

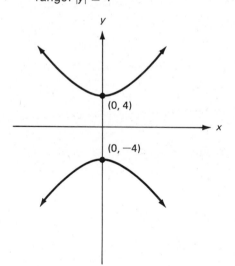

(0, 4)

(0, −4)

17. Hyperbola; domain: $x \neq 0$; range: $y \neq 0$

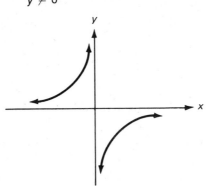

19. One point: (0, 0)

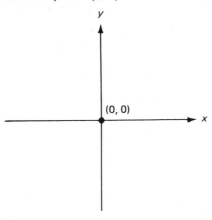

21.

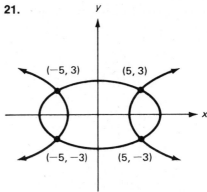

23.

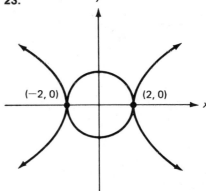

EXERCISES 7.8, pp. 321

1.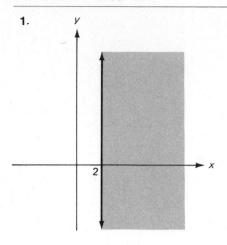

3.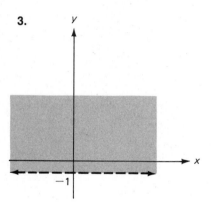

5.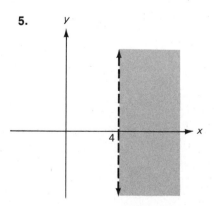

7.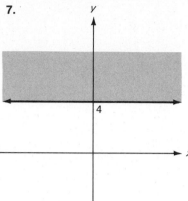

9.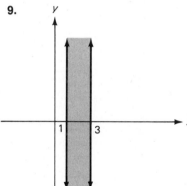

11.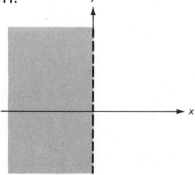

13.

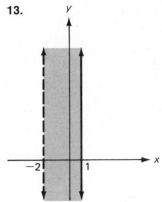

15.

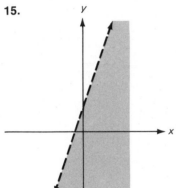

17.

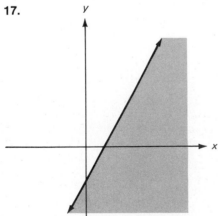

19.

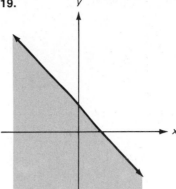

21.

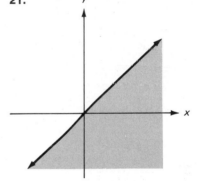

23.

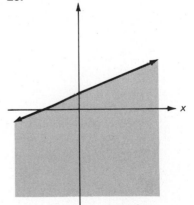

25.

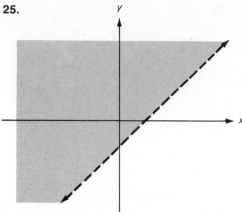

27.

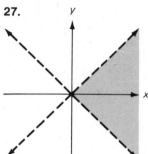

29.

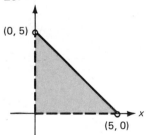

31.

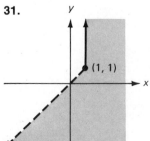

33.

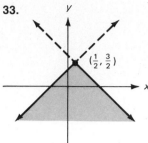

35.

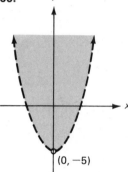

37.

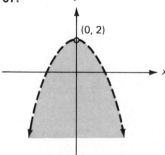

39.

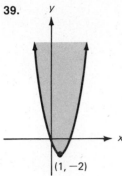

41.

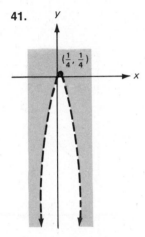

43.

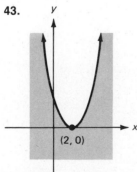

45.

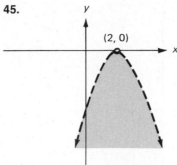

47.

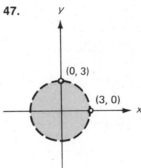

49.

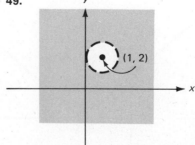

51.

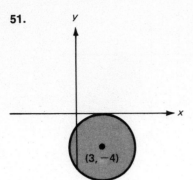

EXERCISES 7.9, pp. 324

1.

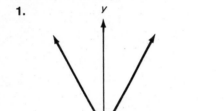

5.

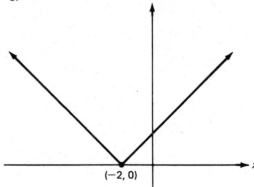

3.

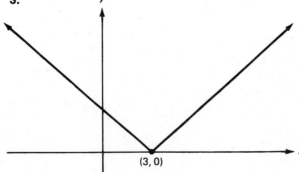

7.

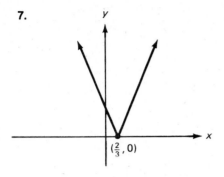

$(\frac{2}{3}, 0)$

13.

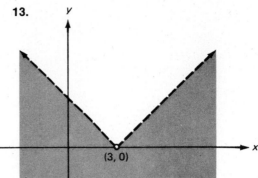

(3, 0)

9.

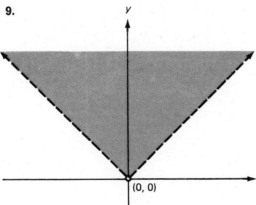

(0, 0)

11.

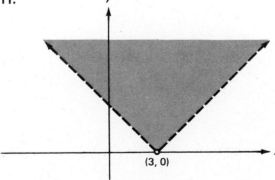

(3, 0)

15.

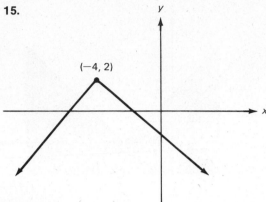

17.

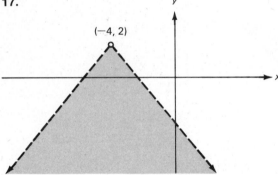

19.

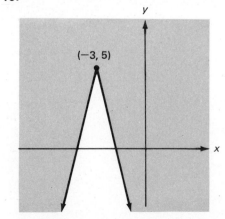

21.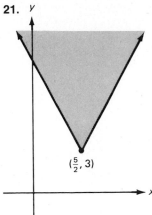

$\left(\frac{5}{2}, 3\right)$

23.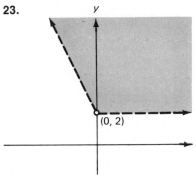

(0, 2)

25.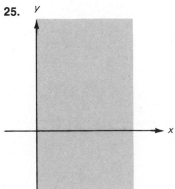

27.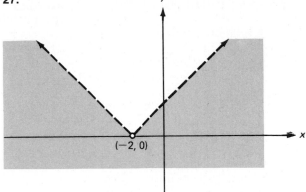

(−2, 0)

29.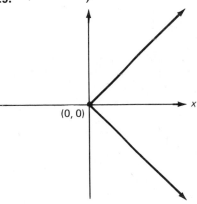

(0, 0)

CHAPTER 7 DIAGNOSTIC TEST, pp. 325

1. A relation
2. A function
3. 12, 9; $f(2) = 3(2)^2 - 2(2) + 4 = 12 - 4 + 4 = 12$
   $ f(-1) = 3(-1)^2 - 2(-1) + 4 = 3(1) + 2 + 4 = 9$
4. $3a^2 + 10a + 12$, $12a + 8$;
   $f(a + 2) = 3(a + 2)^2 - 2(a + 2) + 4$
   $ = 3(a^2 + 4a + 4) - 2a - 4 + 4$
   $ = 3a^2 + 12a + 12 - 2a - 4 + 4$
   $ = 3a^2 + 10a + 12$
   $f(a) = 3a^2 - 2a + 4$
   $f(a + 2) - f(a) = (3a^2 + 10a + 12) - (3a^2 - 2a + 4)$
   $ = 3a^2 + 10a + 12 - 3a^2 + 2a - 4$
   $ = 12a + 8$
5. -1; $m = \dfrac{5 - (-2)}{-4 - 3} = \dfrac{7}{-7} = -1$
6. $y = -x + 1$; Using the point-slope form to find the equation
   $y - (-2) = -1(x - 3)$
   $ y + 2 = -x + 3$
   $ y = -x + 1$
7. $(2, -15)$; $y = f(x) = 2x^2 - 8x - 7$; $a = 2$, $b = -8$, and $x = \dfrac{-b}{2a} = \dfrac{-(-8)}{2(2)} = \dfrac{8}{4} = 2$
   $y = f(2) = 2(2)^2 - 8(2) - 7 = 8 - 16 - 7 = -15$
8. Parabola
9. Hyperbola; $x^2 - 4y^2 = 12$
   $\dfrac{x^2}{12} - \dfrac{y^2}{3} = 1$
10. Circle; this is the equation of a circle with center at $(2, -3)$ and radius of $\sqrt{12}$.
11. $\sqrt{34}$; $d^2 = (4 - 1)^2 + (2 - (-3))^2 = 3^2 + 5^2 = 9 + 25 = 34$
    $d = \sqrt{34}$
12. $(x - 3)^2 + (y + 1)^2 = 16$; an equation of a circle with center at (a, b) and radius r is
    $(x - a)^2 + (y - b)^2 = r^2$
13. $-12 \le x \le 12$; this is the equation of an ellipse. Rewriting this equation in the form
    $\dfrac{x^2}{a^2} + \dfrac{y^2}{b^2} = 1$, we write
    $\dfrac{x^2}{144} + \dfrac{y^2}{9} = 1$
    When $y = 0$, $x = -12$ or $x = 12$, so the domain of the ellipse consists of all values
    for x between and including -12 and 12.
14. $-3 \le y \le 3$; when $x = 0$, $y = -3$ or $y = 3$, so the range of the ellipse consists of all
    values for y between and including -3 and 3.
15. An ellipse; see discussion above.
16. Above
17. Inside
18. Hyperbola; all values of x except $x = 0$; all values of y except $y = 0$.

19. Horizontal line; $m = 0$. This is the graph of a constant function.

20. $-\dfrac{2}{3}$; the slope of the line is the same as the slope of the line whose equation is

$2x + 3y = 12$, or $y = -\dfrac{2}{3}x + 4$. Since the slope of this line is $-\dfrac{2}{3}$, we use the

formula for finding the slope of a line to find k.

$$\frac{k - (-2)}{3 - 5} = \frac{k + 2}{-2} = -\frac{2}{3}$$

$$k + 2 = -\frac{2}{3}(-2)$$

$$k = \frac{4}{3} - 2 = -\frac{2}{3}$$

Note: The fact that the slope of the line and the value of k are the same is pure chance.

CHAPTER 7 REVIEW EXERCISES, pp. 326

1. Domain: $\{-3, 3, 5\}$; range: $\{2, 4\}$; function
2. Domain: $\{-1, -4, 4, -9\}$; range: $\{-1, -4, 4, -9\}$; function
3. Domain: $\{2, 4\}$; range: $\{-3, 3, 5\}$
4. Domain: $\{1, 2, 3, 4\}$; range: $\{2\}$; function
5. Domain: $\{2\}$; range: $\{1, 2, 3, 4\}$
6. $\{x \mid x \neq -2\}$; function
7. $\{x \mid x \neq 0, -2\}$; function
8. $\{x \mid x \text{ is any real number}\}$; function
9. $\{x \mid x \text{ is any real number}\}$; function
10. $\{x \mid |x| \geq 4\}$; function
11. (8) range: $\{y \mid y \geq 0\}$; (9) range: $\{y \mid y \geq -1\}$; (10) range: $\{y \mid y \geq 0\}$
12. $f(1) = 0$
13. $f(0) = 2$
14. $f(3 + h) = (3 + h)^2 - 3(3 + h) + 2 = h^2 + 3h + 2$
15. $f(3 + h) - f(3) = h^2 + 3h$
16. $f(x + 1) = (x + 1)^2 - 3(x + 1) + 2 = x^2 - x$
17. $f(5) = \dfrac{1}{5}$

18. $f(x + 5) = \dfrac{1}{x + 5}$

19. $f(x + 5) - f(5) = \dfrac{-x}{5(x + 5)}$

20. $\dfrac{f(x + 5) - f(5)}{x} = \dfrac{-1}{5(x + 5)}$

21. $m = 2, b = -3$

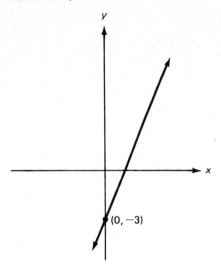

(0, −3)

22. $m = 0, b = 5$

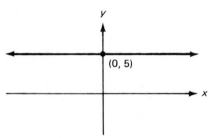

(0, 5)

23. $m = -1, b = 4$

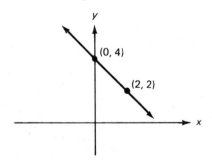

(0, 4)

(2, 2)

24. $m = \dfrac{1}{2}, b = 0$

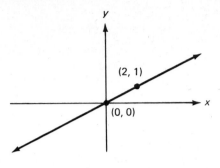

(2, 1)

(0, 0)

25. $m = -\dfrac{3}{7}$

26. $3x + 7y - 5 = 0$

27. $k = \dfrac{1}{8}$

28. $k = \dfrac{3}{8}$

29.

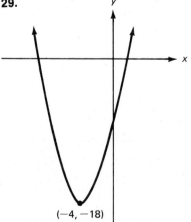

(−4, −18)

30.

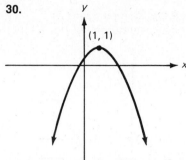

32.

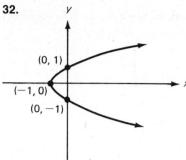

31.

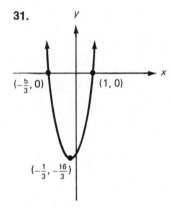

33.

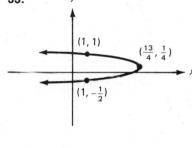

34. $2\sqrt{10}$

35. $(x + 1)^2 + (y - 3)^2 = 16$

36. $(x - 2)^2 + (y + 2)^2 = 18$

37. Center: $(3, -5)$; radius: 5

38. Ellipse; domain: $|x| \le 10$; range: $|y| \le 2$; not a function

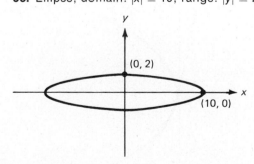

39. Parabola; domain: all real numbers; range: $y \geq -16$; function

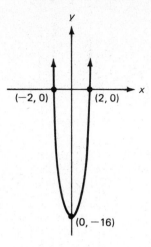

40. Line; domain: all real numbers; range: $y = -1$; function

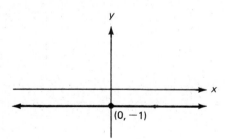

41. Line; domain: all real numbers; range: all real numbers; function

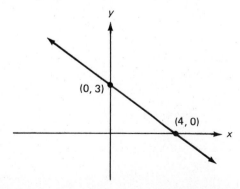

42. Hyperbola; domain: $|x| \geq 3$; range: all real numbers; not a function

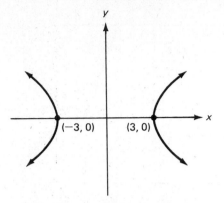

43. Ellipse; domain: $|x| \leq 3$; range: $|y| \leq 2$; not a function

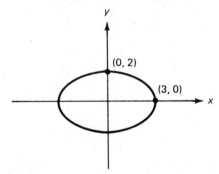

44. Parabola; domain: $x \geq 0$; range: all real numbers; not a function

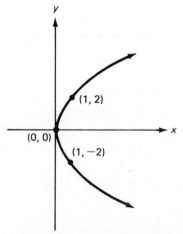

ANSWERS

45. Circle; domain: $|x| \leq 3$;
range: $|y| \leq 3$; not a function

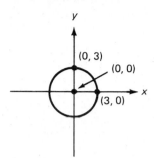

46. Line; domain: all real numbers;
range: all real numbers; function

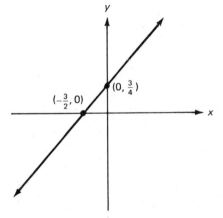

47. Hyperbola; domain: all real numbers;
range: $|y| \geq 4$; not a function

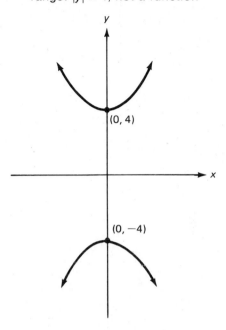

48.

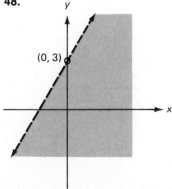

49.

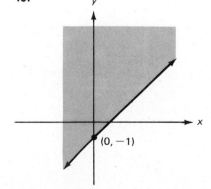

50.

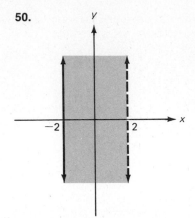

53.

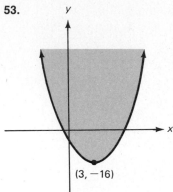

(3, −16)

51.

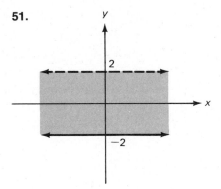

54.

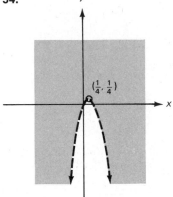

$(\frac{1}{4}, \frac{1}{4})$

52.

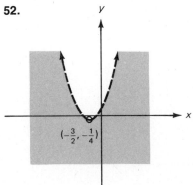

$(-\frac{3}{2}, -\frac{1}{4})$

55.

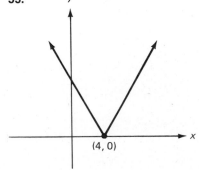

(4, 0)

56.

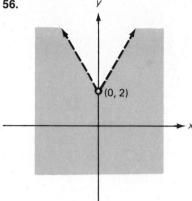

(0, 2)

57.

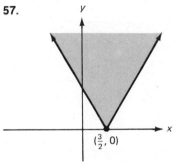

$(\frac{3}{2}, 0)$

58.

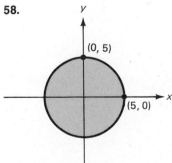

(0, 5)

(5, 0)

59.

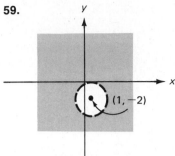

(1, −2)

60. (a)

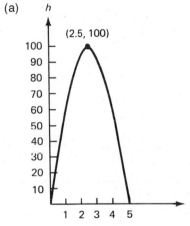

(2.5, 100)

(b) 100 ft   (c) 5 sec

EXERCISES 8.1, pp. 335

1. $r^{-1} = \{(x, y) \,|\, x = 2y - 1\}$
  Domain r = all real numbers = range r^{-1}
  Range r = all real numbers = domain r^{-1}

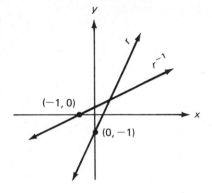

3. $r^{-1} = \{(x, y) \,|\, 3y + 2x = 4\}$
  Domain r = all real numbers = range r^{-1}
  Range r = all real numbers = domain r^{-1}

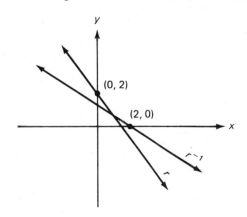

5. $r^{-1} = \{(x, y) \,|\, x = y^2 + 1\}$
  Domain r = all real numbers
  Range r = $\{y \,|\, y \geq 1\}$
  Domain r^{-1} = $\{x \,|\, x \geq 1\}$
  Range r^{-1} = all real numbers

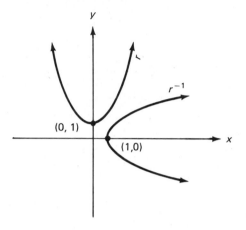

7. $r^{-1} = \{(x, y) \,|\, x = y^2 + 3y - 1\}$
  Domain r = all real numbers
  Range r = $\left\{y \,|\, y \geq \dfrac{-13}{3}\right\}$
  Domain r^{-1} = $\left\{x \,|\, x \geq \dfrac{-13}{4}\right\}$
  Range r^{-1} = all real numbers

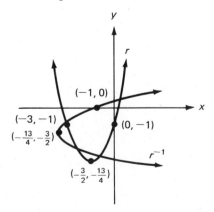

9. $r^{-1} = \{(x, y) \mid x = y^3 + 2\}$
Domain r = all real numbers = range r^{-1}
Range r = all real numbers = domain r^{-1}

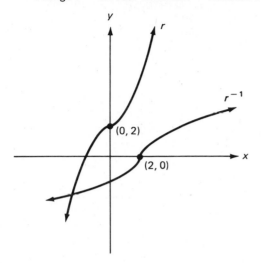

11. $r^{-1} = \{(x, y) \mid x = |y|\}$
Domain r = all real numbers
= range r^{-1}
Range $r = \{y \mid y \geq 0\}$
Domain $r^{-1} = \{x \mid x \geq 0\}$

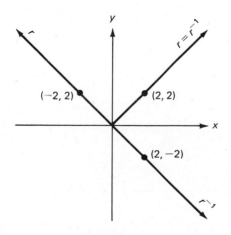

13. $f^{-1}(x) = \dfrac{x + 1}{2}$

15. $f^{-1}(x) = \dfrac{5 - x}{2}$

17. $f^{-1}(x) = \sqrt{x - 2}$

19. $f^{-1}(x) = x^2; \ x \geq 0$

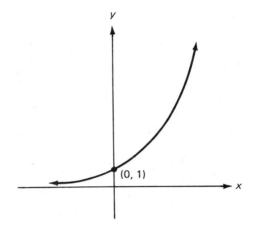

21. $f(-3) = \dfrac{1}{8}$

$f(-2) = \dfrac{1}{4}$

$f(-1) = \dfrac{1}{2}$

$f(0) = 1$

$f(1) = 2$

$f(2) = 4$

$f(3) = 8$

23. $f^{-1}(2) = 1$; $f^{-1}(1) = 0$; $f^{-1}(8) = 3$; $f^{-1}\left(\dfrac{1}{4}\right) = -2$; $f^{-1}\left(\dfrac{1}{8}\right) = -3$

25. $f\left(-\dfrac{1}{2}\right) =$ approx. 0.6; $f\left(\dfrac{1}{2}\right) =$ approx. 1.7; $f\left(\dfrac{3}{2}\right) =$ approx. 5.2

27. $f^{-1}\left(\dfrac{1}{9}\right) = -2$; $f^{-1}(1) = 0$; $f^{-1}(27) = 3$; $f^{-1}\left(\dfrac{1}{27}\right) = -3$; $f^{-1}(9) = 2$

29.

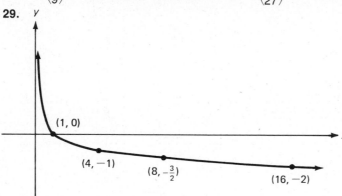

EXERCISES 8.2, pp. 341

1.   **3.**

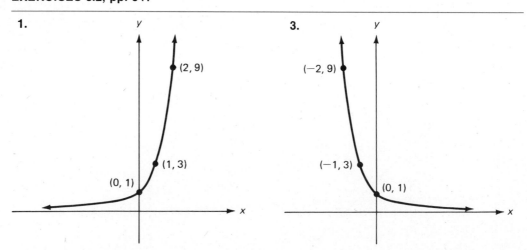

5.

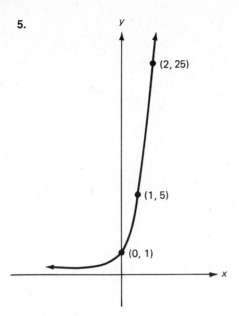

7.

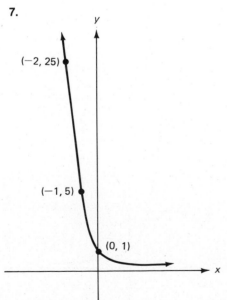

9.

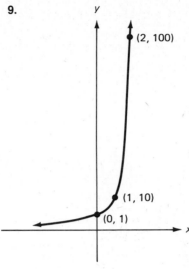

11.

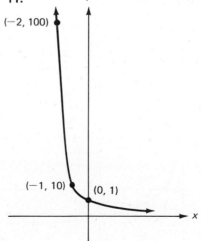

13.

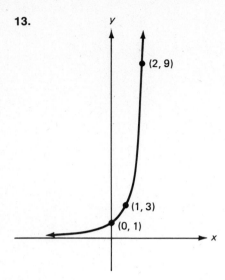

15.

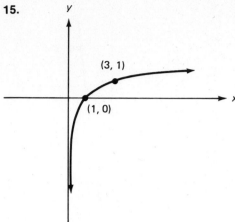

17.

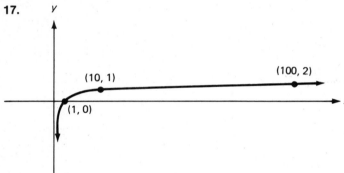

19.

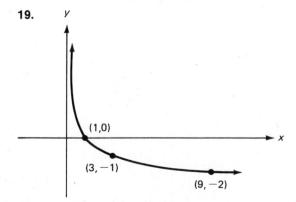

21. 3

23. -2

25. $\dfrac{3}{2}$

27. 4

29. 3

31. -4

33. -2

35. 0

37. $\dfrac{7}{2}$

39. $\dfrac{1}{2}$

EXERCISES 8.3, pp. 345

1. $\log_{10} 10,000 = 4$

3. $\log_9 3 = \dfrac{1}{2}$

5. $\log_3 3 = 1$

7. $\log_9 \left(\dfrac{1}{27}\right) = -\dfrac{3}{2}$

9. $\log_5 1 = 0$

11. $\log_4 \left(\dfrac{1}{64}\right) = -3$

13. $\log_{64} 16 = \dfrac{2}{3}$

15. $\log_{1/6} 36 = -2$

17. $3^2 = 9$

19. $10^1 = 10$

21. $36^{1/2} = 6$

23. $10^{-1} = 0.1$

25. $10^0 = 1$

27. $2^{-3} = 0.125$

29. $10^5 = 100,000$

31. $8^{-2/3} = 0.25$

33. 0

35. -3

37. -3

39. 2

41. 3

43. -4

45. $-\dfrac{3}{2}$

47. -2.5

49. 6

51. 1

53. 5

55. 5

57. 10,000

59. 3

61. 6

63. $\dfrac{1}{3}$

65. -3

67. 10

69. $\dfrac{1}{4}$

71. 1

73. 0.1

75. 5

77. $\dfrac{1}{6}$

79. 10

EXERCISES 8.4, pp. 349

1. $\log_b 5 + \log_b 3$

3. $5 \log_b 3$

5. $\log_b 3 + \log_b 2$

7. $6 \log_b 2$

9. $\log_b 1 - \log_b 3 - \log_b 2 = -\log_b 3 - \log_b 2$

11. $3 \log_b 3 - 3 \log_b 4$

13. $2 \log_b 2 + \dfrac{1}{2} \log_b 3$

15. $\dfrac{1}{2} (\log_b 5 - \log_b 3)$

17. $4 \log_b 3 - 2 \log_b 5$

19. $\dfrac{3}{2} \log_b 5 - \log_b 7$

21. $\log_b 450$

23. $\log_b \sqrt{5} \sqrt[3]{17} = \log_b \sqrt[6]{36,125}$

25. $\log_b 72$

27. $\log_b 243$

29. $\log_b 2025$

31. $\log_b \sqrt[3]{\dfrac{56}{45}}$

33. $\log_b \dfrac{b^3}{5}$

35. $\log_b 2$

37. $\log_b \sqrt[5]{75}$

39. $\log_b \dfrac{9 \sqrt[4]{5}}{343}$

EXERCISES 8.5, pp. 353

1. (a) 0.6776   (b) 0.6776
3. (a) $0.5378 - 1$ (b) -0.4622
5. (a) $0.0899 - 1$ (b) -0.9101
7. (a) 0.8007   (b) 0.8007
9. (a) $0.9768 - 2$ (b) -1.0232
11. (a) 0.1761   (b) 0.1761
13. (a) $0.8451 - 3$ (b) -2.1549
15. (a) $0.9680 + 2$ (b) 2.9680
17. (a) $0.7513 - 1$ (b) -0.2487
19. (a) 0.4346   (b) 0.4346
21. (a) $0.5647 + 4$ (b) 4.5647
23. (a) $0.4249 - 1$ (b) -0.5751
25. (a) $0.1790 - 4$ (b) -3.8210
27. (a) $0.8762 - 2$ (b) -1.1238
29. (a) 0.4969   (b) 0.4969
31. (a) $0.6590 - 7$ (b) -6.3410

33. (a) 6.93×10^1 (b) 69.3
35. (a) 1.17×10^{-1} (b) 0.117
37. (a) 6.83×10^{-3} (b) 0.00683
39. (a) 1.95×10^{-2} (b) 0.0195
41. (a) 3.90×10^2 (b) 390
43. (a) 1.75×10^{-1} (b) 0.175
45. (a) 2.65×10^{-2} (b) 0.0265
47. (a) 1.09×10^0 (b) 1.09
49. (a) 1.71×10^{-2} (b) 0.0171
51. (a) 5.88×10^5 (b) 588,000
53. (a) 2.09×10^{-1} (b) 0.209
55. (a) 7.08×10^{-3} (b) 0.00708
57. (a) 1.54×10^{-4} (b) 0.000154
59. (a) 3.02×10^0 (b) 3.02
61. (a) 7.08×10^{-2} (b) 0.0708
63. (a) 1.27×10^{-4} (b) 0.000127

EXERCISES 8.6, pp. 357

1. 0.4972
3. $0.6391 - 1$
5. 2.3498
7. 3.7224
9. $0.9275 - 1$

11. 1.4707
13. 3.7773
15. $0.6593 - 3$
17. 5.9547
19. $0.9706 - 2$

21. 3.356
23. 0.4966
25. 59.84
27. 0.01903
29. 0.1202

31. 0.001517

33. 16.52

35. 0.1815

37. 704.7

39. 0.0240

41. 27.8

43. 34.9

45. 24.7

47. 3.62

49. 0.000412

51. 1.34

53. 2.07

55. 0.768

57. 1909

59. -1.27

EXERCISES 8.7, pp. 360

1. 7

3. $\dfrac{\log 2}{\log 5} = 0.43$

5. $3 + \dfrac{\log 4.5}{\log 6} = 3.84$

7. $-\dfrac{3}{2}$

9. -2

11. $\dfrac{7}{3}$

13. $\dfrac{1}{3} - \dfrac{\log 2}{3 \log 5} = 0.19$

15. $\dfrac{\log 2}{\log 1.05} = 14.21$

17. 25

19. $2 + \dfrac{1}{\log 3} = 4.10$

21. $\dfrac{1}{8}$

23. $\dfrac{\log 26}{\log 2} = 4.70$

25. 13, 5

27. -2

29. $\dfrac{25\sqrt{3}}{6}$

31. 6, -4

33. No solution

35. $\dfrac{1}{7}$

EXERCISES 8.8, pp. 363

1. 2.303

3. 0.4448

5. -1.610

7. -4.424

9. 1

11. 3.227

13. -2.495

15. 2.276

17. 1.177

19. 0.7014

21. 3.533

23. -2

25. 6.580

27. -3

29. 1.792

EXERCISES 8.9, pp. 367

(*Note:* Most answers are approximate.)

1. (a) 5.3   (b) 1.5   (c) 7.2

3. (a) 1.0×10^{-7} (b) 2.5×10^{-3} (c) 2.0×10^{-9}

5. 24.5 years; 94.5 g

7. $0.000\ 436 = 4.36 \times 10^{-4}$

9. (a) 0.1833   (b) $0.0016 = 1.6 \times 10^{-3}$ g

11. \$5353

13. 7.2%

15. (a) 65 amps   (b) 22 amps

17. $2749

19. (a) 6.32 decibels   (b) 23.2 decibels ($G = -23.2$, which indicates a loss)

21. 261 ft

23. (a) 610   (b) 6765

CHAPTER 8 DIAGNOSTIC TEST, pp. 371

1. True

2. False; only if the function is one-to-one.

3. True; by interchanging x and y in the rule of the relation, the inverse relation is found.

4. True; if $f = \{(x, y)\,|\,y = b^x, b > 0 \text{ and } b \neq 1\}$, then $f^{-1} = \{(x, y)\,|\,x = b^y\}$. But $x = b^y$ if and only if $y = \log_b x$

5. True; see Answer 4 above.

6. False; $\dfrac{1}{81} = 3^{-4}$, so $3^{x-2} = 3^{-4}$ and $x - 2 = -4$, $x = -2$.

7. False; $4^{2 \log_4 x} = 4^{\log_4 (x^2)} = x^2$ because $b^{\log_b z} = z$.

8. True; $\log_b x^2 y = \log_b x^2 + \log_b y = 2 \log_b x + \log_b y$

9. False; $\log \dfrac{17}{6} = \log 17 - \log 6$. The log of a quotient is the difference of the logs.

The problem $\dfrac{\log 17}{\log 6}$ is the quotient of two logarithms, and is log 17 divided by

$\log 6 = \dfrac{1.2304}{0.7782}$

10. True; $\log_3 x = -4$ if and only if $x = 3^{-4} = \dfrac{1}{81}$.

11. True; since $\log_b b = 1$, b must equal 8.

12. True; $\log_2 64 = 6$ if and only if $64 = 2^6$.

13. True; $\log_{10} 0.000835 = \log_{10} (8.35 \times 10^{-4}) = \log_{10} 8.35 - 4$. But $\log_{10} 8.35 = 0.9217$; therefore

$\log_{10} 0.000835 = 0.9217 - 4 = -3.0783$

14. True; the only change in the two antilogarithms is the characteristic. A characteristic of 2 means that 10 is raised to the 2.8089 power, resulting in a number between 100 and 1000. The characteristic of 1 merely moves the decimal point in the answer to a number between 10 and 100.

15. True; $\log_b x = \dfrac{\log_a x}{\log_a b}$. ln stands for $\log_e$, so $b = e$, $a = 10$, and $x = 4.7$.

16. (c); $\dfrac{1}{2} (\log x - \log y) = \dfrac{1}{2} \log \dfrac{x}{y} = \log \left(\dfrac{x}{y}\right)^{1/2} = \log \sqrt{\dfrac{x}{y}}$

17. (b); $\log_{10} 0.075 = \log_{10} (7.5 \times 10^{-2}) = \log_{10} 7.5 - 2 = 0.8751 - 2$

18. (e); the exponent, 1/4, is the logarithm, and the base, 16, is the base of the logarithm.

19. (c); $2^x = 3$; taking the log of both sides, $\log 2^x = \log 3$

$$x \log 2 = \log 3$$

$$x = \frac{\log 3}{\log 2}$$

20. (a); $2 \log x - \log(x + 3) + \log 5 = \log 4$

$$\log x^2 - \log(x + 3) + \log 5 = \log 4$$

$$\log \frac{5x^2}{x + 3} = \log 4$$

$$\frac{5x^2}{x + 3} = 4$$

$$5x^2 - 4x - 12 = 0$$

$$(5x + 6)(x - 2) = 0$$

$$5x + 6 = 0 \quad \text{or} \quad x - 2 = 0$$

$$x = -\frac{6}{5} \quad \text{or} \quad x = 2$$

But since $\log x$ appears in the original problem and $\log x$ is only defined for $x > 0$, we discard the solution $-\frac{6}{5}$. The solution $x = 2$ checks.

CHAPTER 8 REVIEW EXERCISES, pp. 372

1. $r^{-1} = \{(x, y) \mid x = 3y + 2\}$
   Domain r = all real numbers
   Range r = all real numbers
   Domain r^{-1} = all real numbers
   Range r^{-1} = all real numbers

2. $r^{-1} = \{(x, y) \mid y + 5x = 2\}$
   Domain r = all real numbers
   Range r = all real numbers
   Domain r^{-1} = all real numbers
   Range r^{-1} = all real numbers

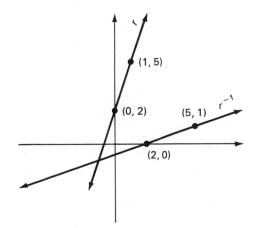

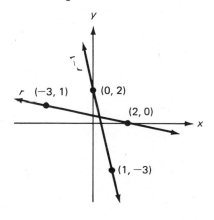

3. $r^{-1} = \{(x, y)\,|\,x = y^2 + 2y + 3\}$
Domain r = all real numbers
Range $r = \{y\,|\,y \geq 2\}$
Domain $r^{-1} = \{x\,|\,x \geq 2\}$
Range r^{-1} = all real numbers

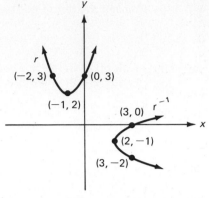

4. $r^{-1} = \{(x, y)\,|\,x = 5 - y^2\}$
Domain r = all real numbers
Range $r = \{y\,|\,y \leq 5\}$
Domain $r^{-1} = \{x\,|\,x \leq 5\}$
Range r^{-1} = all real numbers

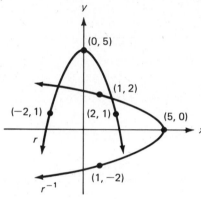

5. $r^{-1} = \{(x, y)\,|\,x = |y - 3|\}$
Domain r = all real numbers
Range $r = \{y\,|\,y \geq 0\}$
Domain $r^{-1} = \{x\,|\,x \geq 0\}$
Range r^{-1} = all real numbers

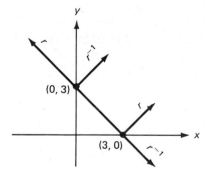

6. $f^{-1}(x) = \dfrac{x - 2}{3}$

7. $f^{-1}(x) = 4 - x$

8. $f^{-1}(x) = \sqrt{x + 1}$

9. $f^{-1}(x) = (x - 1)^2,\ x \geq 1$

10. $f^{-1}(x) = \sqrt{x}$

11. -3

12. $\dfrac{9}{2}$

13. -1

14. $\dfrac{3}{4}$

15. $\dfrac{3}{4}$

16. $\log_{10} 1000 = 3$

17. $\log_4 64 = 3$

18. $\log_{64} 0.125 = -\dfrac{1}{2}$

19. $\log_{10} 10 = 1$

20. $\log_5 1 = 0$

21. $\log_{64} 4 = \dfrac{1}{3}$

22. $\log_{1/2} 8 = -3$

23. -4

24. -2

25. $\dfrac{7}{2}$

26. No answer; not defined

27. -2

28. $\dfrac{1}{2}$

29. 7

30. 5.6

31. 1.2

32. -0.4

52. 0.0159

33. 0.2

34. $-\dfrac{4}{15}$

35. -0.4

36. $\dfrac{2\sqrt{3}}{15}$

37. $\log_b 2 + \log_b 3 + \log_b 5$

38. $3 \log_b 2 + \log_b 3$

39. $4 \log_b 2 + \dfrac{1}{2} \log_b 3$

40. $\dfrac{1}{2} \log_b 3 + \dfrac{3}{2} \log_b 2 - \log_b 5$

41. $0.7589 + 1,\ 1.7589$

42. $0.5403 - 4,\ -3.4597$

43. $0.3692 + 5,\ 5.3692$

44. $0.6596 + 2,\ 2.6596$

45. $0.5541 - 3,\ -2.4459$

46. $0.3012 + 3,\ 3.3012$

47. 253

48. 8760

49. 0.0532

50. 0.8320

51. 0.000 477

53. (a) $\log N = \dfrac{1}{3}(\log 42.21 + \log 23.50 - \log 5650)$ (b) 0.5600

54. $0.000\ 609 = 6.09 \times 10^{-4}$

55. 0.6825

56. 2.944

57. 1.610

58. 40

59. -2

60. 2.4949 or approx. 2.5

61. 3.98×10^{-7}

62. \$3209

63. \$3266

64. 14,640

EXERCISES 9.1, pp. 382

1.

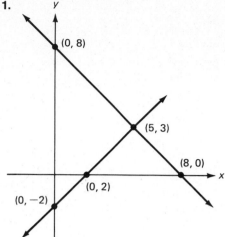

3.

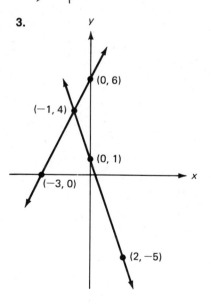

5.

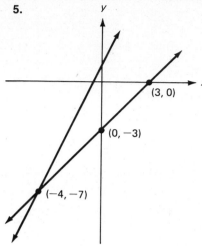

7.

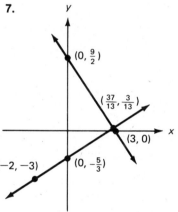

9. $\left\{\left(\dfrac{13}{2}, -\dfrac{7}{2}\right)\right\}$

11. $\left\{\left(-\dfrac{15}{2}, -10\right)\right\}$

13. $\{(-5, -19)\}$

15. $\left\{\left(\dfrac{1}{2}, \dfrac{1}{2}\right)\right\}$

17. $\left\{\left(\dfrac{4}{5}, \dfrac{4}{5}\right)\right\}$

19. $\{(-2, -6)\}$

21. $\left\{\left(\dfrac{1}{3}, 1\right)\right\}$

23. $\left\{\left(\dfrac{124}{25}, \dfrac{48}{25}\right)\right\}$

25. $\{(x, y) \mid 4x + 10y - 3 = 0\}$

27. No solution; $\varnothing$

29. $\left\{\left(\dfrac{5a + 8b}{a^2 + b^2}, \dfrac{-8a + 5b}{a^2 + b^2}\right)\right\}$

31. 21, 5

33. Frame: \$12

Plaque: \$5

35. Boat: 6 mph

Current: $1\dfrac{1}{2}$ mph

37. A: 65% silver; B: 45% silver

39. 10 items of A

9 items of B

EXERCISES 9.2, pp. 389

1. $\{(0, 0, 0)\}$

3. $\{(x, 11 - 6x, 7 - 5x) \mid x$ is any real number$\}$

5. $\{(-1, 1, -2)\}$

7. $\left\{\left(\dfrac{1}{6}, \dfrac{5}{6}, \dfrac{2}{3}\right)\right\}$

9. $\{(x, -4x + 13, 3x - 8) \mid x$ is any real number$\}$

11. $\left\{\left(x, 2x, \dfrac{5x}{2}\right) \,\middle|\, x$ is any real number$\right\}$

13. No solution; $\varnothing$

15. $\{(6, y, 3 - y) \mid y$ is any real number$\}$

17. $\{(5, -2, 1)\}$

19. $\{(5, 3, 6)\}$

21. $\{(6, 1, 2)\}$

23. $\left\{\left(\dfrac{a + b}{2}, \dfrac{a - c}{2}, \dfrac{b - c}{2}\right)\right\}$

25. $\left\{\left(\dfrac{29}{7}, -\dfrac{3}{7}\right)\right\}$

27. $\{(2, -1, 3)\}$

29. $\{(5, 1, -2, -3)\}$

31. $\{(1, -2, -1, 3)\}$

EXERCISES 9.3, pp. 397

1. 7

3. -2

5. 17

7. -14

9. 42

11. $\begin{vmatrix} -2 & 7 & 7 \\ 1 & 0 & 0 \\ 2 & -1 & -3 \end{vmatrix} = 14$

13. $\begin{vmatrix} 1 & 2 & 0 \\ 1 & 4 & 0 \\ 1 & -1 & 2 \end{vmatrix} = 4$

15. $\begin{vmatrix} 0 & 0 & -1 \\ 8 & 1 & 2 \\ 5 & -3 & -1 \end{vmatrix} = 29$

17. $\begin{vmatrix} -13 & 2 & -3 \\ 11 & 1 & 2 \\ 0 & 0 & 1 \end{vmatrix} = -35$

19. $\begin{vmatrix} 1 & 0 & 0 & 0 \\ 2 & -3 & 5 & -10 \\ 4 & -5 & 3 & -11 \\ 3 & -9 & 8 & -11 \end{vmatrix} = 185$

21. $\begin{vmatrix} 1 & 1 & 2(x-y) \\ 0 & x & x(x-y) \\ x & y & (x+y)(x-y) \end{vmatrix} = (x-y)\begin{vmatrix} 1 & 1 & 2 \\ 0 & x & x \\ x & y & x+y \end{vmatrix} = (x-y)\begin{vmatrix} 1 & 1 & 1 \\ 0 & x & 0 \\ x & y & x \end{vmatrix} = 0$

because two columns are identical.

EXERCISES 9.4, pp. 403

1. $\{(4, -2)\}$

3. $\{(-2, 1)\}$

5. $\{(-3, -4)\}$

7. $\{(-2, 3)\}$

9. $\{(-5, 10, 5)\}$

11. No solution; $\varnothing$

13. $\left\{\left(\dfrac{9-5z}{8}, \dfrac{-11+23z}{16}, z\right) \middle| z \text{ is any real number}\right\}$

15. $\{(1, 6, 3)\}$

17. $\left\{\left(\dfrac{3}{4}, -\dfrac{1}{2}, -\dfrac{2}{3}\right)\right\}$

19. $\{(4-2y, y, 4y+2) | y \text{ is any real number}\}$

21. $\{(0, 5, 3, -1)\}$

23. $\{(-2, 3, 5, -4)\}$

EXERCISES 9.5, pp. 409

1.

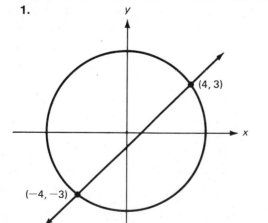

3.

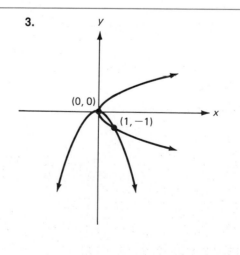

5.

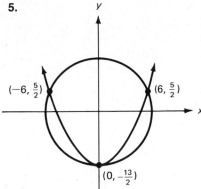

7.

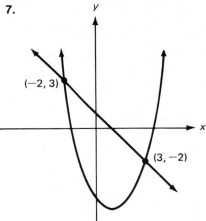

9.

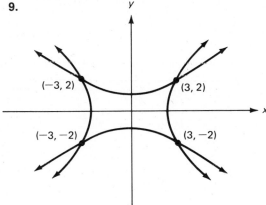

11. $\{(2, -3), (-2, -1)\}$

13. $\{(3 + i, -1 - 3i), (3 - i, -1 + 3i)\}$

15. $\{(6, -2), (-3, 4)\}$

17. $\left\{\left(\dfrac{7}{3}, 1\right), (4, 6)\right\}$

19. $\left\{(-2, -2), \left(\dfrac{22}{9}, \dfrac{2}{9}\right)\right\}$

21. $\left\{(2, 3), (2, -3), \left(-\dfrac{2}{9}, \dfrac{1}{3}\right), \left(-\dfrac{2}{9}, -\dfrac{1}{3}\right)\right\}$

23. $\{(-5, 2), (-8, -1)\}$

25. $\{(0, -4), (\sqrt{3}, 2), (-\sqrt{3}, 2)\}$

27. $\left\{\left(2, -\dfrac{3}{2}\right), \left(-2, -\dfrac{3}{2}\right), \left(\dfrac{3}{2}, 2\right), \left(-\dfrac{3}{2}, 2\right)\right\}$

29. $\{(8, 6), (-8, 6), (0, -10)\}$

EXERCISES 9.6, pp. 412

1. $\{(2, 4), (-2, 4), (2, -4), (-2, -4)\}$
3. $\{(\sqrt{6}, 0), (-\sqrt{6}, 0), (\sqrt{5}, 2), (-\sqrt{5}, 2)\}$
5. $\{(2, \sqrt{5}), (2, -\sqrt{5}), (-2, \sqrt{5}), (-2, -\sqrt{5})\}$
7. $\{(\sqrt{2}, 1), (\sqrt{2}, -1), (-\sqrt{2}, 1), (-\sqrt{2}, -1)\}$
9. $\{(1, 4), (-1, -4), (4, -14), (-4, 14)\}$
11. $\{(1, 1 + 2\sqrt{3}), (1, 1 - 2\sqrt{3})\}$
13. $\{(3, 16), (-8, -6)\}$
15. $\{(4, 2), (-4, -2), (\sqrt{6}, -2\sqrt{6}), (-\sqrt{6}, 2\sqrt{6})\}$
17. $\{(\sqrt{3}, 2\sqrt{3}), (-\sqrt{3}, -2\sqrt{3})\}$
19. $\{(-20, -6), (24, 5)\}$

EXERCISES 9.7, pp 415

1. 15 nickels, 25 dimes, 7 quarters
3. 400 student tickets, 50 faculty tickets, 100 general public tickets
5. P: 2 items; Q: 1 item; R: 3 items
7. $I_1 = 8$; $I_2 = 5$; $I_3 = 3$
9. 12 min by taxi, 36 min by bus, 2 hr 24 min by plane
11. Let a be any real number for which none of the quantities is negative.

$$I = 8 - 5a, \text{ II} = a, \text{ III} = \frac{2a + 4}{3}, \text{ IV} = \frac{8a + 10}{3}$$

No; for $a = 3$, I $= -7$, which is impossible.
Yes; for $a = 1$, I $= 3$, II $= 1$, III $= 2$, IV $= 6$
13. 10 in. $\times$ 20 in.
15. 50 ft right and left of center: 10 ft
100 ft right and left of center: 28 ft
150 ft right and left of center: 58 ft
17. $k = 10$ or $k = -10$
19. $x = 9$, $y = 5$

CHAPTER 9 DIAGNOSTIC TEST, pp. 419

1. (b); the equations of the system graph as lines in a plane. Since the solution set of the system is empty (no solutions) the graphic interpretation is that the lines do not intersect and have no common points (they are not coincident), so the lines are parallel.

2. (c); there are infinitely many solutions for this system, one of which is (5, 1, 2) because if $x = 5$,

$$y = \frac{x-3}{2} = \frac{5-3}{2} = 1 \text{ and } z = \frac{x+3}{4} = \frac{5+3}{4} = 2.$$

3. (d); by first eliminating z from all three equations, we get

$3x + 4y = 30$
$5x + 4y = 42$

By multiplying the second equation by -1 and then adding the two equations, we get

$-2x = -12$ and $x = 6$

If $x = 6$ and $3x + 4y = 30$, then $3(6) + 4y = 30$ and $y = 3$. Replacing x by 6 and y by 3 in any of the three original equations yields $z = 9$. The solution (6, 3, 9) checks for all three equations, and $\{(6, 3, 9)\}$ is the solution set.

4. (d); the solution set of the system is infinite, so we write the solution in terms of one variable. To check, let x stand for any real number, for example, $x = 2$. Then, from (a) we find $2x = 4 = y$ and $\frac{5x}{2} = 5 = z$. Checking (2, 4, 5) in the three equations shows that this solution works. In (b) we let $y = 4$ and find that $x = \frac{y}{2} = 2$, and $z = \frac{5y}{4} = 5$, so we again have (2, 4, 5). Verifying (c) is left for the student.

5. (c); a determinant is the answer obtained from performing the operations on a square matrix according to a rule. For example, for the matrix $\begin{pmatrix} a_1 & b_1 \\ a_2 & b_2 \end{pmatrix}$ the determinant, $\begin{vmatrix} a_1 & b_1 \\ a_2 & b_2 \end{vmatrix} = a_1 b_2 - a_2 b_1$

6. (a); $D = \begin{vmatrix} 1 & 2 & 3 \\ 4 & -3 & 1 \\ 3 & -1 & 2 \end{vmatrix} = 1 \begin{vmatrix} -3 & 1 \\ -1 & 2 \end{vmatrix} - 2 \begin{vmatrix} 4 & 1 \\ 3 & 2 \end{vmatrix} + 3 \begin{vmatrix} 4 & -3 \\ 3 & -1 \end{vmatrix}$

$= 1(-6 + 1) - 2(8 - 3) + 3(-4 + 9)$
$= -5 - 10 + 15 = 0$

7. (e); from Theorem 3, we know that if two rows of a determinant are interchanged, the resulting determinant is the negative of the original one, and from Theorem 4 we know that if each element of a row is multiplied by a constant, then the determinant is multiplied by that constant. In B, rows 1 and 3 of A were interchanged, and each row of A was multiplied by 2. Therefore $B = -8A$.

8. (d); since the number of real solutions corresponds to the number of intersections of the two graphs, it is possible that the graphs do not intersect at all (0 solutions); that the line and the ellipse have only one point in common, that is, the line is tangent to the ellipse (1 solution); or that the line passes through the ellipse, cutting it in two places (2 solutions).

9. (d); $x^2 - 4y^2 = 5$ and $xy = 3$; thus $x = \frac{3}{y}$ and $x^2 = \frac{9}{y^2}$

Substituting, we get $\frac{9}{y^2} - 4y^2 = 5$

$4y^4 + 5y^2 - 9 = 0$
$(4y^2 + 9)(y^2 - 1) = 0$

$$4y^2 + 9 = 0 \quad \text{or} \quad y^2 - 1 = 0$$
$$4y^2 = -9 \qquad\qquad y = 1 \quad \text{or} \quad -1$$
$$2y = 3i \quad \text{or} \quad -3i$$
$$y = \frac{3}{2}i \quad \text{or} \quad y = -\frac{3}{2}i \quad \text{or} \quad y = 1 \quad \text{or} \quad y = -1$$

Since $x = \dfrac{3}{y}$, when $y = \dfrac{3}{2}i$, $x = \dfrac{3}{\dfrac{3}{2}i} = \dfrac{6i}{3i^2} = \dfrac{6i}{-3} = -2i;$

when $y = -\dfrac{3}{2}i$, $x = -(-2i) = 2i$

when $y = 1$, $x = \dfrac{3}{1} = 3$

when $y = -1$, $x = -3$

Therefore the solution set is $\left\{(3, 1), (-3, -1), \left(2i, -\dfrac{3}{2}i\right), \left(-2i, \dfrac{3}{2}i\right)\right\}$

10. (c)

11. $\{(-3, -2), (-1, 2)\};$
$$x^2 + y^2 + 4x = 1$$
$$x^2 + y^2 + 2y = 9$$

Eliminate the second-degree terms by multiplying the second equation by -1 and adding the two equations.
$$4x - 2y = -8$$

Solve for y:
$$y = 2x + 4$$

By substitution in the first equation:
$$x^2 + (2x + 4)^2 + 4x = 1$$
$$x^2 + 4x^2 + 16x + 16 + 4x = 1$$
$$5x^2 + 20x + 15 = 0$$
$$x^2 + 4x + 3 = 0$$
$$(x + 3)(x + 1) = 0$$
$$x + 3 = 0 \quad \text{or} \quad x + 1 = 0$$
$$x = -3 \quad \text{or} \qquad x = -1$$

When $x = -3$, $y = 2x + 4 = 2(-3) + 4 = -6 + 4 = -2$
When $x = -1$, $y = 2x + 4 = 2(-1) + 4 = -2 + 4 = 2$

12. Two intersecting circles; $x^2 + y^2 + 4x = 1$
$$x^2 + 4x + 4 + y^2 = 1 + 4$$
$$(x + 2)^2 + y^2 = 5$$

This is the equation of a circle with center at $C: (-2, 0)$ and radius $\sqrt{5}$.
$$x^2 + y^2 + 2y = 9$$
$$x^2 + y^2 + 2y + 1 = 9 + 1$$
$$x^2 + (y + 1)^2 = 10$$

This is the equation of a circle with center at $C: (0, -1)$ and radius $\sqrt{10}$.

13. $D = \begin{vmatrix} 1 & 1 & -1 \\ 2 & -4 & 2 \\ 3 & -2 & 1 \end{vmatrix} = \begin{vmatrix} 1 & 0 & 0 \\ 2 & -6 & 4 \\ 3 & -5 & 4 \end{vmatrix} = \begin{vmatrix} -6 & 4 \\ -5 & 4 \end{vmatrix} = -4$

14. $X = \begin{vmatrix} 2 & 1 & -1 \\ 3 & -4 & 2 \\ 1 & -2 & 1 \end{vmatrix} = \begin{vmatrix} 0 & 1 & 0 \\ 11 & -4 & -2 \\ 5 & -2 & -1 \end{vmatrix} = -1 \begin{vmatrix} 11 & -2 \\ 5 & -1 \end{vmatrix} = -(-1) = 1$

15. $x = \dfrac{X}{D} = \dfrac{1}{-4} = -\dfrac{1}{4}$

$$Y = \begin{vmatrix} 1 & 2 & -1 \\ 2 & 3 & 2 \\ 3 & 1 & 1 \end{vmatrix} = \begin{vmatrix} 1 & 0 & 0 \\ 2 & -1 & 4 \\ 3 & -5 & 4 \end{vmatrix} = \begin{vmatrix} -1 & 4 \\ -5 & 4 \end{vmatrix} = 16$$

Therefore $y = \dfrac{Y}{D} = \dfrac{16}{-4} = -4$

By substitution we find that $z = \dfrac{-25}{4}$. Therefore the solution set is

$$\left\{\left(-\frac{1}{4},\, -4,\, -\frac{25}{4}\right)\right\}.$$

CHAPTER 9 REVIEW EXERCISES, pp. 421

1. $\{(-4, -8)\}$

2. $\{(-3, 0)\}$

3. $\left\{\left(x, \dfrac{1-5x}{2}\right) \,\middle|\, x \text{ any real number}\right\}$

4. $\left\{\left(\dfrac{71}{2}, \dfrac{17}{2}\right)\right\}$

5. $\{(-1, 2, 3)\}$

6. $\{(3, -7, -3)\}$

7. $\{(1, 2, -3)\}$

8. $\left\{\left(x, \dfrac{-24x}{10}, \dfrac{-13x}{10}\right) \,\middle|\, x \text{ any real number}\right\}$

9. $\{(-4, 2, 3)\}$

10. $\{(0, 2, -2, 1)\}$

11. $\begin{vmatrix} 1 & 3 & 2 & -2 \\ 0 & -3 & -5 & 5 \\ 0 & 11 & 7 & 0 \\ 0 & 7 & 7 & 0 \end{vmatrix} = 5\begin{vmatrix} 1 & 3 & 2 \\ 0 & 11 & 7 \\ 0 & 7 & 7 \end{vmatrix} = 5\begin{vmatrix} 11 & 7 \\ 7 & 7 \end{vmatrix} = 140$

12. $\left\{\left(-\dfrac{11}{4}, \dfrac{3}{2}, -\dfrac{33}{4}\right)\right\}$

13. $\left\{\left(\dfrac{7}{2}, \dfrac{9}{2}, -\dfrac{7}{2}\right)\right\}$

14. $\left\{\left(\dfrac{2}{3}, \dfrac{1}{2}, \dfrac{1}{3}\right)\right\}$

15. $\left\{\left(\dfrac{14}{5}, -\dfrac{2}{5}, -2\right)\right\}$

16. $\{(-5, 0), (4, 3), (4, -3)\}$

17. $\left\{\left(2i, \dfrac{3}{2}\right), \left(-2i, \dfrac{3}{2}\right), \left(2\sqrt{6}, -\dfrac{11}{2}\right), \left(-2\sqrt{6}, -\dfrac{11}{2}\right)\right\}$

18. $\{(2\sqrt{2}, \sqrt{7}), (2\sqrt{2}, -\sqrt{7}), (-2\sqrt{2}, \sqrt{7}), (-2\sqrt{2}, -\sqrt{7})\}$

19. $\{(3, 8), (-3, -8), (4, 6), (-4, -6)\}$

20. $\{(3, 0), (-5, 2), (-5, -2)\}$
21. $\{(2, 3), (-2, 3), (i\sqrt{6}, -7), (-i\sqrt{6}, -7)\}$
22. $\{(1, 2), (-1, -2), (2i, -i), (-2i, i)\}$
23. $\{(2, -1), (-2, 1), (9i, 3i), (-9i, -3i)\}$
24. $\{(2, 6), (-2, 6), (2\sqrt{7}, 18), (-2\sqrt{7}, 18)\}$
25. $\{(1, 1), (-3, -1)\}$
26. $\left\{(1, 1), (-1, 1), \left(\dfrac{\sqrt{21}}{3}, -\dfrac{5}{3}\right), \left(-\dfrac{\sqrt{21}}{3}, -\dfrac{5}{3}\right)\right\}$
27. $\{(15, 32), (-10, -48)\}$
28. 50 units of P; 35 units of Q; 40 units of R

CHAPTER 10, EXERCISES , pp. 427

1. $x^3 + 3x^2 + 3x + 1$

2. $x^4 + 4x^3 + 6x^2 + 4x + 1$

3. $x^3 - 3x^2 + 3x - 1$

4. $x^4 - 4x^3 + 6x^2 - 4x + 1$

5. $x^5 + 10x^4 + 40x^3 + 80x^2 + 80x + 32$

6. $y^5 - 10y^4 + 40y^3 - 80y^2 + 80y - 32$

7. $x^7 + 14x^6 + 84x^5 + 280x^4 + 560x^3 + 672x^2 + 448x + 128$

8. $y^8 - 8y^7 + 28y^6 - 56y^5 + 70y^4 - 56y^3 + 28y^2 - 8y + 1$

9. $b^6 - 6b^5 + 15b^4 - 20b^3 + 15b^2 - 6b + 1$

10. $x^5 + 5x^4y + 10x^3y^2 + 10x^2y^3 + 5xy^4 + y^5$

11. $x^6 + 9x^4 + 27x^2 + 27$

12. $1 - 5y^2 + 10y^4 - 10y^6 + 5y^8 - y^{10}$

13. $x^4 + 2x^3 + \dfrac{3}{2}x^2 + \dfrac{1}{2}x + \dfrac{1}{16}$

14. $a^6 - 3a^4b^2 + 3a^2b^4 - b^6$

15. $8x^3 + 60x^2 + 150x + 125$

16. $81x^4 - 216x^3 + 216x^2 - 96x + 16$

17. $x^{34} - 34x^{33}y + 561x^{32}y^2 - 5984x^{31}y^3$

18. $t^{100} + 100t^{99} + 4950t^{98} + 161{,}700t^{97}$

19. $y^{30} - 15\sqrt{2}y^{28} + 210y^{26} - 910\sqrt{2}y^{24}$

20. $\dfrac{x^5}{3125} + \dfrac{x^4}{250} + \dfrac{x^3}{50} + \dfrac{x^2}{20}$

21. $14{,}080y^9$

22. $-\dfrac{969}{2}x^{30}$

23. $\dfrac{84r^{18}}{s^9}$

24. $\dfrac{-12870y^7}{x^8}$

25. $17010x^6$

26. $\dfrac{35x^{13}}{512}$

27. $(-1.792 \times 10^{-7})y^5$

28. 924

29. 1.268

30. 0.85

31. 0.73726

32. 1.105

33. 16

34. $\left(\dfrac{1}{2} + \dfrac{\sqrt{3}}{2}i\right)^9 = \left[\left(\dfrac{1}{2} + \dfrac{\sqrt{3}}{2}i\right)^3\right]^3$

$(a + b)^3 = a^3 + 3a^2b + 3ab^2 + b^3$

$\left(\dfrac{1}{2} + \dfrac{\sqrt{3}}{2}i\right)^3 = \left(\dfrac{1}{2}\right)^3 + 3\left(\dfrac{1}{2}\right)^2\left(\dfrac{\sqrt{3}}{2}i\right) + 3\left(\dfrac{1}{2}\right)\left(\dfrac{\sqrt{3}}{2}i\right)^2 + \left(\dfrac{\sqrt{3}}{2}i\right)^3$

$= \dfrac{1}{8} + \dfrac{3\sqrt{3}}{8}i - \dfrac{9}{8} - \dfrac{3\sqrt{3}}{8}i = -1$

$(-1)^3 = -1$

INDEX